National Audubon Society

WILDFLOWERS
OF NORTH AMERICA

National Audubon Society

WILDFLOWERS
OF NORTH AMERICA

The complete identification
reference to wildflowers—with
full-color photographs; updated
range maps; common names; and
authoritative notes on flowering
season, usages, scent, habitat,
and conservation status

Audubon

TABLE OF CONTENTS

NATIONAL AUDUBON SOCIETY
WILDFLOWERS OF NORTH AMERICA

6   Introduction
6   About National Audubon Society
6   What Is a Wildflower?
6   Scope of This Book

7   Identification of Wildflowers
7   How to Identify Wildflowers
7   Leaves
11   Flowers
14   Fruits

15   Names and Classification
15   Families
15   Common Name vs. Scientific Name
15   Changes in Classification

16   Native and Non-Native Species
16   Rare and Threatened Species
16   Invasive Species

17   Threats to Wildflowers
17   Climate Change
17   Human Activities

18   How to Use This Guide
18   Species Description
18   Habitat and Range
18   Toxic Plants
18   Conservation Status

19   Wildflower Biology Topics
19   How Growing Blueberries Can Save the World
20   Seeing the Flowers for the Genes—The Power of Hidden Genetic Diversity

THE WILDFLOWERS OF NORTH AMERICA

22   Nymphaeales
22   Water-shield family
23   Water Lily family

26   Piperales
26   Lizard's-tail family
27   Birthwort family

29   Acorales
29   Sweetflag family

30   Alismatales
30   Arum family
35   Water Plantain family

38   Liliales
38   Bunchflower family
47   Flame Lily family
49   Catbrier family
50   Lily family

65   Asparagales
65   Orchid family
95   Star Grass family
96   Iris family
103   Asphodel family
104   Amaryllis family
106   Onion family
111   Asparagus family

127   Commelinales
127   Spiderwort family
131   Water Hyacinth family

132   Poales
132   Cattail family
134   Bromeliad family
135   Yellow-eyed Grass family
136   Pipewort family

137   Ranunculales
137   Poppy family
147   Moonseed family
148   Barberry family
151   Buttercup family

184   Proteales
184   Lotus family

AMERICAN YELLOW LADY'S SLIPPER
(*CYPRIPEDIUM PARVIFLORUM*)

185	**Saxifragales**
185	Peony family
186	Saxifrage family
197	Ditch-Stonecrop family
198	Stonecrop family
204	**Zygophyllales**
204	Caltrop family
206	**Fabales**
206	Pea family
259	Milkwort family
263	**Rosales**
263	Rose family
280	Nettle family
283	**Cucurbitales**
283	Cucumber family
286	**Celastrales**
286	Staff Vine family
288	**Oxalidales**
288	Wood Sorrel family
289	**Malpighiales**
289	St.-John's-wort family
293	Spurge family
298	Violet family
308	Passionflower family
309	Flax family
312	**Geraniales**
312	Storksbill family
315	**Myrtales**
315	Loosestrife family
318	Evening Primrose family
335	Melastomes family
336	**Sapindales**
336	Soapberry family
337	**Malvales**
337	Mallow family
350	Rock-rose family
352	**Brassicales**
352	Caper family
354	Mustard family

379	**Santalales**
379	Sandalwood family
381	**Caryophyllales**
381	Leadwort family
383	Buckwheat family
400	Sundew family
402	Carnation family
417	Amaranth family
422	Ice Plant family
424	Pokeweed family
425	Four-o'clock family
430	Miner's Lettuce family
437	Purslane family
438	Cactus family
449	**Cornales**
449	Hydrangea family
451	Stickleaf family
454	Dogwood family
455	**Ericales**
455	Balsam family
456	Phlox family
478	Theophrasta family
479	Myrsine family
486	Primrose family
492	Diapensia family
493	Pitcher-Plant family
495	Heath family
506	**Gentianales**
506	Madder family
513	Gentian family
526	Dogbane family
540	**Boraginales**
540	Borage family
557	Waterleaf family
563	Heliotrope family
564	**Solanales**
564	Morning Glory family
575	Nightshade family
583	False Fiddleleaf family
584	**Lamiales**
584	Olive family
585	Juniper-Leaf family
586	Plantain family

616	Figwort family
619	Unicorn Plant family
620	Acanthus family
624	Bignonia family
626	Bladderwort family
630	Verbena family
635	Mint family
665	Lopseed family
671	Broomrape family
686	**Asterales**
686	Bellflower family
695	Buckbean family
697	Aster family
851	**Dipsacales**
851	Honeysuckle family
857	**Apiales**
857	Carrot family
876	**Wildflower Families**
892	**Glossary**
897	**Index**
906	**Photography Credits**
911	**Acknowledgments**

SCARLET GLOBE-MALLOW
(*SPHAERALCEA COCCINEA*)

About National Audubon Society

The National Audubon Society protects birds and the places they need, today and tomorrow. Audubon works throughout the Americas using science, advocacy, education, and on-the-ground conservation. State programs, nature centers, chapters, and partners give Audubon an unparalleled wingspan that reaches millions of people each year to inform, inspire, and unite diverse communities in conservation action. A non-profit conservation organization since 1905, Audubon believes in a world in which people and wildlife thrive. The organization introduces children, family, and nature lovers of all ages to the world around them, while Audubon experts, including scientists and researchers, guide lawmakers and agencies in developing conservation plans and policies. Audubon also works with several domestic and international partners, including BirdLife international, based in Great Britain, to identify and protect bird habitat. From urban centers to rural towns, each community can provide important habitats for local and migrating birds. In turn, birds offer us a richer, more beautiful, and healthy place to live. To survive, native birds need native plants and the insects that have co-evolved with them. Audubon promotes the use of native plants and maintains a native plants database as part of its Plants for Birds program. This database allows people to search for plants that are native to their region by entering their ZIP code. You can learn more at **audubon.org/nativeplants**.

FOXGLOVE (*DIGITALIS PURPUREA*)

What Is a Wildflower?

Wildflowers are wild plants that grow and spread naturally without cultivation or human intervention. Wildflowers can be found in just about any environment, from cities, to seashores, to forests and open fields. Some flowering plants bloom for just one day, some bloom during a specific season, and others bloom all year.

Experts estimate there are more than 20,000 wildflower species in North America alone, including grasses, forbs, trees, and shrubs. Although it's not possible to cover every wildflower in this guide, we focus on a selection of representative species from across the continent, including the most common species you're likely to encounter in the wild.

Scope of This Book

This guide covers wildflowers in continental North America north of Mexico. Ecologically, Mexico is often considered part of the more tropical Latin American ecoregion, which has its own unimaginably rich diversity of plants and animals. Likewise, Hawaii—with its distinct and rich biosphere—contains many native species not found anywhere else, and is best considered separately.

The species chosen for inclusion in this book are meant to be a representative selection of the countless wildflowers you might encounter throughout the continent. Although there's no way to include every known species, this book showcases some of the most commonly encountered and characteristic species you might find growing in the wild.

If you're interested in identifying wildflowers near you, consider choosing a field guide specific to your state or province that features the species you're most likely to encounter in your area.

COLORADO BLUE COLUMBINE
(*AQUILEGIA CAERULEA*)

PINK LADY'S SLIPPER (*CYPRIPEDIUM ACAULE*)

How to Identify Wildflowers

The color of the plant is often the first characteristic beginner wildflower enthusiasts notice, but there are many more features to examine when observing a wildflower. The shape and growth patterns of the leaves and flowers, for example, the habitat, and the region the wildflower is growing in are useful to take note of. Pink Lady's Slipper (*Cypripedium acaule*), for example, tends to grow in shady woods, while Marsh Marigold (*Caltha palustris*) prefers wetlands.

There are a series of questions you can ask yourself as you examine each major part of the plant—namely, the flowers, stem, leaves, and fruit.

Start with the flower. Depending on the time of year, the flowers may be in bloom or they may not be. If there are flowers, what's the width of the flower? What color are they? Are there petals? If so, are they united or separated? Are they solitary, or are they growing in a cluster? What shape is the cluster?

If there are no flowers, the leaf pattern and shape can be telling identification tools. Look at the shape and pattern of the leaves. Do the leaves have veins? Do they have a stalk or are they clasped to the stem? Are the edges smooth, toothed, or wavy? Are the leaves hairy?

The stem can take different shapes. Some stems are very thin, some trail to the ground, while others are erect. How tall is the stem? Is the stem covered in hairs? Is it sticky? Is it smooth?

We explain more detailed plant identification tools in the following sections.

Leaves

Function of Leaves

Leaves may not be as showy as flowers, but they are important to the plant's survival. Leaves grow at nodes along the stem and usually contain a green pigment called chlorophyll, which makes food for plants from sunlight in a process called photosynthesis. Leaves also eliminate excess water from the plant through transpiration and pull water from the surface of the leaves into the plant toward the roots.

Photosynthesis is the process by which a plant converts sunlight, water, and carbon dioxide into nutrients like sugars and

IDENTIFICATION OF WILDFLOWERS

ROUNDLEAF SUNDEW (*DROSERA ROTUNDIFOLIA*)

starches. A by-product of photosynthesis is oxygen, which plants eliminate into the air, making it breathable for humans and virtually all other animals on Earth.

Most plants rely on photosynthesis to survive, but some plants lack chlorophyll and have other methods of obtaining nutrients.

Carnivorous plants use their leaves to trap insects and digest them as food. Roundleaf Sundew (*Drosera rotundifolia*), for example, attracts pollinators with its white or pinkish flowers, but once the insects arrive, they become trapped in the plant's sticky leaves. Pitcher plants (*Sarracenia* spp.) are another kind of carnivorous wildflower that capture prey in their hollow leaves, which fill with water and drown insects to then be broken down by bacteria and enzymes into nutrients for the plant.

GHOST PIPE (*MONOTROPA UNIFLORA*)

Some plants get their food from other plants. These **parasitic** plants typically latch onto the stems or roots of the host plants for nutrients, often killing the host. Ghost Pipe (*Monotropa uniflora*) is an example of a parasitic wildflower that can grow in the understory of forests with very little sunlight by getting its nutrients from the roots of a host tree.

Leaf Arrangement

Being able to distinguish the arrangement pattern of leaves can help you identify a plant. There are three main patterns of leaf arrangements: alternate, opposite, and whorled.

If a plant has an **alternate** leaf arrangement on a branch or stem, then there is only one leaf at each node, and the leaves grow from the stem in alternate directions, as in Carolina Larkspur (*Delphinium carolinianum*).

CAROLINA LARKSPUR (*DELPHINIUM CAROLINIANUM*)

If leaves are **opposite**, then pairs of leaves grow from the same point on opposite sides of the stem. Cardinal Catchfly (*Silene laciniata*), for example, has opposite leaves.

CARDINAL CATCHFLY (*SILENE LACINIATA*)

PAINTED TRILLIUM (*TRILLIUM UNDULATUM*)

While most leaves are opposite or alternate, some are **whorled**. A whorled leaf pattern occurs if you see three or more leaves from a point on a stem. Painted Trillium (*Trillium undulatum*) for example, has whorled leaves.

Leaf Structure

There are two leaf structures, defined by how many leaf blades exist per node. Leaves are either simple or compound.

A **simple** leaf blade is undivided. Common Selfheal (*Prunella vulgaris*), for example, has simple leaves.

COMMON SELFHEAL (*PRUNELLA VULGARIS*)

Compound leaves are divided into separate blades, called leaflets, each sometimes with its own stalk. Dog Rose (*Rosa canina*), for example, has compound leaves that are divided clearly into multiple leaflets. White Clover (Trifolium repens) also has compound leaves, with typically three leaflets, but there may be four leaflets if you're lucky enough to find a four-leaf clover.

If there is doubt in distinguishing between simple and compound leaves, locate the bud where the leaf attaches to the stem. Simple leaves each attach individually to the stem, whereas compound leaves include multiple blades that come together to attach to the stem at a single point.

Basal Leaves

Leaves are said to be basal when they grow in a rosette or cluster at the base of the stem. Basal leaves provide important protection to the plant's roots. White Wood Aster (*Eurybia divaricata*), a native of open eastern woodlands, has heart-shaped basal leaves.

WHITE WOOD ASTER (*EURYBIA DIVARICATA*)

Leaf Texture

Some leaves have a distinguishable texture. Plants like Wild Basil (*Clinopodium vulgare*), for example, have very hairy leaves. If you're in the field, take note of how dense the hairs are, whether they are growing in a pattern along the veins, and if they are long, short, or spreading. Some leaves might have rough or scratchy surfaces, like sandpaper. Other leaves may be sticky, slimy, or Velcro-like.

WILD BASIL (*CLINOPODIUM VULGARE*)

Leaf Venation

The veins of a leaf are another characteristic that might help you identify the plant. Some veins look like webs, others run in parallel lines up and down the leaf, such as in grasses.

Leaf Margins

The edge of a leaf is called the margin. Margins may be smooth (also called untoothed or entire) if there are no indentations or incisions at the tips or edges.

If the leaf edges have small incisions, they are called **toothed**. Margins with large incisions are called **lobed**. The tips of toothed margins may be rounded (often called scalloped), as in Ground Ivy (*Glechoma hederacea*), or they may be pointed.

Some leaf margins are rolled or curled—these are called revolute or involute. Revolute margins roll down, toward the lower surface of the leaf, while involute margins roll up.

PASSIONFLOWER (*PASSIFLORA INCARNATA*)

Leaf Tendrils

Tendrils are thread-like structures that help the plant climb. Most passionflowers (*Passiflora* spp.), for example, are vines that climb various supporting structures using tendrils.

GROUND IVY (*GLECHOMA HEDERACEA*)

Leaf Shapes

The overall shapes of leaves are often characteristic of a plant species. Leaves can take dozens of different shapes, but the most commonly mentioned in this book are **lanceolate** (shaped like a lance); **linear** (long and narrow); **cordate** (shaped like a heart), **oblong** (long and oval), **orbicular** (shaped like a circle), **rhomboid** (star- or diamond-shaped), **ovate** (egg-shaped); **elliptical** (ellipse-shaped); and **reinform** (kidney-shaped).

Modified Leaves

Some plants don't have leaves at all or have modified leaves to adapt to various environments. Most cacti, for example, have adapted to the dry places in which they live and often have spikes in place of leaves, such as Mojave Yucca (*Yucca schidigera*), native to the Mojave Desert. Modified leaves in this book include spines, tendrils, hooks, and scales.

EASTERN PRICKLY-PEAR (*OPUNTIA CESPITOSA*)

Leaf Spines

Spines are needle-like structures that protect the plant from predators and help it retain water in dry climates. Cacti such as prickly-pears (*Opuntia* spp.), for example, have spines.

BLUE FIESTA FLOWER (*PHOLISTOMA AURITUM*)

Leaf Hooks

Leaf hooks are claw-like structures that help the plant climb and support its weak stems. Blue Fiesta Flower *(Pholistoma auritum)*, for example, has hooks.

Scale Leaves

Scales are small, membranous, scale-like leaf structures, such as a bud scale or bract. Leaf scales may function to protect part of the plant, such as a bud, or they may be thick and fleshy to store water or nutrients. Spanish Moss *(Tillandsia usneoides)*, for example, is covered in thick scales that extract moisture from the air.

Flowers

Function of Flowers

The primary function of flowers is reproduction. This is where pollination takes place, and where fruits and seeds are produced.

Flower Parts

Flowers have two main organs—the male organ, called the stamen, and the female organ, called the pistil. The **pistil** is the ovule, or egg-producing part of a flower. The **stamen** is the male part of a flower. Other parts of the flower include the petals, sepals, and the flower stalk.

Stamen

The stamen has two parts—anthers and filaments. The **anther** is usually yellow and contains pollen. Each anther contains many pollen grains, which contain male reproductive cells, also called male gametes. The **filament** is a thread-like, tubular structure that holds the anthers.

Pistil

The pistil contains the stigma, style, and ovary. The **stigma** is the flat, sticky surface at the top of the pistil, which traps and holds pollen. The **style** is the tube-like structure that holds up the stigma. The **ovary** contains the **ovules**, or eggs. After fertilization, the ovules become the seeds.

Petals

Petals are usually the most attractive and scented part of the flower. The main function of colorful and fragrant petals is to attract pollinators, but all petals also function to protect the inner reproductive structures of a flower. The petals are collectively called the **corolla**.

Petals may be separated, or they may be united to each other. Some petals are united from the base to the tip, others are united only at the base. Petals or lobes can take different shapes. Those that are partially united are often cup-shaped or tubular. Petals may also curve back, be widely spread, or appear erect.

Sepals

Sepals are structures on the outer parts of the flower that are often green and leaf-like, but they may also look petal-like. The sepals enclose a developing bud. The sepals are collectively called the **calyx**.

SPANISH MOSS (*TILLANDSIA USNEOIDES*)

Tepals

When the petals and sepals are indistinguishable, as in irises (family Iridaceae), they are called tepals.

Stalk

The stalk is the base part of the flower that connects to the plant's stem or branch.

Flower Arrangement

How the flowers are arranged on the plant can help distinguish among species with similar flowers and foliage.

Flowers may be solitary, where only a single flower grows on a stem, or they may grow in clusters, where several flowers grow together.

Some flowers grow amid the plant's foliage, making them difficult to see, and others appear prominently at the end of the stem.

Clustered flowers may be dense, with many flowers tightly growing together, or more open and spaced farther apart. Some clusters are branched, some are spike-like, and others are umbrella-shaped.

Flower Shape

The shape of the flowers also can be a useful identification tool.

Bell-shaped flowers have a long tube and a small opening at one end, as in Virginia Bluebells (*Mertensia virginica*).

VIRGINIA BLUEBELLS (*MERTENSIA VIRGINICA*)

Funnel-shaped flowers are narrow at the base and flare open, as in morning glories (*Ipomoea* spp.).

Trumpet-shaped flowers are similar to funnel-shaped flowers, but the petals turn back, like Northern Gentian (*Gentianella amarella*).

NORTHERN GENTIAN (*GENTIANELLA AMARELLA*)

Bowl- or cup-shaped flowers are shaped like a bowl. They have a wide opening and petals that flare up, creating a cup, like in Spotted Fritillary (*Fritillaria atropurpurea*).

SPOTTED FRITILLARY (*FRITILLARIA ATROPURPUREA*)

Saucer-shaped flowers are flatter than bowl-shaped flowers and have a wide, round opening, like evening primroses (*Oenothera* spp.).

HOOKER'S EVENING PRIMROSE (*OENOTHERA ELATA*)

GLOBE GILIA (*GILIA CAPITATA*)

Spherical flowers are shaped like a ball and are almost completely round, as in Globe Gilia (*Gilia capitata*).

Tubular flowers have united petals and a long tube, as in Rock Harlequin (*Capnoides sempervirens*).

ROCK HARLEQUIN (*CAPNOIDES SEMPERVIRENS*)

Flowering Season

The flowering season for different plants, even of the same species, varies considerably depending on elevation, climate, and other conditions. Plants in southern locations tend to bloom earlier than those in the north, for example. This guide gives the typical timing for a species to be in bloom, which may be different in certain parts of the plant's range.

Reproduction

Flowers are the reproductive organs of a plant that later develop into seeds and fruits. A flower may be one of two types: perfect or imperfect.

Perfect flowers are bisexual, or monoecious, meaning they contain both male and female reproductive parts.

Imperfect flowers are unisexual, or dioecious, meaning they have either a male or female reproductive organ, but not both.

Plant reproduction begins with **pollination**. Plants with perfect flowers can self-pollinate, whereas those with imperfect flowers must cross-pollinate with other individual plants. Most plants are imperfect and need the help of pollinators—such as hummingbirds, butterflies, moths, bees, wasps, or other insects—to start the reproduction process.

Many plants produce a sugar-rich liquid called nectar, especially in flowers. Pollinators, attracted to the flower's color and smell, visit the plant to feed on the nectar. Pollen from the flower's anther sticks to the pollinators' bodies while they feed and is carried with them as they travel from plant to plant. In this way, the pollen from one plant reaches the pistil of another.

Fertilization is the next step after pollination. Once pollen reaches the pistil, it goes through a slender tube until it reaches the ovary. The pollen then needs to fertilize an ovule (egg) inside the stigma.

The ovules mature into seeds. Often their outer walls become the wall of the fruit, the fleshy or dry, ripened ovary enclosing the seeds.

Fruits are often an attractive food for various animals that eat them, including many birds and mammals. Animals consuming the fruits helps the plants to spread into new areas when the animals expel waste that contains undigested seeds. Some seeds are also distributed by the wind.

WOODLAND STRAWBERRY (*FRAGARIA VESCA*)

DWARF RED RASPBERRY (*RUBUS PUBESCENS*)

Fruits

Some fruits appear only after the flowers, while some grow at the same time as the flowers. Fruits can come in different forms. Some fruits are fleshy and others are dry.

Fleshy Fruits

Fleshy fruits include berries, drupes, and pomes.

Berries, including blueberries, cranberries, and tomatoes, are all fleshy fruits. They have a sweet taste and often rely on animals to eat and disperse their seeds.

Drupes, also called stone fruits, have a fleshy outer layer but a thickened, inner layer that contains the seed. Examples include plums, apricots, and peaches.

Pomes like apples and pears are another type of fleshy fruit, but unlike drupes, they don't have a thick, stone-like inner layer.

INDIAN STRAWBERRY (*POTENTILLA INDICA*)

Dry Fruits

Dry fruits, which include familiar nuts and legumes, can be either dehiscent or indehiscent.

Dehiscent fruits break open at maturity to disperse the seeds. Dehiscent fruits include follicles, capsules, and legumes.

Follicles split along a single seam. Milkweeds, for example, produce follicles.

Capsules are another dry fruit type that open when ripe, splitting open lengthwise.

Legumes split along two seams, such as seed pods, that usually open at maturity to allow for seed release. Peas and beans in the family Fabaceae are especially known for their legumes.

SHOWY MILKWEED (*ASCLEPIAS SPECIOSA*)

Indehiscent fruits don't break open and rely on decomposition or some other outside force to release their seeds. There are three kinds of indehiscent fruits — nuts, achenes, and samaras.

Nuts are distinguished by their thick, durable outer walls, such as acorns.

Achenes are thin and lightweight, allowing them to float through the wind. Dandelions, for example, distribute fluff-like achenes when you blow on them.

Samaras, commonly called helicopters, have wing-like structures that catch the wind, causing the seed to spin and float aloft. This motion aids in wind dispersal.

Families

All plants are grouped into families, which are groups of related plants that share a common ancestor and usually many common characteristics. The family name always ends with -aceae. The mint family Lamiaceae, for example, is known for its strong mint scents. Members of the Lamiaceae family also have square stems and simple leaves.

Common Name vs. Scientific Name

Plants have both a common name and a scientific name. It's not unusual for a plant to have several common names, but the scientific name is almost always agreed upon by the scientific community. Trout Lily (*Erythronium americanum*), for example, goes by several common names — various sources or communities may call the plant American Trout Lily, Eastern Trout Lily, Yellow Trout Lily, Dogtooth Violet, or Yellow Dogtooth Violet. In scientific literature, however, the species is denoted by the Latin name *Erythronium americanum*. Scholars around the world, regardless of different languages, can reference this exact species using its scientific Latin name.

The scientific name is always italicized and consists of two words—a genus name and a specific name, or epithet.

Genus is Latin for "origin" or "type," and refers to common characteristics of related plants. Closely related species are classified in the same genus, and as you become more experienced, it's often possible to tell what genus a plant belongs to just by looking at it.

GLACIER LILY (*ERYTHRONIUM GRANDIFLORUM*)

Take Trout Lily, for example, with the scientific name *Erythronium americanum*. It belongs to the genus *Erythronium*, and *americanum* is its specific epithet. All plants in the genus *Erythronium* have tulip-like flowers and small bulbs. *E. americanum* is one distinct species in this genus of around two dozen relatives that are found around the world.

Some scientific names come from characteristics of the plant, whereas others are named after botanists or a plant's geographical location.

Changes in Classification

For thousands of years, botanists grouped plants and families based on appearance, growth habit, habitat, and other identifying features. That all began to change with the advent of DNA testing in the 1980s and early 1990s.

What botanists thought they knew about plants has changed significantly and many plants have been reclassified as a result of molecular and DNA studies. Some families have been split into multiple ones, others have been combined, and new families with new genera have been formed.

The Lily family Liliaceae, for example, was once a very large family of more than 1,000 species, but DNA studies have taught scientists volumes about the evolutionary relationships between—and differences among—its former members. Many species formerly included in the Liliaceae family are now placed in other families that better reflect their relatedness.

TROUT LILY (*ERYTHRONIUM AMERICANUM*)

NATIVE AND NON-NATIVE SPECIES

Native wildflowers are those that occur naturally in the wild in a given location. Native plants have evolved to the environmental and ecological conditions of the region, and support native wildlife communities that have co-evolved to thrive alongside them. Native plants provide food and shelter for native birds, insects, and other wildlife.

In North America, a native plant is defined as one that was naturally found in a particular area prior to European colonization. Native wildflowers support an untold number of North American wildlife species ranging from songbirds and mammals to pollinators like hummingbirds, butterflies, and bees.

Humans, however, have been cultivating and moving plants around for thousands of years, so in most habitats—especially those inhabited by people—plant communities are not exclusively native.

Non-native plants are those that do not occur naturally in a given location and have been introduced there either through direct human transplantation—whether intentional or unintentional—or have spread into new territory due to human activities.

Non-native wildflowers may be casual, meaning they generally stay where you put them, or they can become naturalized, meaning they are able to self-sustain their populations in their new landscape without human cultivation. Some non-native plants can even become invasive, which is described in more detail below.

Rare and Threatened Species

It's important to note that a plant that grows profusely in one state or province may be endangered in another. Many plants are threatened as a result of people picking them and transplanting them into their own gardens. In some states and provinces, picking wildflowers or digging them from public property is prohibited.

Governments protect many threatened species by prohibiting their collection or destruction. The Endangered Species Act in the United States and the Canada Endangered Species Protection Act (CESPA), widely known for protecting rare and endangered animals, also include protection of rare wildflowers. These laws make it illegal to kill, harm, or collect certain endangered or threatened species or to destroy or damage their habitat. Most states and provinces maintain their own lists of locally endangered or threatened wildflower species that are protected by law as well.

JAPANESE KNOTWEED (*REYNOUTRIA JAPONICA*)

Invasive Species

While most of the plants in this guide are native species, many others are introduced exotic plants that have spread to North America through human activity. Invasive species are non-native plants that not only establish a robust wild-growing population, but also expand and spread across long distances in their new location once introduced.

The most destructive of these are noxious weeds, which are weeds that are harmful to native habitats, wildlife, people, or property. Noxious weeds can displace native vegetation, sicken native wildlife or livestock, cause soil erosion, disrupt agricultural landscapes, or have adverse effects on the food chain.

Some exotic species may be prohibited or on a quarantine list in certain states. In these cases, transporting, buying, selling, and distributing these plants is illegal. Reputable seed companies won't sell you seeds if the plant is prohibited in your state. You may be fined for growing a prohibited plant in your garden or landscape. Some particularly destructive or invasive plants are even targeted for removal or eradication by government agencies or conservation organizations.

There are several plants in North America that are considered a weed just about everywhere. Queen Anne's Lace (*Daucus carota*), for example, is a commonly prohibited wildflower that grows in fields and along roadsides—often taking over and crowding out native plants. Japanese Knotweed (*Fallopia japonica*) is another common noxious weed in the East. The tall, bamboo-like plant spreads rapidly and blocks native vegetation from getting sunlight. Once established, populations of Japanese Knotweed are hard to eradicate, with the smallest piece of plant left behind able to rapidly regrow and spread.

It's important to note, however, that a plant that's a weed in one state may be useful or threatened in another. The United States Department of Agriculture's National Invasive Species Information Center is a good place to start to learn about noxious weeds.

Wildflowers are beautiful to look at, but they are also important parts of the ecosystem. They improve soil nutrients and provide food for pollinators and other animals. Some plants also provide shelter and nesting areas for birds. Wildflowers, like many other native plants, animals, and ecosystems in North America, face a variety of threats.

ALPINE SHOOTINGSTAR (*PRIMULA TETRANDRA*)

Climate Change

The delicately balanced ecosystems of North America face an unprecedented threat in the form of climate change. Earlier snowmelt, increased forest fires, and harsh weather are shifting entire ecosystems as we know them.

Some wildflowers are declining or slowly shifting from their native ranges as a result of climate change. Plants with small ranges or specialized habitats are especially vulnerable. For example, many alpine species that rely on the climate at specific altitudes may find their habitats are drifting to ever higher altitudes where there is less available land area—or vanishing entirely.

While climate change studies and models are ongoing, it's clear that a changing climate puts incredible pressure on native plants, which aren't able to migrate or simply walk away from a habitat that's changing or disappearing around them.

HARSH POPCORN-FLOWER
(*PLAGIOBOTHRYS HISPIDULUS*)

WESTERN WATERLEAF
(*HYDROPHYLLUM OCCIDENTALE*)

Human Activities

At the same time, mature forests, prairies, and other native habitats for wildflowers have declined significantly due to human development, including for resource extraction, large-scale agriculture, and urban sprawl. Industrial agriculture's widespread use of powerful insecticides has impacted populations of many native pollinators such as butterflies, moths, and bees, as well as some birds.

Several organizations have formed to research and restore native plants and their habitat. The Native Plant Trust, headquartered in Massachusetts, and the Lady Bird Johnson Wildlife Center in Texas are two such organizations that are researching, conserving, and protecting rare plants.

COBAEA BEARDTONGUE (*PENSTEMON COBAEA*)

This guide includes a representative selection of the countless wildflowers you might encounter throughout the continent, including the most common wildflower species in North America. They are grouped by family and genus. We list both the scientific name and common names for each species. Many plants have more than one common name. In these cases, we provide a list of alternate names.

Species Description

The species description includes a brief discussion of the plant, including its use and history, followed by a detailed description of the plant that gives readers identification clues. We include the geographic range, the habitat in which the plant prefers to grow, and the flowering season. The reader should note that the plant's flowering season fluctuates between regions it grows in—this guide describes the most expansive range possible for each plant. Similar species are also included here to help the reader distinguish between two or more similar plants. In some cases, plants are so similar that they are nearly impossible to differentiate without DNA sampling.

Photos of each plant from different angles are also included to provide a useful visual.

Habitat and Range

The distribution maps provided alongside each species give the reader a visual of where in North America a plant is known to occur. These maps reflect the most current data available as of the time of publication, including distribution at the county level for most species records in the U.S. and at the provincial level in Canada. However, one should note that plant populations can and do change over time.

Most plants are also highly dependent on having the correct habitat or growing conditions, so don't expect a species to appear everywhere inside its possible range. Some plants have a different appearance depending on their location or local environment. The flowers of some species may be a different color at higher elevations, for example, or the plant may be smaller in drier conditions.

Toxic Plants

Plants that are harmful to humans or animals, including pets or farm animals, are noted under the Caution section. Some plants are toxic to consume, others may have sharp prickles or spines, and some can irritate the skin if touched. Unfortunately, some wildflowers produce colorful but poisonous fruits that may be alluring to children or pets, so use caution. Wild plants should never be consumed unless you're certain of their identity and that it's legal to collect the species.

Conservation Status

This guide includes the global conservation status ranking for each wildflower. The conservation status notes whether a species' wild population is secure or in danger of becoming extinct. We use the most up-to-date information available, but the reader should note that conservation statuses are constantly changing.

The information here is sourced from NatureServe, a nonprofit organization based in the United States that works with various networks and more than 1,000 conservation scientists to provide comprehensive wildlife data.

Rankings range from G1 to G5:

G5 (Secure) G5

At very low risk of extinction or collapse. Species ranked G5 tend to have an extensive range, abundant populations, and little to no concern about declines or threats.

G4 (Apparently Secure) G4

At fairly low risk due to a large population and/or wide range, but with some concern due to local recent declines, threats, or other factors.

G3 (Vulnerable) G3

At moderate risk of extinction due to a limited range or widespread decline or threat.

G2 (Imperiled) G2

At high risk of extinction due to a restricted range and/or a steep population decline or severe threat.

G1 (Critically Imperiled) G1

At very high risk of extinction due to a highly restricted range, very few populations or occurrences, catastrophic declines/population collapse, or other factors

In several instances, plants have a global G5 ranking, but may be threatened in specific states or provinces. In those cases, we list the states and provinces that classify the plant as threatened.

Some plants are exotic in North America and therefore are not ranked based on their populations in the wild.

How Growing Blueberries Can Save the World

Over the past decade, a new trend in landscaping has been growing steadily. For many folks the days of pretty for pretty's sake are gone—people are planting for pollinators, for air and water quality, and with human and wildlife health in mind. With authors and conservationist figures such as Doug Tallamy, Rick Darke, and Heather Holm taking center stage on the horticultural scene, there has never been more public interest and concern over landscaping and its direct impacts on the environment.

From backyard pollinator gardens to supporting birds through Audubon's Plants for Birds program to community gardens, ecological landscaping allows anyone who interacts with the land to directly affect their own local ecosystem.

There are still plenty of ecological issues that our landscape choices cannot affect directly. Climate change, the overuse of pesticides and fertilizers, or habitat loss can be tough challenges for any one person to tackle. There are, however, other indirect ways we can tackle some of these issues.

One of the largest drivers of habitat loss worldwide is large-scale, industrialized agriculture. Industrial agriculture clears entire rainforests for palm oil production, destroys prairie for corn or cattle, and eliminates important "weed" plants that insects like Monarchs (*Danaus plexippus*) use to reproduce. Annually we lose 1.88 million hectares of forest, an area the size of Costa Rica, to these highly unsustainable agricultural practices.

This is not a simple or a small problem. Many of us play an unintentional role in this system. Unless you are the sort of individual who reads the ingredients on every supermarket purchase, researches the source of those ingredients, and makes choices based on that research, you are unwittingly involved. As potentially depressing as this bit of information is, it also means each

and every one of us has the ability to impact this issue through our personal choices.

There are numerous ways to go about lessening our role in this form of habitat loss. There are certain ingredients that are worth either avoiding or, better yet, worth sourcing with care. There are also numerous organizations willing to help us make better purchasing decisions.

Look for certification from the Forest Stewardship Council to ensure sustainable forest products. The Monterey Bay Aquarium's Seafood Watch program has an app that tells you which fish come from sustainable sources. Fair trade certification for coffee ensures that coffee farmers get a fair share of the coffee industry proceeds and supports sustainable farming practices. Some certifications go a step further: coffee that is certified Smithsonian Bird Friendly®, including Audubon® coffee, contributes to protecting critical habitat for birds and wildlife, fighting climate change, protecting biodiversity, and supporting farmers committed to conserving bird and wildlife habitat by farming sustainably.

Another strategy is to buy local. Support your local organic, multi-crop farms, or join a community-supported agriculture (CSA) or crop-sharing program. Ask questions about the practices happening in your local food-production systems and put your money toward those farms that are doing things well. Not all of us can grow our own food, but all of us put money into the food economy, and that gives us the power to support those farmers that we feel are worth supporting.

This becomes much more difficult in the numerous areas where food choices are limited (often referred to as food deserts), and this is where our final strategy can be even more effective.

The final strategy—and one that doesn't get enough attention—is growing your own food. Each tomato, ear of corn, or blueberry grown at home or in the community garden is one fewer of those foodstuffs coming from industrial agriculture sources. Although you may not be building habitat in the same manner or scale as a native plant restoration project, you are helping to stop the spread of habitat destruction due to industrial agriculture.

Your part may be small and indirect, but it is no less important. There is great power in large numbers of people making small changes!

And while tomatoes and corn do indeed help to reduce habitat loss, there is little argument in favor of them building habitat for anything other than us humans (and possibly the local rabbit populations). When we start looking at planting native edible species, we not only help to reduce habitat loss but we actually start to build habitat for wildlife locally.

HIGHBUSH BLUEBERRY (*VACCINIUM CORYMBOSUM*)

ELDERBERRIES (*SAMBUCUS CANADENSIS*)

We don't tend to think of the blueberries in our gardens as being a part of the habitat, especially when so many of them are planted under bird netting or some other protective cover. Sometimes we forget that the berries are only a piece of the blueberry plant's function in an ecosystem. While the birds do indeed enjoy the berries, the native bees are much more interested in the flowers.

The flowers are only part of the blueberries' value. The leaves are immensely important, supporting around 275 different species of native caterpillars. Not only are these caterpillars important for the butterflies and moths they will mature into, but the caterpillars are also an important source of protein for birds.

There are a number of great native crops that we all know and love that can play a similar role to the blueberry in our gardens. The genus *Rubus* contains our favorite raspberries, blackberries, black raspberries, and thimbleberries. There are also serviceberries (*Amelanchier* spp.), elderberries (*Sambucus canadensis*), currants (*Ribes* spp.), hazelnuts (*Corylus americana*), beach plums (*Prunus maritima*), and the list goes on. Perhaps the most interesting native edibles are those that are not as common, offering new flavors for enthusiastic chefs to play with. Pawpaws (*Asimina triloba*), fiddleheads (*Matteuccia struthiopteris*), ramps (*Allium tricoccum*), groundnuts (*Apios americana*), and sunchokes (*Helianthus tuberosus*) all fall into this category.

Underplant some of the aforementioned shrubs with our native wild strawberries (*Fragaria virginiana*, a vigorous spreader) and your need to weed drops to a few hours per season.

So can growing blueberries save the world? Perhaps not, but it is certainly a step in the right direction.

Want to see these ideas in action? Come visit Norcross Wildlife Sanctuary in Wales, MA and check out our Native Edibles Test Garden.

■ *Dan Jaffe Wilder*

Seeing the Flowers for the Genes—The Power of Hidden Genetic Diversity

Biodiversity is defined at three scales—ecosystems, species, and genetics. Ecosystem diversity encompasses variation among plant communities and physical environments, while species diversity evaluates the number and relative abundance of the constituent species in those communities. The finest scale of biodiversity—genetic diversity—quantifies variation in genetic characters within an individual species. While historically a comparatively understudied component of biodiversity, recent advances in molecular genetics methods and technology has ushered in a new era of genetic diversity research.

Practically speaking, genetic variation underlies the potential of populations and species to adapt when faced with environmental change. The genetic code of an organism, or its genotype, interacts with the environment to create the expression of these genes as a living organism, or phenotype.

High genetic diversity increases the likelihood that wildflower populations will contain favorable genotypic variation for environmental selection to act upon, driving adaptation of a population to a new environmental context. High genetic diversity is maintained by gene flow, or more specifically, the movement of wildflower pollen and seeds across the landscape.

GRAY GOLDENROD (*SOLIDAGO NEMORALIS*)

Different wildflowers have adapted different means of dispersal, such as producing nectar to lure pollinators close to their pollen-laden anthers and seeds with attachments that help them catch a ride on the wind. Continuous, unfragmented habitat supports long-distance dispersal and increases the likelihood of a pollen grain finding a receptive stigma or a seed finding a suitable germination site.

Gene flow can introduce novel genetic material, increasing the variation present in the gene pool for environmental selection to act upon. When wildflower populations express phenotypic differences as a result of unique selective pressures, this is called local adaptation. While a high level of gene mixing can be occurring among these populations, biotic and abiotic factors such as herbivores, precipitation, and soils are applying different, forceful genetic filters. Consequently, the characteristics that provide maximum fitness can vary among populations. Over long periods of time, environmental selection can drive two wildflower populations further and further from each other, potentially resulting in the generation of two new species, where previously there had only been one.

Throughout North America, the landscape has been heavily developed, resulting in smaller, more fragmented natural areas. Fragmentation and large areas of inhospitable land cover, such as cities, limit gene flow between populations of wildflowers. Many pollinators cannot travel long distances to transport pollen between habitat patches, and small mammals that may disperse seeds via ingestion or stuck to their coats are discouraged from venturing through hostile environments to potentially reach another refuge.

As gene flow becomes more restricted, genetic diversity can be lost due to acts of random chance, such as a tree falling down and eliminating half the local population. Loss of genetic diversity over generations due to chance is called genetic drift. Over time, no or limited gene flow can result in inbreeding, or mating among closely related individuals. Inbreeding can result in inbreeding depression, whereby inbred individuals are less fit, and hence the population size may decline rapidly. Genetically depauperate populations are less resilient to sudden environmental change, increasing their likelihood of extirpation or extinction under rapid, anthropogenic climate change.

To reverse the effects of genetic declines, gene flow can be artificially re-created through the addition of plants, pollen, and/or seeds from a wild population or seed bank collection with high genetic diversity, otherwise known as genetic augmentation. Another approach can be the reestablishment of gene flow among populations through the creation of habitat corridors connecting previously isolated habitat fragments. Investing in greater connectivity among natural areas is a better long-term strategy for maintaining genetic diversity in wildflower populations as natural dispersal vectors like bees and squirrels take care of gene flow all on their own.

In addition to supporting adaptation in response to environmental change, high genetic diversity has been found to increase multiple population, community, and ecosystem metrics. For example, research on goldenrods (*Solidago*) has found that plants grew larger when planted in plots with greater genotypic diversity, as compared to plots containing only a single genotype. Additionally, genetically diverse goldenrod plots contained more insects and a greater number of insect species. Another study on the humble dandelion (*Taraxacum officinale*) found that population performance clearly and consistently increased with genetic diversity.

A potential explanation for these effects is decreased competition among divergent phenotypes, or niche complementarity. A recent summary of the ecological importance of intraspecific variation found that the effects of genetic variation within species are often comparable to, and sometimes stronger than, those between different species.

Future research will continue to explore the relationship between genetic diversity and population, community, and ecosystem performance, as well as the mechanisms underlying such relationships, with the benefit of rapidly developing genomic methods.

■ *Jessamine Finch*

COMMON DANDELION (*TARAXACUM OFFICINALE*)

CAROLINA FANWORT *Cabomba caroliniana*

NATIVE
RARE OR EXTIRPATED
INTRODUCED

ALTERNATE NAMES: Purple Cabomba, Carolina Water Shield, Gray Fanwort, Green Cabomba, Cabomba, Fanwort, Fish Grass, Green Grass, Washington Grass, Washington Plant, Water Shield, Grass *Cabomba aquatica; Cabomba pulcherrima; Cabomba australis; Cabomba caroliniana* var. *pulcherrima*

An oftentimes unwanted aquatic plant native to the southeastern U.S., Carolina Fanwort can be considered weedy in its native area and can be highly invasive in areas to which it is not native. It forms extremely thick stands which can interfere with recreational activities such as swimming and boating, as well as clog drainage systems. A common plant sold for use in aquariums, its sale and transport is now banned in several northern and western states.

DESCRIPTION: An aquatic, herbaceous perennial with feathery, fan-shaped leaves; green, sometimes reddish-brown, attached to long, branching stems that may reach 6.5' (200 cm) long. Large, submerged leaves are in pairs, opposite each other; narrow, fairly inconspicuous floating leaves form near stem tips, alternately. Flowers are typically white (sometimes tinged purple), ½" (1.2 cm) wide, with three petals and three sepals and float just above the water's surface on short stalks.

FLOWERING SEASON: May–September

HABITAT: Stagnant to gently flowing freshwater tidal creeks, rivers, marshes, ponds, lakes, reservoirs, and ditches.

RANGE: Native in the southeastern U.S. through South America; invasive in Northeast and Pacific Northwest.

SIMILAR SPECIES: *Limnophila sessiliflora*, known as Dwarf Ambulia, Ambulis, and Asian Marshweed, which has three or more leaves per node (whorled) instead of two that are opposite.

CONSERVATION: G5
This plant is invasive in areas outside its native range.

■ NATIVE
■ RARE OR EXTIRPATED
■ INTRODUCED

ALTERNATE NAMES: Spatterdock, Cow Lily, Yellow Cow Lily, Wakas, Broadleaf Pond-lily

A common sight across freshwater ponds and slow moving streams of the eastern U.S. and Canada, all parts of this plant were used by numerous Native American tribes for both food and medicine. The Menominee—the native people from the area which is now Wisconsin and upper Michigan—call this plant "woka'tamo" and say the Yellow Pond-lily belongs to the "Underneath Spirits" and creates the fog that hovers over lakes.

DESCRIPTION: Typically found in large clusters in waters less than 10' (300 cm), the aquatic perennial has large, green heart-shaped leaves, 3–15" (7.5–38 cm), with a V-notched base. Leaves are variable and may be wide or narrow and either submerged, floating or raised above the surface of water. Single, yellow cup-shaped fleshy flowers, 1–2" (2.5–5 cm), appear at or above the water surface with yellow, lobed stigma with green outer sepals.

FLOWERING SEASON: March–October.

HABITAT: Slow-moving freshwater, typically in full sun.

RANGE: Eastern U.S. from Texas north to Southern Ontario. Limited populations exist in New England and Canada.

SIMILAR SPECIES: Historically, all plants within the *Nuphar* genus were considered a single species, *N. lutea*, with separate subspecies. Further comparisons of morphology and genetic analysis have recognized eight distinct species. May be confused with any other *Nuphar* species: *N. microphylla, N. orbiculata, N. polysepala, N. rubrodisca, N. sagittifolia, N. ulvacea, N. variegata*. Differentiation between species is difficult, but can be made with a taxonomic key based on detailed study of flower and leaf structures.

CONSERVATION: G4
The species is imperiled in New Jersey and threatened in Kansas, Maine, Wisconsin, Illinois, and throughout its range in Eastern Canada. It is believed extirpated in Connecticut.

ROCKY MOUNTAIN POND-LILY *Nuphar polysepala*

NATIVE
RARE OR EXTIRPATED
INTRODUCED

ALTERNATE NAMES: Alternate names: Great Yellow Pond Lily, Wokas, Spatterdock, Western Yellow Pond-lily

This plant was used by Native Americans for food as well as medicine. Seeds were roasted until cracked similar to popcorn and were also ground to make flour. The seeds formed a large portion of carbohydrates in the diet of the native Klamath and Modoc peoples in what is now Oregon.

DESCRIPTION: Typically in shallow, muddy waters, this aquatic perennial has large, leathery, heart-shaped leaves, 4–18" (10–45 cm), floating on the water with rounded tips and a notch at one side of the stem. The plant produces single, bright yellow cup-shaped fleshy flowers, 2½–4" (6.5–10 cm). They may be found floating or up to 3" (7.5 cm) above the water surface with red anthers and green outer sepals.

FLOWERING SEASON: April–September.

HABITAT: Ponds, lakes, sluggish streams.

RANGE: Western North America—Utah north to Montana, west to Alaska and California.

SIMILAR SPECIES: May be confused with other *Nuphar* species: *N. advena, N. microphylla, N. orbiculata, N. rubrodisca, N. sagittifolia, N. ulvacea, N. variegata.* Differentiation between species is difficult, but can be made with a taxonomic key based on details of flower and leaf structures.

CONSERVATION: G5
The species is threatened in Arizona, Montana, Utah, Wyoming.

■ NATIVE
■ RARE OR EXTIRPATED
■ INTRODUCED

ALTERNATE NAMES: *Castalia odorata*, American White Water-lily, Sweet-scented Water-lily, Beaver Root

A classic native, Fragrant Water-lily is very common throughout much of Central America to Canada with its iconic leaves and flowers floating on the surface of lakes and ponds. The fragrant flowers open each morning, closing typically at noon each day, typically for three days. Afterward, the flower closes and sinks below the water surface while the fruit matures. The spongy leafstalk has four channels to move oxygen and other gases from the leaf surface to the submerged rhizome; it is so porous it can be used as a straw. All parts of this plant are edible and used by native peoples for both food and medicine.

DESCRIPTION: A floating aquatic plant featuring flat, round, bright green leaves that are red to purple underneath, up to 10" (25 cm). Large, 3–6" (7.5–10 cm), showy, and fragrant white flowers can be seen floating or raised just above the water surface. Its long leaf and flower stalks are submerged 2–4' (60–120 cm).

FLOWERING SEASON: July–October.

HABITAT: Shallow, still water of ponds.

RANGE: It is native to all U.S. states except Alaska, Hawaii, North Dakota, and Wyoming.

SIMILAR SPECIES: May be confused with both Pygmy Water-lily (*N. tetragona*), a circumpolar species only found in Canada in the Western Hemisphere, and the Small White Water-lily (*N. leibergii*), a rare, threatened species from the most northern extremes of the continental U.S. and southern Canada. As their name suggests, both will be smaller in average size than the Fragrant Water-lily, but clear differentiation between these species is difficult and primarily made through detailed observation of specific sepal characteristics.

CONSERVATION: G5
This plant is considered invasive in California and Washington.

LIZARD'S-TAIL *Saururus cernuus*

■ NATIVE
■ RARE OR EXTIRPATED
■ INTRODUCED

ALTERNATE NAMES: Breastweed, Water Dragon, Swamp Root

A wetland plant native to much of eastern North America, Lizard's-tail grows in wet areas and shallow water. Crushing the leaves provides a scent of sassafras. The name is descriptive of the shape of the drooping, white flowers that bloom in the summer. The vegetation is a preferred food source for beavers and other wetland animals. Many Native American tribes, including Cherokee, Choctaw, and Seminoles, used the roots in a poultice to treat swelling in the body. It is commercially available and attractive in wetland gardens.

DESCRIPTION: Grows 2–5' (60–150 cm) with heart-shaped leaves 3–6" (7.5–15 cm) long. When in bloom, it features numerous tiny, white flowers on a bottlebrush-like spike which droops at the tip—up to 6" (15 cm) in total length.

FLOWERING SEASON: June–September.

HABITAT: Wet soils and mud, in standing water up to 4" (10 cm). Found at the edges of ponds and streams, swamps, and ditches.

RANGE: Native to Southern Ontario, Quebec and New England to Michigan and Illinois, south to Florida and Texas.

CAUTION: If overeaten, the plant may be toxic to foraging livestock.

CONSERVATION: G5

The species is listed as threatened in Connecticut, Illinois, Kansas, Rhode Island, Ontario, and Quebec.

■ NATIVE
■ RARE OR EXTIRPATED
■ INTRODUCED

ALTERNATE NAMES: Canadian Wild Ginger, Canadian Snake-root, Broad-Leaved Asarabacca, Woodland Ginger, Coltsfoot, Indian Ginger, Namepin and Sturgeon Potato.

This wildflower is known more by its attractive, heart-shaped leaves than its hidden spring flowers which grow low to the ground—perfect for a plant that is pollinated mainly by beetles and has seeds which are dispersed by ants. While not related to the spice found in many kitchens, the roots of Wild Ginger have a similar scent and were once used by early settlers as a substitute for ginger. Its dense, shallow rhizomatous root system can prevent surface erosion and even fend off invasive species.

DESCRIPTION: Large, low-growing leaves with fine hairs which occur in pairs from the ground with virtually no stem, each 2–6" (5–15 cm) wide and 6–12" (15–30 cm) tall. The petalless, cup-shaped flower will be concealed under leaves and consists of 3 pointed, dark red (perhaps purple to brown) sepals roughly 1½" (4 mm) wide.

FLOWERING SEASON: April–June.

HABITAT: Partial to full shade in rich, deciduous forests.

RANGE: Eastern U.S. from the eastern edges of Arkansas to North Dakota to the Atlantic Ocean. In Canada, from Manitoba east to New Brunswick.

CAUTION: Handling the leaves may cause dermatitis in some people. Species of *Asarum* may contain aristolochic acid which is used in rat poison and has been found to be nephrotoxic and carcinogenic in large quantities or regular use. The United States Food and Drug Administration (FDA) and Health Canada warn against consuming Wild Ginger.

SIMILAR SPECIES: In the South, could be confused with Little Brown Jugs (*A. arifolium*), which has hairless, evergreen leaves which are more arrow shaped and with brownish, ground-level flowers that resemble jugs.

CONSERVATION: G5

Threatened in Delaware, Louisiana, Maine, South Dakota, Kansas, Mississippi, Illinois, and Manitoba.

■ NATIVE
■ RARE OR EXTIRPATED
■ INTRODUCED

ALTERNATE NAMES: British Columbia Wild Ginger, Western Wild Ginger

A native of the western U.S., Long-tailed Wild Ginger is the most common of the western wild gingers. It spreads mainly by its mats of rhizomes forming a thick, understory ground cover. Ants, attracted to the seed's fatty appendage called an elaisome, will also spread its seeds. Flowers are hidden, forming under the plant's leaves. The roots smell of ginger if rubbed, but this plant is not related to the culinary herb.

DESCRIPTION: Large, low-growing leaves with fine hairs which occur in pairs from the ground with virtually no stem, each 4" (10 cm) wide and 6" (15 cm) tall. The single, petalless, cup-shaped flower is concealed under leaves and consists of three lobes, each tapering to long, slender, brown-purplish to yellow or greenish tips roughly 1½–5" (4–12.5 cm) wide.

FLOWERING SEASON: April–July.

HABITAT: Moist, shady coniferous woodlands.

RANGE: From near the coast of central California north to British Columbia east to western Montana.

CAUTION: All species of *Asarum* may contain aristolochic acid which is used in rat poison and has been found to be nephrotoxic and carcinogenic in large quantities or regular use. The U.S. FDA and Health Canada warn against consuming Wild Ginger.

SIMILAR SPECIES: It is fairly easy to differentiate the three other western wild ginger species. Lemmon's Wild Ginger (*A. lemmonii*) is similar in leaf appearance, but with a more compact flower without the long, trailing sepal tips. The leaves of both Hartweg's Wild Ginger (*A. hartwegii*) and Marbled Wild Ginger (*A. marmoratum*) have blotchy white veins or patches that follow the major veins.

CONSERVATION: 🄶5
Long-tailed Wild Ginger is common in its range and is not listed as a species of concern.

■ NATIVE
■ RARE OR EXTIRPATED
■ INTRODUCED

ALTERNATE NAMES: Single-vein Sweetflag, Calamus Root, Beewort, Bitter Pepper Root, Flag Root, Sweet Cinnamon, Sweet Myrtle, Sweet Grass

This tall, European wetland plant can be found naturalized and growing in water or wet soils. Sweetflag is often found growing among blue flag (*Limniris* spp.), cattails (family Typhaceae), and other wetland plants. Muskrats are often found to be in the vicinity. This species is a sterile triploid—hosting an extra full set of chromosomes—which can only reproduce clonally by rhizomes. The thick rhizomes were candied to form an old-fashioned confection called "calamus." Today, all *Acorus* species are considered toxic and the use of the root or derivatives is banned in the U.S. and Canada due to documented medical impacts.

DESCRIPTION: Found in or near marshy or wetland areas, tall stiff blade-like leaves with a single mid-vein extend 1–4' (30–120 cm). Slightly conical spikes, 2–3½" (5–9 cm), covered with inconspicuous green, yellow, or brownish flowers occur among the leaves.

FLOWERING SEASON: May–August.

HABITAT: Swamps, marshes, meadows, riverbanks, and small streams.

RANGE: Eastern U.S. from Texas, Colorado and Manitoba to Atlantic seaboard as well as populations in northern California, Oregon, and Washington.

CAUTION: *A. calamas* is considered carcinogenic and toxic, and the use of the root or derivatives is banned in foods in the U.S. and Canada due to documented medical impacts.

SIMILAR SPECIES: In early botany, the North American native *A. americanus* was treated as the same species, and later as a subspecies of *A. calamus*. This confusion has led to the lack of clarity of which species of *Acorus* within North America was being addressed in historic texts, range maps, and other references. Now, botanists widely recognize *A. calamus* as a European species introduced to the U.S. by settlers, likely in the 1600s. Today, this confusion persists with some areas of the U.S. and Canada listing *A. calamus* as an exotic or invasive while others list it as threatened, and oftentimes commercial sale of roots or plants will incorrectly identify the species. The difference between the two species is slight—in *A. calamus* there is one raised vein running the length of the leaves just offset from the center. In the native *A. americanus*, there will be two or more raised veins.

CONSERVATION: G4

At risk in Colorado, North Carolina, Iowa.

■ NATIVE
■ RARE OR EXTIRPATED
■ INTRODUCED

ALTERNATE NAMES: Indian Turnip, Bog Onion, Brown Dragon

Jack-in-the-pulpit is an iconic spring ephemeral (a group of plants which emerge early in spring and go dormant by summer). Its peculiar flowering structure has led to its unusual common name. The "pulpit" refers to the spathe, a green to purple leafy cone topped with a hood. "Jack" is the spadix—the flower spike; a pale cream, club-like structure that holds the true flowers—and can be found hidden inside the spathe.

Jack-in-the-pulpit is a long-lived, slow-growing plant, often taking five years before flowering, and may live over 100 years. Once they begin to flower, these newer plants will only have small male flowers on the spadix. As they mature, the plant will instead produce female flowers at the base of the spadix inside the spathe. Its specialized flower is perfectly adapted for its unique method of pollination.

The flower emits an odor similar to woodland fungi. Fungus gnats, attracted to this smell, enter the spathe and are lured deep into the bottom of the slippery, cup-shaped structure. Unable to climb out or fly straight up, they search for an escape, brushing against the small male flowers which cover them in pollen.

Fortunately, male flowers have a small hole at the base of the spathe which allows the pollinated gnats to escape. However, when these gnats are lured into the spathe with female flowers, no such hole is provided, trapping the gnat forever, ensuring any pollen on the gnat will pollinate the female flowers as it struggles to escape. Once pollinated, female flowers will ripen in late summer into a cluster of bright red berries, a favorite of woodland birds including turkeys and grouse, which will help distribute their seeds.

DESCRIPTION: Leaves are typically a pair of three leaflets, 1–3' (30–90 cm) tall, dull green on top while matte silvery green underneath. Beneath or at leaf height, an upright cone-shaped spathe, green or purplish-brown, often striped on the interior and "hood," surrounding a creamy-white finger-like spadix. A 2" (5 cm) cluster of bright red berries, each ¼" (1 cm) in late summer.

FLOWERING SEASON: April–June.

HABITAT: Moist, rich woodlands, along creeks, marshes and swamps.

RANGE: Texas north to Manitoba, east to Nova Scotia, south to Florida.

CAUTION: The berries, foliage, and roots of this plant contain calcium oxalate crystals and will cause painful irritation of the mouth and throat if ingested. The roots can cause blisters on skin if touched.

SIMILAR SPECIES: The genus *Arisaema* has four species within North America. *A. dracontium* (Green Dragon) is the easiest to differentiate and features 7–15 leaflets per leaf with a spadix (the "Jack" inside the cone-shaped spathe) which rises multiple times higher than the spathe. The other species are sometimes treated as just one entry in many field guides, or as subspecies in taxonomic guides, but recent genetic analysis supports their separation. Field identification is possible. Whereas the underside of *A. triphyllum* leaves are a matte silvery green, both of the following species will have a brighter, glossy green underside to their leaf. *A. stewardsonii* will have a strongly fluted spathe tube (the pulpit) while the spathe hood is green with white or purple stripes. On the more northern *A. pusillum*, the spathe tube is not fluted and the spathe hood will be green or purple and not starkly striped. Both of these species are thought to be fairly rare, imperiled in many areas, and may be protected.

CONSERVATION: ⓖ⑤
Threatened in Manitoba.

■ NATIVE
■ RARE OR EXTIRPATED
■ INTRODUCED

ALTERNATE NAMES: Water-dragon, Wild Calla

Water Arum is a native marsh plant from the arum family which occurs more to the north than all others in North America and follows a circumpolar range. Its attractive white flower is pollinated mainly by syrphid flies which leads to clusters of pear-shaped red berries in late summer. Rarely, some individuals will have a genetic mutation that forms two or three spathes (the white, showy "petal"), which botanists refer to as *C. palustris* forma *polyspathacea*. The Potawatomi, who called the plant *wabasi'pini'bag*, meaning "white potato leaf" or "swan potato root," found the root of the Water Arum could reduce swelling once pounded and applied as a poultice.

DESCRIPTION: The Water Arum has a showy white spathe, 2" (5 cm), that partially surrounds a shorter, upright flower spike which is known as the spadix, 1" (2.5 cm). It is from the spadix that the tiny, true flowers erupt, yellow and petalless. Numerous dark green and heart-shaped oblong leaves, 6" (15 cm) long, with 6–12" (15–30 cm) petrioles that grow above shallow water or mucky soils.

FLOWERING SEASON: Late May–August.

HABITAT: Bogs, wetlands, and pond shores.

RANGE: Northern North America, Alaska east to Newfoundland, south to western mountains of Maryland, northwest to Minnesota and North Dakota.

CAUTION: The berries, foliage, and roots of this plant contain calcium oxalate crystals and will cause painful irritation of the mouth and throat if ingested.

CONSERVATION: G5

Threatened in Newfoundland, Labrador, Illinois, Indiana, Maryland, North Dakota, Rhode Island, New Jersey, and Ohio.

YELLOW SKUNK CABBAGE *Lysichiton americanus*

■ NATIVE
■ RARE OR EXTIRPATED
■ INTRODUCED

ALTERNATE NAMES: Western Skunk Cabbage, American Skunk-Cabbage, Swamp Lantern

An early-blooming western native, the descriptive common name refers to the unpleasant odor of the crushed leaves and sap as well as the musky scent and color of the flowers. The odor attracts its primary pollinator, a species of rove beetle, which feeds on the pollen and uses the flowers as a mating location. The plant was used widely by indigenous peoples from more than 25 different tribes as medicine, as well as using the large, waxy leaves to make cups, cooking tools, and storage containers for food. Because extensive preparation was needed to make the plant edible, it was rarely eaten and then only as a famine food. After hibernation, bears often will dig up and eat the roots. This is believed to be done in part to help remove the fecal plug that was formed prior to denning.

DESCRIPTION: A showy, yellow spathe up to 8" (20 cm) long, partially surrounding a shorter, upright, flower spike known as the spadix. It is from the spadix that the tiny, true flowers erupt, yellow and petalless. Often the flower emerges before the cluster of enormous oval leaves, 1–5' (30–150 cm) long on short, thick petioles, which hide the berries later in the season.

FLOWERING SEASON: April–July, often as soon as the snow melts.

HABITAT: Swampy soil.

RANGE: From Alaska to near the coast in central California, east to Montana.

CAUTION: The berries, foliage, and roots of this plant contain calcium oxalate crystals and will cause painful irritation of the mouth and throat if ingested.

CONSERVATION: G5

Least concern in North America; this species is considered invasive throughout the European Union.

■ NATIVE
■ RARE OR EXTIRPATED
■ INTRODUCED

ALTERNATE NAMES: Green Arrow Arum, Tuckahoe

This aquatic perennial is common throughout the East Coast, sometimes forming large colonies along shallow streams and marshes. It has continued to spread outside its historical range in the past 40 years, with new populations now occurring in Iowa, Kansas, Minnesota, West Virginia, and Wisconsin. The spread is believed to be caused by seed dispersal from migratory birds; however, occurrences in Oregon and California are believed to be solely from human introduction.

Notably, Arrow Arum has a symbiotic relationship with a chloropid fly, upon which it relies for pollination. Not only are the male and female flowers produced at different times within the same plant, but the leaf-like spathe gradually closes to fully encompass the spadix, preventing it from being pollinated by the wind. The flies are drawn to chemicals released by the spathe; they eat the pollen and cross-pollinate the flowers as they move between flowers to lay their eggs on the spadix. As the flower matures, its smell changes from sweet to fetid, which helps cue the flies as to the most opportune time to deposit their eggs.

DESCRIPTION: Showy, large, arrow-shaped leaves from long stalks with prominent veins, 1–2' (30–60 cm) in length. It has a greenish spathe, 4–7" (10–17.5 cm), perhaps with cream to yellow edges, which may be closed or open. Inside is a long, tapering rod-like spadix, green to yellowish. The berry cluster may be green to very dark purple-green.

FLOWERING SEASON: May–June.

HABITAT: Edges of ponds and slow-moving rivers, bogs, swamps, and marshes.

RANGE: Ontario east to the Gulf of St. Lawrence, south to Florida, west to Texas, and north through Missouri to Minnesota. Naturalized in California and Oregon.

CAUTION: The berries, foliage, and roots of this plant contain calcium oxalate crystals and will cause painful irritation of the mouth and throat if ingested.

CONSERVATION: G5
Threatened in Ontario, Quebec, Iowa, Oklahoma, West Virginia, and Vermont.

■ NATIVE
■ RARE OR EXTIRPATED
■ INTRODUCED

ALTERNATE NAMES: Eastern Skunk Cabbage, Swamp Cabbage, Clumpfoot Cabbage, Meadow Cabbage, Foetid Pothos, Polecat Weed

One of the first plants to sprout in late winter, the shell-like spathe of the Skunk Cabbage can often be seen appearing directly through snow cover. This emergence is made possible by the plant's ability to create its own heat, helping it melt through snow and ice. This also helps to spread its namesake odor, which attracts cold-blooded insects and provides them a warm, sheltered location as they help pollinate the plant.

All parts of the plant will emit a skunk-like odor if bruised or cut, leading to its common name. The massive leaves disappear by late summer as the plant becomes dormant. The roots of the plant contract each year, pulling the plant down deeper into the boggy earth, making it nearly impossible to uproot mature specimens.

DESCRIPTION: The first sign of this plant is a large, 3–6" (7.5–15 cm) long, brownish-purple and green shell-like spathe, enclosing a knob-like, yellowish to dark red-purple spadix covered with tiny flowers. In later spring, a tight roll of fresh green leaves emerges beside the spathe, unfolding to form huge, dark green, deeply veined leaves, 1–2' (30–60 cm) long and to 1' (30 cm) wide, on stalks rising directly from the ground. Overall height: 1–2' (30–60 cm).

FLOWERING SEASON: February–May.

HABITAT: Open swamps and marshes, wet woodlands, and stream sides.

RANGE: North Carolina, west to Tennessee, north to Minnesota and Ontario, east to Nova Scotia.

CAUTION: The berries, foliage, and roots of this plant contain calcium oxalate crystals and will cause painful irritation of the mouth and throat if ingested.

CONSERVATION: G5

Threatened in New Brunswick, Nova Scotia, Tennessee, Iowa, North Carolina, and Illinois.

■ NATIVE
■ RARE OR EXTIRPATED
■ INTRODUCED

ALTERNATE NAMES: Northern Water Plantain, Large-Flowered Water Plantain

Emerging from shallow, still water, the Water Plantain is a somewhat tall, leggy plant with small, white (rarely pink) flowers at the ends of branching stems. Leaves formed underwater are ribbon-like and soon rot; these are typically only found on new plants. Mature plants have broad, elliptical leaves emerging above the water on slender stalks.

The plant was used widely by Native American tribes for dozens of medicinal purposes as well as a food source. The Mohegan Nation is said to have used the rough side of fresh leaves to draw out poisons from snake and insect bites.

Some texts may list North American species of *Alisma* as the European Water-plantain, *A. plantago-aquatica* (sometimes with var. *americanum* added). Since 1970, however, botanists have recognized clear differences between the species, most notably the more pointed leaves of the European variety compared to the more rounded North American species.

DESCRIPTION: Leaves are up to 7" (18 cm) long and 3½" (9 cm) across, usually elliptical with smooth edges. Flowers are small, in open clusters of branching stems, rising from the center of the basal leaves. Its total height is 6"–3' (15–91 cm).

FLOWERING SEASON: July–August.

HABITAT: Shallow water; muddy shores, marshes and ditches.

RANGE: Throughout much of North America except in the Southeast, from Texas northeast to Virginia. It is suspected that West Virginia has a suitable climate for the plants, however the species has not been recorded in the state.

SIMILAR SPECIES: Very similar to Southern Water Plantain (*A. subcordatum*), which has smaller petals roughly half the length, ¹⁄₁₆–⅛" (1–3 mm) compared to ⅛–¼" (3.5–6 mm) on *A. triviale*.

CONSERVATION: **G5**

Threatened in Illinois, New Jersey, Oklahoma, Ohio, Pennsylvania, and Wyoming in the U.S., and Yukon, Newfoundland, and the Northwest Territories in Canada.

■ NATIVE
■ RARE OR EXTIRPATED
■ INTRODUCED

ALTERNATE NAMES: Arum-leaf Arrowhead, Wapato, Northern Arrowhead, Duck Potato, Wapatum Arrowhead

A native wetland plant, often seen with attractive, green foliage when exposed above the water, the leaves are highly variable. Underwater leaves may be dramatically different between locations. While generally having more narrow leaves than the closely related Broadleaf Arrowhead (*S. latifolia*), true differentiation can be made between these plants by comparing the length of the side lobes to the main lobe on leaves that are emerging from the water. Narrowleaf Arrowhead will have side lobes significantly shorter than the main lobe, whereas on Broadleaf Arrowhead, the side lobes will be as long as the main lobe.

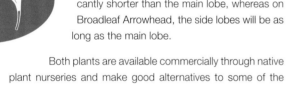

Both plants are available commercially through native plant nurseries and make good alternatives to some of the more invasive plants that overtake our wetlands. Within their shared ranges, Broadleaf Arrowhead is typically the more common of the two, both commercially and in nature, and many older texts do not differentiate between these two species.

DESCRIPTION: A widespread aquatic plant, usually emergent from water. Leaves are variable, typically 2½" (6.5 cm) wide, with two-pointed side lobes shorter than the main lobe and on long, erect stalks from the base. It grows to 1–3' (25–76 cm) above the water. Male flowers are 1¼" (3 cm), with three white, round petals, in whorls of three high on the stem above female flowers, which are green, ball-like, and whorled lower on the stem.

FLOWERING SEASON: June–September.

HABITAT: Ponds, swamps, quiet streams, rivers, and banks and shores.

RANGE: Across the southern tier of Canadian provinces, north into Alaska and the Northwest Territories, and south across the northeastern United States and throughout much of the West south to northern Texas and central New Mexico, Arizona, and California.

SIMILAR SPECIES: Broadleaf Arrowhead (*S. latifolia*).

CONSERVATION: **G5**

Threatened in Connecticut, Iowa, Massachusetts, New Hampshire, New Jersey, Ohio, Oklahoma, Pennsylvania, and Wyoming in the U.S. and Labrador and Yukon in Canada.

■ NATIVE
■ RARE OR EXTIRPATED
■ INTRODUCED

ALTERNATE NAMES: Bigleaf Arrowhead, Duck-potato, Wapato

Unlike the similar Narrowleaf Arrowhead (*S. cuneata*), which is not found in the Southeast, Broadleaf Arrowhead enjoys a wide range across nearly all of North America except for the most northern extremes as well as Nevada. A highly variable species, there have been attempts by taxonomists at various points to subdivide this plant into many different species based on specific characteristics. For instance, when found in the Southeast, the entire plant is often covered in fuzzy hairs not seen on this species from other locales. However, the most recent evidence suggests these minor differences alone are not enough to be recognized separately.

This plant produces edible starchy tubers that float on the water once loosened from the muck in which it grows. These are often uprooted and eaten by beavers and muskrats and are sometimes called "duck potatoes." Ironically, ducks do not eat the tubers, but are fond of the seeds. In late fall in Oregon, Lewis and Clark noticed Native Americans uprooting these from the Deschutes River delta and soon tried it themselves, roasting the tubers in the embers of their fire. Clark said it reminded him of a "small Irish potato" and the plant helped sustain the party through the winter, often boiled in a stew with elk.

DESCRIPTION: A widespread aquatic plant, usually emergent from water. It has long-stalked arrow-shaped leaves, 2–16" (5–40 cm) long, with two backward pointing side lobes nearly as long as the main lobe. it grows to 1–4' (30–120 cm) above water. Flowers 1¼" (3 cm), with three white, round petals in whorls of three high on the stem.

FLOWERING SEASON: July–September.

HABITAT: Wet sites or shallow water along lake and stream edges; ditches, ponds, marshes, and swamps.

RANGE: Throughout North America, except the Far North (above 60°N latitude) and deserts.

SIMILAR SPECIES: Narrowleaf Arrowhead (*S. cuneata*). Broadleaf Arrowhead has some resemblance to the Arrow Arum (*Peltandra virginica*), which is toxic. The Arrow Arum leaf is veinless and the flower is very different.

CONSERVATION: G5
Threatened in Arizona, Wyoming, Newfoundland, and Alberta.

MOUNTAIN DEATHCAMAS *Anticlea elegans*

NATIVE
RARE OR EXTIRPATED
INTRODUCED

ALTERNATE NAMES: *Zigadenus elegans,* Elegant Death-camas, Alkali Grass

Members of the bunchflower family typically have bunches of little white or greenish, lily-like flowers each with 3 sepals and 3 petals, identical in size and color, as well as six stamens, and a three-parted pistil. Most plants in the family are toxic, and the Mountain Deathcamas earned its foreboding common name due to the severe illness and death it can cause to both humans and livestock. Even bees have been poisoned by the nectar and pollen of some bunch-flower species.

Unfortunately for many of its victims, the bulbs of this plant can look very similar to blue camas, wild onion, and other plants that are considered quite tasty and sought after. As the tubers were often dug in the late fall, well after the flowering season, mistakes were often fatal. Native Americans were reported to use sticks to dig up these plants and weed them out when they bloomed because they knew they could be lethal.

DESCRIPTION: Long, grass-like leaves, 6–12" long (15–30 cm) alternately arranged below a raceme of 10 to 50 greenish-white, bowl-shaped flowers, each ¾" (2 cm) wide, the total inflorescence up to 12" (30 cm) long. Overall height: 6–28" (15–70 cm).

FLOWERING SEASON: June–August.

HABITAT: Sunny prairies and meadows, mountain valleys, and forests.

RANGE: Western Canada; south to western Washington, eastern Oregon, Arizona, New Mexico, and Texas.

CAUTION: All parts of the plant are poisonous. Eating it can cause headache, vomiting, tremors, loss of muscle control, heart failure, coma, and death. The toxicity of the plant is due to one of several different alkaloids—primarily a steroidal alkaloid called zygacine—which are contained in all parts of the plant.

SIMILAR SPECIES: Mountain Deathcamas was formerly included in the lily family, an enormous family that included over 3,700 species. Efforts to divide the family have resulted in the suggestion of over 70 different families. Recently there has been consensus to adopt some of those divisions and many plants that were formerly included in the lily family are now included in the bunchflower, amaryllis, and asparagus families.

CONSERVATION: G5

Endangered in Nebraska, Vermont, West Virginia, and New Brunswick. Threatened in New York, Ohio, Virginia, and Quebec.

- ■ NATIVE
- ■ RARE OR EXTIRPATED
- ■ INTRODUCED

ALTERNATE NAMES: Eastern Featherbells

Featherbells is a native, perennial wildflower that is becoming rare to find in the wild, currently listed as endangered in three states (Kentucky, Illinois, Indiana) and threatened or at-risk in several others. Featherbells has attractive, showy, and fragrant flowers that often attract bees. The genus name is based on the Greek words *stenos* for "narrow" and *anthos* for "flower," which refers to the narrow, pointed shape of each sepal and petal as well as to the general shape of the whole flowering structure.

DESCRIPTION: A long, narrow, feathery plume of small, white, nodding flowers on a stem rising from the middle of narrow, folded, grass-like leaves, 8–16" (20–40 cm) long, up to 3–5' (90–150 cm) high. Individually, flowers are about ½" (1.5 cm) wide; each with 3 sharply pointed petals and 3 petal-like sepals.

FLOWERING SEASON: June–September.

HABITAT: Open, rocky woodland and sandy bogs.

RANGE: Pennsylvania south to Florida, west to Texas, and northeast to Missouri and Michigan.

CAUTION: Information on the toxicity of Featherbells is scant; however, as a member of the Bunchflower Family, all plants should be presumed to be toxic.

SIMILAR SPECIES: When in flower, it is nearly impossible to mistake this plant for anything else. However, based only on the leaves, this plant can be difficult to identify and may even be mistaken for a grass by a casual observer.

CONVERSATION: G4
This plant is not particularly common in any part of its range. It is threatened in Arkansas, Illinois, Indiana, Kentucky, Louisiana, Maryland, Mississippi, North Carolina, Ohio, Oklahoma, and West Virginia.

■ NATIVE
■ RARE OR EXTIRPATED
■ INTRODUCED

ALTERNATE NAMES: Painted Wakerobin, Painted Lady

Trilliums in general are very attractive native spring ephemerals. These long-lived, woodland, perennial flowers are known for having a trio of most of their flower parts, leading to both their common names and genus. All species of Trilium are slow growing and fragile, taking years to generate energy needed to flower and nearly always dying if leaves are picked. No species of trillium should be picked, no matter how common the plants may be.

Painted Trillium is a colorful species, producing a single white flower with a splash of bright pink around the center of the petals. While the plants themselves are immediately recognizable, their taxonomy has been unsettled. There are over 50 species within the *Trillium* genus, 39 of them native to North America. Some sources treat *Trillidium* as a distinct genus, while others treat the genera as synonymous. In the past, these were included in their own family, Trilliaceae, and later with the lilies. In 2016 based on molecular genetics, taxonomists moved them into the Melanthiaceae family, subdivided into a tribe called Parideae.

DESCRIPTION: Grows 8–20" (20–50 cm) tall, with three blue-green leaves, 2½–5" (6.5–12.5 cm) long. At the top of the stem has a single white, 3-petaled flower with bright pink at its center. Foliage dies back as the plant is dormant by mid-summer.

FLOWERING SEASON: April–June.

HABITAT: Moist, acidic, humus-rich woods and swamps.

RANGE: Ontario east to Nova Scotia, south to Georgia, and north to Tennessee, Ohio, and Michigan.

CAUTION: Berries and roots are poisonous.

CONSERVATION: **G5**

Endangered in Ohio and Michigan. Threatened in Kentucky, New Jersey, Rhode Island, South Carolina, and Georgia.

■ NATIVE
■ RARE OR EXTIRPATED
■ INTRODUCED

ALTERNATE NAMES: Whip-poor-will-flower, Northern Nodding Trillium, Nodding Wakerobin

Nodding Trillium is the northernmost-occurring trillium, with populations reaching into southern Canada. The flowers are often difficult to spot because they hang down from short stems below the plant's broad leaves. As with all trilliums, individual plants will only produce one set of leaves and a single flower each year. If a plant is foraged by a deer or if the seed or flower is picked, the plant will become dormant for the rest of the year. While the first time it happens it may not kill the plant, the reduction in energy produced in the leaves will likely prevent flowering the following year. Repeated browsing will quickly exhaust the limited energy reserves in the rhizome and the plant will die.

While not a preferred food source for wildlife, overpopulation of deer will quickly eat all the tastier springtime growth. With limited choices, they will preferentially browse large reproductive individuals, reducing the capacity of the plant population to reproduce, eventually leading to site extirpations. No species of trillium should be picked, no matter how common the plants may be.

DESCRIPTION: The iconic trio of 2½–4" (6.5–10 cm) long, diamond-shaped, green leaves with whitish (rarely pale pink) flowers (3–5 cm) with pinkish-purple anthers, bent down ("nodding") and hidden by the leaves. Overall height: 6–24" (15–60 cm).

FLOWERING SEASON: April–May.

HABITAT: Moist, acidic woods and swamps.

RANGE: Saskatchewan east to Newfoundland, south to Virginia, west to Illinois, and north to the Dakotas.

SIMILAR SPECIES: Nodding Trillium is very similar to Bent Trillium (*T. flexipes*), which has flowers with white anthers, rather than pink.

CONSERVATION: G5

Endangered in Illinois and West Virginia. Threatened in Delaware, Iowa, Maryland, New York, Pennsylvania, South Dakota, Vermont, Virginia, and Saskatchewan. It is presumed extirpated in Ohio.

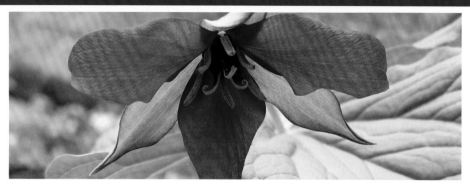

■ NATIVE
■ RARE OR EXTIRPATED
■ INTRODUCED

ALTERNATE NAMES: Purple Trillium, Stinking-Benjamin, Wake-robin, Wet Dog Trillium, Ill-scent Trillium, Birthwort

The colorful alternate names for this plant refer to the unpleasant odor, color, and appearance of the flower, which is said to resemble rotting meat and helps attract flies and carrion beetles as pollinators. A fairly common plant throughout its range, this plant was widely used by many different Native American tribes for various purposes, including as an aid in childbirth, as a poultice to treat tumors and ulcers, and as a remedy for coughs. The Cherokee used it for everything from bowel complaints to asthma. Recent science has now isolated a chemical termed "steroidal saponins" from trilliums, and Red Trillium in particular, and purified forms of these have been found to have some efficacy in combating cancers, fungal infections, and inflammatory disorders.

DESCRIPTION: Single stem that grows 6–20" (15–50 cm) with a whorl of 3 widely spreading, diamond-shaped, dark green leaves, up to 7" (17.5 cm) long. Stems are purplish-green and smooth, 8–16" (20–40 cm). Flowers are purple or maroon, sometimes white, about 2½" (6.5 cm) wide.

FLOWERING SEASON: April–June.

HABITAT: Rich woodlands.

RANGE: Manitoba east to Nova Scotia, south to the northern extremes of Georgia and Alabama, and north, sparsely in Illinois and Michigan.

SIMILAR SPECIES: Vasey's Wake Robin (*T. vaseyi*), found in the southern Appalachians, is larger in all respects and has pleasant-smelling flowers.

CONSERVATION: **G5**

Endangered in Delaware, Illinois, and Rhode Island. Threatened in New Jersey. Today, collection of wild populations of this plant for medicinal use is considered one of the greatest contributors to its population decline. As is the case with many *Trillium* species, Red Trillium is a slow-growing plant which may take years to flower from seed, limiting cultivation of the plant as well as exacerbating the effects of harvesting. No species of trillium should be picked, no matter how common the plants may be.

■ NATIVE
■ RARE OR EXTIRPATED
■ INTRODUCED

ALTERNATE NAMES: Large-flower Wakerobin, White Trillium, Large-flowered Trillium, Great White Trillium, White Wake-robin

This is the largest, showiest trillium in North America, and is popular as a native garden flower. As with all trilliums, seeds are naturally dispersed by ants (a process called myrmecochory), which typically will only spread the seeds about 30' (10 m). Occasionally deer or other wildlife will eat the fruits and spread the seeds farther, leading to patchwork patterns of trillium throughout the woods. Height of *Trillium* plants has been used to estimate deer population and browse pressure, as deer prefer to eat the tallest individual plants.

The white flowers of the Large-flower Trillium turn a distinctive pink and remain so for several days just before the wilting of the flowers. This species of *Trillium* is the most likely to form double flowers from a genetic mutation; however, these are almost always sterile. Other alterations of the plant, including green marks within the typically stark white petals, as well as forming up to 30 petals within a single plant, are not genetic variations, but instead caused by a bacterial infection called a phytoplasma. As the infection progresses, the mutations increase and slowly the infection will kill the plant.

DESCRIPTION: One large, 2–4" (5–10 cm), odorless, waxy-white, 3-petaled flower, turning pink with age, on an erect stalk above a whorl of three broad, diamond-shaped leaves, 3–6" (7.5–15 cm). Height is dependent on deer browse, typically 8–18" (20-45 cm).

FLOWERING SEASON: April–June.

HABITAT: Rich woods, thickets, usually basic or neutral soils.

RANGE: Ontario east to Nova Scotia, south to Georgia and Alabama, and north to Illinois and Minnesota.

CONSERVATION: (G5)

Presumed extirpated in Maine. Endangered in Alabama, New Jersey, South Carolina, and Nova Scotia. Threatened in Georgia, Illinois, and Quebec. Due to their slow-growing nature—when planted by seed, it takes a minimum of five years before the first flower appears and more commonly 10—it is often considered impractical to raise trilliums from seed commercially. As a result, the majority of plants and rhizomes in commerce are collected from the wild. This heavy collecting, combined with other pressures such as habitat destruction and grazing, has led to dramatic reduction of these plants in some areas. No species of trillium should be picked, no matter how common the plants may seem to be.

■ NATIVE
■ RARE OR EXTIRPATED
■ INTRODUCED

ALTERNATE NAMES: Western Wakerobin, Coast Trillium, Pacific Trillium

Western Trillium was initially defined based on a specimen growing along the Columbia River and gathered in 1806 by Meriwether Lewis during the return trip of the Lewis and Clark expedition. It is the most widespread of the trilliums occurring in western North America. It was widely used by Native Americans for medicine including the Lummi and Paiute tribes, which used juice from smashed plants to make eye drops. The Skagit tribe considered the plant poisonous, and today we still recognize that the berries, roots, and other parts are indeed poisonous.

DESCRIPTION: A low-growing plant with 1 white flower, 1½–3" (3.8–7.5 cm), which usually opens white and becomes pink with age. In parts of California and Oregan, it can become deep red. Single stem that grows 4–16" (10–40 cm) with a whorl of 3 broad, oval leaves, 2–8" (5–20 cm) long.

FLOWERING SEASON: February–June.

HABITAT: Coniferous and mixed forests, along stream banks and around alder shrubs. Along the California coast, it can be found often under redwoods.

RANGE: British Columbia to central California; east to northwestern Colorado, Montana, and Alberta.

CAUTION: All parts are poisonous; poison symptoms include burning of the mouth and throat, headache, stomachache, and convulsions.

SIMILAR SPECIES: This species closely resembles the Largeflower Trillium (*T. grandiflorum*) of the eastern U.S. There is only one other Trillium species in the West that has a stalk between the flower and the leaves: Klamath Trillium (*T. rivale*), found in southwestern Oregon and northwestern California. Other western Trilliums include the Giant Wake Robin (*T. chloropetolum*), which grows in dense patches west of the Cascade Range and in the Sierra Nevada. It has stalkless, mottled leaves and its petals vary from white to maroon. Roundleaf Trillium (*T. petiolatum*) is found in eastern Washington and Oregon; it has long-stalked leaves and dark red-brown petals.

CONSERVATION: G5
Endangered in Alberta. Threatened in Colorado and Wyoming. No species of trillium should be picked, no matter how common the plants may seem to be.

Veratrum californicum **CALIFORNIA CORN LILY**

■ NATIVE
■ RARE OR EXTIRPATED
■ INTRODUCED

ALTERNATE NAMES: California False Hellebore, Wild Corn, Cow Cabbage

The California Corn Lily is a tall plant topped with feathery clusters of white to greenish flowers. But their bright green, pleated leaves and all other parts are extremely poisonous. Eaten carelessly, the plant will cause heart conditions which could easily kill people and livestock if medical care is not promptly received. And the flowers can even be toxic to insects—while native bees can pollinate the plant without ill effect, it can wipe out colonies of domesticated European Honey Bees (*Apis mellifera*) used in commercial honey production.

The plant is host to multiple native moth caterpillars, which likely sequester the toxic chemicals and become toxic and distasteful to potential predators. Native Americans primarily used the plant for a variety of external medicinal and cosmetic uses. For instance, the Karok tribe used the inner white stem torn into ribbons as decorations braided into girls' hair. Today, the plant and extracts from it are used to make insecticides as well as blood pressure and heart medications.

DESCRIPTION: A tall plant reaching 4–8' (1.2–2.4 m) with a stout, unbranched, leafy stem topped with long, branched, dense clusters of relatively small, whitish or greenish flowers, each ½–¾" (1.5–2 cm) long. Upward slanted leaves are 8–12" (20–30 cm) long, numerous, and pleated without stalks.

FLOWERING SEASON: June–August.

HABITAT: Swamps and creek bottoms, wet meadows and moist forests.

RANGE: Western Washington to southern California; east to New Mexico, western Wyoming, and Montana.

CAUTION: All parts of the plant are poisonous from alkaloids. Eating it can cause headache, vomiting, tremors, loss of muscle control, heart failure, coma, and death.

CONSERVATION: G5
Endangered in Wyoming. Threatened in Montana.

■ NATIVE
■ RARE OR EXTIRPATED
■ INTRODUCED

ALTERNATE NAMES: American False Hellebore, Indian Poke, Indian Hellebore, False Hellebore, Green False Hellebore, Giant False Helleborine

This species occurs in two distinct populations: one eastern and the other western. This divide is believed to have been caused by glaciation of the middle continent. The pleated, yellow-green leaves of this wetland plant are unmistakable in spring, which is beneficial as all parts of the plant are highly toxic. Fortunately, attempting to eat the foliage provides a strong burning taste—and soon thereafter vomiting—so it is rarely eaten in sufficient quantities to be lethal. It was used to create blood pressure medicines in the 1950s and 1960s, but fell out of favor due to their strong side effects, including seizures and temporary paralysis.

Several poisonous alkaloids are present, especially in young shoots. There are anecdotes about careless people bathing with

hellebore who experienced violent seizures and even death. In Chinook, a historical trade language in the Pacific Northwest, Green Corn Lily is known as skookum root, which is translated to mean "strong and powerful."

DESCRIPTION: Large, 6–12" (15–30 cm) long, pleated leaves are spirally arranged on a 6' (180 cm) tall, stout stem topped by branching clusters of greenish, star-shaped, hairy flowers, each about ½" (1.5 cm) wide. Overall height: 2–7' (60–210 cm).

FLOWERING SEASON: July–August.

HABITAT: Wet soils in swamps, wet woods, and meadows.

RANGE: Quebec east to Nova Scotia, south to Georgia and Alabama, and north to Tennessee, West Virginia, and Ohio; also from Alaska and Northwest Territories south to Wyoming, Oregon, and California.

CAUTION: Plants of the genus *Veratrum* are toxic to humans and animals if ingested. Both the rootstock and foliage of this species are poisonous. Although the latter has a burning taste and is usually avoided by animals, it can be lethal. Toxicity can vary in a plant according to season, the plant's different parts, and its stage of growth.

SIMILAR SPECIES: The two other false hellebores within its range have hairless flowers. These are also toxic. Small-Flowered False Hellebore (*V. parviflorum*), with leaves stalked and mostly from the ground, occurs in drier woods from Virginia and Kentucky south to Georgia and Alabama. Wood's False Hellebore (*V. woodii*), with greenish to blackish-purple flowers, is found in dry woods from Iowa and Missouri east to Ohio and south to Florida, Alabama, Tennessee, and Oklahoma.

CONSERVATION: (G5)

Endangered in Nova Scotia. Threatened in Georgia and Labrador.

■ NATIVE
■ RARE OR EXTIRPATED
■ INTRODUCED

ALTERNATE NAMES: Merrybells

A clump-forming plant that eventually grows up to 2' (60 cm) tall (but in bloom is typically half that) with droopy yellow, bell-shaped flowers, often with twisted petals. The flowers provide nectar to a variety of native bees, including a rare mining bee, *Andrena uvulariae,* which is a specialist, collecting pollen only from flowers in the genus *Uvularia.*

This is a slow-growing species. From seed, the first year is focused on root development, and it often takes 2–3 years for foliage to appear and 5–8 years before the first flowers will bloom. Due to these factors, commercial cultivation is difficult and oftentimes these unique and attractive plants are harvested from the wild. A fairly sensitive species, the presence of this flower in woodlands is a good indication that much of the original flora is still intact. However, it is heavily grazed by deer and is unlikely to be found in woodlands with high numbers of deer present.

The Potawatomi prescribed an infusion of the root for backaches and added it to a lard-based salve to massage sore backs and tendons. Current studies are investigating chemicals from this plant for their effectiveness as a treatment for boils, wounds, arthritis pain, and backaches.

DESCRIPTION: Pale green, oval, and droopy leaves that look to be pierced by the stem (perfoliate) and are alternate, each up to 6" (15 cm) long and 2" (5 cm) across; underside is greenish-white with fine hairs. Stems will be pulled over from their own weight and topped with a bell-shaped, pale yellow flower, possibly hidden by the leaves.

FLOWERING SEASON: March–May. A 30-year comparison shows this plant is now blooming 21 days earlier in New York due to the warming climate.

HABITAT: Rich, deciduous woods; thickets; floodplain woods.

RANGE: Southern Quebec, east to Vermont, south to Georgia and Oklahoma, north to Minnesota and the Dakotas.

SIMILAR SPECIES: Perfoliate Bellwort (*U. perfoliata*) has leaves that are more upright than the droopy leaves of the Large-flower Bellwort and is slightly smaller, typically only 8" (20 cm) when in flower and reaching 20" (50 cm). The main differentiation is the orange to dark yellow or golden granule-like bumps on the petals, which are not present on Large-flower Bellwort.

CONSERVATION: G5

Endangered in Connecticut and New Hampshire. Threatened in Kansas, Maryland, Massachusetts, and Oklahoma.

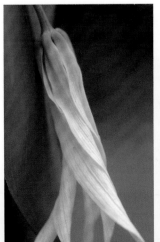

- ■ NATIVE
- ■ RARE OR EXTIRPATED
- ■ INTRODUCED

ALTERNATE NAMES: Sessile-leaf Bellwort, Wild Oats, Merrybells, Spreading Bellwort, Straw Lily

This is the most common of the bellworts, especially in New England. After flowering, ants are attracted to a small, fatty appendage on the seed called an elaiosome. An ant carrying the seed will realize the seed itself is not edible, dropping it near its nest while taking the elaiosome. Clusters of flowers are typically clonal, spreading asexually from its horizontal roots (rhizomes).

Based on the Doctrine of Signatures—a belief from the 16th–18th centuries that supposed that plants resemble the parts of the body which their use could heal—these plants were thought to be good for treating throat diseases because the drooping flowers resemble the uvula, the soft lobe hanging into the throat. This is the basis for the genus name *Uvularia*.

DESCRIPTION: A drooping stem with alternate, erect leaves, each 1¾–3" (4.5–7.5 cm), topped by a 1" (2.5 cm) long creamy-yellow to white flower with 6 petals, possibly obscured by the leaves. Total height: 6–12" (15–30 cm).

FLOWERING SEASON: April–June.

HABITAT: Woods and thickets.

RANGE: Manitoba east to Nova Scotia, south to Florida, west to Louisiana, and north to Oklahoma, Missouri, Iowa, and the Dakotas.

SIMILAR SPECIES: Mountain Bellwort (*U. pudica*) has shiny leaves and stems in clumps. Florida Bellwort (*U. floridana*) has a small, leaf-like bract on the flower stalk.

CONSERVATION: G5

Threatened in Illinois, Iowa, Kansas, Louisiana, North Dakota, and Manitoba.

■ NATIVE
■ RARE OR EXTIRPATED
■ INTRODUCED

ALTERNATE NAMES: Smooth Carrionflower, Jacob's-ladder, *Smilax herbace*

A tall, attractive herbaceous vine which uses its tendrils to climb over other plants. The often unpleasant-smelling spherical flower clusters attract many different insects, especially carrion flies (hence the common name), which serve as pollinators. When first sprouting, it can resemble asparagus, but quickly grows vertically up to 8' (2.4 m) before it must find support or fall over. Closely related to greenbriars, many catbriers are woody, thorny vines; however, carrionflowers are thornless. The Cherokee used powdered leaves for burn dressings and the Iroquois used powdered root as a deodorant. The berries are eaten by game birds and deer.

DESCRIPTION: A thornless vine, 3–9' (90–270 cm), with light green to purple stems and broad pale green leaves 5" (12.5 cm) long and pale green underneath. Spherical clusters of small, ½" (1.5 cm) wide, green or yellowish flowers extended on stalks.

FLOWERING SEASON: May–June.

HABITAT: Rich moist forests and meadows, prairies.

RANGE: Ontario east to New Brunswick, south to Georgia, west to Louisiana and Oklahoma, and north to Kansas, Iowa, and Minnesota.

SIMILAR SPECIES: Common Carrion Flower (*S. lasioneura*) leaves are pale green with fine hairs along the veins of its leaf undersides. Powdery Carrion Flower (*S. pulverulenta*) leaves are medium green with fine hairs along their veins.

CONSERVATION: **G5**
Threatened in Illinois, Kansas, Michigan, Texas, Wisconsin, and Quebec.

■ NATIVE
■ RARE OR EXTIRPATED
■ INTRODUCED

ALTERNATE NAMES: Elegant Mariposa Lily, Star Tulip, Elegant Sego Lily

A western U.S. native plant, the common name refers to each petal's resemblance to a kitten's ear, and the alternate name Star Tulip refers to the overall shape of the flower. The bulb is edible raw or cooked, and reportedly the raw bulb tastes like a raw new potato. They were eaten by many Native American tribes and were widely used by settlers in Utah when food was scarce. The Acoma and Laguna tribes used tea made from the plant to treat rheumatic swelling and by the Ramah Navajo to ease the delivery of the placenta after birth.

DESCRIPTION: A small plant with a cluster of greenish-white to pale purple flowers, each 1" (2.5 cm) with 3 sepals and 3 broad, densely hairy petals, which have a fringed membrane on the lower side. The 4"–8" (10–20 cm), grass-like, green, hairless leaves typically extend above the flowers.

FLOWERING SEASON: May–June.

HABITAT: Grassy hillsides, bedrock meadows and open coniferous woods.

RANGE: Northern California to Washington and western Montana.

CONSERVATION: G3

This plant is vulnerable with a limited range, and harvesting should be avoided. Flowers within the genus may be very similar; petal shape, gland shape, and locations of hair are highly variable within a species, but careful attention to these details are often important in identifying species.

- ■ NATIVE
- ■ RARE OR EXTIRPATED
- ■ INTRODUCED

ALTERNATE NAMES: Bigpod Mariposa

Found in the arid Northwest, if you catch this plant in bloom, you'll find a handsome cluster of up to three lilac-colored flowers atop a tall stem. Any other time, its short, grass-like leaves are likely to be overlooked. The flowers are pollinated by over a dozen species of native bees as well as beetles and other insects.

The bulb of this plant was used as food by the Nlaka'pamux (the First Nation peoples of what is now British Columbia) and many other tribes. A variation of this species is found in southeastern Washington and west-central Idaho. There, it is a white flower with a reddish stripe on each petal.

DESCRIPTION: An unbranched, erect stem, 8–20" (20–50 cm) long, with a loose cluster of 1–3 bell-shaped, lilac flowers, each 1½ –2½" (3.8–6.3 cm) wide, with 3 broad, fan-shaped petals. Leaves are 2–4" (5–10 cm) long, with several on the stem, very narrow, usually curled at the tip.

FLOWERING SEASON: May–August.

HABITAT: Loose, dry soil on plains, among sagebrush, and in yellow pine forests.

RANGE: The eastern side of the Cascade Mountains from southern British Columbia to northern California; east to northern Nevada and across central and northern Idaho to western Montana.

CONSERVATION: G5
This species is not of special conservation concern.

■ NATIVE
■ RARE OR EXTIRPATED
■ INTRODUCED

ALTERNATE NAMES: Yellow Clintonia, Yellow Bluebead Lily

The common name of this shade loving perennial refers to the wide-spaced cluster of small, round, true-blue fruits at the top of the stalk. While these berries may look appealing, they are foul tasting and mildly poisonous. The genus name *Clintonia* honors the former mayor, governor, and senator of New York, DeWitt Clinton (1769–1828), who was largely responsible for the completion of the Erie Canal. The species name *borealis* means "of the north," which refers to the fact that the flower tends to be found in the northern U.S.

The young girls and women in the Chippewa tribe, especially from the region which is now Ontario, use this plant, as well as birch bark, to create a unique artwork they called *mazinashkwemaganjigan*—which is two words meaning "picture" and "he bites or gnaws." It is created by folding the leaves or bark and then crushing it with the points of their eye teeth, sometimes with a grinding or twisting motion, to make small indents in the leaves. These indents create attractive patterns and shapes, including flowers, insects, people, animals, and geometric shapes, and are especially visible when held up to light. Sometimes these are later used as patterns for beadwork, and the finest examples, typically made on bark, are kept, traded and admired.

DESCRIPTION: This plant produces 3–6 greenish-yellow to whitish, drooping, bell-like flowers, each ¾–1" (2-2.5 cm) long, on a central stalk rising from a set of shiny, bright green, oblong leaves, 5–8" (12.5–20 cm) long. Overall height: 6–15" (15–38 cm). Later on the stalk there are blue berries, ¼" (8 mm) wide, with few to many seeds.

FLOWERING SEASON: May–August.

HABITAT: Found in moist woods in acidic soil and shade.

RANGE: Manitoba east to Newfoundland, south to the higher elevations of North Carolina and Tennessee, and northwest to Illinois and Minnesota.

CAUTION: Berries are bad tasting and mildly poisonous if eaten.

SIMILAR SPECIES: A less widespread species, White Clintonia (*C. umbellulata*), is found from New York south to Georgia and Tennessee. It has numerous erect white flowers and black, fewer-seeded berries.

CONSERVATION: G5
Endangered in Indiana and Ohio. Threatened in Maryland, New Jersey, Tennessee, and West Virginia.

■ NATIVE
■ RARE OR EXTIRPATED
■ INTRODUCED

ALTERNATE NAMES: Bride's-bonnet, Bead Lily, Single-flowered Clintonia

If you come across this western native, you will likely find several clusters of 2–3 leaves in a patch as this plant spreads asexually from an extensive system of underground stems (rhizomes). It is typically found in healthy coniferous forests and has been used by scientists as an indicator species for the health of some woodlands.

Unlike all other *Clintonia* species, this plant usually only produces a single flower (rarely two) instead of a cluster, leading to the species name *uniflora*. The bright, blue bead-like berry produced after the flower is pollinated is primarily eaten by Varied Thrush (*Ixoreus naevius*) and grouse, as most species find the berry to be unpalatable and mildly poisonous.

DESCRIPTION: One or rarely two white, star-like flowers, 1–1½" (2.5–3.8 cm) wide, with 6 petals will bloom on a short, leafless stalk growing from a cluster of 2–3 oblong, shiny leaves, 2½–6" (6.3–15 cm) long. Overall height: 2½–6" (6.3–15 cm). Berry is round and blue, approximately ¼–½" (6–13 mm) in diameter.

FLOWERING SEASON: May–July.

HABITAT: Moist, coniferous forests.

RANGE: Alaska to northern California and inland to the southern Sierra Nevada; east to eastern Oregon and western Montana.

CAUTION: Berries are bad tasting and mildly poisonous if eaten.

CONSERVATION:
Threatened in Alberta.

■ NATIVE
■ RARE OR EXTIRPATED
■ INTRODUCED

ALTERNATE NAMES: American Trout Lily, Eastern Trout Lily, Yellow Trout Lily, Dogtooth Violet

One of our most common spring wildflowers and often growing in large colonies from seeds distributed by ants, the Trout Lily is immediately recognizable by its mottled brown and green leaves, which have a similarity to the spotted pattern of Brook Trout (*Salvelinus fontinalis*). Young plants will have a single leaf while more mature plants will produce two or three. Typically, plants take four to seven years prior to their first flower and the same plant will not bloom in consecutive years in order to collect enough resources to produce a new flower. After seeds mature, the leaves quickly fade in early summer and the plant remains dormant until spring.

Trout Lilies are known best for their shy, nodding, bronze-yellow flowers which bring a welcome splash of color across the woodlands. As evening falls, they close their backswept petals until the morning. Few animals eat Trout Lilies regularly, perhaps because eating of the plant can cause vomiting, but chipmunks will feast on the corms in the early spring when other foods like fruits, nuts, or seeds are not yet available. They must work hard to get this food, as the corms are buried deeply, 6–8" under the surface, which also makes them difficult to transplant. An alternative common name, Dogtooth Violet, refers to the resemblance of the corm to the shape of a tooth.

DESCRIPTION: A pair of brownish-mottled leaves, 2–8" (5–20 cm) long, sheaths the base of the stalk, which bears a bronzy-yellow nodding flower, 1" (2.5 cm) wide. Overall height: 4–10" (10–25 cm).

FLOWERING SEASON: March–May.

HABITAT: Rich woods and meadows.

RANGE: Ontario east to Nova Scotia, south to Georgia, west to Mississippi and Arkansas, and north to Minnesota.

CAUTION: Can cause vomiting if ingested in sufficient quantity.

SIMILAR SPECIES: White Dogtooth Violet (*E. albidum*) has narrower, mottled leaves and white, bell-shaped flowers, often tinged with lavender on the outside. Minnesota Adder's Tongue (*E. propullans*) has pink flowers and is found only in Minnesota.

CONSERVATION: **G5**
Threatened in Illinois, Iowa, Mississippi, and North Carolina.

- NATIVE
- RARE OR EXTIRPATED
- INTRODUCED

ALTERNATE NAMES: Yellow Avalanche-lily, Yellow Fawn-lily, Dogtooth Violet

This western species often blooms as snow recedes. It was first collected during the Lewis and Clark expedition in Idaho, and was mentioned several times in Lewis' journal. At first, Lewis considered it to be the same as the familiar Trout Lily (*E. americanum*) from the east, but later understood it to be distinctly different. There are 14 different species of *Erythronium* in North America.

Glacier Lily is an important forage for grizzly bears, which dig for the corms in spring. Bears will stray from their normal course of travel along ridges to seek out Glacier Lily corms. Deeply set into the ground, the corms were not often used as food by many Native Americans because they were difficult to dig and often had a burning taste, but they formed a portion of the winter diet for some tribes, and the children of the Thomson tribe ate the small root end of the corm as a candy.

DESCRIPTION: Typically a single flower, but occasionally up to 5 pale to golden-yellow flowers, 1–2" (2.5–5 cm) long, on a stalk with petals that curve back behind the base of the flower. The flowering stalk grows between two broad, gradually tapering leaves, 4–8" (10–20 cm) long. Overall height: 6–12" (15–30 cm).

FLOWERING SEASON: March–August.

HABITAT: Sagebrush slopes and mountain forest openings, often near melting snow.

RANGE: Southern British Columbia to northern California; east to western Colorado, Wyoming, northern New Mexico, and western Montana.

CAUTION: Can cause vomiting if ingested in sufficient quantity.

SIMILAR SPECIES: Mother Lode Fawn Lily (*E. tuolumnense*) is difficult to differentiate from casual observation. It is easier to differentiate by range, occurring on the western slope of the Sierra Nevada in central California.

CONVERSATION: G5

CONSERVATION: G5

This species is not of special conservation concern. A rare form with white or cream petal-like segments with a band of golden yellow at the base grows in southeastern Washington and adjacent Idaho. Some consider this a separate species and the National Forest Service recommends its protection.

CHECKER LILY *Fritillaria affinis*

■ NATIVE
■ RARE OR EXTIRPATED
■ INTRODUCED

ALTERNATE NAMES: *F. lanceolata*, Chocolate Lily, Rice Root Fritillary, Mission Bells, Ojai Fritillary

There are more than 20 *Fritillaria* species found in North America, all of them in the West. Checker Lily has a fairly large distribution across the Northwest, but can be challenging to find in the wild. The plant is known to be "shy-flowering," and does not flower regularly in nature. Its flower is also highly variable, which has resulted in the attempted naming of several supposedly distinct species that are too similar to differentiate and currently treated as one. The bulbs grow close to the surface and are easily dug. They are covered in small "bulblets" and resemble grains of rice. These bulblets were often replanted while the main bulb was

harvested for food by nearly all the northwestern Native American tribes. Even when cooked, they are slightly bitter, and some people soak them in water overnight to reduce the bitter flavor.

The taxonomic challenge this species presents is long-standing. After examining a specimen first collected by Lewis and Clark, the noted botanist Frederick Pursh attempted to name this plant *L. lanceolata*. Unfortunately, a portion of his descriptions were mistakenly based on a different species, Kamchatka Lily (*L. camschatcense*), which looks similar but has an unpleasant odor. Even though Josef Schultes caught the mistake in the early 1800s and proposed the species name *L. affine*, the name *L. lanceolata* remained widely used for over 150 years, until the 1980s when its current name was proposed by the work of Josef Sealy, who clarified the taxonomy and credited Schultes.

DESCRIPTION: Nodding flowers with yellow stamen, ⅜–1½" (1–4 cm), with highly variable coloration—yellowish- or greenish-brown with a lot of yellow mottling, purplish-black with little mottling, or yellow-green mottled with purple. There are 1–3 whorls, each with 5–11 leaves that encircle the unbranching stems. Overall height: 8–20" (20–50 cm).

FLOWERING SEASON: April–July.

HABITAT: Moist grassy places, coastal prairies, open woods.

RANGE: Southern British Columbia to northern Idaho, to Oregon, typically west of the Cascade Mountains to southern California.

SIMILAR SPECIES: Kamchatka Lily (*F. camschatcense*) can look similar, but is typically more northern and has a foul smell.

CONSERVATION: G5

This species is not of special conservation concern in its range.

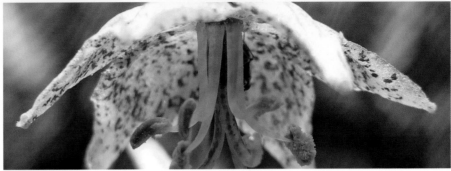

■ NATIVE
■ RARE OR EXTIRPATED
■ INTRODUCED

ALTERNATE NAMES: Spotted Mission Bells, Leopard Lily, Checker Lily, Purple Fritillary.

This is the most widespread fritillary and is known by many different names in different regions. The flowers are nearly camouflaged, speckled in purplish-browns, greens and yellow and pointed down to the ground. These will soon give way to "fritillaria," from the Latin word for "dice box" — unique, nearly rectangular seedpods. There are no records of this plant being used as food; however, the Ute called this plant *Kai'-rumosita-gwiv* and boiled and concentrated the bulbs and roots of the plant to create a medicine. It had to be taken sparingly because in larger quantities the medicine was regarded as highly poisonous.

DESCRIPTION: Linear, grass-like leaves, 4" (10 cm) long, arranged in whorls of 2–3 on a thick, purplish- or brownish-tinted stem topped by 1–4 nodding, bowl-shaped flowers ½–1" (1.5–2.5 cm) long. Flowers may be yellowish-, greenish-, or purplish-brown and mottled or speckled with yellow, white, or light green. Overall height: up to 24" (60 cm).

FLOWERING SEASON: May–July.

HABITAT: Rich wooded or grassy mountain slopes, open forests, shrublands; to mid-elevations.

RANGE: Western Dakotas and Nebraska west through most of Oregon and eastern half of California, south to central Arizona.

SIMILAR SPECIES: This species is very similar to the Pinewoods Fritillary (*F. pinetorum*), which has generally erect, upright blooms and is found in California and Nevada.

CONSERVATION: G5

Threatened in Colorado, Nebraska, and New Mexico.

■ NATIVE
■ RARE OR EXTIRPATED
■ INTRODUCED

Yellow Bells are dainty little flowers that can be found soon after snow melts in sagebrush steppes of western North America. The Okanogan-Colville tribe used the appearance of these flowers as a sign that spring had arrived. As with other *Fritillaria*, the flowers are highly variable but can be mistaken for no other. As the flower ages, the narrow yellow bell becomes rusty red or purplish.

DESCRIPTION: A single, downward-facing, yellow, bell-shaped flower, ½–1" (1.3-2.5 cm) long, each with its own stalk. A plant will typically have just one bloom, but on occasion a plant will sport a small cluster of flowers. It has green leaves, 2–8" (5–20 cm) long, with two or many growing from the middle of the stem. Overall height: 4–12" (10–30 cm).

FLOWERING SEASON: March–June.

HABITAT: Grasslands, among sagebrush, and in open coniferous forests.

RANGE: British Columbia; south on the eastern side of the Cascade Mountains to northern California; east to Utah, western North Dakota; Wyoming, western Montana, and Alberta.

CONSERVATION: G5

Threatened in Colorado, Wyoming, and Alberta. Yellow Bells are becoming rarer as the habitat in which they live is threatened by destruction, fragmentation, and invasive species. Their fate is further at risk due to unethical harvesting for personal and commercial uses.

- ■ NATIVE
- ■ RARE OR EXTIRPATED
- ■ INTRODUCED

ALTERNATE NAMES: *Lloydia serotina*, Common Alplily, Mountain Spiderwort, Snowdon Alplily

This arctic and alpine wildflower is the only native *Gagea* species in North America. Its range includes all of Alaska but becomes more sparse to the south, with several isolated populations following the Rocky Mountains to New Mexico. The same species is also widely distributed around the Northern Hemisphere including northern Asia and Europe.

For most of the year, this plant's long, grass-like leaves would likely be overlooked by most. However, when it flowers in June, it's impossible to confuse with any other plant. With pollinators sometimes in short supply, the flowers will self-pollinate. Molecular studies have shown smaller, isolated populations are exhibiting some genetic drift. The most striking changes can be found on Haida Gwaii, an archipelago off the coast of British Columbia, where the Alplily are larger than the mainland variety and have golden-yellow veins.

DESCRIPTION: Typically one (sometimes two) broad flowers with 6 creamy-white petals and pale-green to purple veins and purple streaks on the outside. Several needle-like leaves, mostly from base, 2–8" (5–20 cm) long. Overall height: 2–6" (5–15 cm).

FLOWERING SEASON: June–July.

HABITAT: Mountain ridges and rock crevices.

RANGE: Alaska; south to northwestern Oregon, northern Nevada, northern Utah, and along the Rocky Mountains to northern New Mexico.

CONSERVATION: G5
Threatened in Wyoming.

CANADA LILY *Lilium canadense*

■ NATIVE
■ RARE OR EXTIRPATED
■ INTRODUCED

ALTERNATE NAMES: Wild Yellow Lily, Meadow Lily

Canada Lily sports a large, showy flower which can be of many different colors, usually yellow to red-orange and often with dark spots. Across the species' range, there are also many differences in leaf width, number of flowers per stalk, and details of the flower morphology. There have been many attempts to define these as specific varieties and subspecies, but there is so much overlap between characteristics that differentiation has proven nearly impossible. Some earlier range maps included populations in the central plains, including Arkansas, Kansas and Nebraska; however it is now believed these were likely misidentified Turk's Cap Lily (*L. superbum*) or Michigan Lily (*L. michiganense*).

Observations of the plant across its flowering range indicate that it is pollinated primarily by Ruby-throated Hummingbirds (*Archilochus colubris*). Overbrowsing by large deer herds is reducing populations near urban and suburban areas where they were once common. Many Native American tribes used the plant medicinally to treat a variety of issues like irregular menstruation, stomach disorders, rheumatism, and snakebites.

DESCRIPTION: A tall lily of up to 6' (1.8 m) with up to 17 stalks topped with large, showy, nodding flowers, each 2–3" (5–7.5 cm), in colors ranging from dirty yellow, orange, to red, sometimes with red to maroon spots. Petals will arch outward but will not point backward. The plants have 4–10 spear-shaped leaves, up to 6" (15 cm) in whorls. Veins on the underside of the leaves have tiny prickles.

FLOWERING SEASON: June–August.

HABITAT: Wet meadows and woodland borders.

RANGE: Ontario east to Nova Scotia, generally following the Appalachian Mountains south to Georgia, west to Alabama, and north to Indiana.

SIMILAR SPECIES: Turk's Cap Lily (*L. superbum*) and Michigan Lily (*L. michiganense*) have similar coloration and general appearance to the Canada Lily, but the petals of Canada Lily do not sweep backward. Turk's Cap Lily and Michigan Lily were considered the same species for 150 years, as they bear a striking resemblance to each other and without flowers they are nearly impossible to tell apart. However, the petals of the Michigan Lily curve backward until they touch the flower tube or near its base. The Turk's Cap Lily petals curve farther backward, well behind the flower tube.

CONSERVATION: G5

Threatened in Alabama, Delaware, Georgia, Indiana, North Carolina, Rhode Island, South Carolina, and Tennessee in the U.S., and in Nova Scotia and Ontario in Canada.

■ NATIVE
■ RARE OR EXTIRPATED
■ INTRODUCED

ALTERNATE NAMES: Wild Tiger Lily, Oregon Lily

This western wildflower was widespread but is now in serious decline, in part because of the popularity of its highly variable, lovely orange blooms. Because they typically take five years to mature from seed to flowering, many people dug them up for personal and commercial sale as a garden plant, where they usually do not survive. This may be due to disruption of their symbiotic mycorrhizal association with fungus in the soil. The Columbia Lily is pollinated primarily by Rufous Hummingbirds (*Selasphorus rufus*) and to a lesser extent by large butterflies, including the Pale Swallowtail (*Papilio eurymedon*). For many Native Americans, Columbia Lily was a staple food, typically steamed or mixed with meat or salmon as well as used as a peppery condiment.

DESCRIPTION: Beautiful nodding flowers, 2–3" (5–7.5 cm) wide, with variable colors of orange to yellow-orange and spotted maroon on the 6 petals on top of a leaf-covered stem. Leaves are 2–4" (5–10 cm) long, narrow and pointed, either in whorls or evenly scattered along the stem. Overall height: 2–5½' (60–170 cm).

FLOWERING SEASON: May–August.

HABITAT: Prairies, thickets, and open forests.

RANGE: Southern British Columbia to northwestern California; east to northern Nevada and northern Idaho.

SIMILAR SPECIES: Leopard Lily (*L. pardalinum*), also called Panther Lily, is similar but grows in more moist locations, along forest streams or near springs in California and southwestern Oregon.

CONSERVATION: **G5**

Threatened in Montana. A small number of occurrences were last reported in the extreme northwest portion of Lincoln County in 1980. This species was once more widespread but is now in serious decline, in part because of collection for personal and commercial sale as a garden plant. Habitat destruction across its range is also a serious threat.

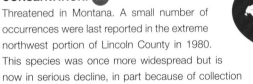

WOOD LILY *Lilium philadelphicum*

■ NATIVE
■ RARE OR EXTIRPATED
■ INTRODUCED

ALTERNATE NAMES: Wood Lily, Philadelphia Lily, Prairie Lily, or Western Red Lily

Once common, this lily has become rare, especially in the southern portion of its range as the meadows and prairies it calls home are rapidly disappearing. While the northeastern areas of its range also have experienced substantial declines, in parts of the eastern U.S, the Wood Lily can often be found at powerline right-of-ways, as the areas cleared of brush creates an ideal habitat.

Wood Lily is pollinated primarily by large swallowtail butterflies, and by a lesser degree by other butterfly species within its wide range. The flower's shape is adapted to force these large butterflies' forewings to contact reproductive structures in order to feed on the nectar. While you may also see hummingbirds visit the flowers, they are merely stealing the nectar without pollination, since these birds can easily get to the nectar without contact. The Dakota tribe said the flowers could provide immediate relief from the pain and inflammation of some spider bites if chewed or pulverized and placed on the affected area.

DESCRIPTION: The plant produces 1–3 funnel-shaped flowers, 2–2½" (5–6.5 cm) wide, each red or red-orange with gently outwardly curved tip. The interior of the flower is yellowish with purple spots at the base. The flowers sit atop an erect, leafy stem with leaves 2–4" (5–10 cm) long, pointed, with lower leaves scattered singly and the upper leaves in 1–2 whorls. Overall height: 12–28" (30–70 cm).

FLOWERING SEASON: June–August.

HABITAT: Dry woods and thickets.

RANGE: British Columbia east to Quebec, south to northern Georgia and Texas, west to Arizona, and north to Montana.

CONSERVATION: **G5**

Threatened in Alabama, Colorado, Georgia, Illinois, Montana, Nebraska, New Mexico, North Carolina, Ohio, Tennessee, Vermont, Virginia, West Virginia, and Wyoming. It is believed extirpated in Delaware and in Washington D.C. Habitat destruction, particularly of meadows and prairies, has threatened this species in much of its U.S. range. Picking of the flowers by mountain visitors and intensive grazing have also caused rapid declines.

■ NATIVE
■ RARE OR EXTIRPATED
■ INTRODUCED

As the common name suggests, the Indian Cucumber's roots' taste and smell are reminiscent of a sweet cucumber. Native Americans historically used the roots for food, both raw and cooked, but due to its scarcity, that is not recommended today. The Indian Cucumber's roots create a series of horizontal rhizomes at 45-degree angles. This produces colonies of these plants in an octagonal pattern that may resemble rings of plants and is thought to help prevent competition with other species.

After flowering, bluish-purple berries form and the leaves below them turn red at the base at the same time. Birds are attracted to these berries and help spread their seeds. The berries are not toxic, but most people wouldn't find them to be tasty. As the only species within the genus, molecular evidence shows the closest relatives to this plant are in the *Clintonia* genus. The genus name was chosen to honor Medea, the sorceress from Greek mythology.

DESCRIPTION: This plant has 2 whorls of 6–10 leaves at mid-stem, each 2½–5" (6.5–12.5 cm) long, and 3 leaves in the whorl above, each 1–3" (2.5–7 cm) long. From the center of the uppermost whorl, several yellowish-green flowers, ½" (1.5 cm) long, grow on a stalk that nods downward, sometimes bending below the leaves. Flowers have 6 tepals curving back and later produce dark bluish-purple berries.

FLOWERING SEASON: May–June.

HABITAT: Moist woodlands.

RANGE: Ontario east to Nova Scotia, south to Florida, west to Louisiana, and north to Illinois and Minnesota.

SIMILAR SPECIES: This species can look like the Starflower (*Trientalis borealis*), and that plant's root is poisonous. The leaf veins in the Indian Cucumber run from the base to the tip—they are parallel to each other. On the Starflower, there is one central vein with side veins coming off the central vein that terminate at the edges of the leaf.

CONSERVATION: 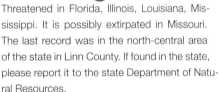 G5

Threatened in Florida, Illinois, Louisiana, Mississippi. It is possibly extirpated in Missouri. The last record was in the north-central area of the state in Linn County. If found in the state, please report it to the state Department of Natural Resources.

■ NATIVE
■ RARE OR EXTIRPATED
■ INTRODUCED

ALTERNATE NAMES: Rough-fruit Fairybells, Rough-Fruited Mandarin

Wartberry Fairybells are known for their delicate nodding flowers and velvety, grape-size berries, which start out yellow and change to orange, then red. This common plant is often found in the same place as chokeberries. In addition to the berries being used for food, the Blackfoot people historically used Wartberry Fairybells to help clean out their eyes. They would insert a fresh seed into their eyelid and rub it until the seed was watered out and any foreign matter in the eye clung to it. They would also place seeds in the eye overnight as a remedy for snow blindness.

DESCRIPTION: This plant has one or two small, creamy-white, narrowly bell-shaped flowers, each ⅜–⅝" (9–16 mm) long which often hang beneath leaves. Numerous oval leaves, 1½–5" (4–12.5 cm) long, occur along the stem, which are round or indented at the base and which have hairs on the edges sticking straight out. The plant produces round berries, ⅜" (9 mm) wide, that start yellow and become red. Overall height: 1–2' (30–60 cm).

FLOWERING SEASON: May–July.

HABITAT: Shady forests, often near streams.

RANGE: British Columbia south to northeastern Oregon, east to eastern North Dakota, and south through the Rocky Mountains to western New Mexico and southern Arizona. Isolated populations exist in Michigan and Ontario.

SIMILAR SPECIES: Smith's Fairybells (*P. smithii*) grows in moist woods on the western side of the Cascade Range and the Sierra Nevada; it has larger flowers ½–1" (1.5–2.5 cm) long. Hooker's Fairybells (*P. hookeri*) is very similar to Wartberry Fairybells, but the upper surfaces of the leaves are hairy; the hairs on the leaf edges point forward. Claspleaf Twisted Stalk (*Streptopus amplexifolius*) is similar in appearance but is differentiated by the heart-shaped leaf base clasping the stem.

CONSERVATION: G5

Endangered in Michigan, Minnesota, and Nebraska. Threatened in Wyoming and Ontario.

■ NATIVE
■ RARE OR EXTIRPATED
■ INTRODUCED

Dragon's Mouth is typically understated and relatively small, and usually doesn't grow to be more than a foot tall. After flowering, it will grow a single, grass-like leaf. In the wild it is often growing with Grass Pink (*Calopogon tuberosus*) and Rose Pogonias (*Pogonia ophioglossoides*). Despite being easy to overlook, the flower is considered by many to be the most beautiful in North America. Unfortunately, this has caused it to be overcollected in the wild, leading to its severe depletion, and collection is often pointless due to the flower's short life. It is not generally sold commercially and is extremely difficult to replant those grown in the wild.

In summer, this rare orchid produces a single pink flower reminiscent of a mouth. The lip is whitish-pink with magenta spots and a yellow center. Although the flower is showy and fragrant, it offers little or no nectar. Pollination is dependent on young, inexperienced bumblebees (*Bombus* spp.), oftentimes by newly

hatched queen bees that emerge throughout its blooming season. The worker bees quickly learn to avoid these flowers that have nothing to offer. As a result of their challenging pollination process, fewer than 1 in 6 flowers develop mature seeds.

DESCRIPTION: Typically pink to magenta flowers, though rarely white or bluish. Three sepals are splayed over the flower while two petals form a hood over the showy lower lip. The lower lip curves out and down revealing wrinkled edges and purple speckling with fleshy white to yellow bristles. After flowering, one grass-like leaf 2–8" (5–20 cm) long and ¼–½" (0.64–1.25 cm) wide will extend. Overall height: up to 6" (15 cm).

FLOWERING SEASON: June–July.

HABITAT: Peat bogs, swamps, wet meadows, lakeshores, and alkaline fens.

RANGE: Newfoundland south to the Appalachian Mountains of North Carolina west to Minnesota and Manitoba.

SIMILAR SPECIES: Grass Pink (*Calopogon tuberosus*) has similarly colored flowers and grows in similar habitats. Unlike Dragon's Mouth, the lip of the flower is at the top of the bloom instead of the bottom.

CONSERVATION: G5
Although globally secure, this orchid is rare and of conservation concern in much of its range. It is extirpated in Connecticut, Delaware, South Carolina, Virginia, and Washington D.C.; possibly extirpated in Indiana and Maryland. It is endangered in New Hampshire, North Carolina, Ohio, Pennsylvania, Rhode Island, Vermont in the U.S. and Alberta, Labrador, and Saskatchewan in Canada. It is threatened in Maine, Massachusetts, Michigan, New Jersey, and New York in the U.S. and in Manitoba, Northwest Territories, Prince Edward Island, and Quebec in Canada.

■ NATIVE
■ RARE OR EXTIRPATED
■ INTRODUCED

ALTERNATE NAMES: Tuberous Grass Pink

Grass Pink boasts 2–10 lightly fragrant flowers that range from pink to magenta on a leafless, grass-like stalk. The flowers of this tuberous plant boast bilateral symmetry. The genus name comes from the Greek for "beautiful beard," which is appropriate because Grass Pink is recognized by its cluster of hairs at its lip, which is at the top of the flower, instead of the bottom like most other orchids. These hairs resemble the pollen-bearing anthers of other flowers, which trick naive, recently emerged bees into landing on them expecting a pollen reward. If the bees are heavy enough, their weight will pull the lip down, dropping with the bee and pressing any pollen attached to the bee onto the stigma. As the bee leaves the flower, a new load of sticky pollen is pressed onto its back as it passes the end of the column.

DESCRIPTION: Fragrant pink flowers with yellow beards, 1½" (4 cm) long, in a spike-like cluster, opening sequentially from a tall, leafless stalk. Solitary, grass-like leaves up to 1' (30 cm) long. Overall height: 6–20" (15–50 cm).

FLOWERING SEASON: March–August, but throughout the year in Florida.

HABITAT: Bogs and bog meadows and acidic, sandy, or gravelly sites.

RANGE: Manitoba east to Newfoundland, south to Florida, west to Texas, and north to Kansas, Missouri, Illinois, and Minnesota.

SIMILAR SPECIES: Two similar species are found from North Carolina southward. Bearded Grass Pink (*C. barbatus*) has pink flowers that open simultaneously, whereas the flowers of *C. tuberosus* open successively. Pale Grass Pink (*C. pallidus*) has pale pink to whitish flowers and is smaller in all respects than *C. tuberosus*.

CONSERVATION: G5

Extirpated in Iowa and Washington D.C. Endangered in Arkansas, Delaware, Maryland, Oklahoma, Virginia, and West Virginia. Threatened in Georgia, Illinois, Missouri, Ohio, Rhode Island, and Vermont in the U.S. and Manitoba and Prince Edward Island in Canada.

■ NATIVE
■ RARE OR EXTIRPATED
■ INTRODUCED

ALTERNATE NAMES: Fairy-slipper Orchid, Calypso

The small Fairy Slipper orchid is the only species within the genus *Calypso* and can be found worldwide throughout the northern latitudes including in Eurasia and Japan. In the fall, it produces a solitary leaf that fades quickly after flowering in late spring. It boasts an attractive, solitary flower that is usually pink, magenta, or white, with a pouch-like lip often spotted with contrasting colors such as yellow. While in most of its range it is found in wet forests and bogs, in the Northwest it can be found in drier, shady coniferous forests. This orchid deceives naive pollinators, mainly newly emerged queen bumblebees (*Bombus* spp.). It relies on its bright colors, anther-like hairs, and a sweet smell to attract them, but provides them with no nectar or pollen.

There are two varieties of this orchid in North America. Eastern Fairly Slippers are widely distributed across Canada and the U.S. and have a white or pinkish lip. Western Fairy Slippers, with white and reddish markings on the lip, are found only in the northwestern regions of Canada and the U.S.

DESCRIPTION: One showy, nodding flower, 1½–2" (4–5 cm) long; typically with a white or pink, slipper-like lip petal blotched with purple or red, bearded with yellow hairs, with two horn-like points at the "toe." A single oval-shaped leaf, 3" (7.5 cm) long, with wavy edges. The leaves wither after flowering, often replaced by an overwintering leaf. Overall height: 3–8" (7.5–20 cm).

FLOWERING SEASON: May–July.

HABITAT: Cool, damp, mossy, mainly coniferous woods.

RANGE: Alaska south and east across Canada to Newfoundland, south to Vermont, and from Michigan west to South Dakota; also in much of the West.

CONSERVATION: (G5)

Extirpated in New York and Vermont. Endangered in Newfoundland and the Navajo Nation. Threatened in Arizona, Maine, Michigan, South Dakota, Wisconsin, and Wyoming in the U.S. and New Brunswick, Nunavut, Saskatchewan, and Yukon in Canada.

■ NATIVE
■ RARE OR EXTIRPATED
■ INTRODUCED

ALTERNATE NAMES: *Dactylorhiza viridis,* Frog Orchid, Long Bracted Green Orchid

Bracted Orchid produces multiple small, inconspicuous green flowers, with the petals and lip often with hints of red or brown. Large, leaf-like bracts extend beyond the flowers, making the flowers hidden as well as camouflaged. An Alaskan subspecies (subsp. *viride*), however, has smaller bracts and a brighter red lip instead of mostly green. Like many of the native orchids, the Bracted Orchid prefers moist habitats, growing in wet coniferous forests, tundras, prairies, meadows, and bogs. Beyond this continent, it has one of the widest global distributions of any orchid and can be found across Eurasia.

The pollinators of this orchid are currently unknown in North America, but there is suspicion that this species may be self-pollinating. In Europe, there is some evidence it is pollinated by beetles and small wasps. The taxonomy is currently in debate as molecular analysis shows the genus *Coeloglossum* is extremely similar to the genus *Dactylorhiza*, but the plants are very different in structure and shape. As *Coeloglossum* is composed of a single species, some experts attempt to include this plant within the *Dactylorhiza* genus, while others maintain separation on the basis of the distinct plant morphology.

DESCRIPTION: A cluster of camouflaged green to reddish or purple flowers from a stalk. Leaves are alternate, clasping the stem at the base. Low on the stem, leaves are more oval-shaped and shift to become more lance-shaped near the top of the stem. Overall height: 6–24" (15–60) tall.

FLOWERING SEASON: May–September.

HABITAT: Woods, meadows, stream banks, boggy places, tundra.

RANGE: Mid to northern United States and throughout Canada.

SIMILAR SPECIES: Pale Green Orchid (*Platanthera flava*) has a fin-like tubercle (bump) on the lip.

CONSERVATION: G5

Extirpated in Kentucky. Endangered in Arizona, Connecticut, Indiana, Maryland, Nebraska, North Carolina, Ohio, Pennsylvania, Utah, Washington, and West Virginia. Threatened in Illinois, New York, Wyoming, Iowa, Massachusetts, New Hampshire, Vermont in the U.S. and Labrador, New Brunswick, Newfoundland, Nova Scotia, and Yukon in Canada.

■ NATIVE
■ RARE OR EXTIRPATED
■ INTRODUCED

ALTERNATE NAMES: Summer Coralroot

Spotted Coralroot has the largest North American distribution of all the coralroots. It grows in a wide variety of woodlands and forests, but generally prefers open areas with minimal brush or other plants. If the habitat is suitable, plants can form large colonies. Coralroots have no chlorophyll and so they cannot create energy from photosynthesis. Instead, these plants are parasites that steal all the energy they require from fungi. This also means the only aboveground sign of this plant will be its infrequently flowering stalks. These often will take several years between sproutings.

This species is highly variable in color and bloom time across its range, but plants in a particular location will usually have the same characteristics. They can produce red or brown stalks, but may be a light yellow or cream color. Each stalk can have up to 40 flowers, which may have brown, yellow, or reddish petals and typically a white labellum spotted with purple. While there are many questions about how this flower is pollinated, there is strong evidence that self-pollination is typical. There have also been observations of the flowers being pollinated by dance flies and mining bees (*Andrena* spp.). The Iroquois, Paiute, and Shoshone all used the orchid's stems dried and brewed as a tea for such maladies as colds, pneumonia, and tuberculosis.

DESCRIPTION: One or more red or brown or yellowish leafless stems with loose clusters of flowers, about ¾" (2 cm) wide, the same color as the stem, with white, usually spotted purple lips. Overall height: between 8–31" (20–80 cm).

FLOWERING SEASON: May–August.

HABITAT: Moist to dry, upland, deciduous or coniferous forests.

RANGE: Alberta east to Newfoundland, south to Georgia, and north and west to Tennessee, Illinois, Iowa, Nebraska, and North Dakota; also found in Texas and throughout the West.

SIMILAR SPECIES: All coralroots could be confused without careful observation. Early Coralroot (*C. trifida*) typically has green stalks and is smaller overall with an earlier blooming period.

CONSERVATION: G5
Possibly extirpated in Delaware and Georgia. Endangered in Illinois, Iowa, Nebraska, Rhode Island, and Tennessee. Threatened in North Carolina, Ohio, and Wyoming in the U.S. and in New Brunswick, Newfoundland, and Prince Edward Island in Canada.

■ NATIVE
■ RARE OR EXTIRPATED
■ INTRODUCED

ALTERNATE NAMES: Wister's Coral-root

At one time mistakenly believed to occur only in the eastern portion of the U.S., this plant can be found throughout much of the continent. Spring Coralroot's range overlaps others of its genus, which can make positive identification somewhat challenging. For many years, western populations were misidentified as Spotted Coralroot (*C. maculata*). Like other coralroots, it also grows in a wide variety of woodlands and forests, but generally prefers open areas with minimal brush or other plants, but prefers a more moist location than the Spotted Coralroot.

Coralroots have no chlorophyll and so they cannot create energy from photosynthesis. Instead, these plants parasitize their energy from fungi. The only aboveground sign of this plant is often its infrequently flowering stalks. These often will take several years between sprouting.

DESCRIPTION: Plants can produce reddish-purple to yellow stalks. Each stalk can have up to 25 flowers, which are purple-brown or greenish typically with a white labellum spotted with purple. Overall height: up to 15" (38 cm).

FLOWERING SEASON: *C. wisteriana* may begin flowering as early as December in Florida and progressively later in the north, as late as August.

HABITAT: Moist mixed to coniferous woods.

RANGE: Idaho and Montana to Texas; Pennsylvania to Florida; New Jersey to Oregon; North Carolina to Arizona; plus New Hampshire and Washington D.C.

SIMILAR SPECIES: All coralroots could be confused without careful observation. Spotted Coralroot (*C. maculata*) is extremely similar; however, its sepals are curved downward or out instead of sweeping upward as in Spring Coralroot. Early Coralroot (*C. trifida*) has white flowers.

CONSERVATION: **G5**

Extirpated in New Jersey and Washington D.C. Possibly extirpated in Delaware and Oregon. Endangered in Maryland, Nebraska, North Carolina, Oklahoma, Pennsylvania, and Utah. Threatened in Alabama, Arizona, Idaho, Illinois, Ohio, Virginia, West Virginia, and Wyoming.

■ NATIVE
■ RARE OR EXTIRPATED
■ INTRODUCED

ALTERNATE NAMES: Pink Moccasin Flower

Pink Lady's Slipper is a showy, highly attractive orchid widely distributed from Alabama to the Northwest Territories. Both the scientific and common names refer to the resemblance between a shoe and the flower's inflated pouch-like labellum (lip), which is typically a bright pink with darker pink veins. It can be found in sandy or rocky mixed forests, often near pines or conifers, but also occasionally in bogs or swamps.

The unusual shape of the lip is used to trick bumblebees (*Bombus* spp.) into pollinating the flower. Lured in by the bright colors and a sweet scent—classic signals of nectar sources—native bumblebees enter the pouch but find no nectar. They are nearly trapped inside, but hairs help guide the bees to the exit. As it crawls to its escape, the bee rubs against the stigma—if the bee has fallen for the trick before, this releases any pollen stuck to it and pollinates the flower. As it squeezes out of the flower, new pollen is pressed onto the bee in the hopes it falls for the trick again. Most bees quickly learn to avoid these flowers, but newly emerging queens are inexperienced and are often victims of the trap. This process leads to relatively low rates of pollination for this species, but those that do make seed will disperse thousands of dust-sized seeds into the air. This increases the chances they land somewhere with the correct fungus needed to provide the energy required for germination.

DESCRIPTION: A leafless stalk bearing one or rarely two showy, bulbous, pink flowers, 2½" (6.5 cm) long, a slipper-like lip petal, veined with red and with a fissure down the front. On the ground will be two ribbed, oval leaves, each up to 8" (20 cm) long. Leaves are dark green above, and silvery, hairy beneath (see Caution). Overall height: 6–15" (15–38 cm).

FLOWERING SEASON: April–July.

HABITAT: Dry to moist forests, especially pinewoods; often in humus mats covering rock outcrops.

RANGE: Northwest Territories and Alberta east to Newfoundland, south to Georgia, west to Alabama, and northwest through Illinois to Minnesota.

CAUTION: The hairs on the leaves and stems of *Cypripedium* orchids can produce a rash similar to poison ivy if touched.

SIMILAR SPECIES: The leaves of other, similar pink lady's slipper orchids — such as *C. arietinum* and *C. reginae*—are on the stems.

CONSERVATION: **G5**
Endangered in Illinois and Indiana. Threatened in Alabama, Alberta, and Manitoba.

■ NATIVE
■ RARE OR EXTIRPATED
■ INTRODUCED

ALTERNATE NAMES: Greater Yellow Lady's Slipper, Yellow Lady's Slipper, Moccasin Flower

American Yellow Lady's Slipper has an enormous range and is found across nearly all of the United States and Canada, from Alaska to Georgia. These plants can be highly variable, changing their growth habit and leaf characteristics to adapt to their environment. There are three recognized varieties of this species, but differentiating them in the field can be difficult because the characteristics that define them are indistinct.

Northern Yellow Lady's Slipper (*C. parviflorum* var. *makasin*) is the easiest to identify because it has the smallest lip size with minimal hairs and has an intensely sweet scent. Small Yellow Lady's Slipper (*C. parviflorum* var. *parviflorum*) has a relatively small lip with a moderate rose to musty scent. Large Yellow Lady's Slipper (*C. parviflorum* var. *pubescens*) is a larger, more southern variety and its showy lip can be twice as large as the others. Its stem and leaves are covered with densely and conspicuously silvery hairs and the flowers have a moderate to faint scent.

These naturally hybridize with at least two other species in the genus where their ranges overlap. This can lead to extreme difficulties for most observers in determining the species with complete certainty. With such a variety of flower size and scent, this species attracts a large variety of insects that pollinate their flowers. These include many species of bees in the superfamily Apoidea and syrphid flies in the family Syrphidae.

DESCRIPTION: The plant produces one or two bright yellow flowers with an inflated, yellow, pouch-shaped lip up to 2" (5 cm) long on a leafy stalk. Its 3–5 oval leaves may reach 8" (20 cm) long. Overall height: 8–28" (20–70 cm).

FLOWERING SEASON: April–August.

HABITAT: Bogs, swamps, and rich woods.

RANGE: Alaska east to Newfoundland, south to Georgia, west to Mississippi, Arkansas, and Oklahoma, and north to North Dakota; also in most of the West following the Rocky Mountains with isolated populations in California, New Mexico, and Arizona.

CAUTION: The plants of the genus *Cypripedium* have glandular hairs on the leaves and stems that can cause a rash similar to a poison ivy rash upon contact.

SIMILAR SPECIES: Southern Lady's Slipper (*C. kentuckiense*), a rare species, has a lip that is cream-colored rather than yellow, and is 2–2½" (5–6 cm) long.

CONSERVATION: G5

Endangered in Arizona, Idaho, Massachusetts, Mississippi, Nebraska, Rhode Island, and Labrador. Threatened in Alaska, Colorado, Connecticut, Georgia, Montana, North Dakota, South Dakota, Vermont, Washington, West Virginia, Wisconsin, and Wyoming in the U.S. and Prince Edward Island, Northwest Territories, Nova Scotia, Saskatchewan, and Yukon in Canada.

■ NATIVE
■ RARE OR EXTIRPATED
■ INTRODUCED

ALTERNATE NAMES: Chatterbox, Stream Orchid

This large western orchid is typically found along the edges of streams, lakes, and other riparian areas. Some people call it Chatterbox due to the way its lower lip and "tongue" move when the flower is touched or shaken. Syrphid flies are the primary pollinators of this orchid. These flies lay their eggs among nests of aphids, which the larvae eat. The flower has a bumpy surface that resembles an aphid nest, and has a sweet smell similar to the honeydew aphids produce. If the fly is fooled, it will squeeze into the flower, which scrapes off any pollen attached to it onto the stigma, and as the fly exits, new pollen is attached to it.

DESCRIPTION: Typically growing in dense groups, plants have leafy stems each with 4–12 alternating leaves, 2–8" (5–20 cm) long. From 5–25 greenish-brown and pinkish flowers, 1–1½" (2.5–3.8 cm) wide and each appearing to sprout from a short stalk from an upper leaf. Overall height: from 1–3' (30–90 cm).

FLOWERING SEASON: March–August.

HABITAT: Deserts to mountains in springs or seeps, near ponds, along streams.

RANGE: British Columbia to Mexico; from the Pacific Coast east to Texas and the Black Hills of South Dakota.

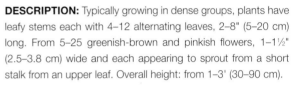

CONSERVATION: G4

Endangered in Oklahoma, South Dakota, and Wyoming. Threatened in Arizona, Colorado, Idaho, Montana, New Mexico, Texas, Utah, and Washington.

NATIVE
RARE OR EXTIRPATED
INTRODUCED

ALTERNATE NAMES: Broad Leaved Helleborine

Helleborine is a non-native, invasive orchid originating in Europe that now has a wide distribution across Canada and the United States after escaping from gardens in New England in the mid-1800s. Capable of self-pollination as well as pollination by wasps and other insects, its dust-like seeds sprout easily from disturbed sites, including urban areas, as well as in woodlands and swamps. This plant is resistant to most herbicides and control can be best accomplished through pulling or digging out the root.

DESCRIPTION: An introduced orchid with up to 50 drooping greenish to purple flowers, each ½" (12 mm), clustered on a long terminal spike. It produces 3–10 oval leaves, 2–6" (5–15 cm) long, with pointed tips and fine hairs, clasping the stem. Overall height: up to 36" (90 cm).

FLOWERING SEASON: July–September.

HABITAT: Woods, thickets, roadsides.

RANGE: Naturalized throughout the Northeast, south to North Carolina and Arkansas, and west to Minnesota. Also found in Montana, New Mexico, Oregon, and primarily coastal California.

CONSERVATION:
This species is not native to North America and can be invasive.

■ NATIVE
■ RARE OR EXTIRPATED
■ INTRODUCED

ALTERNATE NAMES: Western Rattlesnake Plantain, Green-leaf Rattlesnake Plantain, Menzies' Rattlesnake Plantain

The genus *Goodyera* includes all the orchids commonly called rattlesnake plantains. This name stems from the patterns on the leaves that can resemble scales of snakes and the shape of its leaves is similar to plantain. However, the leaf pattern for this specific species typically has less dense scale patterns, with a white midrib area and limited and less defined white patterning elsewhere on the leaf. On the West Coast, this species can have the white veining similar to the eastern Downy Rattlesnake Plantain (*G. pubescens*).

While it blooms between Late May and mid-September, its evergreen leaves can be easily found in the fall and winter when other groundcovers are dormant. In the eastern side of its distribution in North America, it is restricted to formerly glaciated areas, both in coniferous forests and mixed woodlands. Along the Rocky Mountains, it is often located in high-elevation spruce-fir forests. Pollination is provided by bumblebees, which typically start their search for nectar at the bottom of the flower spike where flowers are first to mature and receptive. As the bees work their way up, the flowers will not be mature enough to produce nectar, but will be loaded with pollen. The bumblebee will soon leave the stalk covered in pollen and, with a bit of luck, will soon pollinate the mature flowers of another nearby plant.

DESCRIPTION: This plant has a flattened rosette of three to seven blue-green evergreen leaves, 1–3½" (2.5–9 cm) long, with a prominent white stripe down the middle. (Western coastal populations may have lateral white veining.) It has a tall, woolly, flowering spike from the center of the leaves with a cluster of up to 48 greenish-white flowers, ¼" (6 mm) long, either on one side or in a spiral. Overall height: up to 18" (45 cm).

FLOWERING SEASON: May–September.

HABITAT: Dry or moist, deciduous or coniferous woods and well-drained wooded slopes.

RANGE: Two distinct ranges—Ontario east to Newfoundland, south to Maine, west to Wisconsin. Also, Southern Alaska south to California, west to New Mexico, north to western North Dakota and Alberta.

SIMILAR SPECIES: All four species of rattlesnake plantain can be difficult to distinguish from each other. Downy Rattlesnake Plantain (*G. pubescens*) has dense, net- or scale-like white veining across the leaf. Dwarf Rattlesnake Plantain (*G. repens*) has leaves that are either all green or with a green center with white veining on the outside of the leaf. Checkered Rattlesnake Plantain (*G. tesselata*), which shares the eastern range, does not have a strong white stripe down the middle of its leaves. Also, White-veined Wintergreen (*Pyrola picta*) can appear similar, but the leaves may have red stems and the rosette will not be flattened as in the rattlesnake plantains.

CONSERVATION: Ⓖ5
Possibly extirpated in Vermont. Endangered in Maine, Nebraska, Wisconsin, Newfoundland, Prince Edward Island, and Saskatchewan. Threatened in Arizona, Utah, Wyoming, New Brunswick, Nova Scotia, and Quebec.

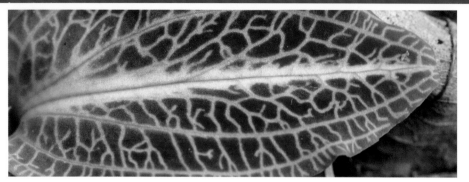

■ NATIVE
■ RARE OR EXTIRPATED
■ INTRODUCED

ALTERNATE NAMES: Rattlesnake Orchid

Downy Rattlesnake Plantain has the largest distribution of this genus in the eastern portion of the U.S., with isolated populations in the Florida panhandle. It grows in coniferous, deciduous, or mixed forests, and occasionally in swamps. While superficially similar, the flowers of this species are very different in exact structure compared to the other rattlesnake plantains. As a result, unlike the other species that are pollinated by bumblebees (*Bombus* spp.), this species attracts metallic-green sweat bees (*Agapostemon* spp.) as its primary pollinator.

The common name may reflect the belief that Native Americans used this plant to treat snakebites; however, ethnobotanists have recorded over 40 uses of these plants in different native groups across the continent, including as eye medicines, to assist childbirth, and to improve appetite. Only one text mentions the possibility of use with snakebites, which was based on a single secondhand account from over 140 years prior, in the late 1700s.

DESCRIPTION: This plant has a flattened rosette of three to seven blue-green evergreen leaves, 1–3½" (2.5–9 cm) long, with a prominent white stripe down the middle and scale or net-like lateral white veining. A tall, woolly, flowering spike from the center of the leaves with a cluster of up to 48 greenish-white flowers, ¼" (6 mm) long, either on one side or in a spiral. Overall height: up to 18" (45 cm).

FLOWERING SEASON: May–September.

HABITAT: Dry or moist, deciduous or coniferous woods and well-drained wooded slopes.

RANGE: Ontario east to New Brunswick, south to Florida, west to Louisiana and Oklahoma, and north through Missouri to Minnesota.

SIMILAR SPECIES: All four species of rattlesnake plantain can be difficult to distinguish from each other. Giant Rattlesnake Plantain (*G. oblongifolia*) has a strong white center with limited veining across the leaf. Dwarf Rattlesnake Plantain (*G. repens*) has leaves that are either all green or with a green center with white veining on the outside of the leaf. Checkered Rattlesnake Plantain (*G. tesselata*), which shares the eastern range, does not have a strong white stripe down the middle of its leaves. Also, White-veined Wintergreen (*Pyrola picta*) can appear similar, but the leaves may have red stems and the rosette will not be flattened as in the rattlesnake plantains.

CONSERVATION: **G5**

Endangered in Florida, Mississippi, and New Brunswick. Threatened in Iowa, Illinois, Maine, Nova Scotia, and Quebec.

- ■ NATIVE
- ■ RARE OR EXTIRPATED
- ■ INTRODUCED

ALTERNATE NAMES: Water-spider False Rein Orchid, Water-spider Bog Orchid

This unusual southern native can be found in swamps and wet ditches as well as floating fully aquatically, forming mats in stagnant pools. As the plant is green with greenish flowers, it can be difficult to locate blended in with the foliage of the wet areas this orchid calls home.

To attract the nocturnal moths that pollinate its flowers, at night the flowers emit a strong fragrance similar to vanilla. The nectar on which moths feed is deep inside a long spur from the rear of the lip, which forces the moths to extend their head and proboscis deep into the flower. This deposits any pollen on the stigma and deposits a new load of pollen as the moth removes its head from the flower.

DESCRIPTION: Leafy stems up to 36" tall, with a dense cluster of spidery, yellowish-green, ½" (12 mm) flowers and spear-like, fleshy leaves sheathing the stalk.

FLOWERING SEASON: Year-round, especially June.

HABITAT: Swamps, marshes, bogs, streams, ponds, ditches, and canals.

RANGE: From eastern North Carolina south to southern Florida and west to Texas.

CONSERVATION: G5

Endangered in Oklahoma. Threatened in Arkansas, Georgia, and North Carolina.

■ NATIVE
■ RARE OR EXTIRPATED
■ INTRODUCED

Large Whorled Pogonia is the far more common of the two native plants of this genus. The other is the rare Small Whorled Pogonia (*I. medeoloides*), an imperiled species that is endangered throughout much of its range and is federally protected.Large Whorled Pogonia has an unusual ring of typically five leaves, whorled around the upper part of the stem. The stem is typically purplish-brown, a key trait to differentiate it from the rarer species. The plant can reproduce clonally as well as through self-pollination, sometimes forming large groups in ideal, acid-rich environments. Although the flowers have no nectar, mining bees, metallic sweat bees, and bumblebees are attracted to the nectar guides, and pollen is transferred as they search for food.

DESCRIPTION: A hairless, purplish to greenish-brown stem with a single whorl of five green to dark-green leaves, 1–4" (2.5–10 cm) long. One or two greenish-yellow flowers, ⅓–2" (1–5 cm) long, with long, purple to brownish sepals, are elevated above the leaves on short, arching stalks.Total height: 2–16" (5–40 cm).

FLOWERING SEASON: April–June.

HABITAT: Acidic woods, bogs.

RANGE: Southern Ontario, east to Maine, south along the Atlantic coast to the South Carolina border, throughout the entire eastern U.S. to Illinois, Missouri, Oklahoma, and Texas. One isolated population occurs in the Florida panhandle.

SIMILAR SPECIES: Small Whorled Pogonia (*I. medeoloides*) is smaller with green stem and sepals. Immature Indian Cucumber (*Medeola virginiana*) with a single whorl of leaves can also appear similar, but has hairy stems.

CONSERVATION: G5

Extirpated in Maine. Possibly extirpated in Ontario. Endangered in Florida, Illinois, Missouri, New Hampshire, Oklahoma, and Texas. Threatened in Alabama, Connecticut, Delaware, Georgia, Indiana, Louisiana, Massachusetts, Michigan, Mississippi, New York, North Carolina, Rhode Island, and Vermont.

■ NATIVE
■ RARE OR EXTIRPATED
■ INTRODUCED

ALTERNATE NAMES: Loesel's Wide-lipped Orchid, Yellow Wide-lip Orchid

Although it is a circumpolar species found across the Northern Hemisphere, Loesel's Twayblade is a rare flower to find anywhere in its range because of its scarcity and its small size and lack of showy features. Typically found in wetland areas, nearly every aspect of this plant's life cycle is highly adapted to water conditions. In spring, the most frequent sign of this plant appears—its two glossy, dark green leaves. Afterward, if conditions are acceptable, the yellowish-green flowers may appear on a stalk. Flowering will not occur if water levels fluctuate too far from normal levels, either too low or too high. In these cases, only leaves will be present for the year.

When flowers are produced, there are no known pollinators. This plant is able to self-pollinate, often with the assistance of rainfall. Droplets of rain help push the pollen-holding anthers down onto the stigma. The moisture can also help hold the pollen packets against the stigma. Rain leads to a fourfold increase in self-pollination success.

DESCRIPTION: Just exposed at the surface will be a bulb-like base that stores water. From it grow a pair of broad, dark green leaves, 1–8½" (2.5–21.5 cm) long and up to 2½" (6.5 cm) wide. A flowering stalk grows from between the leaves with up to 18 small, yellowish-green flowers with narrow sepals but a wide lip that can be translucent. Overall height: 2–8" (5–20 cm).

FLOWERING SEASON: May–August.

HABITAT: Bogs, swamps, wet thickets, meadows, and shores.

RANGE: Northern United States, Washington east to Maine, south to Kansas, Alabama and North Carolina.

CONSERVATION: G5

Extirpated in Washington D.C. Endangered in Alabama, Arkansas, Georgia, Illinois, Kansas, Nebraska, North Carolina, Rhode Island, Tennessee, Washington, and Newfoundland. Threatened in Maryland, South Dakota, Missouri, Montana, New Hampshire, North Dakota, Virginia, Kentucky, Indiana, Iowa, Vermont, West Virginia in the U.S. and in Alberta, British Columbia, Manitoba, New Brunswick, Northwest Territories, Nova Scotia, Prince Edward Island, Quebec, and Saskatchewan in Canada.

■ NATIVE
■ RARE OR EXTIRPATED
■ INTRODUCED

While there are 10 species of adder's-mouth orchids in North America, most have very small ranges and are adapted to the conditions in a highly specific area of the continent. Green Adder's Mouth Orchid, however, has the largest range of them all, occurring from Texas to Newfoundland. As the species name suggests, this orchid typically has just one leaf clasping its stem roughly halfway from the ground. This tiny orchid produces large clusters of up to 160 tiny green flowers on its ribbed stalk. With such tiny flowers, the plant relies on tiny pollinators, including mosquitoes, fungus gnats, and parasitic wasps.

DESCRIPTION: A tiny orchid with a single clasping leaf, up to 3" (7.5 cm) long, and 2" (5 cm) wide, partway up the stem. Overall height: 6" tall.

FLOWERING SEASON: Spring–fall.

HABITAT: Swamps, bogs, sand barrens, heathlands, and dry woods.

RANGE: Eastern North America west to Michigan and Texas.

CONVERSATION: **G5**

Extirpated in Iowa. Possibly extirpated in New Jersey. Endangered in Connecticut, Delaware, Illinois, Indiana, Kansas, Oklahoma, and Rhode Island. Threatened in Florida, Maryland, Massachusetts, Missouri, New Hampshire, Ohio, and Vermont in the U.S. and in Labrador, Manitoba, Newfoundland, and Prince Edward Island in Canada.

■ NATIVE
■ RARE OR EXTIRPATED
■ INTRODUCED

ALTERNATE NAMES: *Listera convallarioides,* Broad-lip Twayblade, Swan Orchid

The common name of twayblades is a reference to having two leaves exactly opposite from each other. This species occurs throughout northern and western North America in sometimes disjunct or scattered populations. Taxonomists have had quite a bit of trouble with this group and have placed twayblades into no fewer than five different genera, including *Bifolium*, *Epipactis*, *Diphryllum*, *Ophrys*, and (until recently) *Listera*. However, current phylogenetic studies have shown them to be closely related to a group of chlorophyll-less plants, the bird's-nest orchids of genus *Neottia*, and all 25 species of *Listera* have been moved into *Neottia* because it was named first. Many older sources, and a few modern ones, continue to use *Listera*.

DESCRIPTION: The small orchid has a pair of green leaves midway up the stem, each ¾–3" (2–7.5 cm) long. Above the leaves is a slender hairy stalk with 5–20 green flowers, each just ⅜–½" (9-13 mm) long. Overall height: 2–14" (5–35 cm).

FLOWERING SEASON: June–September.

HABITAT: Woodlands, bogs, forests, meadows, and swamps.

RANGE: Alaska south and east to Newfoundland, and south and west to southern California, Idaho, northeastern Nevada, northern Utah, and central Colorado; also in southern Arizona, and in the northern portion of the eastern United States.

CONSERVATION: **G5**

Possibly extirpated in Minnesota. Endangered in Arizona, New York, South Dakota, and Wisconsin. Threatened in Alaska, Colorado, Montana, New Hampshire, Vermont, and Wyoming in the U.S. and Alberta, Newfoundland, Prince Edward Island, and Quebec in Canada.

■ NATIVE
■ RARE OR EXTIRPATED
■ INTRODUCED

ALTERNATE NAMES: Northern Green Orchid, North Wind Bog Orchid

A northern native, this orchid is small and easily overlooked. For many years, this species was identified together with *P. hyperborea*, a species name now restricted only to similar plants in Greenland and Iceland. Any historical references to *P. hyperborea* outside of those areas can now safely be assigned to *P. aquilonis*. This plant shows variability in the numbers of flowers and leaves as well as color. Shaded plants typically have whitish-green flowers while those in the sun may show more yellow. Except for populations in the extreme Northwest, the flowers are scentless and all are self-pollinating.

DESCRIPTION: The inflorescense is a raceme of 5–40 yellowish-green flowers, each ¼" (6 mm) long. The stem has 2–6 green, spear-shaped clasping leaves, each ¾–4" (2–10 cm) long . Overall height: 12–24" (30–60 cm).

FLOWERING SEASON: May–August.

HABITAT: Open wet meadows, roadside ditches, seeps, fens, bogs, and river gravel.

RANGE: Alaska south to California, New Mexico, and Iowa, east to Rhode Island.

SIMILAR SPECIES: The Tall White Bog Orchid (*P. dilatata*) has white flowers instead of green. Huron Green Orchid (*P. huronensis*) is similar but has whitish-green flowers.

CONSERVATION: G5

Extirpated in Ohio. Endangered in Pennsylvania. Threatened in Arizona, Indiana, Massachusetts, Nebraska, Labrador, Nunavut, and Prince Edward Island.

■ NATIVE
■ RARE OR EXTIRPATED
■ INTRODUCED

ALTERNATE NAMES: *Habenaria ciliaris,* Orange Fringed Orchid

This is a showy plant, frequently with over 100 yellow-orange blooms in a dense inflorescence. These feathery flowers attract Monarch and Swallowtail butterflies as pollinators. Pressing their head deep into the flowers for nectar, the pollen is often attached to the butterfly's compound eyes, ready to be transferred to other nearby flowers.

DESCRIPTION: A single stem with 2–4 primary leaves, each 3–10" (7.5–25 cm) long, tapering to bracts near the plume of deep orange to bright yellow flowers with a drooping, deeply fringed lip petal. The lip petal is ¾" (2 cm) long, with a long, slender spur projecting downward and backward to 1½" (4 cm) long. Overall height: 1–2½' (30–75 cm).

FLOWERING SEASON: July–September.

HABITAT: Often acidic, peaty, or wet, sandy woods, bogs, and marshes, as well as along roadsides.

RANGE: New York east to Rhode Island, south to Florida, west to Texas, and northeast to Michigan and historically into Ontario.

SIMILAR SPECIES: Yellow Fringeless Orchid (*P. integra*) has orange-yellow flowers with a fringeless lip. Orange Fringed Orchid (*P. cristata*) has a flower ⅜" (9 mm) wide and a spur ½" (1.3 cm) long, shorter than the fringed lip.

CONSERVATION: **G5**

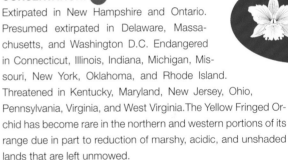

Extirpated in New Hampshire and Ontario. Presumed extirpated in Delaware, Massachusetts, and Washington D.C. Endangered in Connecticut, Illinois, Indiana, Michigan, Missouri, New York, Oklahoma, and Rhode Island. Threatened in Kentucky, Maryland, New Jersey, Ohio, Pennsylvania, Virginia, and West Virginia. The Yellow Fringed Orchid has become rare in the northern and western portions of its range due in part to reduction of marshy, acidic, and unshaded lands that are left unmowed.

WHITE REIN ORCHID *Platanthera dilatata*

ALTERNATE NAMES: Scentbottle, White Bog Orchid, Bog Candle, Tall White Bog Orchid

White Rein Orchid has a very large range reaching both coasts of North America and from the far north into southern California. The Nlaka'pamux tribe, from what is now British Columbia, used this plant to make charms for good luck in hunting and increase their fortunes in wealth and love. The flowers of this plant are very fragrant, with a scent reminiscent of cinnamon or cloves to attract pollinators. The spur that extends behind the flower is filled with glands that excrete nectar. The flower is pollinated by skipper butterflies and owlet moths.

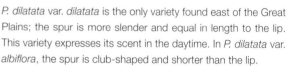

Adaptations to various pollinators across their huge range has led to three distinct varieties, which are largely distinguished by the size of the spur—sized to match the pollinator's tongue. All three varieties overlap in the western portions of the range. *P. dilatata* var. *leucostachys* has a long, strongly curved spur; more than 1.5 times longer than the lip with a primarily nocturnal scent.

P. dilatata var. *dilatata* is the only variety found east of the Great Plains; the spur is more slender and equal in length to the lip. This variety expresses its scent in the daytime. In *P. dilatata* var. *albiflora*, the spur is club-shaped and shorter than the lip.

DESCRIPTION: Tall stems with spear-like leaves clasping the stem, 2–12" (5–30 cm) long, with a cluster of up to 60 fragrant flowers in a spike. Overall height: 6–52" (15–130 cm).

FLOWERING SEASON: June–September.

HABITAT: Wet or boggy ground.

RANGE: Northern New Mexico to southern California; north, through most of the West to central Alaska. In the east, from Manitoba and Minnesota east to Labrador, south to Massachusetts and the Great Lakes.

CONSERVATION: Ⓖ5

Extirpated in Indiana. Possibly extirpated in Connecticut. Endangered in Illinois, Pennsylvania, and South Dakota. Threatened in New Mexico, Alberta, Northwest Territories, Nunavut, Prince Edward Island, Saskatchewan, and Yukon.

■ NATIVE
■ RARE OR EXTIRPATED
■ INTRODUCED

ALTERNATE NAMES: Northern Tubercled Bog Orchid, Southern Rein Orchid

The Pale Green Orchid can thrive in a fairly wide variety of locations, from dry to wet flatwoods to seasonally flooded lakesides. In all cases, a critical factor for survival is the regular clearing of tall plants that may overshadow this species. In the south, this is typically from wildfires while in the north, ice-scouring of lakesides and river edges reduces competition for sunlight. This orchid is pollinated by moths as well as native mosquitoes, which rely on nectar for energy—mosquitoes need a blood meal only for reproduction.

There are two varieties of this orchid, a northern variety (*P. flava* var. *herbiola)* and a southern (*P. flava* var. *flava*). Differentiation can be challenging where their ranges meet. The southern varieties have a more slender flowering spike with short bracts and wider spaces between flowers. The northern variety has a wider plume with bracts extending beyond the flowers, which are denser together. These plants have not been successfully cultivated. Those sold are all wild-collected, leading to diminished populations in some areas.

DESCRIPTION: The plant produces 2–3 alternate leaves with pointed tips, up to 6" (15 cm) long on a smooth stalk topped with a spike of up to 60 small, inconspicuous pale-green flowers about 1/4" (6 mm) wide. On each flower, the lip will have a prominent bump, called a tubercle, in the center. Overall height: 6–24" (15–60 cm).

FLOWERING SEASON: The southern variety flowers from March–October; the northern variety flowers from May–August.

HABITAT: Moist meadows, wetland edges, and floodplains.

RANGE: Eastern North America; Ontario and Quebec south through Florida, west through Minnesota and Texas.

SIMILAR SPECIES: Bracted Orchid (*Coeloglossum viride*), which has similarly colored flowers, does not have the tubercle (bump) on the lip.

CONSERVATION: G5

Extirpated in Washington D.C. Endangered in Arkansas, Delaware, and New Brunswick. Threatened in Alabama, Florida, Georgia, Iowa, Louisiana, Maine, Maryland, Michigan, Minnesota, Missouri, North Carolina, Ohio, Texas, Vermont, and Virginia in the U.S. and Nova Scotia and Ontario in Canada.

■ NATIVE
■ RARE OR EXTIRPATED
■ INTRODUCED

ALTERNATE NAMES: Green Fringed Orchid

There are 32 native North American orchids in this genus and 10 of these have greenish flowers. This species is the most common and widespread of the genus. The flower's lip has three lobes, all of which are deeply divided and is the most delicately fringed of the genus. Numerous moths pollinate this orchid with pollen attached to the proboscis while probing for nectar, typically beginning with the bottom, older flowers with more nectar and working up the cluster. From observations in the wild, pollen from multiple flowers can be found attached to the moth's proboscis at the same time, increasing chances of pollination.

DESCRIPTION: A stem with 1–5 lower leaves, 8" (20 cm) long with smaller leaves higher. The inflorescence is a spike-like clusters of up to 60 whitish-green or creamy-yellow flowers, ½" (1.5 cm) long. The three-lobed lip is highly lacerated with a slender, curved spur, 1¼" (1.5 cm) long. Overall height: 1–2' (30–60 cm).

FLOWERING SEASON: June–September.

HABITAT: Bogs, wet woods, riverbanks, dry to wet meadows and fields.

RANGE: Nova Scotia and New England; south to Florida; west to Texas; north to Tennessee, Minnesota, and Ontario.

CONSERVATION: G5

Endangered in Iowa, Louisiana, Mississippi, Oklahoma, and Manitoba. Threatened in Alabama, Delaware, Georgia, Illinois, Indiana, Kansas, North Carolina, and South Carolina.

■ NATIVE
■ RARE OR EXTIRPATED
■ INTRODUCED

ALTERNATE NAMES: Blunt-Leaved Bog Orchid

A small, fairly inconspicuous plant, the Blunt-leaf Orchid can be an indicator for other wetland orchid species which share the same habitats, like Fairy Slipper (*Calypso bulbosa*), Round-leaved Orchid (*Amerorchis rotundifolia*), the rattlesnake plantains (*Goodyera* spp.), and the twayblades (*Neottia* spp.).

This is the smallest of the native Bog Orchids and has tiny flowers just ⅛" to ¼" (3–6 mm) long. Such small flowers attract small pollinators, mainly native *Aedes* mosquitoes, but occasionally small moths. Once these insects attempt to reach down into the mouth of the flower to the spur to collect nectar, the pollenia spring forward, depositing two large, cone-shaped pollen packets onto the head, and often the eyes of the mosquito, where they will hit the stigma of subsequent flowers. These pollen packets are as large as the mosquito's head itself, which can impair the flight of the insect, and can appear to look like tiny yellow horns when attached to the mosquito.

Fortunately for the plant, the mosquitoes are not attracted to the appearance, but purely to the scent of the flowers, which have a grassy or musky odor. Other close relatives in the genus produce a sweeter scent which has been shown to repel mosquitoes in ways that are similar to DEET, the active ingredient in many commercially sold insect repellents. It is possible to trigger

the pollen attachment manually. Merely poke a small twig under the hood of a flower to see the pollinia spring down and cement themselves to the twig.

DESCRIPTION: Typically grows a single basal leaf, rarely two, 1½–6" (4.25–15 cm) long, broadest above the middle, blunt or rounded tip and tapering at the base. The inflorescence is a loose cluster of 2–15 greenish-white flowers each ⅛–¼" (3–6 mm) long. Overall height: 2–12" (5–30 cm).

FLOWERING SEASON: June–August.

HABITAT: Coniferous bogs and swamps.

RANGE: Circumpolar; Alaska through British Columbia to Newfoundland, south to Oregon, Utah, Montana, Colorado, Minnesota, Wisconsin, and New York.

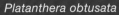

CONSERVATION: G5
Possibly extirpated in Massachusetts. Endangered in Idaho, Oregon, Utah, Vermont, and Prince Edward Island. Threatened in Wyoming and Northwest Territories.

■ NATIVE
■ RARE OR EXTIRPATED
■ INTRODUCED

ALTERNATE NAMES: Lesser Roundleaf Orchid, Flat-leaved Orchid, Dinner Plate Orchid

With a distribution that spans the continent, the Roundleaf Rein Orchid is widespread but never abundant. Growing in moist forests, it is known for its two large, glossy, flat, round leaves on the ground and its flowering spike of relatively large greenish-white flowers. While not nearly as large as some showier orchids, this species' blooms are the largest of all western rein orchids. The flowers are visited by the Large Looper (*Autographa ampla*) and Green-patched Looper (*Diachrysia balluca*) moths to drink nectar, which also pollinates the flowers.

DESCRIPTION: This plant occurs on the ground, growing two broadly round leaves, 2½–6" (6.5–15 cm) long. The leafless stem bears a cluster of up to 25 white or greenish-white flowers about ½" (12 mm) long. Overall height: 8–24" (20–60 cm).

FLOWERING SEASON: June–August.

HABITAT: Moist forest floors.

RANGE: British Columbia east to Labrador and south to northern Oregon, northern Idaho, northwestern Montana, Michigan, northern Ohio, and following the higher elevations of the Appalachian mountains through North Carolina.

SIMILAR SPECIES: In the Northeast, Large-leaved Bog Orchid (*P. macrophylla*) is extremely similar and was previously classified as a subspecies of this plant. It has a longer nectar spur that requires pollination by a larger species of moth. Hooker's Orchid (*P. hookeri*) has greenish or yellowish flowers, and the flower spur is tapered and narrowest at the tip.

CONSERVATION: G5

Extirpated in Indiana. Possibly extirpated in Connecticut, Oregon, and Rhode Island. Endangered in Illinois, Massachusetts, New Jersey, and Wyoming. Threatened in Alaska, Idaho, Montana, North Carolina, South Dakota, and Tennessee in the U.S. and Newfoundland, Northwest Territories, Nova Scotia, Prince Edward Island, and Yukon in Canada.

■ NATIVE
■ RARE OR EXTIRPATED
■ INTRODUCED

ALTERNATE NAMES: Small Purple Fringed Orchid

A beautiful, showy orchid of the wetlands of northeastern North America, the flowers of the Lesser Purple Fringed Orchid resemble butterflies. This gives rise to the species name *psycodes*, named after the Greek goddess Psyche, whose beauty rivaled Aphrodite's and who is represented by a butterfly. Pollinators of this species include numerous moths and butterflies including skippers, swallowtails, and hummingbird moths that are attracted to the unusual odor of the blooms, which is sometimes described as "rank." Pollen is attached to the proboscis as the insects feed on nectar.

DESCRIPTION: The plant grows 2–5 wide-spreading oval leaves with pointed tips, each 8½" (21.5 cm) long, 2¾" (7 cm) wide, and alternate. The purple flowers are ½–¾" (12–18 mm) long with a broad, fringed, 3-lobed lip. Overall height: 1–3' (30–90 cm).

FLOWERING SEASON: June–August.

HABITAT: Meadows, swamps, bogs, stream- and riverbanks, and roadside ditches.

RANGE: Maine south to Georgia and west to Minnesota and Missouri.

SIMILAR SPECIES: Greater Purple Fringed Orchid (*P. grandiflora*) is almost identical but its flowers are larger, 3/4–1" (18–24 mm) long, with a more deeply fringed lip.

CONSERVATION: G5

Extirpated in Missouri. Possibly extirpated in Delaware, Maryland, South Carolina, and Washington D.C. Endangered in Georgia, Illinois, Kentucky, West Virginia, and Manitoba. Threatened in Indiana, Iowa, New Jersey, North Carolina, Ohio, Rhode Island, and Tennessee.

ALASKA REIN ORCHID *Platanthera unalascensis*

- ■ NATIVE
- ■ RARE OR EXTIRPATED
- ■ INTRODUCED

ALTERNATE NAMES: *Piperia unalascensis*

Alaska Rein Orchid is a species whose taxonomy continues to shift and change with our growing understanding. While a few references still have this species under the genus *Piperia*, there is now widespread agreement that all *Piperia* belong in the *Platanthera* genus. Current molecular analysis shows these groups are not distinctly divided and led to merging them into a single genus. The species name comes from Unalaska, an island in the Aleutians, but the species can be found on both sides of the continent, although eastern populations are highly disjointed and limited.

Along the coast, these plants typically are shorter and have more dense flower clusters than more inland populations. When blooming, the small flowers produce a scent that some describe as musky, soapy, or honey-like while others have found it more disagree-able and likened it to ammonia. This scent likely attracts the moths, which pollinate the flowers.

DESCRIPTION: One or a few stems with spear-like leaves near the base, 10" (25 cm) long. Pale green flowers are spaced on long, slender clusters. Each flower is ¼–⅜" (6–9 mm) wide. Overall height: 8–31" (20–80 cm).

FLOWERING SEASON: June–August.

HABITAT: Dry woods, gravelly streambanks, and open slopes.

RANGE: Alaska south and east to Quebec, and south to Baja California, northern Nevada, Utah, Colorado, western South Dakota, and Minnesota; also near the Great Lakes.

CONSERVATION: G5
Endangered in New Mexico, Newfoundland, and Quebec. Threatened in Alaska, Michigan, Wyoming, and Alberta.

■ NATIVE
■ RARE OR EXTIRPATED
■ INTRODUCED

ALTERNATE NAMES: Snake-Mouth Orchid

Rose Pogonia is a beautifully showy wetland orchid that has caught the eye—and nose—of many passersby, including Robert Frost and Henry David Thoreau. Both writers have immortalized their experiences with this plant in their works, although they had very different opinions of the scent, which is said to resemble either violets or raspberries. Having come across a bog clearing in a forest with thousands of blooms, Frost described that, "the air was stifling sweet / With the breath of many flowers." While most find the scent pleasant, in his journal on June 21, 1852, Thoreau said they "smell exactly like a snake."

The Rose Pogonia is the only *Pogonia* found in North America; only two other members of the genus exist, both in East Asia. The genus name is from the Greek word *pogon,* meaning "haired" or "bearded," referring to the bearded appearance of the lip, which appears to have a beard of dark pink, purple or yellow bumps surrounded by a fringe of rose-pink with purplish veins. Both the scent and the appearance attract the many species of bumblebees (*Bombus* spp.) on which it relies for pollination. Fooled into looking for pollen and nectar, the bees push their heads far into the flower where the anthers attach the pollen packet to the head, ready to pollinate the next flower visited.

DESCRIPTION: Produces a single leaf, 4¾" (12 cm) long, midway up a slender, greenish stem topped by a showy light rose-pink flower, 1¾" (4.5 cm) long, featuring a fringed lip with a brightly colored, bumpy center purple or purplish veins. Overall height: 3–24" (7.5–60 cm).

FLOWERING SEASON: May–August.

HABITAT: Wet open woods, meadows, swamps, and sphagnum bogs.

RANGE: Ontario to Newfoundland and Nova Scotia; south on Coastal Plain to Florida; west to Texas; inland to Pennsylvania, Tennessee, Indiana, Illinois, and Minnesota.

CONSERVATION: G5

Extirpated in Washington D.C. Possibly extirpated in Oklahoma. Endangered in Illinois, Kentucky, Missouri, Labrador, and Manitoba. Threatened in Arkansas, Delaware, Georgia, Indiana, Maryland, North Carolina, Ohio, Rhode Island, Tennessee, Vermont, Virginia, and West Virginia in the U.S. and Prince Edward Island and Quebec in Canada.

■ NATIVE
■ RARE OR EXTIRPATED
■ INTRODUCED

ALTERNATE NAMES: White Nodding Ladies'-Tresses

To many people, Nodding Ladies'-Tresses and other flowers in the genus *Spiranthes* do not look like orchids—a fact that may help save them from collectors. To the casual observer, all forms of this species are similar, with grass-like leaves and a spike with 1–4 spirals of white to cream-colored flowers. Taxonomists, however, consider this a "species complex," as there is a large amount of variability in the technical details of the flower shape across the plant's wide range. As a result of extensive molecular and genetic research, it has been proposed to subdivide this species complex into five more geographically restricted species as well as hybrids of related species.

The large variety in forms of this species arises both from hybridization with others in the genus as well as its own asexual reproduction from unpollinated seeds, made possible in part by its multiple sets of chromosomes. Despite the unusual reproduction strategies of the species, bumblebees (*Bombus* spp.) are also known pollinators and provide a route of sexual reproduction between plants.

DESCRIPTION: Lower leaves are lance-shaped, up to 10" (25 cm) long, with a 6' (15 cm) spike of small white to cream-colored flowers in 1–4 spirals; each flower is ½" long with a wavy edge on the lip. Overall height: 6–24" (15–60 cm).

FLOWERING SEASON: August (north)–November (south).

HABITAT: Fields, damp meadows, moist thickets, and grassy swamps.

RANGE: Ontario to Nova Scotia and northern New England; south to Florida; west to Texas; north to South Dakota.

SIMILAR SPECIES: It can be challenging to differentiate between many of the species of *Spiranthes*. The proposed split of the species complex based on molecular studies creates five distinct species: Nodding Ladies'-Tresses (*S. cernua*) found predominantly in the northeastern U.S. and Maritime Canada, Appalachian Ladies'-Tresses (*S. arcisepala*) in the Appalachian Mountains and eastern Great Lakes Basin, Niklas' Ladies'-Tresses (*S. niklasii*) in the Ouachita Mountains in the Southeast, Sphinx Ladies'-Tresses (*S. incurva*) in the Midwestern U.S. and south-central Canada, and Smoky Ladies'-Tresses (*S. kapnosperia*)—a hybrid of *S. cernua* and *S. ochroleuca* found in the Smoky Mountain region. In addition, several other *Spiranthes* species may be visually similar. Great Plains Ladies'-Tresses (*S. magnicamporum*) is quite similar but its leaves wither prior to blooming. Slender Ladies'-Tresses (*S. lacera*) has ovate leaves and a green spot on the lip. Short-lipped Ladies'-Tresses (*S. brevilabris*) has a more downy floral spike. Little Ladies'-Tresses (*S. grayi*) has much smaller, tiny flowers. Fragrant Ladies'-Tresses (*S. odorata*) grows in southern marshes and swamps to 2–3' (60-90 cm) and has spirally arranged clusters of fragrant flowers.

CONSERVATION: G5

Endangered in North Dakota and Prince Edward Island. Threatened in South Dakota, Rhode Island, Illinois, New Brunswick, and Quebec. Recent research has suggested merging the federally endangered Navasota Ladies'-Tresses (*S. parksii*) within this species complex; this plant is endemic to Texas and is classified as G3 (vulnerable).

■ NATIVE
■ RARE OR EXTIRPATED
■ INTRODUCED

This plant is similar in form to Nodding Ladies'-Tresses (*S. cernua*)—until 1973 they were considered to be the same species. However, Great Plains Ladies'-Tresses prefer open prairie and higher pH soils and is found in low, moist areas as well as in higher, drier, gravelly sites. The leaves of this species will wither prior to blooming, sometimes weeks prior. The flowers have a strong almond odor which attracts bumblebees (*Bombus* spp.) for pollination. This orchid is also capable of asexual reproduction; however, this species has only two sets of chromosomes (diploid) and thus is less susceptible to the broad genetic variations seen in the polyploid Nodding Ladies'-Tresses.

DESCRIPTION: The plant has 2–3 slender basal leaves, up to 6" (16 cm) long, which wither before the plant is in flower. The stem is topped with a spike of up to 40 white to yellowish, finely haired flowers arranged in a tight spiral. Overall height: up to 24" (60 cm).

FLOWERING SEASON: September–October.

HABITAT: Meadows, prairies, and fen wetlands.

RANGE: New Mexico and Texas north through North Dakota and east through Pennsylvania, Virginia and Georgia; Manitoba and Ontario.

SIMILAR SPECIES: Nodding Ladies'-tresses (*S. cernua*) will have leaves at flowering time; the flowers are less yellowish, and fairly odorless.

CONSERVATION: G3
Extirpated in Pennsylvania. Endangered in Georgia, Indiana, New York, Tennessee, Virginia, and Manitoba. Threatened in Alabama, Illinois, Iowa, Kansas, Kentucky, Louisiana, Michigan, Mississippi, New Mexico, Ohio, South Dakota, Wisconsin, and Ontario.

■ NATIVE
■ RARE OR EXTIRPATED
■ INTRODUCED

ALTERNATE NAMES: Irish Ladies'-Tresses

A late-summer-blooming orchid, Hooded Ladies'-Tresses is typically found in open, boggy or damp locations. The cluster of up to 60 hooded flowers is arranged in three rows, twisting in a distinctively geometric, spiral pattern. The white, sweetly scented flowers have a tube made by the side petals touching the top sepal. The labellum—the lower, lip-like petal of the flower—is white with green veins. This provides a landing area for its pollinating insects, primarily native bees. This lip has rounded edges and is slightly pinched in the middle, which helps distinguish this orchid from Nodding Ladies'-Tresses (*S. cernua*).

While the plant is highly variable in overall shape and form based on local conditions, the flowers are very consistent throughout most of its range. However, in California and Oregon, the flowers may be yellow and form a less tight spiral due to genetic mixing with Western Ladies'-Tresses (*S. porrifolia*).

This orchid was named in honor of the Russian count who financed a round-the-world scientific expedition in the early 1800s that resulted in the documentation of this and a number of other species. The plant is also found in Europe, albeit more rarely, with populations in England, Scotland, and Ireland. Even today, it is puzzling how this plant is distributed on both sides of the Atlantic Ocean.

DESCRIPTION: This plant has 3–6 spear-like basal leaves, 2–10" (5–25 cm) long and ½–1" (13–25 mm) wide. It grows one stem, or several in a clump, with up to 60 creamy-white flowers, ⅜–½" (9-13 mm) long, blooming in spiraled rows in a dense spike. Overall height: 4–24" (10–60 cm).

FLOWERING SEASON: July–October.

HABITAT: Generally moist open places, but variable, occurring from coasts to high mountains.

RANGE: Alaska to Southern California to New Mexico, north to northern Nebraska and across most of northern North America to Newfoundland.

SIMILAR SPECIES: Western Ladies'-Tresses (*S. porrifolia*) overlaps with the western portion of the range and has a more triangular lip. Nodding Ladies'-Tresses (*S. cernua*) overlaps the eastern portion of the range and features a lip with a dangling tongue.

CONVERSATION: G5

Endangered in Indiana, Iowa, Massachusetts, Nebraska, North Dakota, and Pennsylvania. Threatened in Arizona, Ohio, New Mexico, Vermont, Wyoming in the U.S. and Labrador, Nunavut, Prince Edward Island, and the Northwest Territories in Canada. It is believed extirpated in Connecticut and Illinois.

■ NATIVE
■ RARE OR EXTIRPATED
■ INTRODUCED

ALTERNATE NAMES: Eastern Yellow Star-Grass

When in bloom in the spring, this small plant boasts dainty, yellow, star-shaped flowers that have a faint floral scent. At any other time, it is hardly distinguishable from grass. It is usually no more than 6" (15 cm) tall, and is cultivated for use in native gardens. Yellow Star-Grass thrives in sunny, dry places, such as fields, lawns, prairies, and open woodlands. While the flowers do not produce nectar, they attract small mason and carpenter bees that they rely on for cross-pollination. Other insects like syrphid flies and beetles eat the pollen, and small rodents sometimes eat the corms. *Hypoxis* was formerly in the lily family but now is in its own family, Hypoxidaceae.

DESCRIPTION: The hairy, grass-like stalks, up to 6" (15 cm) long, boast 2–6 yellow, star-shaped flowers, which are ¾" (2 cm) wide. The flowers have 6 softly pointed, oval tepals. The outer 3 tepals are faintly green and hairy on the undersides. The flowers form a loose cluster, diverging from a central point at the top of the stem. The hair-covered leaves are long and thin, only ⅛–½" (2.5–12.5 mm) wide. At 10" (25 cm) long, the leaves are longer than the stem of the plant.

FLOWERING SEASON: March–September.

HABITAT: Open woods and dry meadows.

RANGE: Saskatchewan east to New England, south to Florida, west to New Mexico, and north to North Dakota.

CONSERVATION: **G5**

Extirpated in Maine. Presumed extirpated in Vermont. Endangered in Colorado. Threatened in Delaware, Illinois, New Hampshire, Manitoba, Ontario, and Saskatchewan.

ALTERNATE NAMES: Western Blueflag, *Iris missouriensis*

The characteristics of the Rocky Mountain Iris are varied, as is its ecological range, which can vary from high mountain elevations to sea level in southern California. It seems that the requirement for the flower's success is to be in an area that is extremely wet before it flowers and then to have extremely dry conditions for the remainder of the summer. In the western U.S., this is the only native iris species east of the Sierra Nevada and Cascade Mountains to the great plains. It often forms dense, large populations that can cover hundreds of acres. The blossoms can vary and are popular as landscaping because they require very little maintenance. The native irises provide many seeds and attract hummingbirds and insects. The fibers of the Rocky Mountain Iris are incredibly strong and flexible and are sometimes used to make ropes, strings, hair nets, and fishing nets.

DESCRIPTION: These flowers have large, pale to dark purple flowers with a yellow base. They can have 1–4 blooms per stem. The stem grows 1–2' (30–60 cm) high. The tough, sword-shaped, grayish-green leaves are 8–20" (20–50 cm) long and ¼–½" (6–13 mm) wide.

FLOWERING SEASON: May–July.

HABITAT: Wet meadows and streambanks, where moisture is abundant until flowering time.

RANGE: British Columbia to southern California, east of the Cascade Mountains and Sierra Nevada (and on islands in Puget Sound); east to South Dakota, Nebraska, Colorado, and New Mexico.

CAUTION: This plant is suspected to be poisonous to humans and animals if eaten (especially the rhizome, or root), and it is likely that all irises contain toxins. Plant juices can cause irritation on the skin.

CONSERVATION: G5
Endangered in Alberta.

■ NATIVE
■ RARE OR EXTIRPATED
■ INTRODUCED

ALTERNATE NAMES: Pale-yellow Iris, Water Flag, *Iris pseud-acorus*

The Yellow Flag is a non-native, invasive species that was introduced from Europe, escaped cultivation and is spreading, displacing natural vegetation. The plant is difficult to remove on a large scale. It boasts a large, showy yellow flower that does best in wet soil. It is a hardy plant that can tolerate low oxygen levels due to the air spaces in the cellular tissues. Yellow Flags may be useful in removing heavy metals from wastewater, as it can efficiently and effectively absorb the metals and survive. They can also survive water with high salinity levels.

DESCRIPTION: This perennial grows to 40–60" (100–150 cm) with erect leaves that can be 3' (90 cm) tall, often taller than the flower stalk. The robust stalk has several bright yellow flowers, 3" (7.5 cm) wide with 3 backward-curving petals.

FLOWERING SEASON: June–August.

HABITAT: Marshes and streamsides.

RANGE: The species is invasive throughout the East Coast from Georgia to northern Canada. It is also on the West Coast from California to northern Canada.

CAUTION: The plant is suspected to be poisonous to humans and animals if eaten (especially the rhizome, or root), and it is likely that all irises contain toxins. Plant juices can cause irritation on the skin.

CONSERVATION:

This plant is banned or prohibited in Connecticut, Massachusetts, Montana, New Hampshire, New York, Oregon, and Washington.

■ NATIVE
■ RARE OR EXTIRPATED
■ INTRODUCED

ALTERNATE NAMES: Virginia Blueflag, Southern Blue Flag, *Iris virginica*

Southern Blueflags are small bluish-purple flowers that have a bright yellow base and are commonly found in wetlands and sunny, moist areas. They sometimes grow in both fresh and brackish shallow waters and marshes. The flowers are pollinated by bumblebees and long-horned bees. Butterflies and skippers may also suck nectar from the flower, but are less effective at cross-pollination. Caterpillars and other insects may feed on this plant, including larvae that feed inside the seed capsule, but mammals tend to leave it alone because the foliage and roots can cause irritation of the gastrointestinal tract.

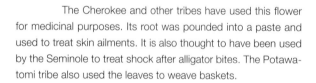

The Cherokee and other tribes have used this flower for medicinal purposes. Its root was pounded into a paste and used to treat skin ailments. It is also thought to have been used by the Seminole to treat shock after alligator bites. The Potawatomi tribe also used the leaves to weave baskets.

DESCRIPTION: This wetland species has 3–4 bright green, ½–1¼" (1–3 cm) wide, lance-shaped leaves that, at 2' (60 cm) long, are sometimes longer than the flower stalk, which is 12–36" (30–90 cm). The slightly fragrant, blue to purple flower is 1½" (3.75 cm) with three upcurving, pointed sepals and three down-curving sepals. They are fuzzy yellow toward the base.

FLOWERING SEASON: April to May.

HABITAT: Marshes, wet meadows, swamps, wet pinelands, lake- and streamsides.

RANGE: Far southern Quebec and Ontario south through Great Lakes to eastern Nebraska and Texas; east to western Pennsylvania and Maryland, and south to central Florida and Louisiana.

CAUTION: The plant is suspected to be poisonous to humans and animals if eaten (especially the rhizome, or root), and it is likely that all irises contain toxins. Plant juices can cause irritation on the skin.

CONSERVATION:
Endangered in Kansas, New York, and Oklahoma. Threatened in Maryland and Pennsylvania.

■ NATIVE
■ RARE OR EXTIRPATED
■ INTRODUCED

This perennial grass-like plant boasts small, white flowers with bright yellow centers. While flowers of grasses are generally wind-pollinated and inconspicuous, these flowers are more conspicuous because they need to attract pollinators. This genus gets its scientific name *Sisyrinchium* from the Greek words meaning "pig snout," which refers to the tubers of this plant being foraged by pigs. The Menominee tribe used to mix the roots with horse feed. The poison in the roots was thought to make the horses vicious and their bites venomous.

DESCRIPTION: The stiff, grass-like leaves can be 8–15" (20–37.5 cm) tall and form clumps. The star-like flowers are white to blue with yellow centers, ½" (13 mm) wide, occurring in loose clusters.

FLOWERING SEASON: March–May in southern locations; April–June in the north.

HABITAT: Dry and sunny habitats like prairies, sand hills, glades, open woods.

RANGE: Gulf Coast states, north to Pennsylvania, southern Ontario, Ohio, southern Michigan, and Wisconsin; a single, introduced population has been reported in Maine.

CAUTION: This plant is suspected to be poisonous to humans and animals if eaten (especially the rhizome, or root), and it is likely that all irises contain toxins. Plant juices can cause irritation on the skin.

SIMILAR SPECIES: All *Sisyrinchium* species can be difficult to differentiate.

CONSERVATION: G5
Extirpated in Michigan. Possibly extirpated in New York and Pennsylvania. Endangered in West Virginia and Ontario. Threatened in North Carolina, Virginia, and Wisconsin.

BLUE-EYED GRASS *Sisyrinchium angustifolium*

■ NATIVE
■ RARE OR EXTIRPATED
□ INTRODUCED

ALTERNATE NAMES: Pointed Blue-eyed Grass, Narrowleaf Blue-eyed Grass

This is most common of all the blue-eyed grasses in the eastern U.S. Although the flowers within this genus can be challenging to differentiate from each other, this species is typically the most likely candidate if the flowers are a deep blue-violet rather than pale blue or white. The foliage of this plant is unique in that the leaves are broader than other blue-eyed grasses.

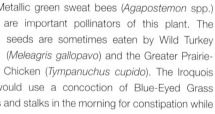

Metallic green sweat bees (*Agapostemon* spp.) are important pollinators of this plant. The seeds are sometimes eaten by Wild Turkey (*Meleagris gallopavo*) and the Greater Prairie-Chicken (*Tympanuchus cupido*). The Iroquois would use a concoction of Blue-Eyed Grass roots and stalks in the morning for constipation while Cherokee would use the roots as an antidiarrheal medicine for children.

DESCRIPTION: The stem of this plant is broadly winged, $\frac{1}{16}$–$\frac{3}{16}$" (2–4 mm) wide, and is taller than the clusters of sword-shaped leaves near its base. The plant is 6–20" (15–50 cm) tall, while the leaves are 2–10" (5–25 cm) long. Generally, only one of the delicate deep blue-violet flowers bloom at a time. The flowers each contain six tipped tepals with a yellow patch in the center.

FLOWERING SEASON: Spring to early summer.

HABITAT: Moist meadows; low, open woods; and shorelines.

RANGE: Ontario, Quebec, and Nova Scotia; Maine south to Florida, west to Texas and Minnesota, and north to Kansas, Iowa, and Minnesota.

CAUTION: The plant is suspected to be poisonous to humans and animals if eaten (especially the rhizome, or root), and it is likely that all irises contain toxins. Plant juices can cause skin irritation.

SIMILAR SPECIES: All *Sisyrinchium* species can be difficult to differentiate. This species can be confused with Eastern Blue-eyed Grass (*S. atlanticum*), which has a very slender stem, whereas this species has a distinct, winged stem.

CONSERVATION: G5
Endangered in Colorado, Wisconsin, and New Brunswick. Threatened in Iowa, Vermont, Wyoming, and Quebec.

■ NATIVE
■ RARE OR EXTIRPATED
■ INTRODUCED

This plant is an evergreen with narrow, erect, grass-like leaves growing from its base. The six-petaled, star-shaped flowers can be light blue to purple with a yellow center. It requires an abundance of moisture and is classified as a wetland indicator in several states. Members of this genus are difficult to differentiate because there is considerable variability both between and within its species. Within this species, there are currently four recognized varieties, many of which have also been proposed as distinct species. Additional molecular and genetic research is likely to expand the understanding of the relationships between the species of this genus.

DESCRIPTION: A clump of small, evergreen, grass-like leaves, 2–10" (5–25 cm) long, with a stem taller than the leaves topped with several delicate, blue or deep blue-violet flowers, ½–1½" (1–3.8 cm) wide. Overall height: 4–20" (10–50 cm).

FLOWERING SEASON: April–September.

HABITAT: Moist areas early in the year, generally in the open, from lowlands and into the mountains.

RANGE: British Columbia south to California, west to Colorado and New Mexico.

CAUTION: The plant is suspected to be poisonous to humans and animals if eaten (especially the rhizome, or root), and it is likely that all irises contain toxins. Plant juices can cause skin irritation.

SIMILAR SPECIES: All *Sisyrinchium* species can be difficult to differentiate.

CONSERVATION:
Threatened in Utah.

NATIVE
RARE OR EXTIRPATED
INTRODUCED

Strict Blue-eyed Grass can be found across a wide swath of northern North America and extending down the Rocky Mountains into New Mexico. This species is sometimes included with Blue-eyed Grass (*S. angustifolium*) due to the similarity of flower structures in eastern populations. However, Strict Blue-eyed Grass is typically differentiated by its lack of the multiple, long-stalked flower clusters seen in *S. angustifolium*.

DESCRIPTION: This species produces small clumps of bright green, grass-like leaves. It has stems with distinct wings topped by a violet-blue, star-like flower at the tip of long flowering stalks. Overall height: 6–15" (15–38 cm) tall.

FLOWERING SEASON: May–July.

HABITAT: Moist, sandy meadows and open woods.

RANGE: Quebec to British Columbia, south to New Jersey, north Indiana, Iowa, Kansas, Texas, and New Mexico.

CAUTION: The plant is suspected to be poisonous to humans and animals if eaten (especially the rhizome, or root), and it is likely that all irises contain toxins. Plant juices can cause irritation on the skin.

SIMILAR SPECIES: All *Sisyrinchium* species can be difficult to differentiate.

CONSERVATION: G5
Endangered in Alaska, Illinois, Indiana, Ohio, Washington D.C, Washington State, and West Virginia. Threatened in Wyoming, Labrador, and Yukon.

■ NATIVE
■ RARE OR EXTIRPATED
■ INTRODUCED

ALTERNATE NAMES: Tawny Daylily, Corn Lily, Ditch Lily

This non-native invasive plant is native to China and Korea. It came to the U.S. during colonial times and was often given as gifts to neighbors, leading it to become a commonly found flower from coast to coast. Its nickname Ditch Lily is a testament to how easily these plants grow, as they are often found growing in ditches. Despite the name, a daylily is not a true lily, which typically grows on stems much taller than the Orange Daylily. These flowers provide little to no value to native pollinators, and larger animals likewise avoid the plants.

The genus name *Hemerocallis* comes from two Greek words for "day" and "beautiful," as the attractive flower only lasts for one day once it blooms. This may not be obvious because as one flower fades, another one on the stalk will bloom, so the period of time that there are blooms on the plant can last for weeks. The flowers, leaves, and tubers of the plant are edible, and the buds have been roasted and eaten for centuries in Asian cultures. The cooked flower buds are said to taste like beans, while the tubers can be a substitute for potatoes.

DESCRIPTION: The Orange Daylily is recognized by its showy, orange flower, which is 2–4" (5–10 cm) across on stems that are 16–59" (40–150 cm) tall. The leaves are basal and linear, 12–59" (50–150 cm) long.

FLOWERING SEASON: Late spring to early summer.

HABITAT: Disturbed sites, prairies, thickets, woodland borders, and along railroads and roadsides.

RANGE: Invasive throughout North America, likely more widespread than official records indicate. Its native range is East Asia.

CONSERVATION:

This is a non-native invasive plant in North America. The waxy coating of the leaves resists penetration by herbicides, making control of this plant more difficult. Digging is the preferred control, but any roots remaining will sprout new plants.

■ NATIVE
■ RARE OR EXTIRPATED
■ INTRODUCED

ALTERNATE NAMES: Rain Lily, Woodland Spiderlily

While most other species of spiderlily are found in wetlands and streambanks, the Carolina Spiderlily is found in moist woodlands and forests. The flowers are distinctive for both their spidery form and their lemonesque scent. The genus name refers to the flower's daffodil-like center crown and is a combination of two Greek words, *hymen* for "membrane," and *kallos* for "beautiful." The flowers are pollinated by moths.

This species has two distinct varieties: *H. occidentalis* var. *occidentalis*, which has the widest range in the entire genus and leaves that remain through flowering, and *H. occidentalis* var. *eulae*, which is only found in eastern Texas and Oklahoma and has leaves that wither before the plant blooms.

DESCRIPTION: The plant produces a small number of spidery, white flowers, up to 7" (17.5 cm) wide, with six narrow petal-like parts attached to a long, slender crown. Flowers appear from a two-edged stalk rising from the ground amid 5–12 deeply grooved, strap-like leaves, each ¾" (2 cm) wide, at least 12" (30 cm) long. Overall height: 18" (45 cm).

FLOWERING SEASON: July–September.

HABITAT: Moist woodlands and forests, swamps, moist fields.

RANGE: Southern Indiana and Illinois to South Carolina west to eastern Texas and Oklahoma.

CAUTION: The bulbs may be toxic if ingested.

CONSERVATION: G4
Possibly extirpated in North Carolina. Threatened in Indiana.

■ NATIVE
■ RARE OR EXTIRPATED
■ INTRODUCED

ALTERNATE NAMES: Crow Poison

The aptly named False Garlic has a striking resemblance to wild garlics (*Allium* spp.); however, the leaves and bulbs of False Garlic do not have the onion-like aroma. The genus name *Nothoscordum* is from the Greek words *nothos* meaning "false" and *skordon* meaning "garlic." Although the colloquial name Crow Poison suggests that this plant is toxic and legend states that Native Americans used this plant to poison crows while growing corn, there is little evidence that either of these ideas is true.

A short, grass-like plant, False Garlic can be found in lawns and along highways as well as in its native woods and prairies, where it is often an early successional plant in disturbed areas. If conditions are favorable, large colonies of plants can eventually form. The plant blooms in the early spring between March and April. Depending on weather, a second blooming from new plants may occur again in autumn.

DESCRIPTION: False Garlic grows one to four grass-like leaves, 4–10" (10–25.4 cm) long. Its flowering stem is erect, with three to six white flowers, each with six tepals, often with a red midvein. Overall height: 8–16" (20–40 cm).

FLOWERING SEASON: March–December.

HABITAT: Open woods, prairies, and disturbed areas.

RANGE: Virginia south through Florida, west through Nebraska and Texas.

SIMILAR SPECIES: Easily confused with a South American non-native invasive, Slender False Garlic (*N. gracile*), which has wider leaves up to ½" (1 cm) wide and tepals that are joined together in their lower third.

CONSERVATION: G4

Endangered in Nebraska. Threatened in Indiana and Ohio.

■ NATIVE
■ RARE OR EXTIRPATED
■ INTRODUCED

ALTERNATE NAMES: Taper-tip Onion

Hooker's Onion is a western variety of wild onion native to areas west of the Rocky Mountains, a large range compared with many of the North American native *Alliums*. This species was used by Native Americans as a food source as well as a bug repellent by the Coast Salish who lived on what is now Vancouver Island. The genus name *Allium* is Latin for "garlic," but the Latin is said to come from a Greek word meaning "to avoid," likely referring to the offensive smell. The species name is Latin for "slender pointed," referring to the leaves.

DESCRIPTION: This plant has a small set of 2–3 narrow leaves, 4–6" (10–15 cm) long and a taller, leafless stalk capped with a round cluster of pink to magenta flowers, each ½" long with 6 tepals.

FLOWERING SEASON: April–July.

HABITAT: Open, often rocky slopes, among brush and pines.

RANGE: British Columbia to central California and southern Arizona; east to southern Wyoming and western Colorado.

CAUTION: Although edible, all *Allium* species are poisonous to some mammals, especially dogs and cats. Additionally, there are several extremely toxic plants these species may be confused with.

CONSERVATION: G5
Endangered in Montana and Wyoming.

■ NATIVE
■ RARE OR EXTIRPATED
■ INTRODUCED

ALTERNATE NAMES: Lady's Leek

The Nodding Onion is prevalent throughout much of the United States, Canada, and Mexico, and it is the most widespread species of its genus in North America. It has distinct, nodding white, pink, or maroon flowers. It boasts a strong onion flavor that lends itself to cooking and historically has been a food staple for many Native American groups, including the Apache and Cherokee. The Cherokee also used its juice as a cold remedy and made an infusion of it to be taken for colic.

DESCRIPTION: This herbaceous plant grows from a conical bulb that tapers into thin, flat, grass-like leaves. It has a bent stalk 4–20" (10–50 cm) tall, with a single, bell-shaped flowering stem coming from each bulb. The pink or white flowers are downward-facing with yellow anthers and are about ³⁄₁₆" (5 mm) across. The flowers mature in fruits, which eventually split to reveal dark, shiny seeds.

FLOWERING SEASON: July–October.

HABITAT: Open woods, prairies, and rock outcroppings.

RANGE: Found throughout the U.S. and Canada.

SIMILAR SPECIES: This species can be confused with Prairie Onion (*A. stellatum*). The flower cluster of Nodding Onion bends downward compared to the more erect Prairie Onion, which also has more star-shaped flowers instead of more rounded blooms.

CONVATION: G5
Threatened in Illinois, Iowa, Minnesota, Nebraska, New York, and South Carolina.

■ NATIVE
■ RARE OR EXTIRPATED
■ INTRODUCED

ALTERNATE NAMES: Wild Chives

Siberian Chives produce edible leaves and flowers that are closely related to onions, leeks, shallots, and garlic. They are widely used in cooking and can commonly be found in grocery stores and home gardens, in addition to the wild. The flowers can be used in salads or to make blossom vinegar, while the green stalks are a common ingredient often used in soups, potatoes, eggs, etc. The Siberian Chives are considered a strong nectar producer, attracting bees and sometimes prolonging their life. They also can be used as pest control—a poultice made from its leaves has insect-repelling properties due to the sulfur compounds they contain. They can also fight fungal infections and mildew. Native peoples used the chives for these purposes as well as for food.

DESCRIPTION: This grass-like plant grows to be 12–20" (30–50 cm) tall and has a faint onion smell. The stems are hollow and tubular and 1/16–1/8" (2–3 mm) across. The pale purple or pink star-shaped flowers have six petals 1/2–1/4" (1–2 cm) wide and grow in dense clusters of 10–30 flowers.

FLOWERING SEASON: April to May in the southern regions. June in the northern regions.

HABITAT: Mountain streamsides, marshes, damp meadows, rocky pastures, gardens and yards.

RANGE: Northern North America south to Virginia, Missouri, Colorado, and Oregon, and throughout much of Canada, north to the Northwest Territories.

CONSERVATION: G5
Endangered in Colorado, Minnesota, New Hampshire, New York, and Vermont. Threatened in Michigan, Nova Scotia, Manitoba, and Seskatchewan.

- ■ NATIVE
- ■ RARE OR EXTIRPATED
- ■ INTRODUCED

ALTERNATE NAMES: Ramp

Wild Leeks are less known for their bloom than their use as a popular food in the rural upland areas of its range. In some regions, especially in central and southern Appalachians, where the plants are known as Ramps, Wild Leeks have become iconic with numerous and increasingly popular food festivals held in early spring when it is harvested. There is minimal commercial propagation and nearly all plants are wild-harvested, leading to significant concerns of overharvest, especially in these areas. To ensure long-term sustainability, it is recommended that only one leaf per plant should be harvested each year.

The springtime arrival of Wild Leeks was celebrated by Native Americans and today's residents of rural Appalachia alike, as it was believed to ward off winter ailments. Indeed, its high vitamin C content was certain to improve health and reduce malnutrition and illnesses like scurvy caused by a lack of fresh vegetables through the winter. Farther west, it is said that the city of Chicago is named after this species, based on a French interpretation of the native Miami-Illinois word *shikaakwa*, meaning "stinky onion."

There are two varieties of Wild Leek that are sometimes treated as separate species. While typically found in moist depressions in rich woodland soil, the rarer, Narrow-leaved Wild Leek (*A. tricoccum* var. *burdickii*) is found growing in drier soils of upland woods.

DESCRIPTION: In spring, this plant produces tall leaves, 8–12" (20–30 cm) long and ¾–3½" (2–9 cm) wide, that wither before flowers appear. Flowers appear atop a leafless stem in a round cluster, 1½" (4 cm) wide, of creamy-white cup-like flowers, each ¼" (6 mm) long. Overall height: 6–20" (15–50 cm).

FLOWERING SEASON: June–July.

HABITAT: Typically in moist woodland depressions. A variety is found in dry upland wooded sites.

RANGE: Northeast of the United States from the eastern Dakotas south to Tennessee and east to Nova Scotia, roughly following the Great Lakes and the Canadian border.

CONSERVATION: G5

Endangered in Alabama, Rhode Island, Tennessee, and Nova Scotia. Threatened in Delaware, Maine, Georgia, Illinois, New Brunswick, and Quebec. The Narrow-leaved variety of Wild Leek is ranked G4 when treated as a distinct species. It is presumed extirpated in New York and Vermont. It is endangered in New Hampshire, Tennessee, Nova Scotia, and Ontario. It is threatened in Illinois, Kentucky, Missouri, and Nebraska.

FIELD GARLIC *Allium vineale*

ALTERNATE NAMES: Wild Garlic, Onion Grass, Crow Garlic, Stag's Garlic

A non-native invasive species, Field Garlic is especially challenging for farmers because the small bulb-like structures, called bulbils, can taint the harvests of grains and milk with a strong garlic flavor and scent. As a bulb-forming plant, it is resistant to pre-emergent herbicides and is also resistant to foliar-applied herbicides because the slick, round leaves do not hold the treatment long enough to absorb the chemicals.

DESCRIPTION: Grows 2–4 hollow, round, linear leaves up to 10" (25.4 cm) long and ¾" (2 cm) across at the base and tapering to a point. A tall stalk emerges topped with a sphere, 2" (5 cm) in diameter, of up to 50 small, purple flowers, each no longer than ³⁄₁₆" (4 mm). These often become bulbils, although in some plants only bulbils occur and no flowers are present. Later, each bulbil may sprout a leaf while still attached in the inflorescence.

FLOWERING SEASON: June–August.

HABITAT: Prairies, degraded meadows, river edges, woodlands, woodland borders, thickets, vacant lots, grassy clay banks, lawns, and waste areas.

RANGE: New Brunswick to Ontario; throughout the eastern United States west to Texas and north to North Dakota. Also occurs on the West Coast from British Columbia to California.

CONSERVATION:

This non-native, invasive species is listed as a noxious weed and may be restricted in many areas.

■ NATIVE
■ RARE OR EXTIRPATED
■ INTRODUCED

ALTERNATE NAMES: Desert Agave, Desert Century Plant

While this slow-growing, desert-dwelling plant is known as the Century Plant, it does not take a century to grow—but it does often take 20 to 40 years for the plant to mature and flower. When conditions are right, a huge, tree-like stem will sprout, growing up to 1' per day with short side branches. While this stem typically reaches 6–10', recorded specimens have reached up to 19' (10 m). At the tips of each short branch, a cluster of relatively small, yellow flowers will bloom.

The rapid growth of the flowering stalk cannot be fully supported by photosynthesis or uptake of water by the roots. Instead, the plant uses the stored resources in its heart and its spiky leaves. However, having used all of its resources to produce this massive flowering structure, the plant quickly dies. Afterward, several clonal offshoots, called "pups," form ring-shaped colonies of genetically identical new plants. Because the plant only flowers once, it is monocarpic, similar to annuals.

The flowers attract numerous birds and insects and the plant is the host of both the Bauer's Giant Skipper (*Agathymus baueri*) and California Giant Skipper (*A. stephens*) butterflies. However, the primary—and some speculate the only—pollinator is the near-threatened Lesser Long-nosed Bat (*Leptonycteris yerbabuenae*), with which the plant has formed a very close symbiotic relationship. The bat is attracted to the musky, melon-like scent of the bloom, whose nectar provides an important food source during the bats' annual migration. The nectar provides two amino acids the plant doesn't need but which are essential to the bat—proline, which supports building muscle tissue, and tyrosine, needed by lactating mothers as a growth stimulator for their young.

There are two varieties: *A. deserti* var. *deserti*, which has numerous rosettes and is primarily found in the Sonoran Desert of southern California, and *A. deserti* var. *simplex*, which typically has a single rosette and a wider range in the north and east into Nevada. The Century Plant is one of the most important plants for Native Americans in the southwest, with its use spanning more than 9,000 years. Different parts of the plant became a dietary staple among many Native American groups, and there is strong evidence of cultivation of many types of *Agave*. The stalks and hearts of the plants were eaten roasted, providing food even during drought. The Cahuilla used stalk ash for tattoo dye and used the dry leaves to make sandals, cactus bags, nets, and women's skirts.

DESCRIPTION: This stout plant typically will have a rosette of thick gray-green leaves, 24" (60 cm) long, with sharp spines along the edges and tip. Up to 50 relatively small, cup-shaped, yellow flowers, 2" (5 cm) long, are found at the tips of each short branch from a tall, woody stem with a tree-like appearance, up to 19' (6 m) tall.

FLOWERING SEASON: June–August.

HABITAT: Rocky slopes and washes in desert scrub.

RANGE: Mojave Desert mountains, Colorado Desert, Sonoran Desert, in California and Arizona.

SIMILAR SPECIES: Agaves of all species can be difficult to differentiate when not flowering.

CONSERVATION: G4
Endangered in Nevada.

PARRY'S AGAVE   *Agave parryi*

■ NATIVE
■ RARE OR EXTIRPATED
■ INTRODUCED

ALTERNATE NAMES: Parry's Century Plant

Many *Agave* species, this plant requires 10 to 25 years to achieve maturity and dies immediately after flowering. As with other *Agaves*, pollination is by bats as well as moths, although self-pollination is also possible. A distinguishing feature of this species is that the protective leaf spines are darker than the leaf, usually dark tan to black. Parry's Agave is named in honor of Dr. Charles C. Parry (1823–1890), who worked as a surgeon and botanist for the U.S. and Mexico border survey of 1848–1855.

An extract made by simmering the chopped-up leaves in water contains saponins and can be used as a soap. A more famous use of *Agave* is the production of liquor. A simple type called "pulque" starts with removal of the central bud, creating a cavity which fills with honey-water or "aguamiel." This liquid is then allowed to ferment and is ready for consumption, but may also be distilled.

DESCRIPTION: This stout plant typically will have a rosette of thick gray-green leaves, 12–18" (30–45 cm) long, with sharp, dark-colored spines along the edges and tip. Up to 50 relatively small, cup-shaped, yellow flowers, each 2" (5 cm) long, are found at the tips of each short branch from a tall, woody stem with a tree-like appearance, up to 19' (6 m) tall.

FLOWERING SEASON: June–August.

HABITAT: Dry, rocky slopes.

RANGE: From central Arizona to western Texas; also found in northern Mexico.

SIMILAR SPECIES: Agaves of all species can be difficult to differentiate when not flowering.

CONSERVATION: G5

This species is not of special concern in its range.

■ NATIVE
■ RARE OR EXTIRPATED
■ INTRODUCED

A California native, Common Goldenstar begins blooming just as many of the early spring flowers begin to fade. Adapted to the dry grasslands where they grow, each plant typically has just a single grass-like leaf, which minimizes moisture loss and grows from corms—bulb-like structures that allow the plant to survive extended periods of drought. Native Americans in the region ate these corms and the Kawaiisu ground them into an adhesive-like paste to seal woven baskets in order to collect seeds.

Three to four years after sprouting, if conditions are favorable, the plant will produce a single leafless stem, crowned with a loose cluster of up to 100 radiating spokes. From this cluster, a few bright yellow star-like flowers will open each day, providing a relatively long total bloom time. With careful inspection of the flowers, one can observe the blue-colored pollen dusting the anthers. The name of the genus *Bloomeria* is in reference to an early San Francisco botanist aptly named Hiram Green Bloomer (1819–1874), who was one of the founders of the California Academy of Sciences. The species name *crocea* is Latin for "saffron colored."

DESCRIPTION: From the ground, one grass-like leaf, 4–12" (10–30 cm) long, ¼–½" (5–13 mm) wide. When in flower, the leafless stem has one cluster of yellow to orange, star-like flow-ers with six tepals accented with dark lines down the middle, ½–1" (1.5–2.5 cm) wide.Overall height: 6–24" (15–60 cm).

FLOWERING SEASON: April–June.

HABITAT: Dry flats and on hillsides in grass, brush, and oak woodlands.

RANGE: Southern third of California to Baja California.

SIMILAR SPECIES: The similar San Diego Goldenstar (*B. clevelandii*) is threatened by urbanization; it has several leaves, each less than ⅛" (3 mm) wide, and the flap at the base of each stamen is not notched.

CONSERVATION: G4
The population is apparently secure in its limited range.

■ NATIVE
■ RARE OR EXTIRPATED
■ INTRODUCED

ALTERNATE NAMES: Small Camas, Swamp Sego

The native Common Camas is widespread in the western U.S. and can be so abundant its blooms will color an entire meadow with shades of blue-violet. This species is highly variable, with eight different subspecies, although these are very difficult to identify due to the minimal differences between them.

Over two dozen Native American groups consumed the bulbs, most frequently after roasting them in a pit for two days, but also as a syrup and beverage. The Blackfoot used the roots to induce labor and many groups used the bulbs for trade. The Nez Perce offered the Lewis and Clark expedition meals of "fish and roots," which likely included these plants, after the party nearly starved in the fall of 1805. Later, the explorers observed women using digging tools able to harvest over 50 pounds of camas per day, with small bulbs and bulblets sown for later harvests.

Parts of the range of Common Camas overlap with the range of Mountain Deathcamas (*Anticlea elegans*). When not blooming,

it can be difficult to determine which species is being collected. Historically, harvests were made in the fall, when plants were not in bloom. Misidentification could be lethal.

DESCRIPTION: A tall cluster of light to deep violet-blue, star-shaped flowers, 1½–2½" (4–6.5 cm) wide, with six tepals. Up to 10 narrow, grass-like leaves near the base, up to 2' (60 cm) long. Overall height: 8–28" (20–70 cm).

FLOWERING SEASON: April–June.

HABITAT: Sandy pinelands and bogs.

RANGE: Southern British Columbia to northern California, east to northern Utah, Wyoming, and Montana.

CAUTION: If not in bloom, this plant may be confused with the poisonous Mountain Deathcamas (*Anticlea elegans*).

SIMILAR SPECIES: Easily confused with Mountain Death-camas (*Anticlea elegans*) when not in bloom.

CONSERVATION: (G5)
Threatened in Montana, Wyoming, and Alberta.

■ NATIVE
■ RARE OR EXTIRPATED
■ INTRODUCED

ALTERNATE NAMES: Amole, California Soaproot

Wavyleaf Soap-plant is a native restricted to the extreme south-west of Oregon and California and is easy to identify through either the unique undulating edges of its leaves or the spindly flowering stalk that can exceed 6' (1.8 m). From seed, this plant typically takes 10 years before its first flowers are produced.Its delicate, wispy flowers open as twilight approaches in the evening, a quality termed vespertine. They bloom for just 6–8 hours to be pollinated by the many bees and moths attracted to them before the tepals wilt and seed production begins.

The genus name *Chlorogalum* is from the Greek for "green milk," referring to the green sap exuded by a broken leaf. An alterna-tive common name, Amole is a Spanish term for the genus; at least five other plant species of the Southwest share this name.

Many Native Americans actively cultivated this plant due to its many uses, especially as food and for making soaps and shampoo, which was said to remove lice and poison oak. One of the most unique uses was crushing the bulbs and throwing the powder into dammed streams or ponds. The saponins that make the plant useful as a soap also prevent the gills of fish from functioning properly, causing all fish and eels to float to the surface for easy collection by hand or net. Cooking removes this toxin and also makes the bulbs palatable.

The bulbs of mature plants are typically 6–8" (15–20 cm) deep and have contractile roots that actively pull the bulb deeper into the earth each year. This makes them extremely difficult to

dig out of hard, compact sandstone or decomposed granite—even with a modern, full-size shovel. Despite these challenges, indigenous peoples dug them up in large quantities due to their exceptional utility.

DESCRIPTION: Delicate, star-like, white flowers, 1–1½" (2.5–4 cm) wide, with six narrow tepals which curve back. Flowers emerge in open clusters from a tall, openly branching stalk up to 10' (300 cm). Several long, narrow, grass-like leaves grow from the base, with strongly undulat-ing, wavy edges, up to 2' (60 cm) long. Overall height: 2–10' (60–300 cm).

FLOWERING SEASON: May–August.

HABITAT: Dry open hills and plains, often in open brush or woods.

RANGE: Southwestern Oregon south to southern California.

CAUTION: The saponins in the bulb are mildly toxic.

CONSERVATION: G5
Two of the three varieties with restricted ranges are threatened in California.

■ NATIVE
■ RARE OR EXTIRPATED
■ INTRODUCED

ALTERNATE NAMES: Desert Pampas Grass, Desert Spoon, Spoonflower

Sotol is a yucca-like plant that grows from a thick stem with sharply toothed sword-shaped leaves that extend in all directions, sometimes giving it the look of a spiky ball. The leaves curve inward at the stem, leading to the common name Desert Spoon. Sotol takes 7–10 years before first flowers appear and then blooms irregularly afterward. This species is dioecious, which means it cannot pollinate itself because each plant has either male flowers that produce pollen or female flowers that can produce seeds—although for Sotol, each female flower produces only one seed. The color of the flower is dependent on the sex of the plant. Male plants produce creamy-yellow flowers while female plants produce purplish-pink flowers.

Native Americans used this plant widely, using the leaves for basket weaving and eating the thick, white leaf ends and the sugary heart. The stalks were dried and split to make the hearths for hand drills, a type of fire starter. Today, the plant is most known for the alcoholic drink of the same name, made from fermented Sotol.

DESCRIPTION: Sotol has up to 100 stout, ridged, spiny, whitish to blue-green leaves all radiating from a central point, up to 38" (96 cm) long and ¾–1¼" (2–3 cm) wide. The teeth on leaf edges curve forward toward the tips. It has large, slender flowering stalks up to 16' (5 m) covered with thousands of tiny flowers that are greenish to yellow with some tinged purplish-pink.

FLOWERING SEASON: May–June.

HABITAT: Dry, rocky slopes of deserts and mountains.

RANGE: Southern Arizona east to western Texas; also northern Mexico.

SIMILAR SPECIES: The smaller Smooth-leaved Sotol (*D. leiophyllum*) has teeth that curve toward the leaf base.

CONSERVATION: G4
The population is apparently secure in its range.

- ■ NATIVE
- ■ RARE OR EXTIRPATED
- ■ INTRODUCED

ALTERNATE NAMES: Covenna, Purplehead, Grass Nuts, Wild Hyacinth, Fool's Onion, Crow Poison

This species was recently moved from the genus *Dichelostemma* and separated back into its own genus as originally suggested in 1912. Currently, this is the only species in the genus, but there is a large amount of variation within the species in habitats, plant and flower shapes and structure, and even in the number of chromosomes. There are three recognized subspecies, which may be separated into individual species with additional research. Bluedicks are an early successional plant well adapted to open, fire-prone woods and grasslands. The underground corms can survive more than 10 years of being shaded over by taller plants and producing no above-ground growth. If a fire burns off this taller growth, there may soon be thousands of Bluedicks blooming where none had been seen for many years.

Native Americans of the Southwest ate the corms of this and many other species to provide the bulk of the starches needed. These corms are also eaten by many animals such as black bears, mule deer, non-native wild pigs, rabbits, and pocket gophers. The process of rooting out the corms often breaks off smaller cormlets and deposits them back in the broken ground.

DESCRIPTION: Bluedicks have two to three narrow leaves, each 4–16" (10–40 cm) long, and emit an onion-like odor when crushed. When blooming in spring atop a leafless stalk, there will be dense clusters of 3 to 15 blue to purplish (rarely white) flowers with six tepals. Overall height: 1–2' (30–60 cm).

FLOWERING SEASON: March–June.

HABITAT: Open woods, grasslands, deserts.

RANGE: Southwestern Oregon, throughout most of California including the southern half of the Baja California peninsula, southeast into New Mexico and Arizona.

CONSERVATION: G5
Threatened in Utah.

■ NATIVE
■ RARE OR EXTIRPATED
■ INTRODUCED

ALTERNATE NAMES: False Lily-of-the-Valley, Wild Lily-of-the-Valley

Until recently a member of the lily family, Canada Mayflower is now classified within the Asparagus family. The genus name *Maianthemum* is from the Latin word *maius*, meaning "May," and the Greek word *anthemon*, meaning "flower." The clusters of white, star-shaped flowers typically open in late May to early June. The delicate stem slightly zigzags up to 6" (15 cm) with one to three shiny, dark green leaves. Plants with only one leaf are immature and will not flower until the second year. After flowering, small berries form, starting green with spots and gradually turning a bright red in fall.

Flowers are pollinated by small syrphid flies and bees. The berries are eaten by Ruffed Grouse (*Bonasa umbellus*) and Canada Jays (*Perisoreus canadensis*) as well as small rodents like Eastern Chipmunks (*Tamias striatus*) and White-Footed Mice (*Peromyscus leucopus*). Native Americans largely did not eat this plant, and it had limited medicinal uses, primarily as treatment for headache and sore throat. The Iroquois used a decoction of the roots for kidney ailments.

DESCRIPTION: The plant grows a single, zigzag stem clasped by one to three glossy, dark green pointed leaves with heart-shaped bases, 1–3" (2.5–7.5 cm) long. It produces an erect cluster of white star-shaped flowers ⅛" (4 mm) long; each flower has four tepals and four stamens.

FLOWERING SEASON: May–June.

HABITAT: Upland woods and clearings.

RANGE: Yukon and British Columbia east to Newfoundland, south along the Appalachian Mountains to Georgia and Tennessee, and west across Ohio to Iowa, and Nebraska to Montana.

CAUTION: The berries may be poisonous.

SIMILAR SPECIES: Three-leaved Solomon's Seal (*M. trifolium*) usually has three elliptic leaves that taper at the base and white star-like flowers with six points instead of four.

CONSERVATION: G5
Possibly extirpated in Montana. Endangered in Illinois and New Jersey. Threatened in Iowa, Kentucky, Washington D.C., Wyoming, and Yukon.

■ NATIVE
■ RARE OR EXTIRPATED
■ INTRODUCED

ALTERNATE NAMES: False Solomon's Seal, Feathery False Solomon's Seal, False Spikenard, Solomon's Plume, Smilacina, Treacleberry

Feathery False Lily-of-the-valley can be found throughout North America except the most northern extremes of the Arctic. Due to this large range, the plant has developed numerous common names and synonyms. In 2010, taxonomists changed the genus name from *Smilacina* to *Maianthemum*, and this former member of the lily family was also moved to the asparagus family.

There are two subspecies of this plant whose ranges have almost no overlap. The western *M. racemosum* ssp. *amplexicaule* has more erect stems, and the eastern *M. racemosum* ssp. *racemosum* has arching stems. The foliage bears a passing resemblance to Lily-of-the-Valley and Solomon's Seals, but the white cluster of flowers at the end of the stem has no similarity to the line of hanging, bell-like flowers of those species. Native Americans used this plant's roots solely for medicinal use, often for pain relief and for calming effects.

DESCRIPTION: The stem is arched in eastern specimens and erect in western regions, with 7–12 elliptical leaves with parallel veins and hairy undersides. A cone-shaped cluster of many tiny, white flowers, ⅛" (3 mm) long, with six tepals grows at the tip of the stem. Flowers give way to berries changing from green with red specks to a translucent ruby red. Overall height: 1–3' (30–90 cm).

FLOWERING SEASON: May–July.

HABITAT: Woods and clearings.

RANGE: Throughout North America, except the Arctic.

CAUTION: Could be confused with various *Veratrum* species, all of which are highly toxic.

SIMILAR SPECIES: The smaller Star-flowered Solomon's Seal (*M. stellatum*) is also found throughout most of North America, excluding the coastal states from North Carolina to Texas. This plant has larger, star-shaped flowers ¼" (6 mm) long and larger berries. If present, berries start striped with blackish-red, eventually becoming completely blackish-red.

CONSERVATION: **G5**
Possibly extirpated in Louisiana. Endangered in Alaska, Manitoba, and Newfoundland. Threatened in Oklahoma, Wyoming, and Saskatchewan.

■ NATIVE
■ RARE OR EXTIRPATED
■ INTRODUCED

ALTERNATE NAMES: Sacahuista Beargrass, Palmilla

While this genus is fairly easy to identify, differentiating to species can be challenging. The common name Beargrass Nolina is confusing because this plant is not a true grass, but a member of the asparagus family. Further complicating issues, all of the 14 North American species of *Nolina* are also commonly called beargrass, which can lead to confusion when scientific names are not used. Even then, the differentiation between species is not well defined, especially in the southwestern U.S. Some *Nolina* species are rare and protected within their range; however, *N. microcarpa* is a dominant species within the dry grasslands of Arizona.

The narrow, grass-like leaves easily catch fire and increase the intensity of wildfires. These slow-growing plants recover from protected woody underground roots, but typically regrow less densely than before. Native Americans used the leaves of this plant to thatch roofs, make brushes, and weave mats and baskets. Young shoots and fruits were eaten fresh, cooked, or ground into flour.

DESCRIPTION: Large clumps of narrow, wiry, grass-like leaves with tips split into fibers, up to 4' (1.2 m) long, without conspicuous stems. Above the center of the leaf cluster, the flowering stalk extends up to 6' (1.8 m) high, bearing a spike of tiny, yellow flowers. Overall height: 6' (1.8 m) tall.

FLOWERING SEASON: May–June.

HABITAT: Arid grasslands, open rocky slopes, dry plains.

RANGE: Arizona and New Mexico.

CAUTION: Beargrass Nolina is poisonous to rats, birds, sheep and goats, leading to impaction of the rumen and liver toxicity. It appears less toxic to cattle and is harmless to deer.

SIMILAR SPECIES: All *Nolina* species are difficult to differentiate.

CONSERVATION: G4
Endangered in Utah.

■ NATIVE
■ RARE OR EXTIRPATED
■ INTRODUCED

ALTERNATE NAMES: Star-of-Bethlehem, Grass Lily, Nap-at-Noon, Eleven o'clock Lady

Considered rare and endangered in its Mediterranean native range, Sleepy Dick is extremely invasive throughout North America. It creates new bulbs every year in densities that crowd out native plants, especially if soil is disturbed, which helps to scatter the bulbs. Folklore has suggested it originally grew from fragments of the star of Bethlehem, hence the alternate common name. Other common names relate to the plant's unique regularity in opening just before noon and closing again before sunset. The attractive blooms have been depicted by Leonardo da Vinci. It is poisonous to livestock, however, and people and all animals should avoid its consumption because it can be lethal. The bulbs resemble onions and, despite their toxicity, have been used as food in ancient times as well as in some regional cuisine.

DESCRIPTION: A cluster of dark-green, shiny, narrow leaves, 4–12" (10–30 cm) long, with a distinct white midrib. It produces attractive clusters of 8–20 white, star-like flowers, each 1¼" (3 cm) wide with six tepals, which close at night and remain closed on cloudy days. When closed, the outer tepals show a green central stripe. Overall height: 8–12" (20–30 cm).

FLOWERING SEASON: May–June.

HABITAT: Pastures, hayfields, open forests, and lawns.

RANGE: Native to Europe and northern Africa. Introduced throughout North America, especially in the eastern United States.

CAUTION: Humans and pets should not consume any parts of this plant due to poisonous compounds found in flowers, bulbs, and possibly all plant parts.

CONSERVATION: G4

This is a non-native invasive species throughout North America. Removal by digging is recommended with generous margins to capture all bulbs and bulblets.

■ NATIVE
■ RARE OR EXTIRPATED
■ INTRODUCED

ALTERNATE NAMES: King Solomon's Seal

Smooth Solomon's Seal is a common plant of woodlands of the eastern U.S. Its arched, zigzagging stems and smooth leaves often conceal the two lines of bell-like flowers hanging underneath. The species name *biflorum,* Latin for "two flowers," is in reference to these pairs of flowers along the stem. The flowers are adapted to a unique form of pollination. After landing on the flowers, native bumblebees (*Bombus* spp.) will attach to the flower and buzz their wings briefly at a higher frequency than used at flight, generally between 385–410 hz. This causes a harmonic that induces the flower to release a plume of dry pollen, coating the bee. The bee then eats most of the pollen as it grooms itself, while any pollen remaining will pollinate the next flower.

With a large range and starchy, edible root, many Native American groups used the two native species of Solomon's Seal as a potato-like food or ground it into a flour. The young shoots and leaves were also eaten as a vegetable similar to asparagus. Many groups also used these plants medicinally for pain relief and for general illness.

DESCRIPTION: Smooth Solomon's Seal has a large degree of variation in size and number of flowers across its range. It has many light green, lance-shaped, parallel-veined leaves, 2–6" (5–15 cm) long, along an arched, zigzagging stem, up to 7' (215 cm) long. Hanging down from the leaf axils are pale green to whitish, bell-like flowers, ½–¾" (1.5–2 cm) long. Flowers are replaced by poisonous blue-black berries.

FLOWERING SEASON: May–June.

HABITAT: Dry to moist woods and thickets.

RANGE: Saskatchewan east to New Brunswick, south to Florida, west to New Mexico, and north to Montana.

CAUTION: The berries contain anthraquinone, which causes vomiting and diarrhea.

SIMILAR SPECIES: Hairy Solomon's Seal (*P. pubescens*) is distinguished by small hairs along veins on the undersides of its leaves.

CONSERVATION: G5
Endangered in Vermont. Threatened in Florida, Illinois, and Texas.

■ NATIVE
■ RARE OR EXTIRPATED
■ INTRODUCED

ALTERNATE NAMES: Soft-leaf Yucca

Arkansas Yucca is smaller and more flower-like than all of the other 28 native species of yucca in North America. As with all *Yucca* species, the Arkansas Yucca has several ways of defending itself from herbivores. In addition to a rosette of tough leaves tipped with a spike, it also contains toxic saponins and cannot be consumed raw.

DESCRIPTION: A sparse rosette of flexible, flattened yellow-green leaves, up to 28" (70 cm) long. From the center of the rosette grows a tall stem topped with a large cluster of greenish-whitish, bell-like flowers, each 1¼–2½" (3.2–6.5 cm) long and attached directly to the main stem. Overall height: to 4' (1.2 m).

FLOWERING SEASON: May–October.

HABITAT: Dry, open areas.

RANGE: Texas north to Kansas and Missouri.

CAUTION: Yucca is toxic to people and animals when consumed raw.

SIMILAR SPECIES: While differentiation between the many species of *Yucca* can be difficult, the Arkansas Yucca is notable for its sparse rosette with spaces between the leaves, the flexibility of the leaves, and the weakness of the spike at the leaf tips. The curly hairs on the edges of the leaves are not twisted, the flower spike is shorter than 4' (1.2 m), and the flowers' stems are directly attached to the main stem (a raceme) and not stemming from a short side branch (a panicle).

CONSERVATION: G5

Endangered in Kansas. Threatened in Missouri.

■ NATIVE
■ RARE OR EXTIRPATED
■ INTRODUCED

ALTERNATE NAMES: Silkgrass, Beargrass

While most species of Yucca are from the southwestern deserts and grasslands, Adam's Needle is a southeastern native that reaches the Atlantic Coast and does not naturally occur farther west than eastern Texas. Adam's Needle has naturalized in areas north of Tennessee and North Carolina, although some records could be confused with the similar *Yucca filamentosa*.

Taxonomists are split on the classification of these two plants, with some regarding them as variations of one species and others finding that molecular markers indicate they are more distantly related than observations of leaf and flower structures may suggest. Yucca have a uniquely mutually symbiotic relationship with Yucca Moths (*Tegeticula yuccasella*), a small whitish moth that is similar in color to Yucca flowers. The plant cannot self-pollinate and does not have any other known pollinators; thus it is entirely dependent on the moth for pollination of its blooms.

DESCRIPTION: A rosette of sword-like evergreen leaves, 30" (75 cm) long, 2½" (6.5 cm) wide, tapering to a spiny point. Along the edges of the leaves are loose, thread-like fibers, less than 2" (5 cm) long. From the center of the leaves, a tall stout stem with a loose cluster of white, nodding, bell-shaped flowers, 1½" (4 cm) wide. Reaches an overall height of 2–10' (60–300 cm).

FLOWERING SEASON: June–September.

HABITAT: Sandy beaches, dunes, and old fields.

RANGE: Ontario; New York and Connecticut south to Florida, west to Texas, and north to Nebraska, Missouri, Illinois, and Wisconsin.

CAUTION: Yucca is toxic to people and animals when consumed raw.

SIMILAR SPECIES: *Y. filamentosa* is extremely similar, bears the same common name, and shares much of the same range. The most obvious differentiating characteristics include *Y. filamentosa*'s wider, more concave leaves, which have longer fibers on the edges, up to 8" (20 cm). Soapweed (*Y. glauca*) is a typical species of the western plains, found east to Iowa, Missouri, and Arkansas. Its rigid leaves have hairy edges, and it has a shorter flowering stalk, reaching a height of just 4' (1.2 m), which does not extend the base of the flower cluster beyond the leaf tips. Spanish Bayonet (*Y. aloifolia*), found from North Carolina south to Florida and Alabama, has toothed leaves with hairless edges.

CONSERVATION: G5

This species is classified as secure but its global status may need to be reviewed once the species status of this plant and *Y. filamentosa* is settled.

■ NATIVE
■ RARE OR EXTIRPATED
■ INTRODUCED

ALTERNATE NAMES: Spanish Dagger

A common yucca in the Mojave Desert, Mojave Yucca can often be found growing near ancient Joshua Trees (*Y. brevifolia*), a tree-like species of yucca that often forms sparse "forests." These species could be confused with each other, if not for the much larger leaves of the Mojave Yucca compared with the short leaves of the Joshua Trees. The species name is from Latin meaning "having a fragment of paper or wood" and referring to the tough, pale fibers peeling from the edges of the leaves.

While Mojave Yucca typically grows up to 16' (5 m), some exceptional specimens have exceeded 30' (9 m). It is a very slow growing plant, adding just ½–1" (1.25–2.5 cm) of growth annually in favorable conditions, with large specimens estimated to be a minimum of 200 years old and up to 500 years. It takes a minimum of 15 years after germination for the first blooms to appear. Seedlings and immature plants are extremely rare because a large percentage of them die off if they do not experience three to five years of favorable moisture levels to become established; these conditions appear just once on average every 40–50 years.

Mojave Yucca is used in numerous industries including as flavoring and foaming agent in soft drinks, as a surfactant and preservative in cosmetics, and as a fertilizer for crops. The stems of species of yucca are commonly ground and added to poultry and livestock feed, as well as to dog and cat foods because they reduce the odor of the animals' waste.

DESCRIPTION: Large, sometimes tree-like trunks, either branched or unbranched, with rosettes of stiff, narrow, spine-tipped, yellow-green leaves which have pale fibers peeling and curling on their edges, each leaf between 1–5' (30–150 cm) long. Within or just beyond leaf tips may be a cluster of white to cream-colored, bell-like, nearly spherical flowers, each 1¼–2½" (3–6.5 cm) long. The flower stalk and cluster are up to 4' (1.2 m). The trunk is typically 4–16' (1.2–5 m).

FLOWERING SEASON: April–May.

HABITAT: Brushy slopes, flats, and open deserts, including hot deserts with Creosote Bush (*Larrea tridentata*) and chaparral on dry, gravelly mountain and valley slopes.

RANGE: Mojave Desert in northwestern Arizona, southern Nevada, southern California, and northern Baja California; at 1000–6000' (305–1829 m), rarely higher.

CAUTION: Yucca is toxic to people and animals when consumed raw.

SIMILAR SPECIES: The leaf size of Joshua Tree (*Y. brevifolia*) is approximately four times smaller, typically 6–12" (15–35 cm).

CONSERVATION: G4
Endangered in Utah. Due to the plant's slow growth and the large numbers of commercial uses, laws exist to regulate commercial harvesting, but conservationists believe that these rules are inadequate for the sustainability of Mojave Yucca.

■ NATIVE
■ RARE OR EXTIRPATED
■ INTRODUCED

ALTERNATE NAMES: Fleshy Yucca, Eve's Needle

The Torrey Yucca may more closely resemble a tree than a wild-flower. This species' oldest specimens reach heights exceeding 20' (6 m), but more commonly are found between 3–10' (1–3 m) tall. The trunk may be branched or single-stemmed. If flowering, the flowering spike will extend an additional 2' (60 cm) above. This plant is named after Columbia University botanist John Torrey, who described this yucca as a new variety in 1859. The Apache ate the fruits raw or roasted and used the juice of the fruit as a beverage, grinding the fruits to make cakes or breads and sometimes pouring the juice over the cakes as a sauce. They also weaved the leaves into baskets and used the roots within the weaving as red accents.

DESCRIPTION: The plant has multiple large, sometimes tree-like trunks, either branched or unbranched, with rosettes

of stiff, yellow-green bayonet-like leaves between 2–3½' (60–110 cm) long with a sharp spine at the tip and coarse, whitish thread-like fibers along the edges. From a long stalk, it grows crowded, upright flower clusters, 16–24" (40–60 cm) long, with many creamy-white, bell-shaped flowers that are sometimes tinged purple, 1½–3" (4–7.5 cm) long with six tepals. The total length of flower stalk and cluster are up to 4' (1.2 m). Overall height: usually up to 13' (4 m), but very old specimens may exceed 20' (6 m).

FLOWERING SEASON: Spring.

HABITAT: Dry soils of plains, mesas, and foothill slopes; in desert grassland and shrub thickets.

RANGE: Southwestern Texas including Trans-Pecos Texas, southern New Mexico, and northeastern Mexico.

CAUTION: Yucca is toxic to people and animals when consumed raw.

SIMILAR SPECIES: Spanish Dagger (*Y. treculeana*) is similar but usually has a single trunk, occasionally with branches instead of multiple trunks.

CONSERVATION: G4
Threatened in New Mexico.

■ NATIVE
■ RARE OR EXTIRPATED
■ INTRODUCED

ALTERNATE NAMES: Mouse Flower

Asiatic Dayflower is a weedy, non-native invasive species origi-nating from East and Southeast Asia and introduced to North America in the 1800s, where it quickly became a weed among gardens in the eastern U.S. Dayflowers are the type genus and namesake of their taxonomic family Commelinacea, commonly also known as the dayflower family.

The common name dayflower refers to the plant's unusual habit of opening its flower for just a single day. It received the spe-cies name *communis* due to the formation of colonies by quickly rooting nodes along the reclined stem. Despite the wide, ovate leaves, Asiatic Dayflower is a monocot with parallel veins. An annual, the seeds germinate throughout the growing season and may survive more than five years before germination. There are at least nine species of dayflower in North America, four of which are native.

DESCRIPTION: Stems often lie along the ground 1-3' (30–90 cm) before turning upright. Wide-spaced branches with wide, fleshy, ovate leaves, 3–5" (8–13 cm) long with pointed tips and rounded bases sheathing the stem. At the tips of stems may be flowers ½" (1.5 cm) wide with two deep blue petals and a third much smaller white petal.

FLOWERING SEASON: June–October.

HABITAT: Open, disturbed areas, roadsides, and woodland borders.

RANGE: Introduced; east of the Rocky Mountains and in the Pacific Northwest.

SIMILAR SPECIES: Slender Dayflower (*C. erecta*) has very similar blooms but typically tall, upright stems. This species shows no rooting from stem nodes. Vir-ginia Dayflower (*C. virginica*) also has an upright stem and three blue petals of equal size.

CONSERVATION: 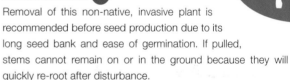 G5
Removal of this non-native, invasive plant is recommended before seed production due to its long seed bank and ease of germination. If pulled, stems cannot remain on or in the ground because they will quickly re-root after disturbance.

NATIVE

RARE OR EXTIRPATED

INTRODUCED

ALTERNATE NAMES: White-mouth Dayflower, Slender Day-flower, Widow's Tears

The Erect Dayflower is the most variable plant within its genus. One example of this variability includes the overall form, which is typically erect especially if some support is provided near other plants; however, stems can be lying across the ground in more open areas. Other variables include the size and specific shapes of leaves, flower petals, and seed capsule.

Erect Dayflower has no nectar and is pollinated by a variety of long-tongued bees, which collect pollen from the stamen. Other visitors include leaf beetles (family Chrysomelidae), whose larvae feed on the stem, as well as Northern Bobwhites (*Colinus virginianus*), Mourning Doves (*Zenaida macroura*), White-tailed Deer (*Odocoileus*

virginianus), and cattle. Stems grow erect or along the ground 1–3' (30–90 cm) before turning upright.

DESCRIPTION: The plant has wide-spaced branches with spear-shaped, fleshy, ovate leaves up to 6" (15 cm) long with slightly rounded tips. At the tips of stems may be flowers ½–1¼" (1.5–3 cm) wide with two deep blue, ear-like petals and a third much smaller white petal.

FLOWERING SEASON: May–October.

HABITAT: In sandy or rocky soil in grassy areas, in open woods, or where weedy.

RANGE: Southern Arizona to eastern Colorado and Wyoming; south to New Mexico and Texas; and throughout the eastern United States.

SIMILAR SPECIES: This native species is extremely similar to the invasive Asian Dayflower (*C. communis*). The native doesn't grow roots from stem nodes, while its invasive relative does.

CONSERVATION: G5

Extirpated in Michigan, New York, and Pennsylvania. Endangered in Delaware, Iowa, and Minnesota. Threatened in Illinois, Maryland, and West Virginia.

- ■ NATIVE
- ■ RARE OR EXTIRPATED
- ■ INTRODUCED

ALTERNATE NAMES: Common Spiderwort, Reflexed Spiderwort, Bluejacket

Ohio Spiderwort is the most common and widespread species of its genus within the U.S. with a range that spans most of the eastern half of the country and into Ontario. The name Spiderwort may refer to leaves, which seem to resemble a squatting spider; the web-like hairs surrounding the stamen; or the thick sap that turns thread-like, white, and silky when dried. The flowers open in the morning and wilt by the afternoon. If the petal is pinched while in the heat of the day, it will wilt immediately in the hand. The tiny hairs on the stamen are typically blue. However, if exposed to low doses of radiation prior to blooming, these hairs mutate and turn pink. This is one of the few known biological indicators of low-dose radiation.

DESCRIPTION: Ohio Spiderwort has pointed leaves, folded lengthwise, up to 15" (38 cm) long and sometimes arching to the ground. It has blue-violet to rose (rarely white) flowers, 1–2" (2.5–5 cm) wide with profusely hairy stamens. Overall height: 8–24" (20–60 cm).

FLOWERING SEASON: March–August in the south; June–August in the north.

HABITAT: Upland forests, rocky open woods, wood edges, and roadsides.

RANGE: Ontario; Maine south to Georgia, west to Louisiana, and north to Missouri, Illinois, and Wisconsin.

CAUTION: All *Tradescantia* species bear clusters of needle-like crystals that may cause minor skin irritations in some people if in contact with the plant.

SIMILAR SPECIES: There are 30 species of *Tradescantia* in North America. Differentiation can be difficult and many hybrids of various species exist in the world and through horticulture. Virginia Spiderwort (*T. virginiana*) is shorter and fuller and is less tolerant of hot, sunny locations. Upon close observation, the sepals of *T. virginiana* are hairier when compared to the smoother *T. ohiensis*.

CONSERVATION: G5

Globally secure but threatened in New Jersey and Ontario.

■ NATIVE
■ RARE OR EXTIRPATED
■ INTRODUCED

ALTERNATE NAMES: Spiderlily

In the early 1600s, Virginia Spiderwort was one of the first plants brought to Europe from the New World by the Elder and Younger John Tradescant, a father-son gardening duo for whom the genus was named and who were also responsible for the earlier introductions of the Larch tree (*Larix* spp.) and Lilac bush (*Syringa persica*) to England. In total, they brought more than 90 North American plants to England before the son returned there to take his father's place as head gardener to the king and queen.

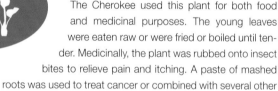

The Cherokee used this plant for both food and medicinal purposes. The young leaves were eaten raw or were fried or boiled until tender. Medicinally, the plant was rubbed onto insect bites to relieve pain and itching. A paste of mashed roots was used to treat cancer or combined with several other ingredients in a medicine for kidney trouble.

DESCRIPTION: The plant has violet-blue (sometimes red or white) flowers, 1–2" (2.5–5 cm) wide, with yellow anthers in a cluster above bearded anthers. It grows long, narrow, pointed leaves, 15" (38 cm) in length, creased along the midrib forming a channel. Overall height: 8–24" (20–60 cm).

FLOWERING SEASON: April–July.

HABITAT: Woodland borders, thickets, meadows, and roadsides.

RANGE: Ontario; Maine south to Georgia, west to Louisiana, and north to Missouri, Illinois, and Wisconsin.

CAUTION: All *Tradescantia* species bear clusters of needle-like crystals that may cause minor skin irritations in some people if in contact with the plant.

SIMILAR SPECIES: Zigzag Spiderwort (*T. subaspera*) has blue flowers and a zigzag stem to 3' (90 cm) high. Ohio Spiderwort (*T. ohiensis*) has blue (sometimes white or rose) flowers with a hairless stem, sepal, and leaves. Hairy-stemmed Spiderwort (*T. hirsuticaulis*) is a hairy plant with light blue flowers.

CONSERVATION: Ⓖ5

Presumed extirpated in Iowa. Endangered in Arkansas and South Carolina. Threatened in Delaware, Michigan, North Carolina, and West Virginia. While populations are secure in portions of their range, in some areas Virginia Spiderwort is becoming more rare. Major threats to the species include habitat destruction as well as collection of wild populations for horticulture.

■ NATIVE
■ RARE OR EXTIRPATED
■ INTRODUCED

Pickerelweed is a marsh native that can be found emerging from lakes and streams throughout eastern North America. It has an extremely large range, reaching from Canada as far south as Argentina. This species also has a large number of possible variations in both the flower structures and leaf shapes, which has led to many rejected attempts to separate these forms as new species and leading to over 40 different synonymous scientific names.

When flowering in summer, the showy flower heads extend up to 2' (60 cm) above the water, the spikes blooming from the bottom to the top. As the season progresses and the blooms fade, the tall stalk bends over, submerging the developing fruits and seeds to mature underwater and out of reach of being browsed by most animals except waterfowl.

The purple flowers typically have a yellow spot with two lobes, which is believed to attract and assist its three known pollinators. One is the native New England Sweat Bee (*Dufourea novaeangliae*), a solitary ground-nesting bee which will use other plants' nectar, but exclusively uses the Pickerelweed's pollen. Two species of long-horned bees (*Melissodes apicata*

and *Florilegus condignus*) also only feed their larvae pollen from Pickerelweed. Some butterflies and moths may be found stealing nectar from the flowers without pollination of the blooms.

DESCRIPTION: The highly variable leaves extend above the water, typically 4–10" (10–25 cm) long and heart-shaped to spear-shaped, tapering to a point. The tall flowering spike is 6" (15 cm) long and bears hundreds of violet flowers, each ⅜" (8 mm) long, with yellow spots on the upper lobe. Flower stalks grow 1–2' (30–60 cm) above water.

FLOWERING SEASON: June–November.

HABITAT: Freshwater marshes and the edges of ponds, lakes, and streams.

RANGE: Ontario east to Nova Scotia; south to Florida; west to Texas; and north to Kansas, Missouri, and Minnesota.

CONSERVATION: G5
Endangered in Kansas, Oklahoma, and Prince Edward Island. Threatened in Illinois and Iowa.

■ NATIVE
■ RARE OR EXTIRPATED
■ INTRODUCED

ALTERNATE NAMES: Broadfruit Bur-reed

The native Giant Bur-reed has a large range across North America as well as in parts of Eurasia. This wetland species frequently occurs in locations where it may be found emerging above spring floodwaters. Its common name refers to its prickly round fruits that resemble a mace or club.

Recently removed from its own family (Sparganiaceae), the Giant Bur-reed is now included within the cattail family Typhaceae.

Bur-reed's large, round, white flowers are wind pollinated, with male flowers blooming separately just above the female flowers. If its large seeds escape predation from waterfowl, they can accumulate in the subsurface muck, helping to form dense stands.

DESCRIPTION: This is a tall, grass-like, oftentimes emergent plant of wetlands and damp places. Light green, narrow, linear leaves reach up to 7' (2.1 m). Flowering stems have clusters of spherical, bur-like heads, ½" (12 mm) wide. Upper, male flowers are greenish while lower, female flowers are bristly white with two stigmas.

FLOWERING SEASON: June–September.

HABITAT: Marshes, wet meadows, lakesides, streams, and fresh to brackish shorelines.

RANGE: From the southern Northwest Territories to the Maritime Provinces, south across the northern half of the U.S., into southwestern California, New Mexico, northern Oklahoma, northern Kentucky, and northern Virginia.

SIMILAR SPECIES: Close observation of the female flowers is needed to see the presence of two stigmas, which distinguishes this species from all other species of bur-reed.

CONSERVATION: **G5**

Possibly extirpated in Oklahoma. Endangered in Kentucky, Wyoming, and Washington D.C. Threatened in Colorado, Maryland, New Hampshire, New Jersey, Utah, Virginia, and West Virginia.

■ NATIVE
■ RARE OR EXTIRPATED
■ INTRODUCED

ALTERNATE NAMES: Bulrush, Common Cattail, Cat-o'-Nine-Tails

Due to its creeping roots, Broadleaf Cattail forms colonies in wetlands. This plant is often seen as a sign of an environment slowly transitioning to a drier habitat. Red-winged blackbirds and other marsh birds find protection in dense cattail stands. Muskrats eat the starchy roots and use the stems to build their huts. All portions of Broadleaf Cattail are edible and it was widely used by Native peoples throughout its range.

There are two other species of cattails (*Typha*) in North America. Southern Cattail (*T. domingensis*) is native to the southern portions of the U.S. Narrow-leaved Cattail (*T. angustifolia*) is a non-native, highly invasive species that can form hybrids with both native species, which can outcompete all other marsh plants and form extremely dense stands.

DESCRIPTION: This is a tall plant with a fan of stiff, flat, sword-like leaves 1" (2.5 cm) wide, which reach taller than the stems, up to 3–9' (1–3 m). The iconic cylindrical spike is ¾" (2 cm) wide and has two sections. Highest on the stalk is a narrow spike of tiny, yellowish male flowers, with a broader brownish spike of female flowers below. There is no gap between male and female spikes.

FLOWERING SEASON: May–July.

HABITAT: Freshwater marshes.

RANGE: Throughout North America, except the Arctic.

CAUTION: Cattails bioaccumulate toxins from their environment which may be poisonous if ingested.

SIMILAR SPECIES: Narrow-leaved Cattail (*T. angustifolia*) is very similar, but both the leaves and flowering spike are more narrow, both just ½" (1.5 cm), and there is an obvious gap between the male and female flowers.

CONSERVATION: G5
Threatened in Wyoming and Nunavut.

■ NATIVE
■ RARE OR EXTIRPATED
■ INTRODUCED

Spanish Moss is a native plant of the southeastern U.S. Despite the common name, it is not a moss, but a type of bromeliad. Like many bromeliads, Spanish moss is an epiphyte, a type of plant that grows on another plant and gathers the resources it needs from the air. It uses photosynthesis to produce energy and is not parasitic on the Southern Live Oak (*Quercus virginiana*) and Bald Cypress (*Taxodium distichum*) on which it often is found. Although it has no roots, Spanish Moss does produce sweet-smelling flowers in the spring that may be green, brown, gray or yellow. The scent attracts the numerous insect pollinators, which help it produce wind-borne seeds. The plant also reproduces vegetatively when pieces are blown by the wind or distributed by animals to new locations.

While the plant has no known natural predators, it provides shelter to numerous species of arthropods, including a species of jumping spider (*Pelegrina tillandsiae*), which has only been found in Spanish Moss. Vesper bats also enjoy the shade the plant provides and many birds use Spanish Moss as nesting material.

DESCRIPTION: Hanging clumps of slender stems up to 13' long, covered with gray scales often on oak trees. It grows inconspicuous pale green to yellow flowers, ½–¾" (1.5–2 cm) long. It has threadlike leaves, 1–2" (2.5–5 cm) long, covered in scales.

FLOWERING SEASON: April–June.

HABITAT: Hanging from branches of Southern Live Oak and other trees and from telephone wires.

RANGE: Virginia south to Florida, primarily along the coast, and west to Texas.

CONSERVATION:
Extirpated in Maryland. Endangered in Virginia. Threatened in Arkansas.

■ NATIVE
■ RARE OR EXTIRPATED
■ INTRODUCED

ALTERNATE NAMES: Carolina Yellow-eyed Grass

Bog Yellow-eyed Grass is one of 21 species of the *Xyris* genus of yellow-eyed grasses in North America. The plants of this botanical family are found in bogs, shorelines, and other wet habitats and, as monocots, have long, narrow, linear leaves with a grass-like appearance. While most species are found in the tropics, a few—like this one—have adapted to temperate climates. In general, *Xyris* species can be told apart from similar species by the unusually large, bulbous head that is made up of a cluster of scales, each protecting an immature flower that will bloom one at a time.

When not flowering, Bog Yellow-eyed Grass is easily overlooked nestled among other wetland plants. When in flower, the small blooms are quite conspicuous. Large plantings can appear showy and attractive. The flowers do not produce nectar, but several insects are attracted to the blossoms, which provides pollination, and the flowers are also capable of self-fertilization. A species of sweat bee, *Lasioglossum zephyrum*, has been observed forcing open *Xyris* flower buds and stealing large quantities of pollen before flowers appear, ensuring they receive the bulk of the pollen before any other pollinators. Flowers produce a large number of seeds that may germinate in the same year.

DESCRIPTION: This plant has narrow to broad fans of long, slender, sword-like leaves, up to 2¾" (7 cm) wide and 20" (50 cm) long. The bases of the leaf are reddish while the rest of the leaf is a deep green. Bright yellow flowers with three petals open in the morning, ⅜" (1 cm) wide. Overall height: up to 36" (90 cm).

FLOWERING SEASON: June–September.

HABITAT: Bogs and swampy areas.

RANGE: Eastern North America from Texas to Prince Edward Island and west to Michigan and Ontario.

SIMILAR SPECIES: Full differentiation of the various species of yellow-eyed grasses is challenging without close, technical study of flowering structures. However, the most common other species within the same range include Broad-scale Yellow-eyed Grass (*X. platylepis*) which has twisting leaves not found on Bog Yellow-eyed Grass. Also Small's Yellow-eyed Grass (*X. smalliana*) appears very similar but blooms in the afternoon instead of the morning.

CONCERVATION: G5

Possibly Extirpated in Vermont. Endangered in Kentucky, Ohio, and New Brunswick. Threatened in Indiana, North Carolina, and Rhode Island.

■ NATIVE
■ RARE OR EXTIRPATED
■ INTRODUCED

ALTERNATE NAMES: Ten-angle Pipewort, Hard-head Pipewort, Large-head Pipewort, Bog Button

Pipewort is a unique and easily recognized plant with a tuft of short, grass-like leaves and longer leafless stalks topped with a hard, button-like knob covered in tiny white flowers. This species occurs primarily along the wet areas of the southeastern coastal plains, but a few isolated populations occur in wetlands within the inland mountains across the northwestern edge of their range. These plants are frequently found where naturally occurring fires limit the amount of shade cast by woody plants.

The relatively inconspicuous flowers are visited by a wide variety of bees and wasps, but some botanists believe these tiny flowers may also be wind-pollinated. Frequently, these plants may be found in proximity to some rare wetland flowers that thrive in similar habitats. Depending on location, examples include Spreading Pogonia (*Cleistes divaricata*), Round-leaved Sundew

(*Drosera rotundifolia*), Hairy-fruit Sedge (*Carex trichocarpa*), and Canada Burnet (*Sanguisorba canadensis*).

DESCRIPTION: A tuft of grass-like leaves up to 16" (40 cm) with one or more ridged, leafless stems up to 44" (110 cm) tall, each of which holds a grayish-white, knob-like cluster of tiny, feathery flowers, with the knob ¾" (2 cm) in diameter and flowers ⁵⁄₆₄"–⁵⁄₃₂" (2–4 mm) long.

FLOWERING SEASON: June–October.

HABITAT: Moist to wet places in pine flatwoods, savannahs, pondsides, and ditches.

RANGE: The coastal plain from New Jersey to Florida and west to Texas; it also occurs in some inland areas, including Arkansas and northern Alabama.

CONSERVATION: G5
Extirpated in Pennsylvania and Washington D.C. Endangered in Arkansas, Delaware, Maryland, Oklahoma, and Tennessee. Threatened in Virginia.

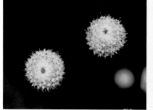

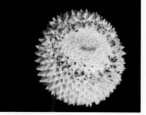

■ NATIVE
■ RARE OR EXTIRPATED
■ INTRODUCED

ALTERNATE NAMES: White Prickly Poppy, Crested Prickly Poppy, Bluestem Prickly Poppy

Annual Prickly Poppies have pretty, white blooms guarded by sharp spines covering their leaves, stems, and fruits. In addition to the physical defenses, 19 different chemical compounds that impact biological activities have been found within the plants' sticky yellow latex sap. Combined, these poisons are highly toxic and distasteful, further protecting the plant from browsing. These defenses are so effective that Annual Prickly Poppy is used as an indicator of overbrowsing by livestock.

These compounds led to wide use among many native groups including Shoshone, Paiute, Tepehuan, Comanche, Hopi, and Aztec peoples to create concoctions for treating sores and burns of the skin as well as the eyes. Other uses include utilizing the yellow sap as a dye, the leaf ash for tattoos, and the prickly stems as a whip used during rite of passage ceremonies.

DESCRIPTION: This plant has branched stems covered with slender yellow spines and deeply lobed, pale blue-green leaves up to 8" (20 cm) long, with prickly spines along the veins. It produces large flowers, about 3" (7.5 cm) wide, with crinkly white petals around stamen-filled, yellow centers. They have two to three sepals covered with spines and tipped with a stout horn, ¼–⅝" (6–15 mm) long. Overall height: 3–4' (1–1.8 m).

FLOWERING SEASON: April–July.

HABITAT: Sandy or gravelly soil on plains or brushy slopes.

RANGE: Northeastern New Mexico and the northern half of Texas; north to South Dakota and eastern Wyoming. Several isolated populations occur farther west.

CAUTION: All parts of the plants, including the seeds, are poisonous if ingested. The spines contain a substance irritating to the skin.

SIMILAR SPECIES: This species of *Argemone* is one of 15 that are native to the southern U.S., mostly through the Southwest. The Mexican Prickly Poppy (*Argemone mexicana*) is an invasive species naturalized along the East Coast and Gulf Coast from New England to Texas and less frequently inland. A unique, identifying trait of the Annual Prickly Poppy is the lack of spines on the top surface of the leaves, unlike others in the genus.

CONSERVATION: 🅖4
Threatened in Wyoming.

■ NATIVE
■ RARE OR EXTIRPATED
■ INTRODUCED

ALTERNATE NAMES: *Corydalis sempervirens*, Tall Corydalis, Pale Corydalis, Pink Corydalis

For most of the past century, Rock Harlequin (often called Tall Corydalis) was classified in the genus *Corydalis*; however, its molecular differences from other flowers in the genus were so large that it was reclassified and is currently the only species in the genus *Capnoides,* returning to its original taxonomic classification from 1763. The genus name comes from Greek for "smoke-like," alluding to the smoky scent of the blooms. Interestingly, these plants can frequently be found following fires, and exposure to heat may increase germination rates of the seeds.

The delicate, dangling pink-and-yellow flowers and bluish-green foliage are unique characteristics. The flowers are capable of self-pollination, but the resulting plants are less vigorous. Research has shown pollination between plants is completed by wind and ants. It is possible

other wildlife also helps to pollinate because the flowers produce nectar that attracts numerous insects, including bumblebees and skipper butterflies.

DESCRIPTION: The plant produces small clusters of yellow-tipped pink, tubular flowers ½" (1.5 cm) long, at the end of a branched stem with intricate, pale blue-green leaves, 1–4" (2.5–10 cm) long, divided into many three-lobed leaflets, each ½" (1.5 cm) long. Overall height: between 5–24" (12.5–60 cm).

FLOWERING SEASON: May–September.

HABITAT: Rocky clearings.

RANGE: Alberta east to Newfoundland, south along the Appalachian Mountains to northern Georgia, and northwest to Minnesota and to Alaska and far western Canada.

SIMILAR SPECIES: Allegheny Vine (*Adlumia fungosa*) has similar foliage. Its flowers are white, biege, or pink but not yellow-tipped. It is a closely related woodland vine that climbs to 10' (3 m).

CONSERVATION: G5

Endangered in Illinois, Indiana, Iowa, Ohio, South Carolina, and Tennessee. Threatened in Georgia, Kentucky, Maryland, Montana, North Carolina, and Rhode Island in the U.S. and Newfoundland and Labrador and Prince Edward Island in Canada.

■ NATIVE
■ RARE OR EXTIRPATED
■ INTRODUCED

ALTERNATE NAMES: Greater Celandine

The common name Celandine may refer to three different species of plants. One is native to North America, the Celandine Poppy (*Stylophorum diphyllum*). The other two are non-native, invasive plants: Lesser Celandine (*Ficaria verna*) and this species, also called Greater Celandine. This early exotic was introduced to North America from Europe before the 1600s and naturalized in many regions, especially in the Northeast.

The genus name is Greek for "swallows," referring to birds of the family Hirundinidae. There are two reasons for this naming. First, the flowers begin to bloom when the swallows arrive and wither around the time when they depart. Second, there is folklore dating back to at least the 1st century A.D. saying that swallows restore sight to blind nestlings by using Celandine. The species name, *majus*, is Latin for "greater," but this is the only species within the genus.

There are numerous medicinal uses of the plant recorded over the centuries, including for eye diseases, warts and other skin ailments, jaundice, and many others. Following the Doctrine of Signatures, the yellow sap was thought to resemble bile, leading to a belief it would treat liver issues, but modern science has shown the plant is especially toxic to the liver. Recent studies show the plant was directly responsible for acute hepatitis in at least 50 people in Germany using this as an herbal remedy.

DESCRIPTION: The plant produces small clusters of deep yellow flowers, ⅝" (16 mm) wide, with four petals. It grows light green, alternate leaves 4–8" (10–20 cm) long, each divided into five to nine lobes. Overall height: 1–2' (30–60 cm). The seed capsule is smooth and slender.

FLOWERING SEASON: April–August.

HABITAT: Disturbed areas, moist soil of roadsides, woodland edges, and around dwellings.

RANGE: Much of North America, especially prevalent in the Northeast; from Ontario northeast to Newfoundland, south to Georgia, northwest to Nebraska, and northeast to Minnesota; also in parts of the West.

CAUTION: All parts of this plant are toxic if ingested. The stems contain sap that can cause skin irritation.

SIMILAR SPECIES: The native Celandine Poppy (*Stylophorum diphyllum*) is very similar and is most easily differentiated by the seedpod. *Stylophorum* has a furry, oval shaped pod while the invasive has a smooth seedpod that is long and slender. *Stylophorum* also has larger, shinier flowers between 1–2" (2.5–5 cm) in diameter, whereas the *Chelidonium's* flowers are typically no more than 1" (2.5 cm).

CONSERVATION:

This widespread non-native invasive plant has seeds that easily germinate and are dispersed by ants. It has a shallow root system and can be controlled through pulling of the plants before they set seed. It can also be controlled through herbicides, especially when the plant is immature.

■ NATIVE
■ RARE OR EXTIRPATED
■ INTRODUCED

ALTERNATE NAMES: Scrambled Eggs, Slender Fumewort

Like many *Corydalis* species, Golden Smoke is an early successional plant well adapted to colonizing areas after wildfires or other disturbances. This plant has been documented flowering just two months after wildfires in the U.S. This yellow-flowered plant develops extremely long-lived seeds that can survive in the soil for more than 160 years. Oftentimes, the plant will only be found on burned sites for two years before being outcompeted by other plants, but this is enough time for their widespread seeds to be scattered over the soil to await the next fire.

The common name refers to the smoke-like scent of the flowers, a common feature among plants of this genus. Along with an unattractive smell, the plant is unpalatable to livestock and is poisonous to people and foraging mammals.

DESCRIPTION: The plant produces clusters of up to 20 tubular, golden-yellow flowers, ½–¾" (2–2.5 cm) long, with weak stems that barely support the leaves and flowers, which sometimes are supported by other vegetation or rocks. It has soft, thinly succulent leaves, each 3–6" (7.5–15 cm) long and fern-like, compoundly divided into five to seven leaflets, each of which is then divided again, with those divisions lobed. Overall height: 4–24" (10–60 cm).

FLOWERING SEASON: February–September.

HABITAT: Gravelly hillsides among rocks or brush, open prairies and flats along rocky creek bottoms, and in open woodlands. It is especially prevalent after wildfires.

RANGE: Throughout much of Canada and the West, east of the Cascade Range and the Sierra Nevada, north to the Yukon and across the Great Plains to Texas and Michigan. It occurs rarely in the Northeast from Ohio and Pennsylvania to New Hampshire.

CAUTION: This plant is poisonous to livestock. Humans should generally avoid ingesting plants that are toxic to animals.

CONSERVATION: G5

Presumed extirpated in New Hampshire. Endangered in Illinois and Pennsylvania. Threatened in Iowa, Kansas, New York, Vermont, and Washington.

■ NATIVE
■ RARE OR EXTIRPATED
■ INTRODUCED

ALTERNATE NAMES: Little Blue Staggers

Dutchman's Breeches is a native spring ephemeral of eastern woodlands and blooms in early spring before the tree canopy fully forms. The unique flowers that give rise to the common name are thought to resemble the Dutch *klepbroek*—a type of men's puffy white pantaloons—if hung upside down to dry. This flower structure is a combination of four flower petals, two outer and two inner. There are nine North American species within the genus *Dicentra*, which includes the bleeding hearts. The genus name is from two Greek words—*dis* meaning "twice" and *kén-tron* meaning "a spur"—referring to the two nectar spurs, the hollow extensions of the flower petals where nectar is produced.

A variety of long-tongued bumblebee (*Bombus* spp.) queens are just emerging from hibernation at the time of flowering and are able to reach deep into these nectar spurs, which pollinates the flowers. Other insects with shorter proboscises are unable to reach the nectar, and instead may chew a hole through the outside of the flower petal to steal the nectar without pollinating the flowers. After pollination, the seeds formed are dispersed by ants. The plants then wither and become dormant until the following spring. After germination, the plants take four to seven years to produce flowers.

It is estimated that the western populations have been separated from the larger eastern range for over 1,000 years. Western plants typically have a somewhat coarser leaf structure than most other populations; however, it is nearly identical to those found in the Blue Ridge Mountains. These variations within the species are believed to be a phenotypic response to the environment.

All parts of this plant are highly toxic and are not eaten by native forging mammals. Cattle that graze on the flowers will start to stumble as if drunk, giving the plant its alternative common name Little Blue Staggers. The blooms should not be picked because they wilt immediately afterward.

DESCRIPTION: Dutchman's Breeches have clusters of 3–12 white, widely-spaced pantaloon-shaped flowers, ¾" (2 cm) long, held aloft by a leafless stalk. The stalk is surrounded by feathery, compound, grayish-green leaves on long stalks, 3–6" (7.5–15 cm) tall, and divided into leaflets that are pale underneath. Overall height: 4–12" (10–30 cm).

FLOWERING SEASON: April–May.

HABITAT: Rich woods.

RANGE: Manitoba east to Nova Scotia, south to Georgia, west to Mississippi and Oklahoma, and north to North Dakota; also found in the Pacific Northwest, in Washington, Idaho, and Oregon.

CAUTION: Can be fatal to animals if eaten. Humans should generally avoid ingesting plants that are toxic to animals.

SIMILAR SPECIES: The closely related Squirrel Corn (*D. canadensis*), often found in the same habitats, has the same pollination story. Its pinkish-white flowers, however, are fragrant and have a narrower heart shape. The frequent occurrence of these two plants together, speckling the woodlands with white and pink blooms, leads to the shared colloquial name "Boys and Girls."

CONSERVATION: G5

Endangered in Mississippi, North Dakota, and South Carolina in the U.S. and in Manitoba and Prince Edward Island in Canada. Threatened in Alabama, Delaware, Georgia, Oklahoma, and Quebec.

■ NATIVE
■ RARE OR EXTIRPATED
■ INTRODUCED

ALTERNATE NAMES: Pacific Bleeding Heart

Western Bleeding Heart is often confused with Eastern Bleeding Heart (*D. eximia*), a similar native wildflower from eastern North America. This is especially true in the nursery trade where both plants are popular and often hybridized. On the Pacific coast, Western Bleeding Heart often blooms in the spring, becomes dormant through the summer, and flowers again in the fall.

The common name refers to the heart-shaped flower with two inner petals that appear at the point of the heart, which are said to resemble a drop of blood. Like the other plants of the genus, Western Bleeding Heart has a symbiotic relationship with ants. After flowering, ants are attracted to a small, fatty appendage on the seed called an elaisome. An ant carrying the seed will realize the seed itself is not edible, dropping it near its nest, helping to distribute the seeds around the forest in a process called myrmecochory.

While the plant is toxic and may be lethal if consumed, Native Americans had several uses for it, including chewing on the roots for toothaches, brewing a root decoction to alleviate worms, and making a wash from the crushed leaves to help hair grow.

DESCRIPTION: The plant has clusters of heart-shaped pink, rose, or purple (rarely yellow) flowers ⅔–1" (16–24 mm) long, hanging down from a leafless stalk. Its long-stalked leaves, up to 20" (50 cm) tall, are soft and pinnately compound and almost fern-like, with leaflets each ¾"–2" (2–5 cm) long. Overall height: 8–18" (20–45 cm).

FLOWERING SEASON: March–July.

HABITAT: Damp, shaded places or, in wetter climates, open woods.

RANGE: Southern British Columbia to central California.

CAUTION: Can be fatal to animals if eaten. Humans should generally avoid ingesting plants that are toxic to animals.

SIMILAR SPECIES: Eastern Bleeding Heart (*D. eximia*) is very similar. Western Bleeding Heart has wider, more rounded flowers with shorter wings on the outer petals, whereas Eastern Bleeding Heart has narrow flowers and larger, widely spread wings. Bleeding Heart (*Lamprocapnos spectabilis,* formerly *Dicentra spectabilis*) is a popular garden plant introduced from Japan. It has larger, rosy-red or white flowers, about 1" (2.5 cm) long, and can become much larger than either of the native species, up to 48" (1.2 m).

CONSERVATION: **G5**
The species is secure in its natural range and also widely cultivated.

■ NATIVE
■ RARE OR EXTIRPATED
■ INTRODUCED

ALTERNATE NAMES: Cup of Gold, Golden Poppy, California Sunlight

California Poppy, the state flower of California, is native to the Pacific slope of North America. This plant was described in the 19th century by German botanist Adelbert von Chamisso after collecting specimens in October 1816 while on a Russian expedition in what is now San Francisco. At that time of the year, it was one of the few plants still in bloom and one of just 30 botanical specimens collected in the month he spent there. He named the genus in honor of J. F. Eschscholtz, the ship's surgeon and entomologist. Just 10 years later, seeds were collected for germination by the Royal Botanical Society of England and were soon thereafter growing in Victorian gardens.

Before Western settlers arrived, on sunny days in spring, entire hillsides would turn orange due to the immense number of California Poppies. This led early Spanish explorers to refer to the California region as *Tierra del Fuego*, or the "Land of Fire." Invasive grasses and excessive browsing by livestock have since greatly reduced the numbers of California Poppies, but experiencing their mass blooms is still spectacular. The flowers are highly responsive to sunlight and will open on sunny days and close at night and on cloudy days. Although they produce no nectar, California Poppies are pollinated by a wide variety of insects. In dry areas where bees are uncommon, beetles are attracted by the spicy fragrance and provide the bulk of pollination.

DESCRIPTION: The plants have feathery, fern-like gray-green leaves that are attached to long stalks by sheaths. Single yellow to orange cup-shaped flowers, each ¾"–2½" (2–6 cm) long, appear with an orange spot at the base. Overall height: up to 8–24" (20–60 cm).

FLOWERING SEASON: February–June and August–October.

HABITAT: Open or disturbed, grassy areas.

RANGE: Southeastern California to southern Washington; often cultivated.

CAUTION: Can be poisonous to animals and people if ingested.

SIMILAR SPECIES: California Poppies are highly variable, and attempts to divide the species have led to more than 90 different suggested taxa. Currently, however, only two subspecies are accepted.

CONSERVATION: G4

California Poppy is considered invasive in many areas of North America outside its native range.

FIRE POPPY *Papaver californicum*

■ NATIVE
■ RARE OR EXTIRPATED
■ INTRODUCED

ALTERNATE NAMES: Western Poppy

The Fire Poppy is native only to central to southwestern California, and even there it is uncommon. Fire Poppies often can be found the first year after wildfires have burned an area. Unlike many other *Papaver* species that are perennials, Fire Poppy is an annual, reliant on new seed germination for each year's blooms. The flowers produce many seeds, which remain viable in the soil for decades until fire facilitates their germination.

DESCRIPTION: The plant produces a single bowl-shaped, reddish-orange flower, 1" (2.5 cm) wide, with four fan-shaped petals that have a greenish area at the base and many bright yellow stamens on top of a tall, hairy or hairless stem. Before the flowers open, the large, immature buds hang down, but turn upright upon flowering. The leaves are short, light green, deeply divided with teeth or lobes, 1¼–3½" (3.1–8.8 cm) long. Overall height: 12–24" (30–60 cm).

FLOWERING SEASON: April–June.

HABITAT: Open brush and woods, especially after fires.

RANGE: The Coast Ranges from San Francisco Bay to southern California; it is not common in the northern half of the range.

SIMILAR SPECIES: Wind Poppy (*P. heterophyllum*) is similar in color and appearance to Fire Poppy, but has a darker circle originating from the center of the flower instead of a lighter color as in Fire Poppy. This species is also related to Opium Poppy (*P. somniferum*), a non-native plant that's commercially cultivated for food (poppy seeds) and poppy-based drugs such as codeine and morphine.

CONSERVATION: ●G4

The species is endemic to California. Although its apparently secure, its limited range and relative lack of abundance are cause for monitoring the health of the population.

■ NATIVE
■ RARE OR EXTIRPATED
■ INTRODUCED

ALTERNATE NAMES: California Creamcups

Creamcups are native to the sandy grasslands of the south-western U.S. The species is highly variable, with more than 57 different taxa having been suggested for it because of its many forms. However, currently it is the only member of the *Platystemon* genus.

Flower colors may range from pure white to pure yellow as well as variations of those, sometimes with more cream-colored petals with yellow centers, and other combinations. While most flowers have six petals, some may have more. Additionally, other parts of the plant may differ depending on the location. Plants from coastal areas often appear semi-succulent with nearly no hairs; those from grasslands are larger with a moderate amount of hair; and those from arid areas are often compact with long, woolly hair. Stems may be fully upright or prostrate with just the tip twisted up to the sky, and leaves may be opposite or whorled with either pointed or rounded tips.

No matter the physical appearance, the blooms can be pollinated by the wind and need no insects to aid their reproduction, but pollination assisted by native solitary bees increases seed production. The flowers open and close with the day, often not opening until mid-morning and closing by early afternoon, although this also can vary by location. Fire assists seed germination, and Creamcups can be prolific in areas after wildfires.

DESCRIPTION: An extremely variable plant, often with soft hairs and several stems, each topped with one small, bowl-shaped flower, ½–1" (1.5–2.5 cm) wide, either pale yellow or cream or a combination of the two. Leaves are attached to the stem, mostly near the base, ¾–3" (2–7.5 cm) long, opposite or whorled, narrow and spear-shaped. Overall height: 4–12" (10–30 cm).

FLOWERING SEASON: March–May.

HABITAT: Open fields with sandy soil, especially abundant after wildfires.

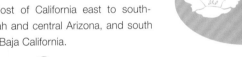

RANGE: Most of California east to south-western Utah and central Arizona, and south to northern Baja California.

CONSERVATION: G5
Endangered in Utah. The sandy, open habitats preferred by Creamcups is extensively developed over the range, as well as encroached upon by non-native, invasive grasses and shrubs. These impacts have led to a general population decline.

■ NATIVE
■ RARE OR EXTIRPATED
■ INTRODUCED

ALTERNATE NAMES: Celandine Poppy

Wood Poppy is a native wildflower that is often found in moist woodlands and along streambanks. In the spring, it features single or small clusters of yellow flowers, each with four wide petals. After the flowers fade, a very hairy, two-part capsule forms filled with seeds, which are spread by ants. This perennial plant spreads easily and has even been found sprouting in trees, having been transported there by ants. The genus name comes from the Greek words *stylos,* which means "style"—the long central tube of the pistil—and *phoros*, which means "bearing," referring to the stoutness of the style. The species name means "having two leaves or leaflets," describing how the blue-green leaves form in pairs. If cut open, the stems exude a bright orange-yellow latex sap that was used as a dye by Native Americans. This sap is toxic and can form welts if left in contact with skin.

DESCRIPTION: Deeply lobed, opposite, blue-green leaves, 4–10" (10–25 cm) long, silvery underneath, occurring at the base of the plant and on the stem. Bright yellow flowers, 1½–2" (4–5 cm) wide, each with four wide, overlapping petals and numerous stamens surrounding a taller, relatively large single pistil. Overall height: 1–1½' (30–45 cm).

FLOWERING SEASON: March–May.

HABITAT: Rich woods and bluffs.

RANGE: Ontario, Michigan, and Pennsylvania south to Georgia, west to Arkansas, and north to Illinois.

CAUTION: All parts of the plant are toxic. The sap can irritate the skin and severely burn the eyes.

SIMILAR SPECIES: Celandine (*Chelidonium majus*) is a non-native, invasive plant that can appear similar. Its flower petals are much narrower and do not overlap, with leaves that appear in an alternate rather than an opposite pattern.

CONSERVATION: Ⓖ5
Endangered in Alabama, Georgia, and Ontario. Threatened in Arkansas and Virginia.

■ NATIVE
■ RARE OR EXTIRPATED
■ INTRODUCED

ALTERNATE NAMES: Canadian Moonseed

A climbing, woody vine that can reach 16' (5 m) high with dark, purple-black berries, Common Moonseed is sometimes confused with Wild Grapes (*Vitis* spp.). This is an unfortunate mistake because the berries are poisonous and can be lethal due to the levels of an alkaloid, dauricine. Fortunately, the berries taste so acrid and bitter it is unlikely anyone would ingest enough to cause harm. In moderation, the Cherokee used the roots and berries as a natural laxative and used the root for sores on the skin. Modern science shows that dauricine can inhibit cancers of the colon, urinary tract, and breast in vitro.

Unlike grapes, which climb by means of tendrils and have fruits with numerous seeds, Common Moonseed climbs by the twining habit of its stem tips and has fruits with a single, crescent-shaped seed. While no mammals feed on this plant, the caterpillars of Moonseed Moth (*Plusiodonta compressipalpis*) sever the leaves and feed on them after they become dry. Many birds eat the berries, including Wild Turkey (*Meleagris gallopavo*), Cedar Waxwings (*Bombycilla cedrorum*), American Robins (*Turdus migratorius*), and several other thrushes (family Turdidae).

DESCRIPTION: This is a woody, climbing vine up to 15' (5 m) with large, shield-shaped leaves, 4–7" (10–17.5 cm). Loose clusters of small, greenish-white flowers, ⅛" (4 mm) wide, grow where the leaves join the stem. The flowers form purplish-black berries that often appear to have a whitish dusting, each berry containing one crescent-shaped seed.

FLOWERING SEASON: June–July.

HABITAT: Woodland edges, thickets, and streambanks.

RANGE: Manitoba east to Quebec, south through Vermont and New Hampshire to Florida, west to Texas, and north to North Dakota.

CAUTION: The fruit of this plant is toxic and may be fatal if ingested in large quantities.

SIMILAR SPECIES: Wild Grapes (*Vitis* spp.) have tendrils and berries with multiple seeds.

CONSERVATION: G5

Presumed extirpated in New Hampshire. Threatened in Georgia, Massachusetts, Mississippi, and South Carolina in the U.S. and Manitoba and Quebec in Canada.

- ■ NATIVE
- ■ RARE OR EXTIRPATED
- ■ INTRODUCED

ALTERNATE NAMES: Blueberry Root

Native east of the Great Plains, Blue Cohosh is a bushy perennial with bluish-green foliage. Each of its two leaves are branched many times, ending with three tulip-shaped leaflets. While the flowers are subtle, the clusters of bright to dark blue, berry-like seeds draw much more attention. While they appear to be berries, these are actually the naked seed of the plant, which burst through the ovary and continue to grow and mature exposed with no protection—a trait that is unique among flowering plants. The seeds contain alkaloids and saponins that are toxic if ingested.

The only other plant of the genus in North America is the Northern Cohosh, also called Early Cohosh (*C. giganteum*), which has a smaller number of purplish blooms that open two weeks earlier. Both plants have six stamens and central pistils that mature at different times, assuring cross-pollination when the nectar glands are visited by the early solitary bees. Native Americans used Blue Cohosh for a wide variety of maladies, including as an anti-convulsive, anti-inflammatory, sedative analgesic, and for complications of pregnancy as well as a general tonic.

DESCRIPTION: The plant produces 5–70 small, inconspicuous yellow-green (fading to purple) flowers, ½" (1.5 cm) wide, in loose clusters among the leaves. Each flower has six large petal-like sepals surrounding six much smaller, hood shaped petals. Each plant has two compound leaves. The lower leaf is larger and divided into 27 leaflets, each 1–3" (2.5–7.5 cm) long. The upper leaf is smaller and divided into 9–12 leaflets each with three to five pointed lobes at the tip. Overall height: 1–3' (30–90 cm). Each flower location develops one or two deep blue, berry-like seeds from a small stalk.

FLOWERING SEASON: April–June.

HABITAT: Moist woods.

RANGE: Manitoba east to Nova Scotia, south to Georgia, west to Oklahoma, and north to North Dakota.

CAUTION: The berries, roots, and leaves of this plant are poisonous and may cause skin irritation if touched. The raw berries are especially poisonous to children if ingested.

SIMILAR SPECIES: Early Cohosh (*C. giganteum*) has a smaller number (4–18) of larger, purplish flowers that bloom earlier.

CONSERVATION: G5

Endangered in Kansas, Nebraska, North Dakota, Rhode Island, and Washington D.C. Threatened in Arkansas, Delaware, South Carolina, and South Dakota in the U.S. and Manitoba and Nova Scotia in Canada. Blue Cohosh is overharvested for medicinal use, with well over 10,000–25,000 lbs. of dry roots traded annually, all of which comes from wild sources, as the plants are not cultivated commercially. Land managers can protect populations through root dyeing, which permanently stains the plant roots and makes them unmarketable.

- ■ NATIVE
- ■ RARE OR EXTIRPATED
- ■ INTRODUCED

Twinleaf is a spring ephemeral of eastern North America. This plant is selective of the locations in which it grows and can be difficult to find; however, it may be prolific in favorable sites. It can often be found on woodland soils above limestone. While both the common name and species name indicates two leaves, the characteristic-shape is actually a single leaf that is so deeply lobed it appears bifurcated into two wing-like halves. The genus only includes one other species, Asian Twinleaf (*J. dubia*), which is found in Japan.

DESCRIPTION: The plant produces a single white flower, 1" (2.5 cm) wide, with eight petals on a naked stalk. The uniquely divided leaves are between 3–6" (7.5–15 cm) long. The leaves are connected to the flowering stalk on an underground rhizome, but the above-ground foliage is separate. Overall height: 5–10" (12.5–25 cm) when in flower and increasing to 1½' (45 cm) when in fruit.

FLOWERING SEASON: April–May.

HABITAT: Rich, damp, open woods; usually in limestone soil.

RANGE: Ontario and New York south to Georgia and Alabama and northwest to Iowa and Minnesota.

CAUTION: All parts of this plant are poisonous to people and animals if ingested.

SIMILAR SPECIES: The white, eight-petal flowers of Bloodroot (*Sanguinaria canadensis*) can appear similar, but the foliage is vastly different. Bloodroot leaves are lobed with three to nine parts instead of the wing-shaped appearance of Twinleaf.

CONSERVATION: G5

Endangered in Georgia, Iowa, New Jersey, and North Carolina. Threatened in Alabama, Michigan, Minnesota, New York, and Wisconsin.

■ NATIVE
■ RARE OR EXTIRPATED
■ INTRODUCED

ALTERNATE NAMES: Mandrake, Ground Lemon

May-apples are a native spring ephemeral with unique, almost umbrella-like leaves. In some areas it is colloquially called a mandrake because the root appears similar to that of a true Mandrake (*Mandragora officinarum*); however, the rest of the plant is vastly different. While the common name hints at having "apples" in May, the name actually refers to the apple-like flowers hidden under the large leaves. The oblong fruits, held just above the ground on a short stem, have a more lemon-like appearance, leading to its other common name: Ground Lemon.

All parts of the plant are poisonous if eaten, including the unripe fruits. Once the fruit ripens and turns fully yellow, it is safe to eat in small quantities. Native Americans used May-apples for the treatment of sores as well as internally to treat worms, constipation, to induce vomiting, and as a spring tonic. It was also used on corn seeds and potato plants as an insect repellent. The toxin podophyllotoxin is unique to the May-apple genus and just one other plant, the Himalayan May-apple (*P. hexandrum*). This toxin is a strong antiviral agent that is used in prescription creams for the treatment of virus-caused skin infections, such as various types of warts. It is less commonly taken internally due to its effects on the central nervous system. Additionally, the rhizomes and roots contain several lignans that exhibit antitumor activity, derivatives of which are used in the treatment of some cancers.

DESCRIPTION: Typically a pair of large, deeply-lobed, circular, glossy, dark-green leaves up to 1' (30 cm) wide, with the leaf stem attached to the middle of the underside of the leaf. A waxy, white flower, 2" (5 cm) wide, with six to nine fragrant petals nods below the leaves. The flower develops into a lemon-shaped berry which turns from green to yellow when ripe.

FLOWERING SEASON: April–June.

HABITAT: Rich woods and damp, shady clearings.

RANGE: Ontario east to Nova Scotia, south to Florida, west to Texas, and north to Nebraska and Minnesota.

CAUTION: Leaves, seeds, and especially the roots (which are used medicinally) are poisonous to humans and animals if ingested in high quantities.

CONSERVATION: G5

Endangered in Florida and Vermont. Threatened in Nebraska and Quebec.

■ NATIVE
■ RARE OR EXTIRPATED
■ INTRODUCED

ALTERNATE NAMES: Columbian Monkshood, Northern Monkshood

Western Monkshood is primarily a western native that is part of a species complex with a highly variable nature, which may include size, flower color, and number of flower parts. One recognized subspecies, *A. columbianum* subsp. *viviparum*, creates reproductive structures known as bulbils from the flowerhead in addition to vegetative bulblets from the roots. This degree of variation is common for all plants of this genus.

While most populations occur in the West, a few small, isolated groups occur—mostly in the midwestern states of Wisconsin and Iowa, but also more easterly, with 11 occurrences in the Catskills of New York and two populations in northeast Ohio. In all of these states, they enjoy both state and federal protections. These populations are believed to be glacial relics, only inhabiting microclimates where ground temperatures remain below 60° F (15° C), which mimics the cooler habitats of the Pleistocene. Often, these areas are where ice forms under talus slopes, where cold air spills out from caves, or situated along high-elevation headwaters and streams. Some taxonomists treat these populations as a separate species, *A. noveboracense*, but the most recent studies show the eastern populations are genetically and molecularly identical to the western plants and therefore are the same species.

Most plants of the genus are poisonous to humans and livestock, although this species is not as highly toxic as many Old World *Aconitum* species. It is related to Wolf's-Bane (*A. lycoctonum*), a European monkshood species with white flowers celebrated in werewolf and vampire lore.

DESCRIPTION: These are usually tall, leafy plants up to 10' (3 m) tall, but can also be found along the ground and twisted together and as short as 8" (20 cm). Leaves are palmately lobed (all veins from a central point) and 2–8" (5–20 cm) wide with jagged, toothy edges. Showy clusters of flowers are typically blue to blue-violet, but may also be white or cream with or without a blue tinge. The uppermost tepal forms a large, arching hood, ⅝–1¼" (1.5–3.1 cm) long. Two additional oval-shaped tepals are found at the sides and another narrow pair are found at the bottom of the flower. Inside the hood, two more true petals typically form.

FLOWERING SEASON: June–August.

HABITAT: Moist woods and subalpine meadows.

RANGE: Alaska to the Sierra Nevada of California; east to New Mexico, Colorado, South Dakota, and western Montana; isolated populations occur in the East.

CAUTION: Plants of the genus *Aconitum* are poisonous to humans and animals if ingested.

SIMILAR SPECIES: All five native North American species of *Aconitum* appear similar, are highly variable, and can be difficult to distinguish. Fortunately, they rarely overlap native ranges of others within the genus. Tall Larkspur (*Delphinium exaltatum*) is more common and can appear similar, but has hollow stems. A more highly poisonous European monkshood species, *A. napellus*, may be found in gardens or as a garden escapee, although it is not considered invasive. This plant has lobes that are divided to the stem, whereas the lobes of most native monkshood species end ½-1" (1.5-2.5 cm) before the stem.

CONSERVATION: G5

Endangered in New York and Ohio. Threatened in Iowa, Wisconsin, and Wyoming.

The bright red berries of the Red Baneberry are attractive but highly dangerous. Eating just six berries is enough to severely poison an adult and just two are likely fatal to a child. Fortunately, the berries are so bitter and acrid that accidental poisoning is rare and no deaths have been documented from this species. However, the black berries of the similar European Baneberry (*A. spicata*) have been implicated in a number of deaths, mostly of children, and are responsible for the common name of these plants. Ingesting the berries induces neurologic effects including optical hallucinations, difficulties speaking, and dizziness; rapid, irregular pulse; and nausea with severe gastrointestinal discomfort. The plant does not affect birds, which eat the fruits but will not ingest the seeds.

DESCRIPTION: This is an upright plant with alternate, compound leaves subdivided into multiple sharply toothed leaflets, each 2½" (6.5 cm) long. The plants have clusters of many small white flowers, ¼" (6 mm), located either at the tops of stems. As the flowers open, the petals fall off, leaving exposed the reproductive flower parts, with the multiple stamen providing a fluffy or hairy appearance. Flowers give way to clusters of red berries (occasionally white), each ⅜" (10 mm), on a slender stalk and with a small black dot opposite the stem.

FLOWERING SEASON: May–July.

HABITAT: Rich woods and thickets.

RANGE: Alberta east to Newfoundland, south to Pennsylvania, west to Illinois, Iowa, and Kansas, and north to North Dakota; also throughout the West.

CAUTION: All parts of this plant are poisonous. Berries are extremely poisonous and potentially lethal.

SIMILAR SPECIES: The only other North American native of the genus, White Baneberry (*A. pachypoda*), is extremely similar. As the name implies, White Baneberry typically has white berries; these feature a large, prominent black dot opposite the stalk, leading to the common name Doll's Eyes. Rarely individuals of this species will produce red berries. Differentiation of White Baneberry with red berries and Red Baneberry with white berries can be made from the stalks that hold the flower and fruit. This stalk is stout in White Baneberry and turns red as the fruit matures. In Red Baneberry, the stalks are slender and remain green or greenish-brown.

CONSERVATION: G5

Possibly extirpated in Kansas. Endangered in Indiana and Rhode Island. Threatened in Arizona, Illinois, Nebraska, Ohio, and Pennsylvania.

■ NATIVE
■ RARE OR EXTIRPATED
■ INTRODUCED

ALTERNATE NAMES: *Anemone canadensis,* Canada Anemone, Meadow Anemone, Windflower

Roundleaf Thimbleweed is a native wildflower often growing in moist locations in central to northeast North America. From 1768 until recently, this plant was in the genus *Anemone*; however, this genus was not monophyletic—that is, it did not include only and all plants related to a common ancestor based on molecular studies. The new genus *Anemonastrum* (which means "somewhat like anemone") includes a monophyletic portion of the species that were formerly included in *Anemone*, which continues on as a smaller genus of plants.

In broad areas of its range, the Roundleaf Thimbleweed is adaptable. In sites with enough moisture, it can be a vigorous grower, spreading by rhizomes as well as seed. Many different insects pollinate the flowers, especially bees. Other woodland animals tend not to browse on this plant due to the presence of toxins held by many plants in the buttercup family. Contact with the sap can irritate the mouth and blister the skin. Ingesting large amounts of the plant can be toxic.

Native Americans used these plants for an extremely wide variety of purposes, which rarely overlapped between groups. The Chippewa used the roots and leaves to treat sores and wounds. The Meskwaki used a root infusion to treat eye issues like crossed eyes and twitching. The Ojibwa ate the root before ceremonies to clear the throat to sing better as well as for back pain.

DESCRIPTION: Between one and five deeply lobed, toothed leaves are attached with a leaf stem (petiole) from the base, the tips of each forming a circular shape. An upright stem, 1–2' (30–60 cm) tall which bears a single whorl of three- to five-part leaves. From the whorl, emerges typically one flower up to 2" (5 cm) in diameter, but occasionally two or three flowers together. Each flower has five white petals surrounding a golden-yellow stamen and a greenish-yellow pistil.

FLOWERING SEASON: May–August.

HABITAT: Moist ditches; damp meadows; sandy streams or lakeshores.

RANGE: Eastern Quebec to British Columbia, south to Virginia, Tennessee, Missouri, and the mountains of New Mexico.

CAUTION: This plant is toxic if ingested in large quantities and may cause dermatitis in sensitive individuals.

SIMILAR SPECIES: Tall Thimbleweed (*Anemone virginiana*) and Long-head Thimbleweed (*Anemone cylindrica*) are somewhat similar to this species but can be distinguished by the stems—Roundleaf Thimbleweed leaves have conspicuous stems, while the others do not. It also grows in damp habitats versus dryer locations for the two true *Anemones*.

CONSERVATION: 🄖⑤

Extirpated in Kentucky and Tennessee. Possibly extirpated in Maryland, Virginia, and Washington D.C. Endangered in New Jersey, West Virginia, Newfoundland, and Prince Edward Island. Threatened in Connecticut, Illinois, Wyoming, British Columbia and Nova Scotia.

■ NATIVE
■ RARE OR EXTIRPATED
■ INTRODUCED

ALTERNATE NAMES: Long-head Anemone, Thimbleweed, Candle Anemone

The greenish-white flowers of the Long-head Thimbleweed are not as showy or attractive as the related Tall Thimbleweed (*A. virginiana*), but otherwise the two plants are very difficult to differentiate. This leads to occasional disappointment for native gardeners who, in their attempts to grow Tall Thimbleweed in their gardens, plant seeds of the Long-headed Thimbleweed and fail to witness the bright white blooms of the preferred plant. For the pollinators, there is no disappointment, but the plant doesn't require their help, as it self-pollinates.

Long-head Thimbleweed is toxic to most animals, with the foliage causing a burning sensation in the mouths of browsers. The chemical behind this burning sensation has antibiotic properties that prevents other plants, fungi, and some bacteria from growing too close to this species. Native Americans found the plant useful to treat ailments including tuberculosis, burns, headaches, and dizziness.

The Ponca burned the woolly fruits and rubbed their hands in the resulting smoke as a good-luck charm.

DESCRIPTION: The plant produces two to eight upright stems, each topped by a greenish-white flower, ¾" (2 cm) across. The center of the flower is round and elongated, loosely resembling a long, narrow thimble that is up to 1.5" (3.75 cm) tall. Deeply lobed, hairy leaves appear from the ground as well as in whorls midway up on flowering stems. Late in the year, after the first frost, the mature seed head begins to open, releasing white, cottony tufts of seeds. Overall height: usually 1–2' (30–60 cm), but rarely to 3' (30–90 cm).

FLOWERING SEASON: May–June.

HABITAT: Dry, open woods and prairies.

RANGE: Maine to British Columbia, south to New Jersey, Nebraska, and in the Rockies to Arizona.

CAUTION: This plant is toxic if ingested in large quantities and may cause dermatitis in sensitive individuals.

SIMILAR SPECIES: Tall Thimbleweed (*A. virginiana*) is very similar but has white flowers. It commonly grows to 4' (1.2 m) and has a slightly shorter "thimble," which reaches just 1" (2.5 cm).

CONSERVATION: G5

Presumed extirpated in New Hampshire. Endangered in Idaho, New Jersey, Pennsylvania, and Rhode Island. Threatened in Missouri, Ohio, Vermont, and Wyoming.

■ NATIVE
■ RARE OR EXTIRPATED
■ INTRODUCED

ALTERNATE NAMES: Nightcaps, Twoleaf Anemone, Windflower

Spreading by creeping rhizomes, Wood Anemone is a spring ephemeral that slowly forms large colonies in moist woodland edges. The alternate common name of Nightcaps refers to the flowers, which close in the evening or on cloudy days. Wood Anemone's scientific name was provided in 1753 by the father of taxonomy, Carl Linnaeus. The species name he provided, *quinquefolia*, means "five-leaved," and indeed his Latin notes describing this plant say, "leaves, five, oval, serrated." However, Wood Anemone actually only has three leaflets, not five. The two outer leaflets are so deeply lobed that, if given a passing glance, the plant may be mistakenly determined to have five leaves.

DESCRIPTION: This small plant has a single stem with a whorl of three compound leaves, each of which is divided into three sharply-toothed leaflets 1¼" (3.2 cm) long and deeply cut at the base. A single white flower 1" (2.5 cm) wide, with 5–7 white, petal-like sepals, often tinged pink on the undersides, is on top of the stem.

FLOWERING SEASON: April–June.

HABITAT: Open woods, clearings, and thickets.

RANGE: Alberta east to Nova Scotia, south to Georgia, west to Mississippi and Arkansas, and north to Iowa, Minnesota, and the Dakotas.

CAUTION: All parts of this plant are poisonous if large quantities are consumed.

SIMILAR SPECIES: This plant is a member of a species complex involving several other native *Anemone* species, including *A. grayi*, *A. lancifolia*, *A. oregana*, and *A. piperi*. All of these plants look extremely similar to each other, but have minute differences in the size of various parts of the plant.

CONSERVATION: G5
Endangered in Mississippi and Utah. Threatened in Arkansas, Illinois, North Dakota, and South Dakota in the U.S. and Alberta, Nova Scotia, and Saskatchewan in Canada.

■ NATIVE
■ RARE OR EXTIRPATED
■ INTRODUCED

ALTERNATE NAMES: *Aquilegia coerulea,* Rocky Mountain Columbine

The Colorado Blue Columbine is a showy species native to the Rocky Mountains and is the state flower of Colorado. The flowers have a long spur, the base of which holds the nectar glands. Only pollinators with very long proboscises can reach the nectar—typically sphinx moths (family Sphingidae) and some bumblebees (*Bombus* spp.).

This species has four varieties. The nectar spurs of each are a slightly different length to match the pollinators in their locales. This plant has also become popular in gardens, oftentimes with a variety of colored blossoms and "doubled" flowers. These cultivated variations often impede most native pollinators from reaching the nectar source, limiting the wildlife value of these plants.

The scientific name was originally spelled with an "o" as the second letter (*coerulea*) — a misspelling of the proper Latin term for "blue" which uses an "a" instead. Later authors have attempted to correct this error. Both spellings have been used interchangeably depending on the author's preference to either reflect the historical spelling or to use correct Latin.

DESCRIPTION: This plant has multiple stems with highly divided leaves, forming leaflets ½–1¼" (1.3–3.1 cm) long and wide, creating a bushy appearance. It typically has white-and-blue flowers, 2–3" (5–7.5 cm) wide, pointing upward at the ends of the stems. Flowers are composed of five broad, blue sepals subtending five white or blue-tinged petals with backward-projecting nectar spurs; the flowers are 1¼–2" (3.1–5 cm) long. It may reach heights up to 3' (90 cm).

FLOWERING SEASON: June–August.

HABITAT: Mountains, commonly in aspen groves.

RANGE: Western Montana to northern Arizona and northern New Mexico.

CAUTION: This plant belongs to a family that contains a number of toxic species.

SIMILAR SPECIES: The similar Alpine Blue Columbine (*A. saximontana*) also has blue sepals and white petal tips but has blue spurs which hook at the tip. Growing at high elevations in the Rocky Mountains, this plant is also much smaller, only 2–8" (5–20 cm) tall.

CONSERVATION: (G5)
Threatened in Arizona.

■ NATIVE
■ RARE OR EXTIRPATED
■ INTRODUCED

ALTERNATE NAMES: Eastern Columbine, Red Columbine, Canadian Columbine

Wild Columbine is a beautiful native woodland wildflower that often can be found in and at the borders of rocky woodlands throughout eastern and central North America. The genus name comes from the Latin word *aquila*, which means "an eagle." This refers to the shape of the backward-pointing petals, which are said to be like an eagle's claw. Wild Columbine is pollinated by hummingbirds, which may depend on the plant as an important nectar source. In addition, at least four bee species have been found to be effective pollinators of Wild Columbine in more northern locales.

Evolutionarily, columbines are thought to have arrived in North America relatively recently, at the end of the Pleistocene. A single species likely crossed into North America over the Bering land bridge that connected to Asia. This common ancestor rapidly spread out of Alaska and throughout the North American continent. New, unique species developed as columbines continuously adapted to new pollinators and ecosystems throughout the continent. Many Native American groups prepared infusions from various parts of Columbine plants to treat a wide assortment of maladies including heart, kidney, and bladder issues, headaches and fever, and as a wash for poison ivy.

DESCRIPTION: The nodding flower, 1–2" (2.5-5 cm) long, has five red petal-like sepals around five rear-pointing petals that are yellow at the face and become redder toward the tip of the spur. Extending out, below the petals and sepals, hang many long yellow stamens. The plant grows long-stalked, compound, light-green leaves, 4–6" (10–15 cm) wide, which are divided into 9–27 tri-lobed leaflets. Overall height: 1–2' (30–60 cm).

FLOWERING SEASON: April–July.

HABITAT: Rocky, wooded, or open slopes.

RANGE: Saskatchewan east to Nova Scotia, south to Florida, west to Arkansas and Texas, and north to North Dakota.

CAUTION: This plant belongs to a family that contains a number of toxic species.

SIMILAR SPECIES: This plant readily hybridizes with other *Aquilegia* species, leading to identification challenges where species ranges overlap. Red Columbine (*Aquilegia formosa*) is a similar western species that does not overlap the native range of Wild Columbine. The non-native European Columbine (*A. vulgaris*) has escaped cultivation and has become well established in many parts of the Northeast. It typically has blue, violet, pink, or white short-spurred flowers.

CONSERVATION: G5
Possibly extirpated in New Brunswick. Endangered in Florida and Mississippi. Threatened in Delaware.

RED COLUMBINE *Aquilegia formosa*

■ NATIVE
■ RARE OR EXTIRPATED
■ INTRODUCED

ALTERNATE NAMES: Crimson Columbine, Scarlet Columbine

Red Columbine is a western species of columbine that occurs from Alaska to Utah and Southern California. The species name is Latin for "beautiful," which aptly describes this large plant. Red Columbine is similar in overall appearance to the eastern Wild Columbine and can be difficult to differentiate. However, the ranges of these species are well separated by the Rocky Mountains. In the more southern area of the range, the flowers are pollinated by hummingbirds, which have long tongues that can reach the nectar located far back in the tube-like petal spurs. Farther north, individuals of this species are likely to have slightly shorter spurs to allow long-tongued insects to reach nectar and support pollination.

Flower species that are adapted to pollination by hummingbirds produce a larger amount and sweeter nectar than other species, and Red Columbine is no exception. Yurok and Haisla children used the flowers like candy, sucking the nectar from the flowers. Other native uses included treatment of cough, gastrointestinal issues, and dermatological maladies, as well as good-luck charms.

DESCRIPTION: The plant produces a downward-facing flower, 2" (5 cm) long, with five red petal-like sepals around five rear-pointing petals that are yellow at the base and red for most of the upward-pointing spur. Extending out, below the petals and sepals, hang many long yellow stamens. Its long-stalked, compound, and light-green leaves are 4–6" (10–15 cm) wide, divided into 9-27 leaflets with either two or three lobes. Overall height: 6–36' (90 cm).

FLOWERING SEASON: May–August.

HABITAT: Open woods, on banks, near seeps.

RANGE: Southern Alaska to Baja California; east to western Montana and Utah.

CAUTION: This plant belongs to a family that contains a number of toxic species.

SIMILAR SPECIES: This plant readily hybridizes with other *Aquilegia* species, leading to identification challenges where species ranges overlap. Wild Columbine (*A. canadensis*) is a similar eastern species that does not overlap the native range of Red Columbine. It is a popular nursery plant and may be present in and near planted locations.

CONSERVATION: G5

Threatened in Wyoming. Endangered in Montana and Alberta.

■ NATIVE
■ RARE OR EXTIRPATED
■ INTRODUCED

ALTERNATE NAMES: Yellow Marsh Marigold, Cowslip

Although the common name suggests otherwise, Marsh Marigold's showy, bright yellow, and cup-shaped flowers belong to the buttercup family. It is not related to marigolds at all, as those are of the genus *Tagetes* in the Asteraceae family. This plant has a large circumboreal distribution and can be found in northern temperate locations across North America, as well as Europe and Asia.

The flowers attract an extremely wide variety of pollinators, which is necessary because the flowers are unable to produce seed if self-pollinated. After the flowers fade, seeds develop inside a small pod. The shape of this pod when open catches raindrops, flinging the seeds into the air. As the seeds float on water, they may be carried a great distance. Stems are hollow, helping to deliver oxygen down to the often submerged roots.

As with many buttercups, all parts of the plant are toxic and handling plants can cause irritation. Prepared properly, the leaves and seeds are edible. Native Americans often used them as a food source as well as to create medicines.

DESCRIPTION: The plant has thick, dark green, glossy leaves, either heart- or kidney-shaped, 2–7" (5-17.5 cm) wide. The stems carrying leaves and flowers are branching and hollow. Bright yellow, glossy flowers, 1–1½" (2.5–4 cm) wide, with five to nine petal-like sepals (no petals are present).

FLOWERING SEASON: April–June.

HABITAT: Swamps, marshes, wet meadows, and along streams and brooks.

RANGE: Alberta east to Newfoundland, south to North Carolina, west to Illinois and Nebraska, and north to North Dakota; also found in parts of the West.

CAUTION: Plant juices can cause blistering or inflammation on skin or mucous membranes on contact, and gastric illness if ingested.

SIMILAR SPECIES: Floating Marsh Marigold (*C. natans*) is smaller and has small white or pinkish flowers. Marsh Marigold is often confused with Lesser Celandine (*Ranunculus ficaria*), yellow flowers with true petals underlain by three yellow-green sepals. Its fruit is seed-like.

CONSERVATION: G5

Endangered in Illinois, Indiana, and Maryland. Threatened in New Jersey, North Dakota, Ohio, Rhode Island, and Newfoundland and Labrador.

SWAMP LEATHER FLOWER *Clematis crispa*

■ NATIVE
■ RARE OR EXTIRPATED
■ INTRODUCED

ALTERNATE NAMES: Marsh Clematis, Curly Clematis, Blue-Jasmine Leatherflower

Swamp Leather Flower is a southeastern U.S. native flowering semi-woody vine that can reach 10' (3 m) if supported. As the common name suggests, this plant can be found in swamps and floodplain forests where the soil is consistently moist. The large, nodding, bell-like flowers are very showy—with petal-like sepals that curve back at the flower opening, often curing into a tight roll—and have ruffled or wavy edges. As with all *Clematis* species, this plant is highly toxic due to the presence of the irritant protoanemonin. Any mammal unfortunate enough to sample the leaves will experience an intensely bitter taste followed by severe mouth pain and possibly ulcers of the mouth. While the plant is poisonous enough to be lethal, it has caused no known deaths, due to the challenges of ingesting enough to be fatal.

DESCRIPTION: This is a semi-woody climbing vine that can grow to 10' (3 m) if supported. Its compound leaves have 4–10 leaflets, each 1–3" (2.5–7.5 cm) long. The last leaflet is tendril-like, helping support the vine. The plant produces violet-blue, lavendar, or white bell-shaped flowers, 2" (5 cm) long, with ruffled edges. The tips of the flower's four sepals curl entirely backward.

FLOWERING SEASON: March–June.

HABITAT: Wet woods and marshes.

RANGE: Atlantic and Gulf coastal plains from Virginia to Florida and Texas, north to Tennessee, Missouri, and southern Illinois.

CAUTION: All parts of the plant can cause blistering or inflammation of mucous membranes on contact and are highly poisonous.

CONSERVATION: G5

Endangered in Illinois and Oklahoma. Threatened in Kentucky and Virginia.

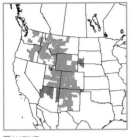

■ NATIVE
■ RARE OR EXTIRPATED
■ INTRODUCED

ALTERNATE NAMES: Hairy Clematis, Leather Flower, Vase Flower, Lion's Beard

Although the genus name from the Greek *klema* means "a vine," Sugarbowls are not viny. This *Clematis* instead is an upright, clump-forming species, generally under 24" (60 cm) tall. Most specimens have a covering of soft, silvery hairs—even on the petal-like sepals—leading to both the species name, which means "very hairy," and another common name for the plant, Hairy Clematis. When seeds are mature, the plant forms a wonderfully feathery seedpod with long plumes dangling out of the fruiting cluster, ready to catch the wind and pull the seeds out. This mane-like feature leads to another alternate name, Lion's Beard. Although it is mildly toxic, Montana and Navajo people used the plant medicinally to seek relief from colds, headaches, and sinus pain.

DESCRIPTION: This is a very hairy plant with several stems, each with a violet to purplish-brown bell-like flower, 1" (2.5 cm) long, nodding down. The plant has silvery green, finely divided, carrot-like leaves up to 5" (12.5 cm) long. Overall height: up to 24" (60 cm).

FLOWERING SEASON: April–July.

HABITAT: Grasslands, among sagebrush, and in open pine forests.

RANGE: British Columbia south to eastern Washington, east to Montana and Wyoming, and south through Utah and Colorado to northern Arizona and New Mexico; it is also found in western South Dakota, western Nebraska, and western Oklahoma.

CAUTION: Members of the *Clematis* genus are generally poisonous, and animals have become ill after grazing on the plants. Humans should generally avoid ingesting plants that are toxic to animals.

CONSERVATION: G4
Endangered in Arizona and South Dakota.

■ NATIVE
■ RARE OR EXTIRPATED
■ INTRODUCED

ALTERNATE NAMES: Western Blue Virgin's Bower, Purple Clematis

Blue Virgin's Bower should be easy to spot with large, drooping, purple-blue flowers with four petal-like sepals. However, it can be challenging to find in the wild and despite the large blooms, you may not see it unless you walk into it as the flower color tends to blend into the shadows of the surrounding leaves. After flowering, fluffy, ethereal tufts catch the wind and distribute the seeds. There are three recognized varieties of *Clematis occidentalis*. Var. *dissecta* only occurs in the Cascades while var. *grossesserata* is found in the West. A third variety, var. *occidentalis* can be found around the Great Lakes region, northeastern U.S., and central Canada.

Only the Blackfoot people were known to use this plant, primarily as protection from ghosts. The Blackfoot believed that fainting was the result of coming near a ghost. The victim could be revived by using a smudge of the stem of this plant. Blackfoot children sometimes wore Blue Virgin's Bower flowers in their hair at night to keep ghosts away. The leaves were used in rituals to remove "ghost bullets"—supernatural objects believed to be shot into people by ghosts. These "bullets" were removed by a diviner, and the leaves were boiled and applied to the spot where the bullets were removed.

DESCRIPTION: This is a vining plant, often creeping over the ground or growing over other plants. It has opposite leaves, each with three leaflets that are 1–2½" (2.5–6.5 cm) long and may be either toothed or deeply lobed. From a leafless stalk, it produces a single pale pinkish-purple to blue-violet, narrowly bell-shaped flower, 1¼–2½" (3–6.5 cm) long and consisting of four petal-like sepals and no petals. Vines may reach 10' (3 m) long. Later, it produces a silvery plume 1¼–2½" (3–6 cm) long, with many seeds in a shaggy, round head.

FLOWERING SEASON: May–July.

HABITAT: Wooded or brushy areas in mountains, often on steep rocky slopes.

RANGE: British Columbia south to northeastern Oregon, east to Wyoming and Montana. Separately, a population occurs from Illinois and Iowa north through Minnesota and Ontario, east to New Brunswick, south throughout all of New England to Pennsylvania, then following the Appalachian Mountains into North Carolina.

CAUTION: Some members of the *Clematis* genus are poisonous, and animals have become ill after grazing on the plants. Humans should generally avoid ingesting plants that are toxic to animals.

SIMILAR SPECIES: Rocky Mountain Clematis (*C. columbiana*), found from Montana south to New Mexico and northeastern Arizona, is similar but each of the three leaflets is divided into three smaller, jaggedly toothed leaflets. Matted Purple Virgin's Bower (*C. columbiana* var. *tenuiloba*), found in Montana, Wyoming, and western South Dakota and south to northeastern Utah and central Colorado, has similar flowers but plants form low mats rather than vines, and the leaves are even more finely divided.

CONSERVATION: **G5**
Presumed extirpated in Delaware and Ohio. Endangered in Illinois, Maryland, North Carolina, Rhode Island, and Nova Scotia. Threatened in Iowa, Maine, Massachusetts, Michigan, Pennsylvania, Vermont, Virginia, West Virginia, and Wyoming in the U.S. and Saskatchewan and New Brunswick in Canada.

- ■ NATIVE
- ■ RARE OR EXTIRPATED
- ■ INTRODUCED

ALTERNATE NAMES: Devil's Darning Needles, Wild Clematis, Woodbine

A vigorously growing native vine, this is the most common native *Clematis* in eastern North America. It is often found trailing over fences and other shrubs along moist roadsides and riverbanks. Unlike other vines, this species has no tendrils to give support as it grows. Instead, the stems of its leaves wrap around thin objects on which it grows, often the stems of other plants. It can reach 23' (7 m) and can easily add much of that height in one growing season in ideal locations.

Each plant produces either male or female flowers. Male plants tend to have larger, showy flowers. After pollination, females produce showy, feathery tufts of seeds in late summer that are released through the fall and early winter. The roots of these plants provide an overwintering home to the larva of Clematis Clearwing Moths (*Alcathoe caudata*), a wasp mimic which lacks wing scales on its rear wing, rendering it transparent. The adult moths can often be found resting under the Eastern Virgin's Bower foliage or feeding on nectar from the flowers from June through September.

DESCRIPTION: This is a quick-growing vine up to 23' (7 m) with alternate leaves that are composed of three sharply toothed leaflets, each 2" (5 cm) long. From the leaf axil, clusters of up to 30 white flowers grow, each 1" (2.5 cm) wide, with four to five petal-like sepals. Female plants may have downy seed tufts into early winter.

FLOWERING SEASON: July–September.

HABITAT: Borders of woods, thickets, and moist places.

RANGE: Manitoba to Nova Scotia; south from New England to Georgia; west to Alabama, Mississippi, and Louisiana; north to eastern Kansas.

CAUTION: Some members of the *Clematis* genus are poisonous, and animals have become ill after grazing on the plants. Humans should generally avoid ingesting plants that are toxic to animals. A few susceptible people may acquire dermatitis from handling the leaves of this plant.

SIMILAR SPECIES: The rarer Coastal Virgin's Bower, (*C. catesbyana*), can be distinguished by the presence of leaves with five to seven ovate leaflets instead of three leaflets. The non-native, invasive Sweet Autumn Clematis (*C. terniflora*) also differs in usually having leaves with five leaflets rather than three; however, the leaves are thick, leathery and lack the teeth seen on either of the native species.

CONSERVATION: G5

Endangered in Oklahoma. Threatened in Manitoba and Prince Edward Island.

ALTERNATE NAMES: *Coptis groenlandica*, Three-leaf Goldthread, Canker-root, Yellow Snake-root

Goldthread is a small, native wildflower adapted to survival in the extremes of the Far North, with a circumpolar range extending into Greenland and Siberia. This plant is found in nutrient-poor swamps, wetlands, and bogs. It has a symbiotic, mycorrhizal association with soil fungi that help provide the plant with soil nutrients—primarily phosphorus—while the plant provides the fungus energy. Several of the plant's common names refer to the slender, golden-yellow creeping rhizome from which it forms colonies. The name Canker-root comes from the use of these roots to treat mouth sores by both Native Americans and colonists.

DESCRIPTION: Goldthread has evergreen leaves, 1–2" (2.5–5 cm) wide, divided into three leaflets with scalloped or toothy edges. Leaves and flowering stems rise from a shallow, underground stem, sometimes forming mats and often under and around mosses. On single, straight stems grows a white flower, ½" wide with five to seven petal-like sepals and tiny, yellow, club-like petals. Overall height: typically 3–6" (7.5–15 cm).

FLOWERING SEASON: May–July.

HABITAT: Cool woods, swamps, and bogs.

RANGE: Alberta east to Newfoundland, south along higher elevations of the Appalachian Mountains into West Virginia, to northern Ohio, Indiana, and much of Wisconsin and Minnesota. A single disjunct population is located in far northwestern North Carolina.

CAUTION: This plant is moderately toxic and may be poisonous if large amounts are consumed.

CONSERVATION: G5

Endangered in Maryland, North Carolina, Oregon, and Washington. Threatened in West Virginia and Alberta.

■ NATIVE
■ RARE OR EXTIRPATED
■ INTRODUCED

ALTERNATE NAMES: Prairie Larkspur, White Larkspur, Plains Larkspur

With vibrant spikes of violet, blue, or white flowers, this midwestern species often carpets acres of prairie before the grasses take over. The genus name comes from the Greek word *delphis*, which means "dolphin"—a reference to the shape of the closed flower buds, which are thought to resemble dolphins in some species. When the flowers open, the long nectar spurs ensure only pollinators with the longest tongues are rewarded. Hummingbirds often can be found feeding on Carolina Larkspurs. Plants within this genus have many variations, and this species is no exception. Variations may include differences in flower color (from dark blue to white), as well as stem pubescence, plant vigor, and leaf size. Larkspurs also hybridize easily across species, increasing the challenges in classification.

The current accepted taxonomy breaks this species complex into four subspecies. Subspecies *virescens* is the most common of these and inhabits the largest range, native to the central grasslands of North America. It was previously considered a separate species and is still commonly referred to as Prairie Larkspur. Subspecies *carolinianum* can be found across the central and southeastern United States. Subspecies *vimineum* is limited to the Gulf Coast areas of the U.S. and Mexico. Subspecies *calciphilum* is native to only a small range from Kentucky to Alabama. All parts of this plant contain a large number of toxic alkaloids. An infusion of the seeds in alcohol has been used medically to treat head lice.

DESCRIPTION: Typically narrow clusters of widely spaced purple to pale blue or white, 1" (2.5 cm) flowers with long spurs extending over ½" long. Alternate leaves, deeply divided with narrow segments each about ¼" (6 mm) wide, attached by leaf stalks which are shorter on leaves higher on the stem. The stem is usually covered by short, soft hairs. Often, the leaves at the base wither before flowering, leaving 5–12 stem leaves. Overall height: 1–3' (30–90 cm).

FLOWERING SEASON: May–July.

HABITAT: Prairies and dry open woods.

RANGE: Saskatchewan to Minnesota southeast to Illinois, Kentucky, South Carolina, and Florida, west to Texas, and north to North Dakota.

CAUTION: Plants of the genus *Delphinium* contain toxins that cause paralysis, and no parts of them should ever be eaten. Cattle, and to a lesser extent horses, find the plant palatable and are susceptible to rapid death after ingestion due to paralysis and bloat. Sheep and goats are resistant to these toxic alkaloids.

SIMILAR SPECIES: Dwarf Larkspur (*Delphinium tricorne*) has a nearly hairless stem. The white-flowered Wooton's Larkspur (*D. wootoni*) overlaps the western edge of this species' range. It usually has leaves mostly at the base, whereas *D. carolinianum* has leafy stems. Trelease Larkspur (*D. treleasei*) is restricted to a very small range in the Ozark Highlands; the flowers of this species are extended from the stem by long stalks (pedicels). Tall Larkspur (*D. exaltatum*) overlaps the eastern edge of Carolina Larkspur's range; the flower spurs of this species are generally horizontal, angled no more than 45 degrees at the extreme. The spurs of Carolina Larkspur are highly inclined, often nearly parallel to the stem.

CONSERVATION: Ⓖ5
Endangered in Florida, Illinois, and Manitoba. Threatened in Georgia.

- NATIVE
- RARE OR EXTIRPATED
- INTRODUCED

ALTERNATE NAMES: Desert Larkspur

Bare-stem Larkspur is a southwestern species avidly sought by bees and hummingbirds for its nectar. The species name *scaposum* means "multiple stems." When there is an abundance of moisture, this species produces several tall stems. However, in average to dry conditions it is common to find only one flowering stem. If the conditions are very dry, only the basal leaves grow and no flowering stems emerge.

A close investigation of the flowers will show that the outermost "petals" are in fact sepals with a very long spur; small white or blue petals are located at the mouth of the sepal spur. The petals also have a spur, which protrudes into the sepal spur. Navajo and Hopi people used the plant, especially the flowers, in religious ceremonies, as well as for a wash following childbirth.

DESCRIPTION: This plant has only basal leaves, which are finely divided and lobed. The unbranched flower stem up to 24" (60 cm) tall with a long column of widely spaced, distinctive, bright blue-and-white flowers. Each flower is 1" (2.5 cm) and has long spurs that often may be bronze-colored near the tip. The upper petals of the flower are usually white or blue-tinged; the lower petals are deep blue.

FLOWERING SEASON: March–May.

HABITAT: Open deserts and gravelly mesas.

RANGE: Southwestern Colorado to Utah, Arizona, and New Mexico.

CAUTION: Plants of the genus *Delphinium* contain toxic alkaloids, and no parts of them should ever be eaten.

CONSERVATION: (G5)
Endangered in California.

- NATIVE
- RARE OR EXTIRPATED
- INTRODUCED

Dwarf Larkspur is the most commonly encountered species of *Delphinium* east of the Great Plains. The plants have very showy spikes of flowers that often bloom in mass. At the most suitable, undisturbed sites, thousands of plants may be present, carpeting the forest with a sea of blue-violet blooms. The long nectar spurs of the flowers are specifically adapted to protect the limited nectar produced from insects that are unable to pollinate the plants. While other species may try, only hummingbirds and the queens of bumblebees (*Bombus* spp.), both of which have very long tongues, are able to reach deep into the nectar spurs to pollinate the flowers. In spite of the flowers' adaptations, some insects rob the nectar by chewing through the outside of the sepals to access the nectar glands.

After pollination, the seeds are dispersed, shaken from their seedpods by the wind. The seeds remain dormant through the summer, not maturing until temperatures fall below 40° F (5° C) for at least 10–12 weeks.

DESCRIPTION: This plant produces a cluster of blue or violet flowers, ¾" (2 cm) wide with a backward-facing spur that extends ½" (1.25 cm). Flowers occur on unbranched, fleshy and hairless stems. Leaves are 2–4" wide and deeply divided into narrow lobes. Overall height: 4–24" (10–60 cm).

FLOWERING SEASON: April–May.

HABITAT: Rich woods.

RANGE: Minnesota southeast to Illinois and Pennsylvania, south to Georgia, west to Oklahoma, and north to Nebraska and Iowa.

CAUTION: Plants of the genus *Delphinium* contain a harmful alkaloid that frequently poisons grazing cattle, and no parts of them should be eaten by humans or animals.

SIMILAR SPECIES: Carolina Larkspur (*D. carolinianum*) has a stem covered in tiny, soft hairs. Tall Larkspur (*D. exaltatum*) reaches a height of 6' (1.8 m) and is leafier, with more flowers; it occurs from Ohio and Pennsylvania south to North Carolina and Tennessee, and in Missouri. The flower structure is roughly similar to that of the annual Giant Larkspur (*Consolida ajacis*), a non-native species that has escaped cultivation in gardens and naturalized onto roadsides and waste places. Close observation of the flower will show the non-native has only one pistil, whereas the native has three.

CONSERVATION: **G5**
Possibly extirpated from Washington D.C. Endangered in Nebraska. Threatened in Georgia, Iowa, Maryland, Mississippi, and North Carolina.

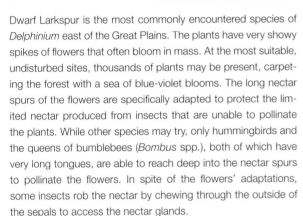

FALSE RUE ANEMONE *Enemion biternatum*

■ NATIVE
■ RARE OR EXTIRPATED
■ INTRODUCED

ALTERNATE NAMES: *Isopyrum biternatum,* Eastern False Rue Anemone

False Rue Anemone is a small, native, early-blooming spring ephemeral. The common name is quite accurate because this plant is easily mistaken for Rue Anemone (*Thalictrum thalictroides*) by all but the most careful observers. Although the small, white flowers do not produce nectar, they frequently attract a wide variety of pollinators to feed on the pollen. Although it does form large populations, this small, somewhat slow-growing plant it is easily overwhelmed by non-native invasive species. Large populations have been extirpated from woodlands after invasion by Garlic Mustard (*Alliaria petiolata*) and non-native shrubs like Japanese Barberry (*Berberis thunbergii*). Garlic Mustard is especially insidious because it excretes chemicals that prevent other plants from growing nearby.

DESCRIPTION: The plant usually produces one to four dainty white flowers, ½" wide, from the end of stems from the axils of the leaves. Each flower has five petal-like sepals. Its compound leaves often are divided into three parts, with each leaflet deeply lobed, approximately 1" (2.5 cm) long. The leaves along the stems are alternate. Overall height: 4–16" (10–40 cm).

FLOWERING SEASON: April–May.

HABITAT: Rich limestone woods and thickets.

RANGE: Ontario southeast to Virginia and Florida, west to Louisiana and Oklahoma, and north to Kansas and Minnesota.

SIMILAR SPECIES: As the common name of this plant suggests, it is similar in form to Rue Anemone (*Anemonella thalictroides*). Rue Anemone can be differentiated by a number of characteristics. Usually Rue Anemone has clusters of three to six flowers, each of which have five to 10 petal-like sepals, instead of only five. Also, only one set of three leaflets appears on its stems and these leaflets may be scalloped or slightly lobed, not deeply lobed. Finally, Rue Anemone does not occur in dense colonies but has greater space between plants.

CONSERVATION: **G5**
Extirpated in New York. Possibly Extirpated in South Dakota. Endangered in Florida, Mississippi, South Carolina, Virginia, West Virginia. Threatened in Alabama, North Carolina, and Ontario.

■ NATIVE
■ RARE OR EXTIRPATED
■ INTRODUCED

ALTERNATE NAMES: *Ranunculus ficaria*, Fig Buttercup, Pile-wort

Lesser Celandine is a non-native, invasive plant that has spread widely throughout large swaths of North America. Emerging earlier than the native spring ephemerals, Lesser Celandine can quickly cover much of the ground in dense, thick mats of vegetation, obstructing the growth of all early-blooming native flowers.

Because the plant sprouts readily from bare dirt and is poisonous to livestock, even in its native Europe many gardeners and farmers consider the plant a nuisance weed. However, several authors have expressed their affection for the plant, including poet William Wordsworth, who dedicated no fewer than three works to this species, writing in one, "There's a flower that shall be mine, 'tis the little Celandine."

In North America, reproduction is most often through small, bud-like bulblets usually produced at the leaf axils. These are easily detached and dispersed by rain, whereupon they readily grow into new plants. Reproduction by seed is less common. The alternate common name Pilewort alludes to the use of the root in folk medicine to treat hemorrhoids.

DESCRIPTION: This is a short perennial herb with a cluster of dark green, glossy, heart-shaped leaves, ¾–1½" (2–4 cm) wide, with smooth or scalloped edges. It produces single yellow flowers, 1–1½" (2.5–4 cm) wide, with 8–10 petals that fade to whitish with age. It also has several greenish sepals that underlie the petals. Overall height: 5–10" (12.5–25 cm).

FLOWERING SEASON: February–May.

HABITAT: Lawns, shaded areas, streamsides, and moist disturbed places.

RANGE: The plant likely exists in more areas than are officially reported. It is introduced from Ontario east to Newfoundland, Vermont south to South Carolina and Alabama, west to Missouri, and north to Wisconsin. It is also introduced in the Pacific Northwest from Oregon north to British Columbia.

CAUTION: These plants contain toxins that can cause poisoning in both humans and animals if ingested, and plant juices can cause blistering or inflammation on skin or mucous membranes on contact.

CONSERVATION:
This is a non-native, invasive species that is restricted for sale, use, or transport in most U.S. states.

ROUND-LOBED HEPATICA *Hepatica americana*

◼ NATIVE
◼ RARE OR EXTIRPATED
◼ INTRODUCED

ALTERNATE NAMES: *Hepatica nobilis* var. *obtusa*, Roundleaf Hepatica, Liverwort, *Anemone americana*

Round-lobed Hepatica is often one of the first wildflowers to bloom within its range in early spring. Many parts of this plant are covered with fine hairs, which are thought to be an adaptation that helps insulate the plant from cold temperatures that may occur early in its growing season. The short, dainty blooms are typically lavender, but may also be shades of pink, purple, or white, and only open on sunny days.

The leaves persist year-round and are very recognizable with three tough, leathery lobes, which overwinter with a deep burgundy-red color. New leaves are a bright green. The leaves were thought to resemble the three lobes of the liver, which led to the genus name as well as several common names. The ancient concept called the Doctrine of Signatures, which suggested medical uses for plants based on the shapes and colors of their parts, has led to use of *H. americana* in folk remedies for liver ailments. However, modern medicine has found no medicinal efficacy of the chemical constituents of this plant.

The taxonomy of this species has been unsettled in recent years, with the results of molecular research changing both the genus from *Anemone* to the closely related *Hepatica* as well as the species name, which has included *A. nobilis* var. *obtusa* or *A. hepatica* depending on the date of publication.

DESCRIPTION: A short plant with 2–2½" (5–6.5 cm) wide leaves which have three rounded lobes and may be burgundy or green. It has several hairy stalks, each with a single pinkish, lavender-blue, or white flower with five to nine petal-like sepals and three green bracts behind sepals. Each flower is ½–1" (1.5–2.5 cm) wide.

FLOWERING SEASON: March–June.

HABITAT: Dry, rocky woods.

RANGE: Manitoba east to Nova Scotia, south to Florida, west to Louisiana and Arkansas, and north to Minnesota.

SIMILAR SPECIES: Sharp-lobed Hepatica (*H. acutiloba*) shares nearly the same range as Round-lobed Hepatica. It has more pointed leaf lobes and bracts.

CONSERVATION: Ⓖ5
Endangered in Manitoba and Nova Scotia. Threatened in Delaware, Florida, Illinois, Mississippi, Rhode Island, and New Brunswick.

■ NATIVE
■ RARE OR EXTIRPATED
■ INTRODUCED

ALTERNATE NAMES: Prairie Crocus, Prairie Anemone, Blue Tulip, American Pulsatilla, Prairie Smoke, Cutleaf Anemone

This species is South Dakota's state flower and is the floral emblem of Manitoba. As many of its common names suggest, the flowers resemble crocus or tulips, emerging in early spring, often when snow is still on the ground. However, this species is not closely related to these plants. The precise taxonomy of this species complex has been long debated and has continued to evolve in recent years. The classification has been complicated due to subtle differences between flowers across its large circumpolar range and efforts to either group or separate populations based on those differences. Depending on the source and time of publication, the scientific name may assign this species to genus *Anemone* or consider it a subspecies or variety of *Pulsatilla patens* (syn: *Anemone patens*). Older names for the same species include *P. hirsutissima* or *P. ludoviciana*.

DESCRIPTION: American Pasqueflower has one to many hairy stems, with a single deeply cup-shaped flower, 1½–2" (4–5 cm) wide. Its petal-like sepals may be lavender, purple, blue, or rarely white. At the base, it grows many leaves 1½–4" (4–10 cm) wide, with three additional leaves in a whorl below the flower. The leaves are divided multiple times into narrow lobes. Overall height: up to 14" (35 cm).

FLOWERING SEASON: May–August.

HABITAT: Well-drained soil from prairies to mountain slopes.

RANGE: Alaska south and east to eastern Canada, south to central Washington, Idaho, Utah, New Mexico, and Texas, and east across northern Great Plains to Illinois and Michigan.

CAUTION: Plant parts are poisonous if eaten. Contact with this plant can cause irritation or inflammation of skin and mucous membranes.

SIMILAR SPECIES: Western Pasqueflower (*P. occidentalis*) is a white-flowered species and has longer, less crowded leaf segments.

CONSERVATION:
Presumed extirpated in Kansas. Endangered in Illinois, Utah, Washington, and Ontario.

■ NATIVE
■ RARE OR EXTIRPATED
■ INTRODUCED

ALTERNATE NAMES: Western Anemone

Western Pasqueflower incorporates an Old French word for Easter due to the springtime blooming and the pure white color of the petal-like sepals. The Western Pasqueflower often grows at high elevations, between 1600–12000 ft (500–3700 m). It blooms later in the season than related species, but is still among the earliest-blooming plants wherever it grows. The curved, white sepals help direct heat from the sun into the center of the flower, helping to warm the reproductive structures as well as any insects that may come to feed—usually small flies feeding on the pollen, since it produces no nectar.

DESCRIPTION: Western Pasqueflower has one to many hairy stems, with a single deeply cup-shaped flower, 1½–2" (4–5 cm) wide. Petal-like sepals may be white to cream-colored. At the base, it produces many leaves 1½–3" (4–7.5 cm) wide, with three additional leaves in a whorl below the flower. The leaves are divided multiple times into narrow, crowded segments. Overall height: up to 24" (60 cm).

FLOWERING SEASON: May–September.

HABITAT: Mountain slopes and meadows.

RANGE: British Columbia to the Sierra Nevada of California; east to northeastern Oregon and western Montana.

CAUTION: Plant parts are poisonous if eaten. Contact with this plant can cause irritation or inflammation of skin and mucous membranes.

SIMILAR SPECIES: American Pasqueflower (*P. nuttalliana*) rarely has white flowers, and also shorter, less crowded leaf segments.

CONSERVATION: G5
The population of this species is secure and is not of special conservation concern.

■ NATIVE
■ RARE OR EXTIRPATED
■ INTRODUCED

ALTERNATE NAMES: Small-flower Buttercup, Early Wood Buttercup

With small flowers and tiny petals, Kidneyleaf Buttercup does not look like most buttercups. This species enjoys a large range from Florida to Alaska. It is abundant in many areas, especially in disturbed sites, but is often overlooked due to its weedy appearance. There are three varieties recognized by some sources based on slight alterations of leaf shape. In the East alone, there are 10 very similar buttercups which may be mistaken for this species. Native Americans mashed the leaves of this plant to treat infections and the roots to treat fainting, seizures, and snakebite.

DESCRIPTION: Kidneyleaf Buttercup is surrounded by stalked kidney-shaped basal leaves erupting from the ground, each ½–1½" (1.5–4 cm) wide with scalloped edges. Leaves occurring on the branched stems are different—they do not have stalks and are typically deeply divided into three to five lobes. The inconspicuous yellow flowers are just ¼" (6 mm) wide, each with five tiny petals and numerous stamens and pistils.

FLOWERING SEASON: April–August.

HABITAT: Fields, open woods, and waste places.

RANGE: Alberta east to Newfoundland, south to Florida, west to Texas, and north to North Dakota; also in parts of the West.

CAUTION: Buttercups contain toxins that can cause poisoning in both humans and animals if ingested. Plant juices can cause blistering or inflammation on skin or mucous membranes on contact.

SIMILAR SPECIES: At least 10 similarly small-flowered species occur in eastern North America.

CONSERVATION: G5

Endangered in New Mexico. Threatened in Wyoming, Labrador, and Yukon.

■ NATIVE
■ RARE OR EXTIRPATED
■ INTRODUCED

ALTERNATE NAMES: Tall Buttercup

The Common Buttercup is generally a non-native, invasive species that is widely spread across North America and is one of our tallest and most common buttercups. This plant can be found in moist soils of nearly every state and territory, especially in the Northeast and Northwest, but is uncommon in the Desert Southwest. Common Buttercup is highly variable in form and division of leaves. Nearly all North American plants were introduced from Eurasia. However, a rare variety, *Ranunculus acris* var. *fridigus,* features larger flowers and is found only in the Aleutian Islands of Alaska and parts of Greenland. Some botanists believe those plants are native to those areas.

The species name refers to the acrid sap of the stems and leaves, which discourages browsing of the plant by most animals. This foul taste also hints at the moderately toxic properties of the plant. Some livestock develop a taste for the plant, leading to consumption of fatal quantities in pastures infested by Common Buttercup.

DESCRIPTION: This is generally an upright, branching plant covered in fine hairs. The leaves are 1–4" (2.5–7.5 cm) long and deeply cut into segments. It produces glossy, bright yellow flowers, 1" (2.5 cm) wide, with five petals. Overall height: 1–3' (30–90 cm).

FLOWERING SEASON: May–September.

HABITAT: Old fields, meadows, and disturbed areas.

RANGE: Throughout much of North America.

CAUTION: Buttercups contain toxins that can cause poisoning in both humans and animals if ingested. Plant juices can cause blistering or inflammation on skin or mucous membranes on contact.

CONSERVATION:

The non-native variety of this plant can be weedy or invasive in many regions or habitats and may displace native vegetation if not properly managed. *Ranunculus acris* var. *fridigus*, native to Alaska, is classified as a vulnerable variety (G3).

■ NATIVE
■ RARE OR EXTIRPATED
■ INTRODUCED

ALTERNATE NAMES: Plantainleaf Buttercup, Meadow Buttercup

Water-plantain Buttercup may have small flowers that grow just ½" wide, but if the conditions are right, they can blanket a meadow with a yellow carpet of thousands of blooms early in spring. This species is highly variable and includes no fewer than six different recognized varieties. The genus comes from the diminutive form of the Latin word *rana*, which literally translates as "tadpole." It is likely this refers to the wet, damp locations where many members of the genus can be found—often near the vernal pools in woodlands, grasslands, and meadows that provide the breeding grounds for frogs.

While the flower's cup shape inhibits direct pollination by insects, the flowers produce large amounts of pollen that a variety of bees and beetles feed upon. Often, these pollinators will become covered in pollen and transport it to other flowers. Instead of directly contacting the female structures, some of the pollen falls off the insect upon landing, dropping onto the slick, glossy petals where it is funneled to the base and into contact with the female reproductive structures, sometimes assisted by floating on drops of rain.

DESCRIPTION: This is a perennial with shiny, green leaves that may be rounded or lance-shaped up to 5½" (14 cm). Waxy, bright yellow flowers, about 1/2–1" (12–25 mm) wide, have 4–14 petals, yellow to greenish centers, and five green sepals. They are very showy in groups. Overall height: may reach 1–2' (30–60 cm).

FLOWERING SEASON: April–June.

HABITAT: Muddy banks and moist mountain meadows.

RANGE: British Columbia to West Montana, south to North California and Colorado.

CAUTION: Buttercups contain toxins that can cause poisoning in both humans and animals if ingested. Plant juices can cause blistering or inflammation on skin or mucous membranes on contact.

CONSERVATION: G5
Endangered in British Columbia. Threatened in Montana.

■ NATIVE
■ RARE OR EXTIRPATED
■ INTRODUCED

ALTERNATE NAMES: White Water Crowfoot, Water Crowfoot

The aptly named Water Buttercup is a fully aquatic species of buttercup that may produce specialized floating leaves that help the flower rise out of the water ½–¾" (1–2 cm). This plant is highly variable and adapts to growing conditions across nearly all of North America. This has led to numerous proposals for the taxonomy of this species, with some sources dividing it into at least four species, all with many varieties. Currently many taxonomists concur on just two varieties within North America: *R. aquatilis* var. *aquatilis*, which produces the specialized floating leaves when in shallow waters, and *R. aquatilis* var. *diffusus* that does not.

The Goshute, indigenous peoples who traditionally resided in what is now called the Salt Lake Valley of Utah, used this plant as a food after boiling, which neutralizes the toxins found within.

DESCRIPTION: This plant sometimes forms dense beds of finely divided, hair-like leaves from submerged stems up to 3' (90 cm) long. For some varieties, less-divided floating leaves may also be present. Above the vegetation, the plant produces small white flowers, ½–¾" (1.5-2 cm) wide, held just above the water's surface on short stalks up to 1" (2.5 cm) tall.

FLOWERING SEASON: May–August.

HABITAT: Ponds and slow streams.

RANGE: Much of North America.

CAUTION: Buttercups contain toxins that can cause poisoning in both humans and animals if ingested. Plant juices can cause blistering or inflammation on skin or mucous membranes on contact.

CONSERVATION: G5

Endangered in Wyoming. Threatened in British Columbia.

■ NATIVE
■ RARE OR EXTIRPATED
■ INTRODUCED

ALTERNATE NAMES: Tufted Buttercup

Unlike most in the genus, this plant is found in drier woods, fields, and prairies. As the name hints, Early Buttercup is often the first native buttercup to bloom across its range. The bright yellow flowers, each atop a hairy stem, often was a welcome sight after a long winter. Today, this plant is becoming a rare find, especially throughout the Northeast where much of its suitable habitat has been developed.

DESCRIPTION: This is one of the earliest spring wildflowers, standing 6–12" (15–30 cm) tall. It has compound, finely hairy leaves with three to five leaflets, each whole or divided with few teeth along the edges. It produces bright yellow, five-petaled flowers, ¾–1" (2–2.5 cm).

FLOWERING SEASON: March–May.

HABITAT: Dry, open woods; prairies.

RANGE: Maine, New Hampshire, southern Ontario, and Minnesota, south to Georgia, Texas, Colorado, and southeastern Nebraska.

CAUTION: Buttercups contain toxins that can cause poisoning in both humans and animals if ingested. Plant juices can cause blistering or inflammation on skin or mucous membranes on contact.

SIMILAR SPECIES: Hispid Buttercup (*R. hispidus*) is quite similar and exists within the same range. It is typically found in wetter habitats, nearly all of its parts are slightly larger, it has more numerous teeth along the edges of leaves, and it blooms slightly later.

CONSERVATION: **G5**

Presumed extirpated in Vermont. Endangered in Maine, Maryland, Nebraska, New Hampshire, New Jersey, North Carolina, Pennsylvania, South Carolina, and Manitoba. Threatened in Massachusetts and Ohio.

This non-native, invasive plant has become naturalized in many areas of the world well beyond its native range in Eurasia and northwestern Africa. In North America, where this plant has historically been used in landscaping, there are many cultivated native species of buttercup that make better plants for the garden and that positively impact the surrounding environment.

Similar to other buttercups, the plant has a bitter, acrid taste that livestock tend to avoid. Herbivores devouring the surrounding plants, cropping them close to the ground, helps provide clear spaces for this plant to quickly spread via runners, soon taking over agricultural fields. Pulling the plant is rarely effective because its fine, filamentous roots are well anchored and the runners often break off the main plant, providing additional spouting locations.

DESCRIPTION: This invasive species spreads by creeping runners along the ground. It has blotchy leaves that are divided into three deeply cut toothy leaflets ¾–1½" (2–4 cm) long. Tall, branching stalks lead to shiny, yellow flowers, ½" (12 mm) wide, with five petals. Overall height: 6–24" (15–60 cm).

FLOWERING SEASON: May–August.

HABITAT: Disturbed open areas, roadsides, meadows, marshes, and lawns.

RANGE: A Eurasian native. It is a non-native invasive plant found throughout North America.

CAUTION: Buttercups contain toxins that can cause poisoning in both humans and animals if ingested. Plant juices can cause blistering or inflammation on skin or mucous membranes on contact.

CONSERVATION:

This non-native, invasive species has escaped from cultivation and is found globally outside of its natural range.

■ NATIVE
■ RARE OR EXTIRPATED
■ INTRODUCED

ALTERNATE NAMES: *R. hispidus* var. *nitidus*, *R. caricetorum*, Bristly Buttercup, Marsh Buttercup

Swamp Buttercup was formerly recognized as a variety of *R. hispidus*, but was recently elevated to a separate species. Like many of its relatives in the genus, and as its common name implies, this plant is often found in the moist soils of swamps and marshes. This species often grows in areas so soggy that even many invasive species like Garlic Mustard (*Alliaria petiolata*) cannot survive there, thus providing some stability to the population. However, this plant only tolerates standing water for a short time.

DESCRIPTION: This plant produces weak, hollow stems, sometimes arching or lying on the ground or across other plants, with bright, glossy, yellow flowers, 1" (2.5 cm) wide with five petals. It has divided leaves with three lobes, each 1½–4" (4–10 cm) long, growing from short stalks. Overall height: 1–3' (30–90 cm).

FLOWERING SEASON: April–July.

HABITAT: Moist woods, thickets, and meadows.

RANGE: Manitoba east to New Brunswick, south to Florida, west to Texas, and north to North Dakota.

CAUTION: Buttercups contain toxins that can cause poisoning in both humans and animals if ingested. Plant juices can cause blistering or inflammation on skin or mucous membranes on contact.

SIMILAR SPECIES: The stems of Hispid Buttercup (*R. hispidus*) have abundant spreading hairs and they are usually more erect. Carolina Buttercup (*R. carolinianus*) tends to have longer fruits (3.5–5 mm in length) than either Swamp or Hispid buttercups. Creeping Buttercup (*R. repens*) leaves are less deeply cleft and are often splotched with pale green or white patterns.

CONSERVATION: **G5**

Endangered in Delaware, Illinois, and Maryland. Threatened in Kansas and North Carolina.

■ NATIVE
■ RARE OR EXTIRPATED
■ INTRODUCED

ALTERNATE NAMES: Tall Meadow-rue

Purple Meadow-rue is an unusual and highly variable plant with tall, purple stems and loose, delicate clusters of white flowers. Each plant is either male or female, with males producing pollen from their stamens, which attracts some insects that feed on pollen. However, because the female flowers produce no nectar, the insects on the male flowers do not visit or pollinate the female plants. Instead, this plant relies only on the wind for pollination.

DESCRIPTION: This is a large, upright perennial with purple stems that begin branching in the upper portion of the stems. Lower leaves are long-stalked and larger, becoming smaller in size and stalk length higher on the plant. Leaves are usually divided into three leaflets, ¾–2" (2–5 cm) long, and most are shallowly lobed and smooth with no or slight hairs on the underside. The plant produces large, loose drooping clusters, up to 2' (60 cm), of creamy to greenish-white flowers with petal-like sepals that fall after opening, leaving many long, thread-like stamens on male plants or slightly smaller clusters of flowers on female plants with similarly thread-like pistils. Overall height: 2–6' (60–180 cm).

FLOWERING SEASON: March–July (southern populations); June–July (northern populations).

HABITAT: Wet meadows, streambanks, and prairies.

RANGE: Ontario to eastern British Columbia, south to Pennsylvania, Ohio, Alabama, Texas, and Arizona.

CAUTION: This plant is in the buttercup family and may contain toxins that can cause poisoning in both humans and animals if ingested.

SIMILAR SPECIES: A few tall *Thalictrum* species are similar and can be difficult to differentiate. Most similar is the Waxy Purple Meadow Rue (*T. revolutum*), which has more densely glandular-hairy leaf stalks and flowering branches, and more prominent leaf veins. Its leaves are foul-smelling if crushed.

CONSERVATION: G5

Possibly extirpated in Pennsylvania. Endangered in Georgia, Idaho, and Washington D.C. Threatened in Arizona, Colorado, Illinois, Washington, Wyoming.

- ■ NATIVE
- ■ RARE OR EXTIRPATED
- ■ INTRODUCED

Early Meadow-rue blooms a little earlier than the others of the genus, as the common name implies—often just as leaves start to form on trees. The species name, from Greek meaning "two households," alludes to the fact that the male and female flowers are on separate plants. The toxicity of the foliage helps deter herbivory by deer and most other woodland animals, but caterpillars of some Noctuid moths (family Noctuidae) feed on the leaves of this genus as their host plant.

DESCRIPTION: This upright perennial has branching, green to purplish-green, smooth stems. Its leaves are long-stalked and divided into three or four lobed leaflets, each ½–2" (1.5–5 cm) wide, medium green on top and pale underneath. The plant produces long-stalked, loose, drooping clusters of greenish to purple flowers with petal-like sepals about ⅛" (3 mm) long, dangling many long, thread-like, pale-yellow stamens on male plants and slightly smaller clusters of flowers on female plants with similarly thread-like purplish pistils. Overall height: 8–30" (20–75 cm).

FLOWERING SEASON: April–May.

HABITAT: Rich, moist woods and ravines.

RANGE: Ontario east to southern Maine, south to Georgia, west to Mississippi and Arkansas, and northwest to the eastern edge of Nebraska and the Dakotas.

CAUTION: This plant is in the buttercup family and may contain toxins that can cause poisoning in both humans and animals if ingested.

CONSERVATION: G5
Endangered in Arkansas, Delaware, and Kansas. Threatened in Illinois and North Carolina.

■ NATIVE
■ RARE OR EXTIRPATED
■ INTRODUCED

Individual plants of Western Meadow-rue are either male or female, like many others in the genus. However, compared to some of the others, which have showier male flowers, the female flowers of Western Meadow-rue steal the show with their deep purple pistils waiting to catch grains of pollen wafting through the air. The common name meadow-rue comes from the strong resemblance to European rues in the genus *Ruta*.

The Blackfoot people had many uses for fruits of this plant, including as a perfume or fragrance as well as an insecticide and spice. The Gitksan, an indigenous people from what is today British Columbia, primarily used the root to create medicines to treat the eyes, legs, and wounds, as well as an analgesic.

DESCRIPTION: This plant has highly divided, soft leaves, each ½–1½" (1.5–4 cm) long, with three leaflets, each notched shallowly into three lobes and sometimes tinged purple at the edges. Flowers occur at the tips of branched stems, ⅜" (9 mm) wide with no petals but greenish, whitish, or purplish petal-like sepals that detach after flowering. Female pistils are typically deep, saturated purple; male stamens are thread-like and brownish-purple. Overall height: 1–3' (30–90 cm).

FLOWERING SEASON: May–July.

HABITAT: Moist ground, often in shady woods.

RANGE: British Columbia south to northern California and east to Montana, Wyoming, and Utah.

CAUTION: This plant is in the buttercup family and may contain toxins that can cause poisoning in both humans and animals if ingested.

CONSERVATION: G5

Threatened in Alaska, Wyoming, Saskatchewan, and Yukon.

■ NATIVE
■ RARE OR EXTIRPATED
■ INTRODUCED

ALTERNATE NAMES: *Anemonella thalictroides*

Rue Anemone is a small but beloved spring ephemeral with showy white blooms on short purple stems. The flower clusters are unlike many within the genus, both in retaining their petal-like sepals and by possessing both male and female parts together. The plant appears to be closer to an anemone and leads some to place it in the anemone-like genus of *Anemonella*. However, the stems and leaves are very similar to others in the *Thalictrum* genus, where most taxonomists agree is the best fit.

DESCRIPTION: The plant produces dainty clusters of several stalked, white or pinkish-tinged flowers, 1" (2.5 cm), with 5–10 petal-like sepals. It has whorls of compound leaves with three round-lobed leaflets, each 1" (2.5 cm) wide. Overall height: 4–8" (10–20 cm).

FLOWERING SEASON: April–June.

HABITAT: Open woods.

RANGE: Ontario; Maine south to Florida, west to Arkansas and Oklahoma, and north to Kansas, Iowa, and Minnesota.

CAUTION: This plant is in the buttercup family and may contain toxins that can cause poisoning in both humans and animals if ingested.

SIMILAR SPECIES: False Rue Anemone (*Enemion biternatum*) has alternate leaves instead of whorls, and usually only five petal-like sepals. Wood Anemone (*Anemone quinquefolia*) has solitary flowers and toothed, jagged leaves.

CONSERVATION STATUS: **G5**
Endangered in Florida, Maine, New Hampshire, Rhode Island, and Vermont. Threatened in Illinois, Kansas, Massachusetts, and Ontario.

■ NATIVE
■ RARE OR EXTIRPATED
■ INTRODUCED

ALTERNATE NAMES: Water Chinquapin, Yellow Lotus

One of only two living species in the lotus family, and the only one native to North America, the American Lotus is an aquatic perennial found in eastern lakes and swamps. The roots, leaves, and seeds are edible and were valued as a food source by Native Americans. Today the American Lotus is widely planted for its foliage and flowers.

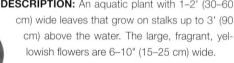

DESCRIPTION: An aquatic plant with 1–2' (30–60 cm) wide leaves that grow on stalks up to 3' (90 cm) above the water. The large, fragrant, yellowish flowers are 6–10" (15–25 cm) wide.

FLOWERING SEASON: July through September.

HABITAT: Found in the quiet waters of lakes, ponds, and marshes.

RANGE: Occurs from southern Ontario and southern New England south to Florida and west to eastern Texas, Oklahoma, Kansas, Nebraska, and southern Minnesota.

SIMILAR SPECIES: The only other extant species in this family is the Sacred Lotus (*N. nucifera*), native to Asia and Australia and introduced to parts of North America.

CONSERVATION: G4

This species appears to be fairly secure globally but is critically imperiled in Delaware and New Jersey; imperiled in Ontario, Michigan, North Carolina, Nebraska, and West Virginia; and vulnerable in Illinois, Indiana, and Virginia.

■ NATIVE
■ RARE OR EXTIRPATED
■ INTRODUCED

ALTERNATE NAMES: Brown's Peony, Native Peony

Typically found at high elevations, Western Peony is one of only two peony species native to North America. The genus name is derived from Paeon, the physician to the gods in Greek mythology. Native Americans used the roots of Western Peony to make a medicinal tea. This species is rarely cultivated.

DESCRIPTION: A somewhat fleshy plant reaching up to 24" (60 cm), with bluish leaves and dark maroon flowers. The leaves are up to 2½" (6.3 cm) long, with five to eight per stem. The flowers are 1–1½" (2.5–3.8 cm) wide, with five to six yellowish or greenish sepals, and maroon petals with yellowish edging.

FLOWERING SEASON: April through June.

HABITAT: Grows on mountain slopes and in sagebrush and open pine forests.

RANGE: Found in the northwestern United States from Washington south to northern California and east to Utah, western Wyoming, and Idaho.

SIMILAR SPECIES: California Peony (*P. californica*), found only in southern California.

CONSERVATION: G5
Although globally secure, this species is critically imperiled in Utah and Wyoming.

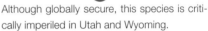

■ NATIVE
■ RARE OR EXTIRPATED
■ INTRODUCED

ALTERNATE NAMES: Coastal Brookfoam

One of several boykinias native to North America, this species can be found growing in moist woodlands along the Pacific coast. The genus name honors the 19th-century naturalist Dr. Samuel Boykin.

DESCRIPTION: A perennial herb with large, heart-shaped leaves with five to seven lobes with toothed edging, borne on stems with brownish or reddish stipules. The small flowers are about ¼" (6 mm) wide and feature five white, narrow petals and the same number of stamens.

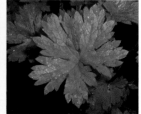

FLOWERING SEASON: June through August.

HABITAT: Found in moist woods and along streams.

RANGE: Occurs along the Pacific coast from British Columbia south to central California.

SIMILAR SPECIES: Mountain Boykinia (*B. major*) and Round-leaf Boykinia (*B. rotundifolia*).

CONSERVATION: G5
The species is globally secure.

■ NATIVE
■ RARE OR EXTIRPATED
■ INTRODUCED

ALTERNATE NAMES: Water-mat, Water-carpet

This small, mat-forming perennial herb favors moist habitats, such as wetlands and seeps, throughout its range. The greenish flowers are rather small and inconspicuous save for the red- or orange-tipped stamens.

DESCRIPTION: A mat-forming herb growing to a typical height of 1–3' (30–90 cm). The leaves are hairless, mostly opposite, and ⅜–⅝" (1–1.5 cm) long and up to ¾" (2 cm) wide. The tiny, solitary flowers are greenish-yellow with typically eight stamens with reddish anthers.

FLOWERING SEASON: Spring.

HABITAT: Grows in springy and muddy places.

RANGE: Found in eastern North America south to northern Georgia and west to Minnesota.

SIMILAR SPECIES: Pacific Golden Saxifrage (*C. glechomifolium*) and Northern Golden Saxifrage (*C. tetrandrum*) are similar but their ranges do not overlap.

CONSERVATION:
This species is globally secure but is critically imperiled in Georgia and South Carolina, imperiled in Indiana and Kentucky, and vulnerable in North Carolina.

NATIVE
RARE OR EXTIRPATED
INTRODUCED

ALTERNATE NAMES: American Alumroot, Common Alumroot

This evergreen wildflower is natively found in rocky woodlands and other habitats of eastern North America. It is often planted as an ornamental for its interesting foliage, and its bell-shaped blooms attract bees and hummingbirds to backyard gardens. The genus name *Heuchera* honors the 18th-century German botanist Johann Heinrich von Heucher.

DESCRIPTION: An evergreen perennial with 3–4" (7.5–10 cm) wide, heart-shaped, lobed leaves. Rising above the basal foliage is a hairy, leafless stalk featuring ⅛–¼" (3–6 mm) long, bell-shaped, drooping, greenish flowers with orange stamen tips, in loose, branching clusters. Typical height is 2–3' (60-90 cm).

FLOWERING SEASON: April through June.

HABITAT: Found in woods, shaded slopes, and rocky places.

RANGE: Occurs in eastern North America from Ontario and southern New England south to northern Georgia and west to Oklahoma, Kansas, Nebraska, and Iowa.

SIMILAR SPECIES: Prairie Alumroot (*H. richardsonii*).

CONSERVATION: G5

Globally secure but critically imperiled in Ontario, imperiled in Louisiana, and vulnerable in Delaware and Iowa.

■ NATIVE
■ RARE OR EXTIRPATED
■ INTRODUCED

ALTERNATE NAMES: Mountain Alumroot, Red Alumroot, Wild Coral Bells

This small member of the saxifrage family is native to the western United States and northern Mexico, preferring dry, rocky places throughout its range. Pink Alumroot can be quite variable in appearance, from the size of its leaves, the height of its stalks, and even the color of its flowers, ranging from white to deep pink.

DESCRIPTION: A small perennial with low basal leaves, ¼–2" (6–50 mm) long, and flowering stalks ranging 4–12" (10–30 cm) in height. The bell-shaped flowers, each ⅛–¼" (3–6 mm) long, are usually pink but can vary in color.

FLOWERING SEASON: May–August.

HABITAT: Grows in dry, rocky places at high elevations.

RANGE: Pink Alumroot is found throughout much of the western United States.

SIMILAR SPECIES: Alumroot (*H. americana*) and other *Heuchera* species.

CONSERVATION: **G5**

Although globally secure, this species is critically imperiled in Colorado.

■ NATIVE
■ RARE OR EXTIRPATED
■ INTRODUCED

ALTERNATE NAMES: Prairie Woodland-star, Small-flower Woodland-star

This flowering plant is a common sight in prairies, open woodlands, and a variety of other open places throughout northwestern North America. Its attractive, star-like flowers are generally white but can be variable. Although this species is sometimes called the Small-flower Woodland-star, its flowers are actually a little larger than those of its close relative, the Bulbous Woodland-star (*L. glabrum*).

DESCRIPTION: This is a perennial herb consisting of deeply lobed, ½–1¼" (1.3–3.1 cm) wide basal leaves and a slender, purplish stem featuring small clusters of white or pale pink ½–1" (1.3–2.5 cm) wide flowers. This plant can grow up to 20" (50 cm) in height.

FLOWERING SEASON: March–June.

HABITAT: Grows in a variety of open habitats including grasslands and prairies, in sagebrush, and in open forests.

RANGE: This species is found in western North America from British Columbia south to northern California and east to Alberta, Montana, western South Dakota, and western Nebraska.

SIMILAR SPECIES: Bulbous Woodland-star (*L. glabrum*) and Bolander's Woodland-star (*L. bolanderi*).

CONSERVATION: G5

Globally secure but critically imperiled in Nebraska and vulnerable in Wyoming.

■ NATIVE
■ RARE OR EXTIRPATED
■ INTRODUCED

ALTERNATE NAMES: Eastern Swamp Saxifrage, Eastern Saxifrage

This is a rather large saxifrage found in bogs, swamps, and other wet areas of southeastern Canada and the northeastern United States. Although somewhat bitter in taste, the young leaves are edible and are sometimes used in salads or cooked.

DESCRIPTION: A large herb growing to 1–3' (30–90 cm) with a sticky, hairy stem and basal, untoothed leaves 4–8" (10–20 cm) long. The greenish-yellow, five-petaled flowers are about ⅛" (4 mm) wide and arranged in initially compact clusters; as the plant ages, these clusters become more open and loose.

FLOWERING SEASON: April–June.

HABITAT: Grows in wet, swampy places, such as areas around bogs, wet meadows and prairies, and along banks.

RANGE: Found in eastern North America from Saskatchewan east to southern Maine, south to North Carolina and west to Missouri.

SIMILAR SPECIES: Oregon Saxifrage (*M. oregana*).

CONSERVATION: G5

Although globally secure, this species is critically imperiled in Manitoba, Ontario, Saskatchewan, Delaware, North Carolina, Rhode Island, and Tennessee; imperiled in West Virginia; and vulnerable in Iowa, Maine, New Jersey, and Virginia.

DIAMOND-LEAF SAXIFRAGE *Micranthes rhomboidea*

■ NATIVE
■ RARE OR EXTIRPATED
■ INTRODUCED

ALTERNATE NAMES: Snowball Saxifrage

Most common in moist mountain meadows, this plant may be found in a wide range of habitats at many elevations in much of the West. It belongs to a complex of several closely related species.

DESCRIPTION: A perennial herb growing up to 12" (30 cm) with a sticky, hairy stem and basal, diamond-shaped, and usually toothed leaves ½–2" (1.5–5 cm) long. The small, white, five-petaled flowers are arranged in close clusters.

FLOWERING SEASON: May–August.

HABITAT: Found in several habitats ranging from moist mountain meadows to sagebrush hills.

RANGE: Western North America from British Columbia and Alberta south to Arizona and New Mexico.

SIMILAR SPECIES: Grassland Saxifrage (*M. integrifolia*) and American Bistort (*Bistorta bistortoides*).

CONSERVATION: G4

This species appears to be fairly secure.

■ NATIVE
■ RARE OR EXTIRPATED
■ INTRODUCED

This widely distributed, early spring wildflower is native to eastern and central North America, favoring rocky places.

DESCRIPTION: A herb growing up to 16" (40 cm) in height and featuring a hairy stalk with basal, egg-shaped and broadly toothed leaves up to 3" (7.5 cm) long. The fragrant white flowers, each about ⅛" (4 mm) wide, are arranged in branched clusters that are initially compact but become more open as the plant matures.

FLOWERING SEASON: April–June.

HABITAT: Grows in rock outcrops, rocky woods, along streambanks, and on dry slopes.

RANGE: Found from Manitoba east to New Brunswick, south to Georgia, and west to Oklahoma and Missouri.

SIMILAR SPECIES: Mountain Saxifrage (*M. petiolaris*) and California Saxifrage (*M. californica*).

CONSERVATION: G5

Globally secure but critically imperiled in New Brunswick and vulnerable in Manitoba, Delaware, and Indiana.

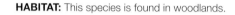

■ NATIVE
■ RARE OR EXTIRPATED
■ INTRODUCED

ALTERNATE NAMES: Two-leaf Bishop's-cap, Two-leaf Miterwort

This woodland wildflower of eastern North America features tiny clusters of intricately fringed white flowers, almost resembling snowflakes. It is sometimes called Two-leaf Miterwort, a reference to the single pair of leaves near the middle of the stem.

DESCRIPTION: This perennial consists of several lobed, stalked leaves at the base and a single pair of stalkless leaves at the middle of the stem, with an elongated cluster of tiny, ⅛" (3.2 mm) wide, white flowers above. The plant grows to 8–18" (20–45 cm).

FLOWERING SEASON: April–June.

HABITAT: This species is found in woodlands.

RANGE: Found from Minnesota, Ontario, and Quebec south to Georgia and west to Arkansas.

SIMILAR SPECIES: Naked Miterwort (*M. nuda*) and other *Mitella* species.

CONSERVATION: G5
Globally secure but critically imperiled in Alabama and South Carolina, imperiled in Arkansas and Delaware, and vulnerable in North Carolina.

■ NATIVE
■ RARE OR EXTIRPATED
■ INTRODUCED

ALTERNATE NAMES: Tufted Alpine Saxifrage, *Saxifraga cespitosa*

This highly variable plant grows in the high Arctic tundra but is also found south into the western United States. The genus name is from two Latin words that mean "stone" and "to break," likely referring to the fact that many of these plants grow in crevices and other rocky places.

DESCRIPTION: A small, often mat-forming perennial that ranges in height from 1–6" (2.5–15 cm). The crowded basal leaves usually have three lobes, and one to several small, often white but sometimes pale yellow flowers blooming from the top of each stem.

FLOWERING SEASON: Late spring through summer.

HABITAT: Grows in rocky places at high elevations.

RANGE: Found in Alaska, in much of Canada, and in scattered mountainous areas from Washington south to California and east to Arizona, Colorado, Wyoming, and Montana.

SIMILAR SPECIES: Wedge-leaf Saxifrage (*S. adscendens*).

CONSERVATION: G5

Globally secure but critically imperiled in Manitoba, Ontario, Arizona, California, Maine, and New Mexico, and vulnerable in Newfoundland and Labrador, Colorado, and Wyoming.

■ NATIVE
■ RARE OR EXTIRPATED
■ INTRODUCED

ALTERNATE NAMES: Heart-leaf Foamflower, False Miterwort, Coolwort

This clump-forming perennial adorns the forest floor with its small, white flowers. These clusters of cream-colored flowers resemble foam, leading to the plant's common name. The heart-shaped leaves account for one of its alternate names, Heart-leaf Foamflower.

DESCRIPTION: A perennial averaging 6–12" (15–30 cm) in height with long-stalked, lobed, heart-shaped leaves generally 2–4" (5–10 cm) long. The tiny white flowers, about ¼" (6 mm) wide, feature five petals and five conspicuous stamens and are arranged in a somewhat elongated cluster with an overall fuzzy or foamlike appearance.

FLOWERING SEASON: April–June.

HABITAT: This species favors wooded areas.

RANGE: Found from Ontario east to Nova Scotia and south through Georgia, Alabama, and Mississippi.

SIMILAR SPECIES: Three-leaf Foamflower (*T. trifoliata*).

CONSERVATION: (G5)

This species is globally secure but critically imperiled in Wisconsin and imperiled in Nova Scotia and Mississippi.

■ NATIVE
■ RARE OR EXTIRPATED
■ INTRODUCED

This plant has the distinction of being the sole member of its genus native to North America. The genus name is derived from two Greek words meaning "five" and "mark," a reference to the five-part pattern of the flower.

DESCRIPTION: A herbaceous perennial that grows to about 12–24" (20–60 cm) in height. The leaves are lanceolate to elliptical, sharply toothed, and 2–4" (5–10 cm) long. The five-petaled flowers are small, about ¼" (6 mm) wide, greenish-yellow, and clustered along the upper side of slender, curved stalks.

FLOWERING SEASON: July–October.

HABITAT: Grows in ditches, along streams, and in other wet or muddy places.

RANGE: This species is found throughout eastern North America from Manitoba east to New Brunswick, south to Florida, and west to Texas.

SIMILAR SPECIES: Although the only native member of its genus, the Ditch Stonecrop resembles and may be confused with members of the genus *Sedum* in the stonecrop family (Crassulaceae).

CONSERVATION: **G5**
Globally secure but critically imperiled in Manitoba and Rhode Island and vulnerable in New Brunswick.

- ■ NATIVE
- ■ RARE OR EXTIRPATED
- ■ INTRODUCED

ALTERNATE NAMES: Canyon Live-forever

Native to the western United States, this attractive succulent grows on cliffs or in other rocky areas, brightening the landscape with its red or yellow flowers. Canyon Dudleya is the larval host plant for the Sonoran Blue butterfly (*Philotes sonorensis*) and is also frequently visited by hummingbirds.

HABITAT: This species favors rocky cliffs, canyons, and outcrops.

RANGE: Found in California and southern Oregon.

SIMILAR SPECIES: Lance-leaf Live-flower (*D. cymosa*).

CONSERVATION: G5
This species is globally secure.

DESCRIPTION: This perennial, succulent plant features yellow to red thimble-shaped flowers, about ¼–½" (7–14 mm) long, clustered atop reddish stalks, with thick, gray-green basal leaves 2–4" (5–10 cm) long. The plant ranges from 4–8" (10–20 cm) in height.

FLOWERING SEASON: April–June.

■ NATIVE
■ RARE OR EXTIRPATED
■ INTRODUCED

ALTERNATE NAMES: Live-forever, Golden Stonecrop, Orpine, *Hylotelephium telephium* ssp. *telephium, Sedum purpureum*

Native to Eurasia, this perennial plant has been introduced in North America, where it is considered invasive. It can be found growing in dry meadows, along roadsides, and in a variety of other places. This wildflower goes by many names, including Live-forever or Life-everlasting, referencing the plant's hardiness. The name Witch's Moneybags comes from the idea that children would separate the outer leaves to form little purses or money bags.

DESCRIPTION: A fleshy plant growing to 10–18" (25–50 cm). The leaves are 1–2½" (2.5–6.5 cm) long, alternate, and coarsely toothed. Small, five-petaled, deep pink to purple-red flowers, about ¼–½" (5–11 mm) wide, are arranged in clusters at the ends of succulent stalks.

FLOWERING SEASON: July–September.

HABITAT: Typically found growing in disturbed areas such as along roadsides and in open woods.

RANGE: Found from Saskatchewan east to Newfoundland, south to North Carolina, and west to Missouri and Kansas. Also found in parts of the northwestern United States.

SIMILAR SPECIES: Allegheny Stonecrop (*H. telephi-oides*) and Woods Stonecrop (*Sedum ternatum*).

CONSERVATION:
The conservation status of this introduced species is not ranked.

■ NATIVE
■ RARE OR EXTIRPATED
■ INTRODUCED

ALTERNATE NAMES: King's-crown

This red-flowered plant is native to Eurasia as well as North America, where it prefers mountainous habitats.

DESCRIPTION: A perennial herb typically growing to 2–12" (5–30 cm). It features small, dark red to maroon-purple flowers atop clustered stems, with broadly lanceolate leaves ¼–1" (6–25 mm) long.

FLOWERING SEASON: June–August.

HABITAT: Grows in moist ground in mountainous areas.

RANGE: Found across western North America. In the West, it occurs from the Arctic south to southern California and New Mexico. Isolated populations (*R. integrifolia* ssp. *leedyi*) are found in Minnesota and upstate New York.

SIMILAR SPECIES: Roseroot (*R. rosea*), a yellow-flowered plant found in parts of the eastern United States, is similar but considered a separate species.

CONSERVATION: G5

This species appears to be fairly secure. An isolated population in southeastern Minnesota is critically imperiled.

■ NATIVE
■ RARE OR EXTIRPATED
■ INTRODUCED

ALTERNATE NAMES: Spearleaf Stonecrop

This low-growing wildflower prefers dry, rocky places with plenty of sunshine. Native to much of western North America, it is sometimes planted as an ornamental in rocky gardens. Lance-leaf Stonecrop is the larval host plant for the Rocky Mountain Parnassian butterfly (*Parnassius smintheus*).

DESCRIPTION: This succulent ranges from 2–8" (5–20 cm) in height. Its thick, lance-shaped basal leaves can be pale green to reddish and are typically about ⅜" (10 mm) long. The yellow starlike flowers, ⅜–¾" (10–18 mm) wide, sometimes tinged with red, are arranged in rounded clusters.

FLOWERING SEASON: June–August.

HABITAT: This species favors sunny, rocky places.

RANGE: Found from Alaska south to Arizona and New Mexico, and east to Alberta, South Dakota, and Nebraska.

SIMILAR SPECIES: Broad-leaf Stonecrop (*S. spathulifolium*) and other *Sedum* species.

CONSERVATION: G5

Although globally secure, this species is critically imperiled in Alaska and vulnerable in Saskatchewan.

ALTERNATE NAMES: Rock Stonecrop, Lime Stonecrop, Pink Stonecrop

This stonecrop is an attractive, hardy little plant that thrives in rocky soils of the south-central United States. There are many common names for this species, including Widow's Cross, a reference to the cross pattern of the flower, and Rock Stonecrop, for this plant's preferred habitat.

DESCRIPTION: This is an annual that grows to 4–12" (10–30 cm) in height. The leaves are narrow, cylindrical, and about ¾" (2 cm) long, with two lobes that grasp the stem. It features pink to pinkish-white flowers, ⅙–¼" (4–6 mm) long, arranged in branched sprays.

FLOWERING SEASON: April–June.

HABITAT: This species favors rocky cliffs and outcrops.

RANGE: Found from eastern Kansas south to Texas and east to Kentucky, Tennessee, and Alabama. It is absent in Louisiana.

SIMILAR SPECIES: The flowers of Elf Orpine (*Diamorpha smallii*) are similar, but its succulent leaves are red.

CONSERVATION:  G5

Globally secure, but critically imperiled in Mississippi and Virginia and vulnerable in Georgia.

- ■ NATIVE
- ■ RARE OR EXTIRPATED
- ■ INTRODUCED

ALTERNATE NAMES: Woodland Stonecrop, Wild Stonecrop, Three-leaved Stonecrop

The Woods Stonecrop or Woodland Stonecrop is the most widespread native stonecrop in the eastern United States. A shade-tolerant species, it is often found growing in rocky areas within forests. Its leaves are arranged in groups of three, hence the alternate common name. This plant is frequently cultivated for ornamental use.

DESCRIPTION: A perennial growing 4–8" (10–20 cm) high with small, rounded green leaves arranged in whorls of three. The star-like white flowers, ½–¾" (12–18 mm) wide, are arranged in branched clusters. It blooms for about a month in late spring or early summer.

FLOWERING SEASON: April–June.

HABITAT: Typically found in rocky woodlands and on cliffs.

RANGE: This species is found in the eastern United States from southern Michigan east to Vermont, and south into Arkansas, Mississippi, Alabama, and Georgia.

SIMILAR SPECIES: This species resembles Mossy Stone-crop (*S. acre*), a European native naturalized in North America.

CONSERVATION:

Globally secure but critically imperiled in Mississippi, imperiled in Delaware, and vulnerable in Arkansas, Missouri, and New Jersey.

■ NATIVE
■ RARE OR EXTIRPATED
■ INTRODUCED

ALTERNATE NAMES: Small-flowered Carpetweed

This is an annual herb found in much of the southwestern United States, where it hugs the dry ground along roadsides and in other disturbed areas. The species name *parviflora* means "with small flowers," and this plant's golden flowers are in fact quite small at about ½" (1.2 cm) wide.

DESCRIPTION: This plant grows along the ground and features compound leaves 1–2.5" (3–6 cm) long and orange to yellow-orange flowers ½–1" (1.2–2.5 cm) across.

FLOWERING SEASON: July–October.

HABITAT: Usually found in dry, open places and disturbed areas including along roadsides and railroads.

RANGE: Occurs in the southwestern United States from California east to Kansas, Oklahoma, and Texas, and in some scattered areas of the Midwest.

SIMILAR SPECIES: The closely related California Caltrop (*K. californica*) and the introduced Puncturevine (*Tribulus terrestris*) have similar growth forms.

CONSERVATION: Ⓖ5

Globally secure, but critically imperiled in Kansas and Utah, imperiled in Nevada, and vulnerable in California.

- NATIVE
- RARE OR EXTIRPATED
- INTRODUCED

ALTERNATE NAMES: Goat's-head, Caltrop

This widespread plant is named for the notoriously sharp spines on its fruit, strong enough to puncture bare feet or even bicycle tires. It is native to the Mediterranean region but widely introduced elsewhere, including North America, where it is considered a noxious weed in several U.S. states. The foliage is poisonous to livestock.

DESCRIPTION: This is a creeping, mat-forming plant with 1–2" (2.5–5 cm) long, opposite leaves pinnately divided into leaflets and small, five-petaled yellow flowers. The bur-like fruit features sharp, paired spines.

FLOWERING SEASON: April–November.

HABITAT: Found in weedy open areas and disturbed places.

RANGE: Occurs throughout much of western North America.

CAUTION: The spines of this plant's woody fruit are sharp enough to puncture bicycle tires and can cause painful injury to hikers and livestock. The foliage is poisonous to livestock if eaten.

SIMILAR SPECIES: The plant resembles the native Warty Caltrop (*Kallstroemia parviflora*).

CONSERVATION:
The conservation status of this introduced species is unranked.

■ NATIVE
■ RARE OR EXTIRPATED
■ INTRODUCED

ALTERNATE NAMES: American Bird's-foot Trefoil

This native plant grows in a variety of habitats and at different elevations, and is often found along roadsides and at other disturbed places. It is a larval host plant to several butterfly species, including the Melissa Blue (*Plebejus melissa*).

DESCRIPTION: This is an annual herb typically growing 6–20" (15–50 cm) and sporting a single, pale pink to cream-colored flower atop a hairy stalk. The leaves, also hairy, are lanceolate-shaped and ¼–1" (6–25 mm) long.

FLOWERING SEASON: May–September.

HABITAT: Grows in most habitats except desert.

RANGE: Occurs throughout North America, but more common in the western states and provinces.

SIMILAR SPECIES: This plant resembles the introduced Bird's-foot Trefoil (*Lotus corniculatus*) from Eurasia.

CONSERVATION: Ⓖ5
This species is globally secure but critically imperiled in Virginia, imperiled in Manitoba and Wyoming, and vulnerable in Saskatchewan.

- ■ NATIVE
- ■ RARE OR EXTIRPATED
- ■ INTRODUCED

ALTERNATE NAMES: Indigo-bush, False Indigo-bush

This flowering shrub, native to North America, is found throughout southeastern Canada, much of the United States, and in northern Mexico. It also occurs on several other continents as an introduced species. It is commonly cultivated as an ornamental plant and serves as a larval host to many native butterfly species.

DESCRIPTION: A shrub growing 5–17' (1.5–5.1 m) in height and often exceeding that in width. It features long racemes of purplish flowers and 10–25 green leaflets, each 1–3" (2.5–7.5 cm) long.

FLOWERING SEASON: May–June.

HABITAT: Grows in open woodlands and along stream- and river-banks, marsh edges, and roadsides.

RANGE: Found from Manitoba east to New Brunswick, and throughout much of the United States.

SIMILAR SPECIES: Leadplant (*A. canescens*) is similar but typically grows only to about 3' (1 m) tall, has gray-green leaflets, and prefers drier habitats.

CONSERVATION: **G5**

Globally secure, but critically imperiled in Manitoba, imperiled in West Virginia and Wyoming, and vulnerable in North Carolina and Virginia.

■NATIVE
■RARE OR EXTIRPATED
■INTRODUCED

This is a flowering vine native to eastern North America. The genus name *Amphicarpaea* is derived from the Greek words for "both kinds" and "fruit," referencing the two kinds of fruit produced by these plants. The seeds of the upper fruit, although inedible to humans, are eaten by several bird species. The seeds of the underground fruit are often eaten by hogs, leading to this plant's common name.

DESCRIPTION: This is a perennial vine growing to 4' (120 cm) long with two kinds of flowers. Those on the upper branches are pale purple or pink to white, arranged in clusters. Those on the lower branches lack petals and are inconspicuous. The leaves are divided into three leaflets, each ¾–3" (2–7.5 cm) long, ovate, and pointed at the tip.

FLOWERING SEASON: August–September.

HABITAT: Favors woodlands, thickets, and moist slopes.

RANGE: Occurs from Manitoba east to Nova Scotia, south to Florida, and west to Texas.

SIMILAR SPECIES: Trailing Wild Bean (*Strophostyles helvola*) and other *Strophostyles* species.

CONSERVATION: Ⓖ5

Although globally secure, this species is critically imperiled in Wyoming.

- ■ NATIVE
- ■ RARE OR EXTIRPATED
- ■ INTRODUCED

Groundnut is a climbing vine of the pea family found through-out eastern North America, preferring low, wet places such as stream banks and moist woodlands and meadows. Its edible tubers were a popular food source among Native Americans and early European colonists. The tubers can be used in soups or fried like potatoes.

DESCRIPTION: A climbing vine growing to 10' (3 m) long with purple-red to maroon flowers produced in compact racemes. The leaves are 4–8" (10–20 cm) long and pinnately compound, with five to seven leaflets.

FLOWERING SEASON: July–September.

HABITAT: This species favors wet thickets, streambanks, woodlands, and meadows.

RANGE: Found in eastern North America from Manitoba east to Nova Scotia, south to Florida, west to Texas, and north to North Dakota.

SIMILAR SPECIES: Price's Groundnut (*A. priceana*) is a rare species with greenish-white, purple-tipped flowers native to Alabama, Mississippi, Kentucky, and Tennessee.

CONSERVATION: **G5**
Globally secure, but critically imperiled in Prince Edward Island and Colorado.

■ NATIVE
■ RARE OR EXTIRPATED
■ INTRODUCED

ALTERNATE NAMES: Purple Milk-vetch, Purple Loco, Cock's-head

This plant, known by several common names, is native to northern and western North America and is representative of many *Astragalus* species with purplish or pinkish flowers, many of which can be difficult to identify. Field Milk-vetch is also native to eastern Asia.

DESCRIPTION: This perennial herb grows to 12" (30 cm) in height. The leaves are narrow and pinnately compound, with 13–21 leaflets, each ¼–¾" (6–20 mm) long. The pink to purple flowers are arranged in crowded clusters of up to 15 flowers, each about ½–¾" (1.5–2 cm) long.

FLOWERING SEASON: May–August.

HABITAT: Grows in moist meadows and prairies.

RANGE: Found across much of Canada and south to the southwestern United States.

CAUTION: All plants in the genus *Astragalus* are potentially toxic to humans and animals if ingested, causing a condition called locoism.

SIMILAR SPECIES: Canada Milk-vetch (*A. canadensis*) and other *Astragalus* and *Oxytropis* species.

CONSERVATION: **G5**
Globally secure, but critically imperiled in Nebraska, imperiled in California, and vulnerable in Yukon, Iowa, and Washington.

■ NATIVE
■ RARE OR EXTIRPATED
■ INTRODUCED

The first *Astragalus* species from North America to be scientifically described, Canada Milk-vetch is common and widespread across Canada and the United States. Although this species includes toxic compounds, the roots have been used medicinally by some Native Americans.

DESCRIPTION: This is a perennial plant growing 12–31" (30–80 cm) in height. The flowers are yellowish-white, arranged in dense racemes and situated atop leafy stems. The leaves are 2½–6" (6.5–15 cm) long and pinnately compound, with 13–29 leaflets. The pods are usually hairless.

FLOWERING SEASON: June–September.

HABITAT: This species grows in thickets, open meadows and woodlands, along roadside ditches, and lakeshores.

RANGE: Found throughout much of Canada and the United States, except for the extreme Southwest, most of Florida, and northern New England.

CAUTION: All plants in the genus *Astragalus* are potentially toxic to humans and animals if ingested, causing a condition called locoism.

SIMILAR SPECIES: Field Milk-vetch (*A. agrestis*) and other *Astragalus* and *Oxytropis* species.

CONSERVATION: G5
Although globally secure, this species is critically imperiled in Alabama, Georgia, Maryland, and Michigan; imperiled in Mississippi, Ohio, Utah, and Vermont; and vulnerable in Colorado, Louisiana, North Carolina, New York, and Wyoming.

NATIVE
RARE OR EXTIRPATED
INTRODUCED

ALTERNATE NAMES: Prairie False Indigo, White False Indigo, *Baptisia leucantha*, *Baptisia lactea*

Long known as *B. leucantha*, this plant is native to eastern and central North America, where it favors grasslands but may also be found in open woods or along roadsides. It is sometimes planted in gardens. The genus name is derived from a Greek word meaning "to dip" or "to dye," a reference to the fact that many of these species were used to make blue dye for fabrics.

DESCRIPTION: A perennial growing 2–5' (60–150 cm) in height with white to cream-colored flowers arranged in long spiky clusters. The smooth leaves are compound, with three leaflets, each 1–2½" (2.5–6.5 cm) long. The hairless pods become black at maturity.

FLOWERING SEASON: May–July.

HABITAT: This species may be found in prairies, open woods, and along roadsides.

RANGE: Found from Ontario and Ohio south to Florida, west to Texas, and north to Minnesota.

CAUTION: This plant may be fatal to livestock and can be irritating to humans if ingested.

SIMILAR SPECIES: Blue False Indigo (*B. australis*), Yellow False Indigo (*B. tinctoria*), and Cream False Indigo (*B. bracteata*).

CONSERVATION: G5

Globally secure, but imperiled in North Carolina.

■ NATIVE
■ RARE OR EXTIRPATED
■ INTRODUCED

ALTERNATE NAMES: Plains Wild Indigo, Cream Wild Indigo, Longbract Wild Indigo

Native to the eastern United States, this species is among the earliest to bloom in spring, and is very important to bumblebees and other pollinators. Its cream-colored flowers can be quite showy. This species can be damaged by wind and it needs support from companion plants.

DESCRIPTION: A perennial with cream-colored flowers growing to 2' (60 cm) tall with a wide, bushy habit. The leaves are alternate, 1½–4" (4–10 cm) long, and divided into three segments; they turn from green to dark gray or black late in the season.

FLOWERING SEASON: March–June.

HABITAT: This species favors prairies and open woods.

RANGE: Occurs throughout much of the eastern and central United States.

CAUTION: Although no human fatalities have been recorded, other plants in this genus are poisonous if ingested, so caution is advised.

SIMILAR SPECIES: Blue False Indigo (*B. australis*), Yellow False Indigo (*B. tinctoria*), and White Wild Indigo (*B. alba*).

CONSERVATION: G4

This species appears to be fairly secure globally but is vulnerable in Indiana and Minnesota.

■ NATIVE
■ RARE OR EXTIRPATED
■ INTRODUCED

ALTERNATE NAMES: Spurred Butterfly-pea, Coastal Butterfly-pea

This showy vine is found in open woods and clearings of the southeastern United States. Details in the flowers can help to separate this species from another plant commonly known as Butterfly Pea (*Clitoria mariana*), also called Atlantic Pigeon-wings. In *Centrosema virginianum*, the calyx lobes are longer than the tubular base of the flower, whereas in *Clitoria mariana*, the lobes are shorter.

DESCRIPTION: A trailing or twining vine growing 2–4' (60–120 cm) long. The lavender flowers are pea-like but upside down, and the leaves are divided into three ovate to lanceolate leaflets, each 1–2½" (2.5–6.5 cm) long.

FLOWERING SEASON: July–August.

HABITAT: Grows in open woods, fields, and clearings.

RANGE: Found from New Jersey south to Florida, west to Texas and Oklahoma, and northeast to Missouri, Illinois, and Kentucky.

SIMILAR SPECIES: See Butterfly Pea (*Clitoria mariana*).

CONSERVATION: **G5**

Globally secure, but critically imperiled in Kentucky and imperiled in Maryland.

■ NATIVE
■ RARE OR EXTIRPATED
■ INTRODUCED

ALTERNATE NAMES: Showy Partridge-pea, Sleepingplant, Sensitive-plant, *Cassia fasciculata*, *Cassia chamaecrista*

There is little wonder why this plant is sometimes called Showy Partridge-pea: its bright yellow flowers stand out in prairies and other open areas throughout the eastern United States. Those bright flowers attract the attention of bees, butterflies, and other pollinators, and the seed pods are eaten by a wide variety of bird species.

DESCRIPTION: A slender-stemmed annual growing 1–3' (30–90 cm) in height. Its leaves are pinnately compound with 12–36 leaflets that fold together when touched. The large, showy flowers are yellow with a reddish mark near the center.

FLOWERING SEASON: June–October.

HABITAT: Grows in prairies, along roadsides, and in open woods.

RANGE: Found throughout much of the eastern United States from Minnesota east to Massachusetts, south to Florida, and west to Texas.

SIMILAR SPECIES: Sensitive Partridge-pea (*C. nictitans*).

CONSERVATION: G5

Although globally secure, this species is vulnerable in New York and West Virginia.

BUTTERFLY PEA *Clitoria mariana*

■ NATIVE
■ RARE OR EXTIRPATED
■ INTRODUCED

ALTERNATE NAMES: Atlantic Pigeon-wings, Maryland Butterfly-pea

Native to much of the eastern United States, this species can be confused with Climbing Butterfly-pea (*Centrosema virginianum*). A key difference is in the flowers: while both species sport the characteristic lavender upside-down flowers, in *Centrosema virginianum*, the calyx lobes are longer than the tube, whereas in *Clitoria mariana*, the lobes are shorter than the tube.

DESCRIPTION: A trailing or sprawling vine that grows 1–3' (30–90 cm) with lavender pea-like but upside-down flowers. The leaves are divided into 3 ovate leaflets, each 1–2½" (2.5–6.5 cm) long.

FLOWERING SEASON: June–August.

HABITAT: Typically found in fields, open woods, and roadsides.

RANGE: Occurs from Kansas northeast to New Jersey, south to Florida, and west to Texas.

SIMILAR SPECIES: See Climbing Butterfly-pea (*Centrosema virginianum*).

CONSERVATION:

Globally secure, but critically imperiled in New Jersey and Pennsylvania, imperiled in Delaware and Ohio, and vulnerable in Indiana.

■ NATIVE
■ RARE OR EXTIRPATED
■ INTRODUCED

ALTERNATE NAMES: Arrowhead Rattlebox

This wildflower, native to the midwestern and eastern United States, is named for its seed pods, which rattle when shaken. It prefers open areas with a history of disturbance, often found growing in gravel prairies or in areas along railroads.

DESCRIPTION: An annual growing 4–16" (10–40 cm) tall with alternate undivided leaves ½–2½" (1–6 cm) long. The pea-like flowers are yellow and the pods are plump, becoming blackish with age.

FLOWERING SEASON: June–September.

HABITAT: Grows in open disturbed places such as fields, small meadows, and along railroads.

RANGE: Occurs from Nebraska northeast to Minnesota, southeast to Kentucky, northeast to Massachusetts bypassing Ohio and much of West Virginia, south to Georgia, and west to Texas. Largely absent in Florida. Also found in southeastern Arizona.

SIMILAR SPECIES: Low Rattlebox (*C. pumila*) is similar, but the two species' ranges do not overlap.

CONSERVATION: **G5**

Globally secure, but critically imperiled in New York, Rhode Island, Vermont, and Wisconsin, and vulnerable in Kentucky, Minnesota, and North Carolina.

■ NATIVE
■ RARE OR EXTIRPATED
■ INTRODUCED

ALTERNATE NAMES: *Petalostemum candidum*, *Petalostemon candidus*

This plant is native to North America, where it may be found growing in a variety of places from prairies to forests. Some Native American groups have used a variety of this species to aid with stomach pain. White Prairie Clover was once placed in the genus *Petalostemon*.

DESCRIPTION: A perennial herb typically growing 1–2' (30–60 cm) in height. The pinnately compound leaves are divided into 5–9 (often 7) light green leaflets, each ½–1½" (1.5–4 cm) long. The small, white flowers are arranged in dense spikes.

FLOWERING SEASON: May–September.

HABITAT: Grows in arroyos, prairies, open woods, and along hillsides.

RANGE: Found in central Canada south to Alabama, west to New Mexico and western Texas, and south to Mexico.

SIMILAR SPECIES: Purple Prairie Clover (*D. purpurea*) and White Dalea (*D. albiflora*).

CONSERVATION: G5

Globally secure, but critically imperiled in Georgia, imperiled in Tennessee, and vulnerable in Alberta and Kentucky.

■ NATIVE
■ RARE OR EXTIRPATED
■ INTRODUCED

ALTERNATE NAMES: Violet Prairie Clover, *Petalostemum purpurea*, *Petalostemon purpureus*

This familiar wildflower is native to the prairies and plains of central North America. It is a nutritious food source for Pronghorn (*Antilocapra americana*) and other wildlife. Like several similar species, Purple Prairie Clover was formerly placed in the genus *Petalostemon*.

DESCRIPTION: A perennial herb growing 1–3' (30–90 cm) in height with pinnately compound leaves divided into three to seven (often five) narrow leaflets, each ½–1" (1.5–2.5 cm) long. The small, purplish flowers are arranged in dense spikes.

FLOWERING SEASON: May–August.

HABITAT: Grows in prairies and along dry hillsides.

RANGE: Occurs from Indiana south to Alabama, west to New Mexico, north to British Columbia, and east to Ontario.

SIMILAR SPECIES: See White Prairie Clover (*D. candida*).

CONSERVATION:  G5

Although globally secure, this species is critically imperiled in Ontario, Georgia, and Tennessee, and vulnerable in Alberta, Indiana, and Kentucky.

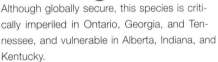

■ NATIVE
■ RARE OR EXTIRPATED
■ INTRODUCED

ALTERNATE NAMES: Prairie Bundle-flower, Illinois Bundle-flower

This common plant is native to the central United States, preferring prairies and other open habitats where it can receive full sun. It is high in protein and quite nutritious, readily eaten by a variety of wildlife; it also attracts bees and butterflies and is often planted in backyard gardens for this purpose.

DESCRIPTION: This is a perennial growing 2–4' (60–120 cm) in height with tall stalks and ball-shaped clusters of small, whitish flowers. The leaves, typically 2–4" (5–10 cm) long, are pinnately divided into many tiny leaflets.

FLOWERING SEASON: June–August.

HABITAT: Typically found in prairies and open woods.

RANGE: Occurs in the middle of the United States from North Dakota southwest to New Mexico, east to Georgia, and northwest to Minnesota.

SIMILAR SPECIES: Prairie Acacia (*Acacia angustissima*).

CONSERVATION: G5

Globally secure, but critically imperiled in North Dakota and vulnerable in Minnesota.

NATIVE
RARE OR EXTIRPATED
INTRODUCED

ALTERNATE NAMES: Canadian Tick-Trefoil

Native to eastern North America, this perennial plant is appropriately named, with its pinkish flowering clusters catching the eye wherever it is found growing. It is a larval host plant for the Eastern Tailed-Blue (*Cupido comyntas*) and several other butterfly species. This plant is often cultivated as an ornamental.

DESCRIPTION: This is an herbaceous perennial growing 2–6' (60–180 cm) in height with pinnately compound leaves divided into three untoothed leaflets, each up to 3" (7.5 cm) long. The numerous pea-like flowers are pinkish-purple, becoming dark blue with age, and arranged in crowded clusters along the upper stems.

FLOWERING SEASON: July–August.

HABITAT: Prefers streambanks, field edges, and moist, open woods.

RANGE: Occurs from Manitoba east to Nova Scotia, south to Virginia, west to Missouri and Texas, and north to North Dakota.

SIMILAR SPECIES: Panicled Tick-Trefoil (*D. paniculatum*) and other *Desmodium* species.

CONSERVATION: G5

Globally secure, but critically imperiled in Nova Scotia and Virginia, and imperiled in Manitoba.

LARGE TICK-TREFOIL *Desmodium glutinosum*

- ■ NATIVE
- ■ RARE OR EXTIRPATED
- ■ INTRODUCED

ALTERNATE NAMES: *Hylodesmum glutinosum*, Pointed-leaved Tick-Trefoil

This plant is native to woodland habitats of eastern America. The genus name is derived from a Greek word meaning a "branch" or "chain," likely a reference to the shape of the seed pods of these species.

DESCRIPTION: This perennial grows 1–3' (30–90 cm) tall with a light green, usually hairless stem. The leaves are divided into three ovately shaped leaflets, each 2–5" (5–12 cm) long and 1½–3" (4–8 cm) across. The pea-like flowers are light to rosy pink in color.

FLOWERING SEASON: July–August.

HABITAT: Grows in woodlands and along woodland edges.

RANGE: Occurs throughout much of eastern North America from Quebec south into northeastern Mexico.

SIMILAR SPECIES: Other *Desmodium* species.

CONSERVATION: G5

This species is globally secure but critically imperiled in New Brunswick and Nova Scotia, imperiled in Delaware and North Carolina, and vulnerable in Quebec.

■ NATIVE
■ RARE OR EXTIRPATED
■ INTRODUCED

ALTERNATE NAMES: Narrow-leaf Tick-Trefoil

This common wildflower is native to eastern North America, found in woodlands, open fields, and various disturbed areas such as roadsides and clearings. Like other species in this genus, the seed pods are segmented and covered with tiny hairs; the pod segments in Panicled Tick-Trefoil are triangular, a helpful clue in identification.

DESCRIPTION: This is a perennial herb growing 1–3' (30–90 cm)

tall with pinnately compound leaves divided into three narrow leaflets. The pea-like flowers are pink to purple in color.

FLOWERING SEASON: July–August.

HABITAT: Found in woods and fields.

RANGE: Occurs from Nebraska northeast to Maine, south to Florida, and west into Texas.

SIMILAR SPECIES: Showy Tick-Trefoil (*D. canadense*) and other *Desmodium* species.

CONSERVATION: (G5)
This species is globally secure but vulnerable in Vermont.

■ NATIVE
■ RARE OR EXTIRPATED
■ INTRODUCED

ALTERNATE NAMES: Wild Licorice

This widespread plant is native to North America and found across much of the continent. Although the root of this plant does indeed have a distinct licorice flavor, most commercial licorice comes from another plant not native to North America.

DESCRIPTION: This perennial plant grows 16–40" (40–100 cm) tall with alternate, pinnately compound leaves divided into 11–19 lance-shaped leaflets. The cream-colored flowers are clustered on a spike.

FLOWERING SEASON: June–July.

HABITAT: Prefers prairies, streambanks, and roadsides.

RANGE: Found from British Columbia east to western Ontario, and southwest to Missouri, Oklahoma, Texas, and Mexico, and California.

SIMILAR SPECIES: Not likely to be confused with other species in its range.

CONSERVATION: G5

Although globally secure, this species is critically imperiled in Wisconsin and vulnerable in British Columbia and Ontario.

Hedysarum boreale **NORTHERN SWEET-VETCH**

■ NATIVE
■ RARE OR EXTIRPATED
■ INTRODUCED

ALTERNATE NAMES: Boreal Sweet-vetch

This highly variable species is found throughout northern and western Canada and the western United States, where it prefers well-drained or sandy soils, often at high elevations. Northern Sweet-vetch is valued by wildlife as a food source as well as for cover; it is a particularly important component of sage-grouse habitat. The roots were used as a food source by some Native American groups.

DESCRIPTION: This is a perennial herb growing 8–24" (20–60 cm) in height, with clusters of purplish pea-like flowers. The leaves are pinnately divided into 7–15 leaflets. The side veins on the leaflets are inconspicuous; these are prominent in the similar Alpine Sweet-vetch (*H. alpinum*).

FLOWERING SEASON: May–August.

HABITAT: This wildflower typically grows in grasslands and other dry open areas.

RANGE: Occurs throughout much of northern and western Canada and the western United States.

SIMILAR SPECIES: Alpine Sweet-vetch (*H. alpinum*) and other *Hedysarum* species. Alpine Sweet-vetch has prominent side veins on the leaflets.

CONSERVATION: G5
Globally secure, but critically imperiled in Newfoundland and Labrador and Texas, and vulnerable in Manitoba and Wyoming.

- NATIVE
- RARE OR EXTIRPATED
- INTRODUCED

ALTERNATE NAMES: Waxy Rush-pea, Camote de Raton, Pig Nut

This plant often grows in low patches in the southwestern United States and Mexico, in grasslands, along roads, and in various disturbed areas. The nutritious roots are readily eaten by small mammals and other wildlife; the common Spanish name is *Camote de Raton*, meaning "mouse's yam."

DESCRIPTION: This is a perennial growing 4–12" (10–30 cm) tall with pinnately compound leaves divided into 5–11 sections, each with another 5–11 pairs of small, oblong leaflets. The flowers are yellow-orange with reddish markings.

FLOWERING SEASON: March–September.

HABITAT: Commonly grows in grasslands and along roads in agricultural areas.

RANGE: The plant's native range extends from southern California east to southern Kansas, and south to Mexico.

SIMILAR SPECIES: James' False Holdback (*Pomaria jamesii*) is a look-alike pea species also native to the American Southwest; it has leaflets with red gland-dots on the underside.

CONSERVATION: G5

Although globally secure, this species is critically imperiled in Kansas.

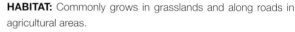

- ■ NATIVE
- ■ RARE OR EXTIRPATED
- ■ INTRODUCED

ALTERNATE NAMES: Sea Vetchling, *Lathyrus maritimus*

This perennial plant is widely distributed across several continents in both hemispheres, growing along seashores and other waters. The seeds can float and remain viable for years, possibly explaining its worldwide distribution. In some cultures, the leaves of this species are used for medicinal purposes.

DESCRIPTION: This trailing vine grows 1–2' (30–60 cm) long with pinnately compound leaves divided into 6–12 oval-shaped leaflets, each up to 2½" (6.5 cm) long. The pea-like flowers are typically pinkish-lavender and arranged in long-stalked clusters.

FLOWERING SEASON: July–August.

HABITAT: Found along seashores continent-wide and along the shores of the Great Lakes.

RANGE: Occurs from Manitoba east to Newfoundland, south to New Jersey, and west to Indiana and Minnesota; it is also found along the Pacific Coast from Alaska south to northern California.

CAUTION: Plants in the genus *Lathyrus*, particularly the seeds, can be toxic to humans and animals if ingested.

SIMILAR SPECIES: Marsh-pea (*L. palustris*) and Silky Beach Pea (*L. littoralis*).

CONSERVATION: G5
This species is globally secure but critically imperiled in Indiana and imperiled in Yukon, California, Ohio, Pennsylvania, and Vermont.

EVERLASTING PEA *Lathyrus latifolius*

■ NATIVE
■ RARE OR EXTIRPATED
■ INTRODUCED

ALTERNATE NAMES: Perennial Pea

Native to Europe, this attractive vine has been widely grown as an ornamental and can now be found as an escapee in much of the United States. It can easily become a nuisance as it grows over other plants and spreads into natural habitats.

DESCRIPTION: A trailing or climbing vine growing to 10' (3 m) long with pea-like pink to purple flowers arranged in clusters. The leaves are alternate, divided into two lance-shaped leaflets, each 2–5½" (5–14 cm) long.

FLOWERING SEASON: June–September.

HABITAT: Typically grows in fields, along roadsides, and in other disturbed places.

RANGE: Native to Europe but can be found in much of the United States.

SIMILAR SPECIES: See native *Lathyrus* species.

CONSERVATION:
The conservation status of this introduced species is unranked.

■ NATIVE
■ RARE OR EXTIRPATED
■ INTRODUCED

ALTERNATE NAMES: Marsh Vetchling

This perennial plant is native to North America, Europe, and Asia, preferring damp meadows and other wet places. It will climb on other plants and is often found growing among reeds. The species name *palustris* means "of the marsh," alluding to its common name and typical habitat.

DESCRIPTION: This is a trailing vine growing 1–4' (30–120 cm) in length with pinnately compound leaves divided into four to eight (often six) leaflets. The stems are often angled or winged; the flowers are reddish-purple.

FLOWERING SEASON: July–August.

HABITAT: Grows in moist or wet places.

RANGE: Found in Alaska and throughout most of Canada, and from Maine south to Georgia, and northwest to North Dakota.

CAUTION: Plants in the genus *Lathyrus*, particularly the seeds, can be toxic to humans and animals if ingested.

SIMILAR SPECIES: Cream-pea (*L. ochroleucus*) and other *Lathyrus* species.

CONSERVATION: G5

This species is globally secure but critically imperiled in Alberta, Alabama, Georgia, Maryland, Pennsylvania, Tennessee, Virginia, and West Virginia; imperiled in California, Kentucky, North Carolina, Nebraska, New Jersey, and Vermont; and vulnerable in Newfoundland and Labrador.

■ NATIVE
■ RARE OR EXTIRPATED
■ INTRODUCED

ALTERNATE NAMES: Round-head Lespedeza

Lespedeza is a genus of clover-like flowering plants, commonly known as bush-clovers. This plant is native to eastern North America, where it may be found in a variety of habitats from prairies to open woods. It is a larval host plant for the Eastern Tailed-Blue (*Cupido comyntas*), Silver-spotted Skipper (*Epargyreus clarus*), and many other butterfly species, and Northern Bobwhites (*Colinus virginianus*) and other birds often feast on the seeds.

DESCRIPTION: This perennial herb grows 2–5' (60–150 cm) in height with alternate leaves divided into three narrow leaflets, each up to 3" (7.5 cm) long and 1" (2.5 cm) wide. The flowers, white with purple inside, are arranged in dense, short-stalked clusters.

FLOWERING SEASON: July–September.

HABITAT: Grows in open woods, prairies, and along roadsides.

RANGE: Found in eastern North America from Ontario and New Brunswick south through Texas and Florida.

SIMILAR SPECIES: Hairy Bush-clover (*L. hirta*).

CONSERVATION: G5

Although globally secure, this species is critically imperiled in New Brunswick, imperiled in North Dakota, and vulnerable in Kentucky and Vermont.

■ NATIVE
■ RARE OR EXTIRPATED
■ INTRODUCED

ALTERNATE NAMES: Creeping Lespedeza

Creeping Bush-clover is native to open woods and clearings of the eastern United States, where its seeds are a favored food source of the Northern Bobwhite (*Colinus virginianus*).

DESCRIPTION: This is a trailing plant growing 6–24" (15–60 cm) long with compound leaves divided into three leaflets, each

about ½" (1.5 cm) long. The flowers are pink to purple and pea-like, arranged in loose clusters.

FLOWERING SEASON: May–September.

HABITAT: Grows in open woods, clearings, and thickets.

RANGE: Occurs from New York and Connecticut south to Florida, west to Texas, and north to Kansas and Wisconsin.

SIMILAR SPECIES: Trailing Bush-clover (*L. procumbens*).

CONSERVATION: **G5**
Globally secure, but critically imperiled in Connecticut, imperiled in Kansas, and vulnerable in New York.

■ NATIVE
■ RARE OR EXTIRPATED
■ INTRODUCED

ALTERNATE NAMES: Bird's-foot Deervetch, Eggs and Bacon

This flowering plant is native to the grasslands of Eurasia and Africa, but can be found throughout much of North America and many other places as an introduced species. Once it becomes established in an area, it often outcompetes native species. The arrangement of the slender pods of this plant resemble a bird's foot, hence the common name.

DESCRIPTION: This trailing plant grows 3–12" (7.5–30 cm) in length. The alternate leaves are divided into five leaflets, with three grouped near the tip, and two near the base of the leaf-stalk. The flowers are bright yellow, becoming red-marked in age, and arranged in clusters.

FLOWERING SEASON: May–September.

HABITAT: In North America, generally found in fields, along roads, and in other disturbed places.

RANGE: Occurs throughout much of North America, except in the Far North.

CAUTION: This plant is reportedly poisonous to animals if ingested.

SIMILAR SPECIES: See native *Lotus* species.

CONSERVATION:
The conservation status of this introduced species is unranked.

■ NATIVE
■ RARE OR EXTIRPATED
■ INTRODUCED

ALTERNATE NAMES: Silver-stem Lupine

This species is native to western North America, occupying a variety of habitats from open pine woodlands to meadows and roadsides. It spreads quickly, forming colonies. This plant's common name refers to its silvery-hairy leaves and stems.

DESCRIPTION: This is a perennial lupine growing 6–24" (15–60 cm) in height, with palmately compound silvery-green leaves. The violet, pea-like flowers are arranged in an elongated cluster atop the stem.

FLOWERING SEASON: June–August.

HABITAT: Grows in plains, meadows, open woods, along roadsides, and on dry slopes.

RANGE: Found from parts of Washington, Oregon, and California southeast to Arizona and New Mexico, and north to western North Dakota and Alberta.

CAUTION: Plants in the genus *Lupinus*, especially the seeds, can be toxic to humans and animals if ingested.

SIMILAR SPECIES: Sundial Lupine (*L. perennis*) is a similar eastern lupine species, found only as far west as the Gulf Coast and the Great Lakes region.

CONSERVATION: **G5**
This species is globally secure but imperiled in Saskatchewan.

NATIVE
RARE OR EXTIRPATED
INTRODUCED

ALTERNATE NAMES: Stemless Dwarf Lupine, *Lupinus lepidus* var. *utahensis*

This is a very low-growing wildflower native to the western United States, found in meadows, sagebrush, and other dry habitats. The leaves, which grow from the base of the plant, are actually taller than the stemless flowers.

DESCRIPTION: This is a low-growing lupine reaching 3–10" (7.5–25 cm) in height with palmately compound, silver-hairy leaves. The blue-purple, pea-like and stemless flowers are whorled in clusters close to the ground.

FLOWERING SEASON: June–August.

HABITAT: Grows in dry meadows, sagebrush, and Ponderosa Pine forest.

RANGE: Occurs from eastern Oregon to Montana, south to western Colorado and northern Nevada, and west into east-central California.

CAUTION: Plants in the genus *Lupinus*, especially the seeds, can be toxic to humans and animals if ingested.

SIMILAR SPECIES: Other *Lupinus* species.

CONSERVATION: G5
This species is globally secure.

■ NATIVE
■ RARE OR EXTIRPATED
■ INTRODUCED

ALTERNATE NAMES: Tailcup Lupine

This common wildflower is native to the western United States, where it grows on open hillsides and in pine woods at mid- to high elevations. The species name *caudatus* means "with a tail," possibly a reference to the long flower cluster of this plant.

DESCRIPTION: This perennial grows 8–24" (20–60 cm) tall with palmately compound leaves divided into five to seven narrow leaflets, each 1–2" (2.5–5 cm) long. The lavender flowers are arranged in an elongated cluster atop the stem.

FLOWERING SEASON: May–August.

HABITAT: Found in sagebrush plains and hills, and in Ponderosa Pine forests.

RANGE: Occurs from California and eastern Oregon and east to Montana, South Dakota, Nebraska, Colorado, and New Mexico.

CAUTION: Plants in the genus *Lupinus*, especially the seeds, can be toxic to humans and animals if ingested.

SIMILAR SPECIES: Other *Lupinus* species.

CONSERVATION: G5
Globally secure, but imperiled in Arizona.

NATIVE
RARE OR EXTIRPATED
INTRODUCED

ALTERNATE NAMES: Chinook Lupine, Seashore Lupine

This low-growing perennial herb is native to western North America from southern British Columbia south to northern California. As the common name suggests, this is a wildflower of coastal areas, thriving in sandy soils. Some Native American groups used the roots of this species as a food source after drying and roasting them.

DESCRIPTION: This mat-forming perennial grows to 12" (30 cm) tall. The palmately compound leaves are divided into six to eight leaflets. The purple flowers are clustered along upright stalks at the top of the stems.

FLOWERING SEASON: May–August.

HABITAT: Grows in sandy coastal areas.

RANGE: Occurs from British Columbia south to northwestern California.

CAUTION: Plants in the genus *Lupinus*, especially the seeds, can be toxic to humans and animals if ingested.

SIMILAR SPECIES: See other *Lupinus* species.

CONSERVATION: G5
This species is globally secure.

■ NATIVE
■ RARE OR EXTIRPATED
■ INTRODUCED

ALTERNATE NAMES: Meadow Lupine, Bog Lupine, Large-leaved Lupine

Native to western North America, this is a wildflower of wet habitats, often found along streams and creeks or in moist meadows and forests. It is one of the tallest western lupine species, reaching up to 5' (1.5 m) in height. It has been crossed with other lupines to produce beautiful horticultural hybrids.

DESCRIPTION: A perennial growing 2–5' (60–150 cm) in height with palmately compound leaves divided into 9–13 leaflets, each 1½–4" (4–10 cm) long. The violet or blue pea-like flowers are arranged in long dense racemes.

FLOWERING SEASON: June–August.

HABITAT: Grows along streams and creeks and in other moist habitats.

RANGE: Occurs from British Columbia south into central California and northeast to Montana and Alberta.

CAUTION: Plants in the genus *Lupinus*, especially the seeds, can be toxic to humans and animals if ingested.

SIMILAR SPECIES: Stinging Lupine (*L. hirsutissimus*) and other *Lupinus* species.

CONSERVATION: G5
This species is globally secure.

■ NATIVE
■ RARE OR EXTIRPATED
■ INTRODUCED

ALTERNATE NAMES: Pursh's Silky Lupine

This wildflower, native to western North America, is found in many types of habitats from forests to grasslands. Although this plant can be toxic to sheep and certain other livestock, parts of it are readily consumed by White-tailed Deer (*Odocoileus virginianus*), Columbian Ground Squirrels (*Urocitellus columbianus*), and various other native wildlife species.

DESCRIPTION: A perennial herb varying from 1–2' (30–60 cm) in height, with palmately compound leaves divided into several narrow leaflets, each covered with silky hairs. The flowers, usually blue to purple, are arranged in dense, terminal racemes.

FLOWERING SEASON: June–August.

HABITAT: Grows in dry to moist valleys and mountains up to 7000' (2150 m) elevation.

RANGE: Occurs from British Columbia east to Saskatchewan and south to northern New Mexico and Arizona.

CAUTION: Plants in the genus *Lupinus*, especially the seeds, can be toxic to humans and animals if ingested.

SIMILAR SPECIES: See other *Lupinus* species.

CONSERVATION: G5

This species is globally secure, but vulnerable in Colorado and Wyoming.

■ NATIVE
■ RARE OR EXTIRPATED
■ INTRODUCED

ALTERNATE NAMES: Nonesuch, Hop Clover

Native to dry Eurasian grasslands, this trailing plant is found throughout North America as an introduced species, growing along sidewalks and in other disturbed places. It is considered an undesirable, invasive weed in many areas.

DESCRIPTION: This is a trailing annual or biennial reaching 6–24" (15–60 cm) in height and 1–2' (30–60 cm) in length. The leaves are ½" (1.25 cm) long and clover-like, alternately divided into three leaflets. The tiny, bright yellow flowers are arranged in dense clusters.

FLOWERING SEASON: April–September.

HABITAT: Grows in various disturbed places, such as along curbs and sidewalks.

RANGE: Native to Eurasia but found throughout North America.

SIMILAR SPECIES: Large Hop Clover (*Trifolium aureum*) is a similar species, also introduced in much of North America.

CONSERVATION:
The conservation status of this introduced species is unranked.

■ NATIVE
■ RARE OR EXTIRPATED
■ INTRODUCED

ALTERNATE NAMES: Lucerne

Native to Eurasia, this species is widely cultivated in North America and can be found growing in fields and along roadsides throughout the continent. It is an important forage crop and commonly used to make hay and silage. Although Alfalfa is the name commonly used in North America, in many other parts of the world this species is known as Lucerne.

DESCRIPTION: A perennial growing 18–36" (45–90 cm) in height with alternate clover-like leaves, divided into three leaflets, each toothed at the tip. The small blue-violet flowers are arranged in short spikes of crowded clusters.

FLOWERING SEASON: April–October.

HABITAT: Grows in fields, along roadsides, in ditches, and other disturbed places.

RANGE: Planted throughout North America and naturalized in many areas.

CAUTION: Alfalfa can be toxic to some animals and mildly toxic to some humans if ingested.

SIMILAR SPECIES: Not likely to be confused with other species.

CONSERVATION:
The conservation status of this introduced species is unranked.

■ NATIVE
■ RARE OR EXTIRPATED
■ INTRODUCED

ALTERNATE NAMES: Honey Clover

This tall wildflower is native to Eurasia but can now be found in many areas of the world, including much of North America. It is especially common in meadows and riparian habitats. This species has a distinct sweet odor, reminiscent of freshly cut grass.

DESCRIPTION: An annual or biennial growing 2–10' (60–300 cm) in height with alternate leaves divided into three leaflets, each ¾–1¼" (2–3 cm) long with toothed edges. The tiny flowers are white in color and arranged in slender racemes.

FLOWERING SEASON: May–October.

HABITAT: Grows in fields, meadows, along roads, and in various other habitat types.

RANGE: Occurs throughout much of North America, except in the Far North.

SIMILAR SPECIES: Yellow Sweet Clover (*M. officinalis*) is a similar but separate Eurasian species that is widely naturalized in North America; it generally prefers drier habitats and features yellow rather than white flowers. Sour Clover (*M. indica*) is also a similar Old World species with yellow flowers and naturalized in North America.

CONSERVATION:

The conservation status of this introduced species is unranked.

PURPLE LOCOWEED   *Oxytropis lambertii*

■ NATIVE
■ RARE OR EXTIRPATED
■ INTRODUCED

ALTERNATE NAMES: Woolly Locoweed, Lambert's Crazy-weed

This wildflower is native to the grasslands of western North America. It is one of the most dangerously poisonous plants on western ranges, lethally toxic to livestock. Plants in this genus can appear quite similar to those in the genus *Astragalus*; a distinguishing difference is in the leaves, which are usually basal in *Oxytropis* species.

DESCRIPTION: A perennial herb growing 4–16" (10–40 cm) in height. The basal leaves are 3–12" (8–30 cm) long and pinnately compound, with 7–17 leaflets ¼–1½" (6–38 mm) long. The flowers are purplish-pink to violet and arranged in dense racemes atop long stems.

FLOWERING SEASON: June–September.

HABITAT: Grows in dry prairies and plains, on slopes, and at limestone sites.

RANGE: Found from British Columbia, Saskatchewan, and Manitoba south to Texas, New Mexico, and Arizona.

CAUTION: This species is lethally toxic to all kinds of livestock. Most plants in the genus *Oxytropis* are potentially toxic to humans and animals if ingested, causing a condition called locoism. The milk from an animal that has ingested these plants may also be toxic.

SIMILAR SPECIES: White Locoweed (*O. sericea*) and various *Astragalus* species.

CONSERVATION: G5

Although globally secure, this species is vulnerable in Manitoba, Saskatchewan, and Iowa.

- NATIVE
- RARE OR EXTIRPATED
- INTRODUCED

ALTERNATE NAMES: Silvery Scurfpea

This species may be found in a variety of dry, open habitats throughout central North America, where it is native. Parts of this plant have been used by Native American groups as a food source, as well as medication for people and horses.

DESCRIPTION: A perennial growing 1½–3' (45–90 cm) in height with leaves palmately divided into three to five silvery white leaflets, each ⅖–1⅖" (1–3.5 cm) long. The flowers are blue or purple, becoming tan in age.

FLOWERING SEASON: June–August.

HABITAT: Grows in prairies and plains, and along hills.

RANGE: Occurs from Alberta east to Manitoba, south to Missouri, and west to New Mexico.

SIMILAR SPECIES: Sampson's Snakeroot (*Orbexilum pedunculatum*).

CONSERVATION: G5

Although globally secure, this species is critically imperiled in Wisconsin, imperiled in Montana, and vulnerable in Alberta.

KUDZU   *Pueraria montana*

■ NATIVE
■ RARE OR EXTIRPATED
■ INTRODUCED

ALTERNATE NAMES: *Pueraria montana* var. *lobata, Pueraria lobata*

Native to eastern Asia, this flowering vine is introduced to North America. It is highly invasive, aggressively spreading throughout the southern United States and beyond. It grows over other plants, covering trees and even buildings as it spreads and takes over new areas.

DESCRIPTION: This is a high-climbing vine growing up to 60' (18 m) long, with alternate leaves divided into three egg-shaped leaflets, each 4–6" (10–15 cm) long and often lobed. The purplish flowers are arranged in dense, elongated clusters.

FLOWERING SEASON: July–September.

HABITAT: Grows along borders of woods and fields, and in various disturbed places.

RANGE: In North America, found from New York and Massachusetts south to Florida, west to Texas, and north to Nebraska and Illinois.

SIMILAR SPECIES: Not likely to be confused with native species.

CONSERVATION:
The conservation status of this introduced species is unranked.

- NATIVE
- RARE OR EXTIRPATED
- INTRODUCED

ALTERNATE NAMES: Purple Crown-vetch, Axseed, *Coronilla varia*

This member of the pea family is native to Africa, Asia, and Europe. It has been introduced to North America and is now widespread, aggressively establishing itself in many areas of the continent.

DESCRIPTION: This is a low-growing vine 1–2' (30–60 cm) in height with compound leaves 2–4" (5–10 cm) long, pinnately divided into 15–25 leaflets, each ½–¾" (1.5–2 cm) long. The pink-and-white flowers are arranged in rounded clusters.

FLOWERING SEASON: June–August.

HABITAT: Grows in fields, along roadsides, and in other disturbed places.

RANGE: Occurs throughout much of North America, except in the Far North.

CAUTION: This plant is toxic to horses and certain other animals if ingested.

SIMILAR SPECIES: Not likely to be confused with any native species.

CONSERVATION:

The conservation status of this introduced species is unranked.

■ NATIVE
■ RARE OR EXTIRPATED
■ INTRODUCED

ALTERNATE NAMES: Maryland Senna, Southern Wild Senna

This wildflower of open woods and roadsides is native to the southeastern United States. It is a larval host plant for several butterfly species, including the Cloudless Sulphur (*Phoebis sennae*). Parts of this plant have been used to treat a variety of ailments, including cramps, heart problems, and pneumonia.

DESCRIPTION: A perennial growing 3–5' (1–1.5 m) tall with alternate leaves pinnately divided into 4–8 pairs of leaflets, each about 1" (2.5 cm) long. The five-petaled flowers are yellow. The pod segments are rectangular; these are nearly square in Northern Wild Senna (*S. hebecarpa*).

FLOWERING SEASON: July–August.

HABITAT: Grows in open woods and along streambanks and roadsides.

RANGE: Occurs from New York south to Florida, west to Texas, and north to Nebraska, Iowa, and southern Wisconsin.

SIMILAR SPECIES: Northern Wild Senna (*S. hebecarpa*) is a similar but separate species.

CONVERSATION: G5

This species is globally secure, but critically imperiled in Wisconsin, imperiled in Nebraska, and vulnerable in Iowa, Maryland, North Carolina, and Pennsylvania. It is possibly extirpated in Delaware.

■ NATIVE
■ RARE OR EXTIRPATED
■ INTRODUCED

ALTERNATE NAMES: Sidebeak Pencil-flower

This perennial wildflower is native to the southeastern United States, where it may be found growing in a variety of open habitat types from dry woods to fields and thickets. The orange-yellow flowers bloom from late spring into fall. White-tailed Deer (*Odocoileus virginianus*) and other mammals readily eat the leaves.

DESCRIPTION: A perennial growing 4–16" (10–40.6 cm) tall with alternate compound leaves divided into three narrow leaflets. The flowers are small and orange-yellow.

FLOWERING SEASON: May–September.

HABITAT: Grows in open woods, along woodland borders, and on roadsides.

RANGE: Occurs from New York south into Florida, west into Texas, and north to Kansas and Missouri.

SIMILAR SPECIES: Viperina (*Zornia bracteata*) is similar but has compound leaves divided into four narrow leaflets instead of three.

CONSERVATION: **G5**
Although globally secure, this species is imperiled in Pennsylvania and vulnerable in New Jersey. It is presumed to be extirpated in New York.

NATIVE
RARE OR EXTIRPATED
INTRODUCED

ALTERNATE NAMES: Devil's Shoestrings, Hoary-Pea

This wildflower is native to open habitats of eastern North America. The common name Goat's-rue refers to the previous practice of feeding this plant to goats to increase their milk production; it is now known that this plant contains rotenone and is therefore potentially toxic to livestock. Another common name, Devil's Shoestrings, refers to the long, stringy roots of this plant.

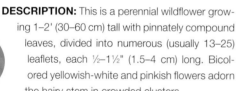

DESCRIPTION: This is a perennial wildflower growing 1–2' (30–60 cm) tall with pinnately compound leaves, divided into numerous (usually 13–25) leaflets, each ½–1½" (1.5–4 cm) long. Bicolored yellowish-white and pinkish flowers adorn the hairy stem in crowded clusters.

FLOWERING SEASON: May–August.

HABITAT: Grows in open woods, roadsides, and clearings.

RANGE: Occurs from Ontario south to Florida, west to Texas, and north to Nebraska and Minnesota.

CAUTION: This plant is poisonous and should never be ingested.

SIMILAR SPECIES: Spiked Hoary Pea (*T. spicata*) and Scurfy Hoary Pea (*T. chrysophylla*).

CONSERVATION: G5

Although globally secure, this species is critically imperiled in Ontario, Nebraska, New Hampshire, and Rhode Island, and vulnerable in Iowa and Minnesota.

■ NATIVE
■ RARE OR EXTIRPATED
■ INTRODUCED

ALTERNATE NAMES: Montane Golden-banner, False Lupine, Buckbean

Native to the western United States, this tall wildflower is sometimes planted in gardens as an ornamental. It produces clusters of bright yellow, lupine-like flowers, which are followed by velvety, brown seed pods. This plant is generally avoided by livestock and suspected to be poisonous.

DESCRIPTION: A wildflower growing 2–4' (60–120 cm) in height with compound leaves divided into three broadly lanceolate leaflets, each 2–4" (5–10 cm) long. The yellow flowers are arranged in long racemes.

FLOWERING SEASON: May–August.

HABITAT: Grows in open woods, meadows, along streambanks, and in openings in coniferous forests.

RANGE: Found from British Columbia south to northern California, and east to Montana and Colorado.

CAUTION: This plant is suspected to be poisonous and should never be ingested.

SIMILAR SPECIES: Prairie Golden-banner (*T. rhombifolia*) and Allegheny Mountain Golden-banner (*T. mollis*) are similar related species. Plants in this genus resemble those in *Lupinus*, but can be separated by the leaves: *Thermopsis* species have only three leaflets on each leaf, whereas *Lupinus* species have more than three.

CONSERVATION: G4

This species appears to be fairly secure globally but is imperiled in Wyoming.

■ NATIVE
■ RARE OR EXTIRPATED
■ INTRODUCED

Native to Eurasia, this wildflower has been introduced to North America and may now be found in a variety of disturbed areas in much of the East and in parts of the West. The common name is derived from the look and feel of the furry flower head.

DESCRIPTION: This wildflower grows 6–18" (15–45 cm) in height with silky-hairy stems, and with leaves divided into three narrow leaflets, each ½–¾" (1.5–2 cm) long and toothed at the tips. The tiny, light pink to whitish flowers are arranged in fuzzy, feathery, cylindrical heads.

FLOWERING SEASON: May–October.

HABITAT: Favors dry, open areas and road-sides.

RANGE: Introduced from Ontario east to Newfoundland, south to Florida, west to Texas, and north to North Dakota.

SIMILAR SPECIES: Crimson Clover (*T. incarnatum*) and Large Hop Clover (*T. aureum*).

CONSERVATION:
The conservation status of this introduced species is unranked.

■ NATIVE
■ RARE OR EXTIRPATED
■ INTRODUCED

ALTERNATE NAMES: Low Hop Clover, Field Clover

This annual wildflower is native to grasslands of Europe and western Asia, introduced to North America. Its yellow flower heads are reminiscent of hop flowers (genus *Humulus*), hence the common name. This species is used to feed livestock and has become naturalized throughout much of the North American continent. Look for it in fields, pastures, and along roadsides.

DESCRIPTION: This annual plant grows 3–12" (10–30 cm) tall with leaves alternately divided into three leaflets. The small, yel-

low flowers are arranged in tight clusters, ½–¾" (12–18 mm) long, and become brown in age.

FLOWERING SEASON: May–September.

HABITAT: Grows in fields and on roadsides.

RANGE: This plant is naturalized throughout much of North America.

SIMILAR SPECIES: Large Hop Trefoil (*T. aureum*).

CONSERVATION:
The conservation status of this introduced species is unranked.

RED CLOVER *Trifolium pratense*

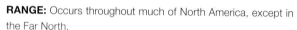

■ NATIVE
■ RARE OR EXTIRPATED
■ INTRODUCED

Native to Europe, western Asia, and parts of Africa, this perennial herb has been planted and naturalized in many other parts of the world, including North America. It grows in a variety of places from lawns to roadsides and is frequently pollinated by bumblebees (*Bombus* spp.).

DESCRIPTION: This is a perennial growing 1–3' (30–90 cm) in height. The compound leaves are divided into three broad leaflets, each ¾–2½" (2–6.5 cm) long, with the upper two leaves usually placed close to the flower head. The deep pink flowers are arranged in round or egg-shaped clusters, ½–¾" (12–18 mm).

FLOWERING SEASON: June–August.

HABITAT: Grows in old fields, lawns, and road-sides.

RANGE: Occurs throughout much of North America, except in the Far North.

CAUTION: This plant can be toxic if eaten and should never be ingested.

SIMILAR SPECIES: White Clover (*T. repens*).

CONSERVATION:

The conservation status of this introduced species is unranked.

■ NATIVE
■ RARE OR EXTIRPATED
■ INTRODUCED

ALTERNATE NAMES: Dutch Clover, Ladino

This Eurasian native is the common clover of lawn seed mixes, also widely cultivated as a livestock forage crop. It can be found growing on lawns and other grassy areas throughout North America. The leaves and flowers of this plant, high in protein, have been used by some people as a food source.

DESCRIPTION: A perennial herb growing 4–24" (10–60 cm) in height with compound leaves divided into three broad leaflets,

each ¾–2½" (1.5–2 cm) long. The tiny white or pale pink flowers are arranged in round heads, ½" (12 mm) wide, above the leaves.

FLOWERING SEASON: April–September.

HABITAT: Grows on lawns, along roadsides, and in fields.

RANGE: Occurs throughout much of North America, except in the Far North.

SIMILAR SPECIES: Alsike Clover (*T. hybridum*) and Red Clover (*T. pratense*).

CONSERVATION:
The conservation status of this non-native species is unranked.

COW CLOVER *Trifolium wormskioldii*

ALTERNATE NAMES: Wormskjold's Clover

From beaches to mountains, this clover is found in a variety of places throughout its native range in western North America. The relatively large, pinkish flower heads help to identify it. Some Native American groups have used the roots and other parts of this plant as a nutritious food source.

DESCRIPTION: This is a perennial growing 4–31" (10–80 cm) in height with compound leaves divided into three rather narrow leaflets, each ½–1¼" (1.5–3 cm) long with tiny teeth on the edges. The tiny individual flowers are deep pink to reddish-lavender in color, clustered in flower heads 3/4–11/4" (20–30 mm) wide.

FLOWERING SEASON: May–September.

HABITAT: Grows in a variety of habitats from coastal dunes to mountain meadows.

RANGE: Occurs from British Columbia south along the Pacific Coast to northern Mexico and east to New Mexico, Colorado, western Wyoming, and Idaho.

SIMILAR SPECIES: White-tip Clover (*T. variegatum*).

CONSERVATION: G5

Although globally secure, this species is critically imperiled in Alaska and Utah.

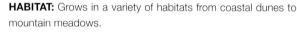

- ■ NATIVE
- ■ RARE OR EXTIRPATED
- ■ INTRODUCED

ALTERNATE NAMES: American Purple Vetch

This slender climbing plant is widespread across North America, where it is native. It commonly clings to other vegetation using tendrils at the tips of the leaves. American Vetch is often used to reclaim land after wildfires.

DESCRIPTION: A climbing perennial growing 2–4' (60–120 cm) in height with compound leaves pinnately divided into 8–12 leaflets, each ½–1½" (1.5–4 cm) long. The deep pinkish-purple flowers, ½–¾" (12–18 mm) long, becoming bluish in age, are arranged in loose racemes.

FLOWERING SEASON: May–July.

HABITAT: Grows in thickets, open woods and meadows, and along fencerows.

RANGE: Occurs throughout much of North America except the southeastern United States.

CAUTION: The seeds of this and some other *Vicia* species contain compounds that produce toxic levels of cyanide when ingested.

SIMILAR SPECIES: Wood Vetch (*V. caroliniana*).

CONSERVATION: G5

Globally secure, but critically imperiled in Missouri, Ohio, and Virginia; imperiled in Alaska; and vulnerable in New York. It is possibly extirpated in Maryland.

■ NATIVE
■ RARE OR EXTIRPATED
■ INTRODUCED

ALTERNATE NAMES: Purple-white Tufted Vetch, Tinegrass

This vetch is native to Europe and Asia, introduced to North America and other continents. It is now commonly found growing in old fields and roadside ditches throughout much of the United States and Canada. As the common name suggests, this species is widely used as a forage crop for cattle.

DESCRIPTION: A climbing plant growing 4–7' (1.2–2.1m) with compound leaves pinnately divided into 19–29 leaflets, each ¾–1½" (2–4 cm) long. The reddish-lavender to bluish-purple pea flowers, ½" (12 mm) long, are closely clustered in racemes.

FLOWERING SEASON: May–July.

HABITAT: Grows in fields and along roadsides.

RANGE: Occurs from Alaska and British Columbia south and east across Canada to Newfoundland, south to Georgia and Alabama, northwest to South Dakota and Minnesota, and west and southwest to Washington and California.

CAUTION: The seeds of some *Vicia* species contain cyanide compounds and are toxic to humans and animals if eaten.

SIMILAR SPECIES: Hairy Vetch (*V. villosa*).

CONSERVATION:
The conservation status of this introduced species is unranked.

■ NATIVE
■ RARE OR EXTIRPATED
■ INTRODUCED

ALTERNATE NAMES: Garden Vetch, Common Vetch

This Eurasian native is widely cultivated and naturalized across the United States. Like many other *Vicia* species, this plant is commonly used as a forage crop for livestock. It is often found growing as a weed along roadsides and fencerows.

DESCRIPTION: A climbing annual vine with many branches and variable height. The leaves are pinnately compound, with four to eight pairs of narrow leaflets, ¾–1¼" (18–30 mm) long. The reddish-purple flowers are arranged in clusters of one to three in the leaf axils.

FLOWERING SEASON: April–July.

HABITAT: Grows along roadsides, woodland edges, fencerows, and in fields and clearings.

RANGE: Occurs across the United States, north to Alaska.

CAUTION: The seeds of some *Vicia* species contain cyanide compounds and are toxic to humans and animals if eaten.

SIMILAR SPECIES: See other *Vicia* species.

CONSERVATION:
The conservation status of this introduced species is unranked.

■ NATIVE
■ RARE OR EXTIRPATED
■ INTRODUCED

This climbing vine is native to the southeastern United States, where it grows in wet woodlands and along streams and rivers. It is widely introduced and may be found in areas outside its native range. This species is similar to but less aggressive than Chinese Wisteria (*W. sinensis*), an invasive species also found in parts of the eastern United States.

DESCRIPTION: A deciduous vine growing 25–30' (8–10 m), with pinnately compound leaves divided into 9–15 shiny, dark green leaflets. The fragrant flowers are blue-purple and ⅔–¾" (15–20 mm) long.

FLOWERING SEASON: April–May.

HABITAT: Grows in moist woods and along riverbanks.

RANGE: Occurs from Maryland and Virginia south to northern Florida and west to Texas; widely introduced elsewhere.

CAUTION: Parts of this plant are toxic to humans if ingested.

SIMILAR SPECIES: Chinese Wisteria (*W. sinensis*) is a related plant native to China that has been grown as a fragrant ornamental but which has become invasive in the eastern United States. Kentucky Wisteria (*W. macrostachya*) is a native relative found in the southeastern United States. It has been classified as a variety of *W. frutescens* and as a separate species.

CONSERVATION: G5

Globally secure, but critically imperiled in Michigan and Oklahoma, and vulnerable in Indiana.

■ NATIVE
■ RARE OR EXTIRPATED
■ INTRODUCED

White Milkwort is native to North America, where it grows at rocky or sandy sites, from prairies to roadsides. The genus name *Polygala* comes from a Greek word meaning "much milk," a reference to the idea that these plants stimulated the flow of milk in cattle.

DESCRIPTION: This perennial averages 8–14" (20–35 cm) in height with hairless stems and very narrow leaves. The tiny white flowers are arranged in slender, cone-shaped clusters, 1–3" (2.5–7.5 cm) long.

FLOWERING SEASON: March–October.

HABITAT: Grows in prairies, plains, and along roadsides.

RANGE: Occurs from eastern Montana and North Dakota south to Texas, and west to central Arizona; also found in Mexico.

SIMILAR SPECIES: Seneca Snakeroot (*P. senega*).

CONSERVATION: G5

Although globally secure, this species is imperiled in Wyoming and vulnerable in Saskatchewan and Montana.

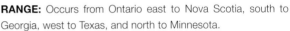

■ NATIVE
■ RARE OR EXTIRPATED
■ INTRODUCED

ALTERNATE NAMES: Field Milkwort

This milkwort is native to eastern North America, where it prefers wet soils in prairies, fields, and other open areas. Despite the name, the flowers of this species are more pink than purple. The roots exude a distinct wintergreen odor when crushed.

DESCRIPTION: This annual averages 5–15" (13–38 cm) in height with narrow, alternate leaves ½–1½" (1.5–4 cm) long. The tiny, pinkish-purple flowers are arranged in a dense, cylindrical cluster about 1" (2.5 cm) long.

FLOWERING SEASON: June–October.

HABITAT: Grows in wet fields, meadows, and open woods.

RANGE: Occurs from Ontario east to Nova Scotia, south to Georgia, west to Texas, and north to Minnesota.

SIMILAR SPECIES: Sandfield Milkwort (*P. polygama*), Seneca Snakeroot (*P. senega*), and Cross-leaf Milkwort (*P. cruciata*).

CONSERVATION: G5
This species is globally secure but critically imperiled in Prince Edward Island and vulnerable in New Brunswick, Nova Scotia, Ontario, Quebec, North Carolina, Rhode Island, and Vermont. It is possibly extirpated in Georgia.

■ NATIVE
■ RARE OR EXTIRPATED
■ INTRODUCED

This wildflower is native to eastern North America and may be found in prairies and other open habitats, blooming from summer into fall. Milkworts do not have milky juice; they are named after the folk belief that they increased milk production in grazing animals.

DESCRIPTION: This is a slender-stemmed annual wildflower with narrow, mostly whorled leaves up to 1" (2.5 cm) long. The tiny flowers are white or greenish-white, clustered in narrow, conical spikes up to ¾" (18 mm) long.

FLOWERING SEASON: July–October.

HABITAT: Grows in prairies and open woods.

RANGE: Occurs in much of the East from Saskatchewan east to Quebec, south to Florida, and west to Texas.

SIMILAR SPECIES: White Milkwort (*P. alba*) and Seneca Snakeroot (*P. senega*).

CONSERVATION:
Globally secure, but critically imperiled in New Brunswick, Rhode Island, Utah, and Wyoming. Imperiled in Massachusetts and Vermont in the U.S. and Manitoba and Saskatchewan in Canada. Vulnerable in Ontario and North Carolina.

ALTERNATE NAMES: Gaywings, Flowering-wintergreen, *Polygala paucifolia*

This species is considered a "false milkwort" due to its resemblance to members of genus *Polygala*. It is a low-growing plant, barely reaching half a foot in height, with an almost orchid-like appearance. It grows in moist wooded habitats in northeastern North America.

DESCRIPTION: This perennial grows 3–6" (7.5–15 cm) tall with ¾"–1½" (2–4 cm) long, oval leaves clustered at the top of the stem. The plant features one to four pink or purple flowers ¾" (18 mm) wide with noticeably large wings.

FLOWERING SEASON: May–June.

HABITAT: Grows in moist woodlands.

RANGE: Occurs from Manitoba east to New Brunswick, south through New England and the Appalachians, and in Minnesota, Wisconsin, and Michigan.

SIMILAR SPECIES: See true milkworts (*Polygala* spp.).

CONSERVATION: **G5**

Although globally secure, this species is critically imperiled in Indiana, Kentucky, and Ohio; imperiled in Alberta, New Brunswick, North Carolina, and South Carolina; and vulnerable in Saskatchewan and Georgia. It is possibly extirpated in Newfoundland and presumed extirpated in Delaware.

■ NATIVE
■ RARE OR EXTIRPATED
■ INTRODUCED

ALTERNATE NAMES: Tall Hairy Agrimony, Hooked Agrimony, Tall Hairy Grooveburr

This flowering plant, native to North America, is commonly known by many names. The species name *gryposepala* is derived from Greek words meaning "hooked sepals." This wildflower shares its native range with several other very similar *Agrimonia* species, including Roadside Agrimony (*A. striata*) and Southern Agrimony (*A. parviflora*).

DESCRIPTION: This is a perennial ranging from 1–6' (30–180 cm) in height with five to nine leaves pinnately divided into bright green, coarsely toothed leaflets, each 2–4" (5–10 cm) long. The small, yellow, five-petaled flowers 6–8 mm wide are arranged in long, narrow clusters.

FLOWERING SEASON: July–August.

HABITAT: Grows in thickets, marshy areas, and open woods.

RANGE: Occurs from Ontario east to Nova Scotia, south to South Carolina, and northwest to Nebraska and North Dakota; also in parts of the West.

SIMILAR SPECIES: Roadside Agrimony (*A. striata*), Southern Agrimony (*A. parviflora*), and other *Agrimonia* species.

CONSERVATION: G5

Globally secure, but critically imperiled in Manitoba, Kansas, Kentucky, and Oregon; imperiled in Prince Edward Island, Georgia, and Wyoming; and vulnerable in Nova Scotia, Delaware, Montana, North Carolina, and North Dakota.

■ NATIVE
■ RARE OR EXTIRPATED
■ INTRODUCED

ALTERNATE NAMES: Common Silverweed, *Potentilla anserina*

With a native range that covers much of the Northern Hemisphere, this plant may be found growing as far north as the edge of the Arctic. It favors shorelines, marshy areas, and wet meadows. Pacific Silverweed (*A. egedii*), treated as a subspecies of *A. anserina* by some botanists, is a similar plant that grows in coastal saltmarshes.

DESCRIPTION: This is a hairy, prostrate perennial plant with runners averaging 1–3' (30–90 cm) long. The compound leaves are up to 1' (30 cm) long and pinnately divided into many toothed leaflets, with smaller leaflets interspersed between larger ones. The plant features one solitary yellow flower per leafless stalk.

FLOWERING SEASON: June–August.

HABITAT: Grows along shorelines, in wet meadows and marshy areas, and on roadsides.

RANGE: Occurs throughout much of Canada, the northernmost United States, and in much of the West.

SIMILAR SPECIES: Pacific Silverweed (*A. egedii*) and Silvery Cinquefoil (*P. argentea*).

CONVERSATION: **G5**

Globally secure but locally rare or threatened at the southern edge of its range in the northern United States. It is critically imperiled in Iowa, Nebraska, and New Hampshire; imperiled in Indiana; and vulnerable in Ohio, Pennsylvania, and Wyoming.

■ NATIVE
■ RARE OR EXTIRPATED
■ INTRODUCED

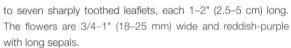

ALTERNATE NAMES: Purple Marshlocks, Swamp Cinquefoil, *Potentilla palustris*

The only cinquefoil with purple flowers, this is a wildflower of wet habitats, found in swamps, marshy areas, and along lakeshores, sometimes partially submerged in the water. It is distributed throughout northern North America, Europe, and Asia.

DESCRIPTION: This is a shrub-like plant ranging 6–20" (15–50 cm) in height, with long-stalked leaves pinnately divided into five

to seven sharply toothed leaflets, each 1–2" (2.5–5 cm) long. The flowers are 3/4–1" (18–25 mm) wide and reddish-purple with long sepals.

FLOWERING SEASON: June–August.

HABITAT: Grows in swamps and bogs.

RANGE: Occurs throughout Canada and south to California, Iowa, Ohio, and New Jersey.

SIMILAR SPECIES: See other cinquefoils.

CONSERVATION: G5

Although globally secure, this species is critically imperiled in Utah and Wyoming, imperiled in North Dakota and Ohio, and vulnerable in Colorado and Iowa.

■ NATIVE
■ RARE OR EXTIRPATED
■ INTRODUCED

ALTERNATE NAMES: White Cinquefoil, Prairie Cinquefoil, *Potentilla arguta*

This rather tall wildflower is native to meadows, prairies, and other open areas of Canada and the northern United States. The white flowers are visited by small bees and hoverflies, and White-tailed Deer (*Odocoileus virginianus*) and other mammals feed on the foliage.

DESCRIPTION: This perennial averages 12–40" (30–100 cm) in height, with mostly basal leaves up to 15" (40 cm) long, divided into 3–11 leaflets. The white or cream-colored, five-petaled flowers are arranged in a tight cluster atop the stem.

FLOWERING SEASON: June–September.

HABITAT: Grows in open woods, meadows, prairies, and along roadsides.

RANGE: Occurs from Alaska to New Brunswick, south to Virginia, Indiana, Arkansas, Oklahoma, and Arizona.

SIMILAR SPECIES: Sticky Cinquefoil (*D. glandulosa*) is a similar species that occurs in the West.

CONSERVATION: G5

Globally secure, but critically imperiled in Ohio, Oklahoma, and Virginia; imperiled in Connecticut and New Jersey; and vulnerable in New Brunswick and Vermont. It is possibly extirpated in Maryland and presumed extirpated in West Virginia.

■ NATIVE
■ RARE OR EXTIRPATED
■ INTRODUCED

ALTERNATE NAMES: Alpine Strawberry, Wild Strawberry

This low-growing, creeping plant is found in a variety of habitats, from open forests to meadows and roadsides, across the Northern Hemisphere. It produces sweet red fruits and is often planted in gardens as a ground cover.

DESCRIPTION: This is a perennial growing up to 9" (22 cm) in height with toothed leaves divided into three oval leaflets. The five-petaled flowers are white, 1/3–1/2" (10–12 mm) wide, with many yellow stamens.

FLOWERING SEASON: March–July.

HABITAT: Grows in open forests, along streams, in clearings, and along woodland edges.

RANGE: Widespread throughout much of North America.

SIMILAR SPECIES: Wild Strawberry (*F. virginiana*).

CONSERVATION:  G5
Although globally secure, this species is critically imperiled in Missouri and imperiled in Virginia.

■ NATIVE
■ RARE OR EXTIRPATED
■ INTRODUCED

ALTERNATE NAMES: Virginia Strawberry, Mountain Strawberry

This native strawberry plant grows in meadows and open woodlands across southern Canada and the United States. The edible portion of the strawberry is in fact the central part of the flower, which enlarges with maturity and is covered with the fruit.

DESCRIPTION: This creeping perennial averages 3–6" (7.5–15 cm) in height with basal leaves divided into three toothed leaflets, each 1–1½" (2.5–3.8 cm) long. The five-petaled flowers are white ½–¾" (12–18 mm) wide, with numerous stamens and pistils.

FLOWERING SEASON: April–June.

HABITAT: Grows in open fields and along woodland edges.

RANGE: Occurs throughout North America, except the Arctic islands and Greenland.

SIMILAR SPECIES: Woodland Strawberry (*F. vesca*).

CONSERVATION: G5

Globally secure, but critically imperiled in Louisiana and vulnerable in Labrador.

■ NATIVE
■ RARE OR EXTIRPATED
■ INTRODUCED

This leafy woodland plant is native to eastern North America. Its small starlike flowers attract a variety of butterflies and other pollinators. This species often naturally hybridizes with the introduced Wood Avens (*G. urbanum*). The resulting hybrid is commonly known as Catling's Avens (*Geum* x *catlingii*).

DESCRIPTION: A perennial growing up to 24" (60 cm) in height with leaves pinnately divided into three to five leaflets; the upper- most leaves may be toothed or lobed. The five-petaled flowers are white, ⅓–⅔" (10–15 mm) wide, and usually solitary.

FLOWERING SEASON: April–June.

HABITAT: Grows in woods, along woodland edges, and in swamps.

RANGE: Occurs from eastern Canada to Minnesota, south to northern Georgia and Texas.

SIMILAR SPECIES: Rough Avens (*G. lacini- atum*) and Virginia Avens (*G. virginianum*).

CONSERVATION: G5
Globally secure, but imperiled in Prince Ed- ward Island and Wyoming.

NATIVE
RARE OR EXTIRPATED
INTRODUCED

ALTERNATE NAMES: Appalachian Barren Strawberry, *Waldsteinia fragaroides*

This evergreen mat-forming perennial is native to eastern North America, typically found in woods and thickets. The fruit of this strawberry-like plant is neither fleshy nor edible. Some Native American groups have used parts of the plant to treat snakebites.

DESCRIPTION: This is a strawberry-like plant growing 3–8" (7.5–20 cm) in height. The long-stalked leaves are divided into three wedge-shaped leaflets, each 1–2" (2.5–5 cm) long. The showy flowers are yellow in and ⅓–⅔" (10–15 mm) wide.

FLOWERING SEASON: April–June.

HABITAT: Grows in woods, thickets, and clearings.

RANGE: Occurs from Ontario east to New Brunswick, south to Georgia, west to Tennessee and Arkansas, and north to Illinois, Wisconsin, and Minnesota.

SIMILAR SPECIES: Lobed Strawberry (*Waldsteinia lobata*).

CONSERVATION: G5

Globally secure, but critically imperiled in, Arkansas, Connecticut, and Maine in the U.S. and New Brunswick in Canada. It is vulnerable in Quebec and in Indiana, Kentucky, Massachusetts, and Minnesota. It is possibly extirpated in Illinois.

■ NATIVE
■ RARE OR EXTIRPATED
■ INTRODUCED

ALTERNATE NAMES: Old-man's-whiskers, *Erythrocoma triflora*

This perennial is native to North America and may be found growing in prairies and other open habitats. It is one of the earliest wildflowers to bloom in much of its range. Some Native American groups have made a tea from the roots of this plant.

DESCRIPTION: This is a hairy perennial plant ranging 6–16" (15–40 cm) in height with pinnately divided leaves with 7–17 toothed or lobed leaflets. The flowers are brownish-purple or pinkish, ½–1" (12–25 mm) long, and often arranged in groups of three.

FLOWERING SEASON: April–August.

HABITAT: Grows in open woods and prairies.

RANGE: Occurs from Alberta east to Ontario and south into New York, Michigan, Illinois, Iowa, South Dakota, and through much of the western United States.

SIMILAR SPECIES: Water Avens (*G. rivale*).

CONVERSATION: G5

Although globally secure, this species is imperiled in Michigan and New York in the U.S. and Yukon in Canada. It is vulnerable in Iowa.

■ NATIVE
■ RARE OR EXTIRPATED
■ INTRODUCED

ALTERNATE NAMES: Alpine Mousetail

This tuft-forming member of the rose family grows in rocky mountainous areas of the western United States. Ivesias are named for Joseph Christmas Ives, an American soldier and explorer of the 19th century. This particular species is named in honor of Alexander Gordon, an English horticulturist of the same period.

DESCRIPTION: This perennial grows 2–8" (5–20 cm) tall. The 5-mm-wide yellow, five-petaled flowers are arranged in dense clusters, and the basal leaves are divided into numerous leaflets, each with three to five rounded segments.

FLOWERING SEASON: June–August.

HABITAT: Grows in alpine meadows and rocky areas.

RANGE: Occurs in scattered areas from Washington east into Montana, south to northern Colorado, and west to central California.

SIMILAR SPECIES: Clubmoss Ivesia (*I. lycopodioides*) and Alpine Avens (*Geum rossii*).

CONSERVATION: G4

This species is apparently secure globally, but vulnerable in Wyoming.

■ NATIVE
■ RARE OR EXTIRPATED
■ INTRODUCED

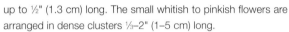

ALTERNATE NAMES: Rocky Mountain Rockmat, Rock Spiraea, Dwarf Spiraea, Petrophytum, Tufted Rockmat

This is a rather unusual plant that grows almost exclusively from the crevices of barren rock. It is mat-forming and shrub-like, with grayish leaves that resemble stone. Rockmat is native to mountainous areas of the western United States.

DESCRIPTION: A mat-forming plant growing to about 3" (7.5 cm) thick and 3' (90 cm) wide, with crowded, narrow leaves up to ½" (1.3 cm) long. The small whitish to pinkish flowers are arranged in dense clusters ⅓–2" (1–5 cm) long.

FLOWERING SEASON: June–August.

HABITAT: Grows from rocky crevices and ledges.

RANGE: Occurs from southeastern Washington south to southern California, and east into Montana, South Dakota, Colorado, New Mexico, and western Texas.

SIMILAR SPECIES: Chelan Rockmat (*P. cinerascens*) and Olympic Mountain Rockmat (*P. hendersonii*).

CONSERVATION: **G5**

This species is globally secure, but vulnerable in Wyoming.

■ NATIVE
■ RARE OR EXTIRPATED
■ INTRODUCED

ALTERNATE NAMES: Canadian Cinquefoil, Running Five-fingers

This small plant with toothed, wedge-shaped leaves and solitary flowers grows in alpine regions of the eastern United States. The stems may be tinged reddish. Some Native American groups used this plant for medicinal and religious purposes.

DESCRIPTION: This spreading plant grows 2–6" (5–15 cm) tall and features single yellow, five-petaled flowers about ½" (12 mm) wide on a long stalk. The leaves are palmately divided into five leaflets, each toothed along the upper half and growing up to 1½" (4 cm) long.

FLOWERING SEASON: March–June.

HABITAT: Widespread in fields and woods; also grows in disturbed sites such as roadsides.

RANGE: Occurs from Ontario east to Nova Scotia, south to Georgia, and west to Arkansas, Missouri, and Ohio.

SIMILAR SPECIES: Common Cinquefoil (*P. simplex*) is a similar but more widespread species. Dwarf Cinquefoil can be identified by its smaller size and more wedge-shaped leaves.

CONSERVATION: **G5**

This species is globally secure, but critically imperiled in New Brunswick, and imperiled in Nova Scotia and Ontario.

■ NATIVE
■ RARE OR EXTIRPATED
■ INTRODUCED

ALTERNATE NAMES: Graceful Cinquefoil

This attractive wildflower, native to meadows and open woods of western North America, is variable in appearance. It is a larval host plant for the Two-banded Checkered Skipper (*Pyrgus ruralis*), a butterfly that ranges from British Columbia to California.

DESCRIPTION: This is a perennial herb ranging 8–40" (20–100 cm) in height. The bright yellow flowers 2/3–3/4" (16–20 mm) wide are situated at the end of branching stems, and the leaves are palmately divided into five to nine toothed leaflets, each 1–3" (2.5–7.5 cm) long.

FLOWERING SEASON: June–August.

HABITAT: Grows in dry meadows and open woods.

RANGE: Occurs in western North America from Alaska south to California, Arizona, and New Mexico.

SIMILAR SPECIES: Mountain-meadow Cinquefoil (*P. diversifolia*).

CONSERVATION: **G5**
Although globally secure, this species is vulnerable in Yukon.

■ NATIVE
■ RARE OR EXTIRPATED
■ INTRODUCED

ALTERNATE NAMES: False Strawberry, Mock Strawberry, Backyard Strawberry, *Duchesnea indica*

This flowering strawberry-like plant is native to eastern and southern Asia. It is introduced to North America, where it has naturalized and is spreading as an invasive weed, growing in backyard lawns and various disturbed places. The species is commonly known by many names, including Mock Strawberry and False Strawberry. The fruit, although tasteless, is not toxic.

DESCRIPTION: This is a trailing plant with flowering stalks 1–3" (2.5–7.5 cm) tall. The solitary yellow flowers, ¾" (18 mm) wide, each have five petals, and the leaves are divided into three leaflets. The fruits are white or red.

FLOWERING SEASON: April–June.

HABITAT: Grows in old fields, lawns, and disturbed areas.

RANGE: Occurs throughout much of North America.

SIMILAR SPECIES: Common Cinquefoil (*Potentilla simplex*) is similar but has five leaflets.

CONSERVATION:
The conservation status of this introduced species is unranked.

■ NATIVE
■ RARE OR EXTIRPATED
■ INTRODUCED

ALTERNATE NAMES: Sulphur Cinquefoil

This perennial herb is native to Eurasia but grows throughout North America as an introduced species. It can be found in a wide variety of habitats, particularly weedy fields and disturbed places. This species has been used medicinally in some parts of the world.

DESCRIPTION: This hairy plant averages 1–2' (30–60 cm) in height. The pale yellow, usually five-petaled flowers are about ¾" (18 mm) wide and arranged in sparse clusters, and the com-

pound leaves are divided into five to seven blunt-tipped and toothed leaflets, each 1–3" (2.5–7.5 cm) long.

FLOWERING SEASON: May–August.

HABITAT: Grows along roadsides and in dry fields and disturbed places.

RANGE: Introduced and widespread throughout North America.

SIMILAR SPECIES: The similar Norwegian Cinquefoil (*P. norvegica*) sports leaves with only three leaflets.

CONSERVATION:
The conservation status of this introduced species is unranked.

■ NATIVE
■ RARE OR EXTIRPATED
■ INTRODUCED

ALTERNATE NAMES: Oldfield Five-fingers

This is a familiar plant of woodlands and fields of eastern North America. It is quite similar to Dwarf Cinquefoil (*P. canadensis*), another native species that co-occurs with *P. simplex* in much of its range. Common Cinquefoil is a taller plant with narrower leaves. The flowers of this species attract a variety of small bees and flies, including mason bees.

DESCRIPTION: This plant averages 8–12" (20–30 cm) in height, with single yellow flowers about ½" (12 mm) wide on long stalks. The leaves are palmately divided into five leaflets, each toothed along most of the edge, and growing to 3" (8 cm) long and ¾" (2 cm) across.

FLOWERING SEASON: April–June.

HABITAT: Grows in open woods, prairies, and along roadsides.

RANGE: Occurs from Newfoundland and Quebec south to Georgia, west to Texas, and north to Minnesota.

SIMILAR SPECIES: Dwarf Cinquefoil (*P. canadensis*) is a shorter plant with wider, wedge-shaped leaves.

CONSERVATION: G5

Globally secure, but critically imperiled in Nebraska and vulnerable in Quebec.

■ NATIVE
■ RARE OR EXTIRPATED
■ INTRODUCED

ALTERNATE NAMES: Dewberry

This widespread plant is native to wet habitats of Canada and the northernmost United States. It serves as an important food source for many wildlife species, including American Black Bear (*Ursus americanus*), Ruffed Grouse (*Bonasa umbellus*), and various small rodents.

DESCRIPTION: This thornless shrub ranges 4–12" (10–30 cm) in height. The white to pinkish, five-petaled flowers are ⅓–½" (8–12 mm) wide and loosely clustered at the ends of the branches, and the leaves are alternately divided into three toothed, diamond-shaped leaflets. The fruits are dark red.

FLOWERING SEASON: May–June.

HABITAT: Grows in bogs, swamps, and other wet places.

RANGE: Occurs throughout much of Canada and the northern United States, south to Colorado and West Virginia.

SIMILAR SPECIES: Nagoonberry (*R. arcticus*).

CONCONSERVATION: 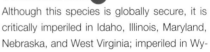 **G5**

Although this species is globally secure, it is critically imperiled in Idaho, Illinois, Maryland, Nebraska, and West Virginia; imperiled in Wyoming; and vulnerable in Iowa.

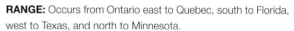

NATIVE
RARE OR EXTIRPATED
INTRODUCED

ALTERNATE NAMES: Small-spike False Nettle, False Nettle

Found in wet areas throughout eastern North America, this plant is an unusual member of the nettle family in that it lacks stinging hairs. The other nettle species with this distinction, Clearweed (*Pilea pumila*), has translucent stems.

DESCRIPTION: This flowering herb grows 1½–3' (45–90 cm) in height, with 2–4" (5–10 cm) long, elongated clusters of greenish flowers. The leaves are long-stalked, opposite, and coarsely toothed, generally 1–3" (2.5–7.5 cm) long.

FLOWERING SEASON: July–October.

HABITAT: Grows in bogs and other wet places.

RANGE: Occurs from Ontario east to Quebec, south to Florida, west to Texas, and north to Minnesota.

SIMILAR SPECIES: Clearweed (*Pilea pumila*) is similar but has translucent stems and much shorter influorescenses.

CONSERVATION: G5

Globally secure, but critically imperiled in Nova Scotia and Arizona, and vulnerable in New Brunswick and Quebec.

■ NATIVE
■ RARE OR EXTIRPATED
■ INTRODUCED

ALTERNATE NAMES: Canada Nettle

Native to open woodlands of eastern and central North America, this plant is known to cause severe pain to those who touch it. Tiny stinging hairs irritate the skin, sometimes resulting in blisters that last several days.

DESCRIPTION: This herbaceous plant averages 1½–4' (45–120 cm) in height. The small greenish flowers are arranged in feathery clusters to 4" (10 cm) long, and the alternate leaves are thin, coarsely toothed, and range from 2½–8" (6.5–20 cm) in length.

FLOWERING SEASON: July–September.

HABITAT: Grows in low woods and along streambanks.

RANGE: Found from Saskatchewan east to Nova Scotia, south to Florida, west to Mississippi and Oklahoma, and north to North Dakota.

CAUTION: This plant's tiny stinging hairs can cause a painful skin irritation.

SIMILAR SPECIES: The similar Stinging Nettle (*Urtica dioica*) is generally a taller plant with opposite leaves and flowers arranged in tighter, more slender clusters.

CONSERVATION: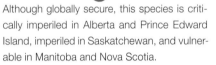
Although globally secure, this species is critically imperiled in Alberta and Prince Edward Island, imperiled in Saskatchewan, and vulnerable in Manitoba and Nova Scotia.

■ NATIVE
■ RARE OR EXTIRPATED
■ INTRODUCED

ALTERNATE NAMES: Slender Stinging Nettle, *Urtica dioica* ssp. *gracilis*

Found in fields and woodlands throughout much of North America, this plant is covered with stinging hairs that can cause severe skin irritation when touched. Nevertheless, the top leaves are sometimes cooked or steamed to destroy the hairs and then eaten as greens or used in soups.

DESCRIPTION: This plant grows 1–6' (30–182 cm) in height. The small greenish flowers are arranged in slender clusters 1–3" (2.5–7.5 cm) long, and the opposite toothed leaves are generally 2½–8" (6–20 cm) long.

FLOWERING SEASON: June–August.

HABITAT: Grows in fields, woodlands, and alongside roads.

RANGE: Occurs throughout much of North America except in the southeastern United States.

CAUTION: This plant's tiny stinging hairs can cause a painful skin irritation.

SIMILAR SPECIES: Some sources consider this plant a subspecies of Stinging Nettle (*U. dioica*), found in the Old World. Wood Nettle (*Laportea canadensis*) is similar but generally shorter and has alternate leaves rather than opposite.

CONSERVATION: **G5**

Globally secure, but imperiled in Labrador and vulnerable in Yukon.

■ NATIVE
■ RARE OR EXTIRPATED
■ INTRODUCED

ALTERNATE NAMES: Balsam Apple, Wild Mock Cucumber

This species grows along rivers and streams and in other wet (mainly freshwater) areas of Canada and the United States. It is an annual herb with broad, simple leaves and a climbing habit. The fruits, although cucumber-like in appearance, are inedible. The roots of the plant have been used medicinally by many Native American groups.

DESCRIPTION: This annual climbing vine features hairless stems and small, whitish, six-petaled flowers about ½" (12 mm) wide. The leaves are somewhat maple-like, with three to seven mostly triangular lobes. The fruits are covered with weak prickles and grow up to 2" (5 cm) long.

FLOWERING SEASON: June–October.

HABITAT: Grows in moist woods and along streambanks.

RANGE: Occurs from Saskatchewan east to New Brunswick, south to Georgia, and west to Oklahoma.

SIMILAR SPECIES: Two similar plants include One-seeded Bur Cucumber (*Sicyos angulatus*), which has smaller fruits and five-petaled flowers, and Creeping Cucumber (*Melothria pendula*), a species of the southeastern United States featuring smooth fruits and yellow flowers.

CONVERSATION: 🄶🄵

CONSERVATION: G5

Globally secure, but critically imperiled in North Carolina and Oklahoma, and imperiled in Kentucky and Wyoming.

■ NATIVE
■ RARE OR EXTIRPATED
■ INTRODUCED

ALTERNATE NAMES: Guadeloupe Cucumber

This vine is native to the southeastern United States, Mexico, and South America. It resembles One-seeded Bur Cucumber (*Sicyos angulatus)* and Wild Cucumber (*Echinocystis lobata*) but tends not to occur as far north as these look-alikes. Creeping Cucumber can be identified by its yellow, five-petaled flowers and smooth fruits, unlike the prickly or hairy fruits of other species.

DESCRIPTION: This perennial climbing vine features small, yellow, five-petaled flowers less than ¼" (5 mm) wide. The heart-shaped leaves are palmately lobed, and the fruits are smooth and grow to ¾" (2 cm) long.

FLOWERING SEASON: June–September (March–December in Florida).

HABITAT: Grows in woods, marshes, floodplains, roadsides, and disturbed areas.

RANGE: Found from Maryland south to Florida, west to Texas, and north to southeastern Kansas. Also occurs in Mexico and South America.

SIMILAR SPECIES: Creeping Cucumber is very similar to two other plants: One-seeded Bur Cucumber (*Sicyos angulatus)* and Wild Cucumber (*Echinocystis lobata*). Creeping Cucumber can be identified by its yellow, five-petaled flowers and smooth fruits that lack hairs or prickles. It is also not found as far north as these other plants.

CONSERVATION: G5

Globally secure, but critically imperiled in Illinois and West Virginia, imperiled in Indiana and Kansas, and vulnerable in Missouri.

Sicyos angulatus **ONE-SEEDED BUR CUCUMBER**

■ NATIVE
■ RARE OR EXTIRPATED
■ INTRODUCED

ALTERNATE NAMES: Star-cucumber

Native to generally wet habitats of the eastern United States, this vine bears cucumber-like but inedible fruits covered in hairs or bristles. It is a mat-forming plant and also a climbing vine that features whitish, five-petaled flowers.

DESCRIPTION: This annual vine features hairy stems and whitish, star-shaped, five-petaled flowers about ½" (12 mm) wide. The leaves are maple-like, with three to five shallow lobes and growing up to 8" (20 cm) long and wide. The fruits, covered with bristles and growing about ½" (1.25 cm) long, are arranged in clusters.

FLOWERING SEASON: July–September.

HABITAT: Grows along streambanks, in thickets, at edges of lowland woods, roadsides, and disturbed places.

RANGE: Occurs from Maine south to the western panhandle of Florida, west to Texas, and north to North Dakota.

SIMILAR SPECIES: This species can be confused with Creeping Cucumber (*Melothria pendula*) and Wild Cucumber (*Echinocystis lobata*), but note that One-seeded Bur Cucumber features whitish, five-petaled flowers and bears hairy or bristly fruits.

CONSERVATION: **G5**
This species is globally secure but vulnerable in North Carolina.

ALTERNATE NAMES: Climbing Bittersweet

This vine is native to central and eastern North America, growing in woodlands and along rivers. It is facing tough competition from Asiatic Bittersweet (*C. orbiculatus*), an aggressive invasive taking over the native bittersweet's range in the Northeast. The colorful fruit of American Bittersweet is an important food source for many bird species.

DESCRIPTION: This is a woody vine climbing up to 56' (17 m), with ovately shaped, finely toothed leaves averaging 2–4" (5–10 cm) in length. The plant features green flowers about ¼" (6 mm) wide with four or five petals, arranged in terminal clusters, and bicolored fruit.

FLOWERING SEASON: May–June.

HABITAT: Grows in thickets, woods, and along riverbanks.

RANGE: Occurs from Saskatchewan east to New Brunswick, south to Georgia, west to Texas, and north to North Dakota.

SIMILAR SPECIES: Asiatic Bittersweet (*C. orbiculatus*), an invasive species with showy, yellow-orange and scarlet fruit.

CONSERVATION: G5

Although globally secure, this species is critically imperiled in Saskatchewan, Alabama, Louisiana, Montana, Rhode Island, Texas, and Wyoming; imperiled in Connecticut, Georgia, and North Carolina; and vulnerable in Quebec, Massachusetts, Mississippi, New Jersey, and Vermont. It is possibly extirpated in Delaware and South Carolina and presumed extirpated in New Brunswick.

■ NATIVE
■ RARE OR EXTIRPATED
■ INTRODUCED

ALTERNATE NAMES: Fen Grass-of-Parnassus

This flowering plant is native to wetlands and other such places of southeastern Canada and the northeastern United States. The common name Grass-of-Parnassus comes from the idea that a similar species was discovered on Mount Parnassus in Greece.

DESCRIPTION: This perennial wildflower averages 6–20" (15–50 cm) in height, featuring solitary, cream-colored, five-petaled flowers ¾–1" (18–24 mm) wide. The mostly basal leaves are thick and broadly heart-shaped, growing to 2" (5 cm) wide.

FLOWERING SEASON: July–September.

HABITAT: Grows along shores of rivers and lakes and edges of wetlands.

RANGE: Occurs from southeastern Canada south to Iowa, Illinois, Ohio, New York, and New Jersey.

SIMILAR SPECIES: Marsh Grass-of-Parnassus (*P. palustris*) and Fringed Grass-of-Parnassus (*P. fimbriata*).

CONSERVATION: G5

Globally secure, but imperiled in Newfoundland, New Hampshire, Pennsylvania, and South Dakota. It is vulnerable in New Brunswick, Quebec, Saskatchewan in Canada and in Iowa, Maine, and New Jersey in the U.S. It is possibly extirpated in Rhode Island.

YELLOW WOOD SORREL *Oxalis stricta*

■ NATIVE
■ RARE OR EXTIRPATED
■ INTRODUCED

ALTERNATE NAMES: Sour Grass, Lemon Clover

This plant has a global distribution that ranges from North America to Eurasia, and is considered an undesirable weed in many areas. It is informally known by many names, including Lemon Clover, referencing the tangy-tasting leaves. All parts of the plant are edible, and the leaves are sometimes used to make tea and other flavored drinks.

DESCRIPTION: This perennial spreading plant grows 6–15" (15–38 cm) tall and features one to several yellow, five-petaled flowers ½–⅔" (12–15 mm) wide. The leaves are palmately divided into three heart-shaped leaflets, each ½"–¾" (1.5–2 cm) wide.

FLOWERING SEASON: May–October.

HABITAT: Grows along roadsides and in fields.

RANGE: Occurs from Saskatchewan east to Newfoundland, south to Florida, west to Texas, and north to North Dakota.

SIMILAR SPECIES: Upright Yellow Wood Sorrel (*O. dillenii*) and Large Yellow Wood Sorrel (*O. grandis*).

CONVERSATION: **G5**

CONSERVATION: **G5**

This species is globally secure but imperiled in Wyoming.

- ■ NATIVE
- ■ RARE OR EXTIRPATED
- ■ INTRODUCED

ALTERNATE NAMES: Bog St.-John's-wort

Native to western North America, this wildflower grows in meadows, along streams, and in other wet areas. It is quite similar to Small-flower St.-John's-wort (*H. mutilum*) found in eastern North America. For centuries, these plants were thought to ward off evil and were hung in homes or carried in pockets for this purpose.

DESCRIPTION: This mat-forming plant averages 1–3" (2.5–7.5 cm) in height. The yellow, five-petaled flowers are small, only about ¼" (5 mm) wide. The leaves are ovately shaped and ¼–⅝" (5–16 mm) long.

FLOWERING SEASON: June–August.

HABITAT: Grows in wet places from sea level to high elevations.

RANGE: Occurs from British Columbia south to Baja California and east to Montana.

SIMILAR SPECIES: Small-flower St.-John's-wort (*H. mutilum*) is a similar Eastern equivalent.

CONSERVATION: Ⓖ⑤

Although globally secure, this species is critically imperiled in Utah and imperiled in Arizona.

ALTERNATE NAMES: Orange-grass

This plant may be found growing in open areas throughout much of eastern North America, identified by its tiny flowers, scaly leaves, and wiry branches that resemble pine needles. The leaves give a citrus scent when crushed.

DESCRIPTION: This flowering plant grows 4–20" (10–50 cm) tall. The small, yellow, five-petaled flowers up to ¼" (6 mm) wide are situated on wiry stems, and the scale-like leaves are only about ¼" (5 mm) long.

FLOWERING SEASON: July–October.

HABITAT: Grows in open woods, fields, and rocky areas.

RANGE: Occurs from Ontario and Nova Scotia south to Florida, west to Texas, and north to Minnesota.

SIMILAR SPECIES: Drummond's St.-John's-wort (*H. drummondii*), also commonly called Nits-and-lice, is a similar species in the same range, but it is especially prominent in the Lower Mississippi River Basin from Iowa to Ohio and south. By contrast, Pineweed populations are more dense in the Atlantic states.

CONSERVATION: G5

Globally secure, but critically imperiled in Ontario and Oklahoma, imperiled in Vermont, and vulnerable in Michigan. It is presumed extirpated in Iowa.

■ NATIVE
■ RARE OR EXTIRPATED
■ INTRODUCED

ALTERNATE NAMES: Klamath-weed, Perforate St.-John's-wort

Native to Eurasia, this introduced plant is now widespread throughout North America, growing in fields, on roadsides, and in various disturbed areas. It can easily be identified by the tiny translucent dots that give the leaves a perforated appearance. This species also has little black dots that appear mostly along the edges of the flower petals. The native North American equivalent is Spotted St.-John's-wort (*H. punctatum*), which is widespread in the East and has black dots on both the leaves and flowers.

DESCRIPTION: This leafy wildflower grows 1–3' (30–90 cm) in height. The bright yellow, star-like, five-petaled flowers are ¾–1" (18–24 mm) wide, arranged in a round-topped, terminal cluster, and the opposite leaves, featuring distinct, translucent dots, are generally ½–1½" (1.5–4 cm) long.

FLOWERING SEASON: June–September.

HABITAT: Grows in fields, along roadsides, and in waste places.

RANGE: Occurs throughout much of North America except in the Far North.

CAUTION: Often used as an herbal remedy, St.-John's-wort can cause extreme sun sensitivity in some individuals. It can also cause poisoning in animals if consumed in high quantities.

SIMILAR SPECIES: Spotted St.-John's-wort (*H. punctatum*) is a close relative native to eastern North America that has dark spots on the leaves and the interior of the flower petals.

CONSERVATION:

The conservation status of this introduced species is unranked.

- NATIVE
- RARE OR EXTIRPATED
- INTRODUCED

ALTERNATE NAMES: Virginia Marsh St.-John's-wort, Purple St.-John's-wort, *Triadenum virginicum*

This perennial wildflower of eastern North American wetlands features light pink flowers, setting it apart from most other St.-John's-worts with yellow flowers. The closely related and very similar Bog St.-John's-wort (*H. fraseri*) also favors wet habitats, but is more common from the Great Lakes north and not as likely to be found in the southeastern United States.

DESCRIPTION: A marsh herb growing 8–24" (20–60 cm) in height with 1–2½" (2.5–6.5 cm) long, opposite, heart-shaped leaves. The five-petaled pinkish flowers, ½–¾" (12–18 mm) wide are clustered atop the stem and feature long acute sepals.

FLOWERING SEASON: July–August.

HABITAT: Grows in swamps, bogs, and other wet places.

RANGE: Occurs from Ontario, Quebec, and Nova Scotia, south to Florida, west to Texas, and northeast to Indiana and Illinois.

SIMILAR SPECIES: Bog St.-John's-wort (*H. fraseri*) and Lesser Marsh St.-John's-wort (*H. tubulosum*).

CONSERVATION: G5

Globally secure, but critically imperiled in New Brunswick, Arkansas, and Oklahoma. It is vulnerable in Quebec and North Carolina. It is possibly extirpated in Illinois.

■ NATIVE
■ RARE OR EXTIRPATED
■ INTRODUCED

ALTERNATE NAMES: White-margin Sandmat, *Chamaesyce albomarginata*

This small but showy member of the spurge family is native to open desert and grassland areas of the southwestern United States and northern Mexico. What appears to be a single, white-and-maroon flower is actually a cluster of numerous tiny flowers. The common name comes from the idea that this species was useful for treating snakebites; however, there is no medical evidence that this is the case. In fact, this plant secretes a milky sap that is poisonous to humans.

DESCRIPTION: This mat-forming plant features branches barely ½" (1.3 cm) high, and stems 2–10" (5–25 cm) long. Numerous tiny flowers are clustered to appear as one small, white flower with a maroon center and four to five petals. The small, roundish leaves are ⅛–⅜" (3–9 mm) long.

FLOWERING SEASON: April–November.

HABITAT: Grows in deserts, grasslands and woodlands, and alongside roads.

RANGE: Occurs from southern California and Nevada east to Oklahoma, and south to Mexico.

CAUTION: Most members of this family are poisonous, and their milky sap can irritate the eyes, mouth, and mucous membranes.

SIMILAR SPECIES: Smallseed Sandmat (*E. polycarpa*).

CONSERVATION: **G5**
This species is globally secure.

■ NATIVE
■ RARE OR EXTIRPATED
■ INTRODUCED

This flowering plant is native to dry open areas of eastern North America. The common name "spurge" comes from a Latin word meaning "to purge." This species has been used as a laxative, but can be poisonous in large doses. Many bird species feed on the seeds, including Wild Turkey (*Meleagris gallopavo*), Mourning Dove (*Zenaida macroura*), and Horned Lark (*Eremophila alpestris*).

DESCRIPTION: This perennial grows 10–36" (25–90 cm) tall. A group of tiny flowers are clustered to appear as a single, small, five-petaled white flower about ¼" (6 mm) wide. The narrow, mostly alternate leaves are about 1½" (4 cm) long.

FLOWERING SEASON: June–September.

HABITAT: Grows in dry open woods, fields, and alongside roads.

RANGE: Occurs from Ontario and Maine south to Florida, west to Texas, and north to South Dakota and Minnesota.

CAUTION: Although sometimes used as a laxative, large doses of Flowering Spurge can be poisonous, and contact with the plant—especially its milky sap—can cause irritation of skin, eyes, and mucous membranes.

SIMILAR SPECIES: Snow-on-the-mountain (*E. marginata*).

CONSERVATION: G5

This species is globally secure, but vulnerable in Maryland.

■ NATIVE
■ RARE OR EXTIRPATED
■ INTRODUCED

ALTERNATE NAMES: Wild Poinsettia, Painted-leaf

This wildflower is commonly called Wild Poinsettia and is reminiscent of the popular holiday plant with its red floral leaves. Fire-on-the-mountain is native to the central and southern United States as well as to parts of Central and South America. It grows in ravines, woods and various other habitat types.

DESCRIPTION: This flowering plant averages 2–3' (60–90 cm) in height. Many tiny, greenish flowers lacking petals are clustered above what are usually red and green floral leaves. The leaves are narrow and generally grow to about 3" (7.5 cm) long.

FLOWERING SEASON: August–September.

HABITAT: Grows in wooded areas, ravines, and alongside roads.

RANGE: Occurs from Nebraska east to Indiana, southeast to Virginia and North Carolina, south to Florida, and west to Arizona.

CAUTION: Contact with this plant's milky sap can cause irritation of skin, eyes, and mucous membranes, and can be poisonous if ingested.

SIMILAR SPECIES: This species resembles Toothed Spurge (*E. dentata*) as well as true Poinsettia (*Poinsettia pulcherrima*), which is native to Mexico and Central America. There is much confusion in the literature between this species and Fireplant (*E. heterophylla*); some sources treat Fire-on-the-mountain as a subspecies of this tropical and subtropical plant.

CONVERSATION: **G5**
Globally secure, but critically imperiled in Utah.

■ NATIVE
■ RARE OR EXTIRPATED
■ INTRODUCED

This European native was introduced to North America as an ornamental plant in the 19th century and can now be found growing in dry meadows and along roadsides throughout much of the continent. It is still widely used as an ornamental and frequently escapes from cultivation and can be highly invasive.

DESCRIPTION: This introduced wildflower grows 6–12" (15–30 cm) in height. It features tiny flowers situated atop bright yellow, saucer-shaped bracts each about ¼" (6 mm) wide. The crowded leaves are very narrow, almost needle-like. The plant produces milky sap that will irritate skin and should be avoided.

FLOWERING SEASON: March–June.

HABITAT: Grows along roadsides and in other disturbed places.

RANGE: Naturalized throughout much of North America except northern Canada and the southern United States.

CAUTION: Although the roots of this plant are sometimes used as a purgative, large doses can cause poisoning; in fact, all members of this genus can be toxic if ingested. Contact with this plant, especially its milky sap, can cause irritation of skin, eyes, and mucous membranes.

SIMILAR SPECIES: This plant can be separated from other *Euphorbia* species by its very narrow leaves. It is similar to Leafy Spurge (*E. esula*), another introduced Old World species.

CONSERVATION: G5

Although introduced in North America, this species is globally secure.

This showy plant is native to prairies and pastures of the mid-western United States. It is similar to Flowering Spurge (*E. corol-lata*) but can be easily distinguished by its floral leaves, which are bordered with white. This plant is sometimes planted as an ornamental for its attractive flowers and foliage.

DESCRIPTION: This small annual grows 1–3' (30–90 cm) tall. A group of tiny flowers are clustered to appear as a single, small white flower. The floral leaves are small and green with white borders. The oval stem leaves are up to 3½" (8.5 cm) long and 1½" (3.7 cm) wide.

FLOWERING SEASON: August–October.

HABITAT: Grows on dry slopes, in prairies and pastures, and alongside roads.

RANGE: Occurs from Montana east to Minnesota, south to Texas, and west to New Mexico.

CAUTION: Parts of this plant and extracts made from them can be toxic if ingested. Contact with the plant can cause irritation of the skin, eyes, and mucous membranes.

SIMILAR SPECIES: Flowering Spurge (*E. corol-lata*) lacks the white borders on the floral leaves.

CONSERVATION: 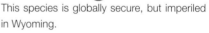 G5
This species is globally secure, but imperiled in Wyoming.

■ NATIVE
■ RARE OR EXTIRPATED
■ INTRODUCED

ALTERNATE NAMES: Hook-spur Violet, Sand Violet

This violet is found in meadows and forests throughout Canada and the western and northern United States. The leaves and flowers are sometimes eaten in salads or brewed as tea. It is a larval host plant for Myrtle's Silverspot (*Speyeria zerene myrtleae*), an endangered butterfly.

DESCRIPTION: This small perennial grows 2–4" (5–10 cm) in height. The pansy-like, five-petaled purple flowers are about ½" (12 mm) long and hang at the tips of slender stalks. The heart-shaped leaves are finely scalloped and generally ½–1¼" (1.3–3.1 cm) long.

FLOWERING SEASON: April–August.

HABITAT: Grows in meadows, open woods, and along road-sides.

RANGE: Occurs throughout much of Canada south to California, Arizona, New Mexico, the northern Great Plains, and New England.

SIMILAR SPECIES: Common Blue Violet (*V. sororia*).

CONSERVATION: ⬤G5

Although this species is globally secure, it is critically imperiled in Connecticut, Iowa, and Nebraska, and vulnerable in Massachusetts and New Brunswick.

■ NATIVE
■ RARE OR EXTIRPATED
■ INTRODUCED

ALTERNATE NAMES: Wild Pansy

This small wildflower grows in open woods and fields through-out much of the eastern United States, blooming in early to mid-spring. It is quite similar to Johnny-jump-up (*V. tricolor*), a European wildflower introduced to North America.

DESCRIPTION: This is a small annual plant growing to about 6" (15 cm) tall. It has bluish-white or blue flowers about ½" (12 mm) wide, and both circular basal leaves and oval to lance-shaped stem leaves.

FLOWERING SEASON: March–April.

HABITAT: Grows in open woods, fields, and along roads.

RANGE: Found from Nebraska east to southern New York, south to Florida, and west to Texas.

SIMILAR SPECIES: Can be difficult to distinguish from the introduced Johnny-jump-up (*V. tricolor*), which usually has some dark blue on the upper petals.

CONSERVATION: G5

Globally secure, but critically imperiled in New York and Ontario and vulnerable in Iowa and New Jersey.

CANADA VIOLET *Viola canadensis*

■ NATIVE
■ RARE OR EXTIRPATED
■ INTRODUCED

ALTERNATE NAMES: Canadian White Violet

This violet is native to woodlands of Canada and mostly the western and northeastern United States. The attractive white flowers feature yellow bases and purple streaking. It is sometimes planted in backyard gardens, and the blossoms can be used to make jelly.

DESCRIPTION: This flowering plant is 8–16" (20–40 cm) in height. The fragrant, five-petaled flowers are white and 1" (2.5 cm) wide, and the leaves are heart-shaped, finely toothed, and generally 2–4" (5–10 cm) long.

FLOWERING SEASON: May–July.

HABITAT: Grows in wooded areas.

RANGE: Occurs throughout much of Canada from Alaska to Newfoundland, and south to Georgia and South Carolina in the East and Arizona and New Mexico in the West.

SIMILAR SPECIES: Cream Violet (*V. striata*), also called Striped Violet, is a similar species found only in eastern North America. Its flowers have a purple-veined lower petal.

CONSERVATION: G5

Although globally secure, this species is critically imperiled in Illinois and New Brunswick. it is imperiled in Alabama and Arkansas in the U.S. and Yukon in Canada. It is vulnerable in Connecticut, Iowa, and Wyoming. It is possibly extirpated in Maine.

- ■ NATIVE
- ■ RARE OR EXTIRPATED
- ■ INTRODUCED

ALTERNATE NAMES: *Viola blanda* var. *palustriformis*

This violet is native to eastern North American forests and has fragrant white flowers on reddish stems and dark green basal leaves. It is a self-supporting perennial plant found in cool wood-lands. The upper flower petals may be twisted or bent back-ward.

DESCRIPTION: This violet grows to 3–5" (7.5–12.5 cm) in height. The white flowers are about ½" (12 mm) long, situated on reddish stems rising from an underground stem, separate from the basal leaves, which are shiny dark green, heart-shaped and sharp-pointed, and grow up to 2½" (6.5 cm) wide.

FLOWERING SEASON: April–May.

HABITAT: Grows in wooded areas.

RANGE: Occurs from Saskatchewan east to Newfoundland, south to Georgia, west to Alabama, and northwest to Illinois, Iowa, and Minnesota.

SIMILAR SPECIES: The very similar Northern White Violet (*V. macloskeyi*) does not have reddish stems and ranges farther west.

CONSERVATION: G5

Globally secure, but critically imperiled in Mani-toba and Illinois, imperiled in Saskatchewan and North Dakota, and vulnerable in Delaware and Indiana.

■ NATIVE
■ RARE OR EXTIRPATED
■ INTRODUCED

This species grows in well-drained woodlands of the central United States and is one of the most common violets in its range. It can be recognized by its bluish flowers, somewhat triangular leaves, and lack of stems. Missouri Violet is also one of the earliest bloomers in the central states.

DESCRIPTION: This stemless perennial grows 3–8" (7.5–20 cm) tall and features blue-violet, five-petaled flowers ½–¾" (12–18 mm) wide with light-colored centers. The leaves, which grow on separate stalks from the flowers, are triangular in shape and have scalloped edges.

FLOWERING SEASON: March–May.

HABITAT: Grows in well-drained woodlands.

RANGE: Occurs from North Dakota and Minnesota southeast to Ohio, and southwest to Texas and New Mexico.

SIMILAR SPECIES: Common Blue Violet (*V. sororia*).

CONVERSATION: G5

This species is globally secure, but imperiled in North Dakota, and vulnerable in Iowa and New Mexico.

■ NATIVE
■ RARE OR EXTIRPATED
■ INTRODUCED

ALTERNATE NAMES: Mountain Pansy

This relatively large violet is native to dry open areas of central and eastern North America. The common name is inspired by this plant's distinctive leaves, which are arranged like a bird's foot. The showy, blue-violet flowers attract a variety of bees and butterflies.

DESCRIPTION: This perennial violet grows 4–10" (10–25 cm) tall and features five-petaled, blue-violet flowers 3/4–11/4" (18–30 mm) wide, with a conspicuous cluster of orange stamens at the center. The leaves are fan-shaped with linear, toothed segments, and generally 1–2" (2.5–5 cm) long.

FLOWERING SEASON: March–June.

HABITAT: Grows in open woods, fields, and along roads.

RANGE: Occurs from Ontario and parts of New England south to Georgia, west to Texas, and north to Nebraska, Iowa, and Minnesota.

SIMILAR SPECIES: Prairie Violet (*V. pedatifida*).

CONSERVATION: **G5**

Globally secure, but critically imperiled in Ontario, Delaware, and Nebraska. It is vulnerable in New York, Ohio, and Rhode Island.

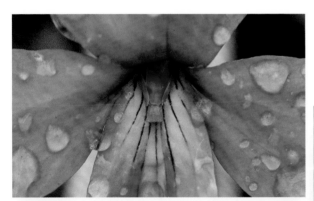

■ NATIVE
■ RARE OR EXTIRPATED
■ INTRODUCED

ALTERNATE NAMES: Crowfoot Violet, Prairie Violet, Coastal Violet

This small flowering plant is native to North America, where it generally prefers dry, open habitats. The flowers and leaves of this plant bear a resemblance to those of larkspurs (of the genus *Delphinium*), hence the common name. This species is known to hybridize with Common Blue Violet (*V. sororia*).

DESCRIPTION: This stemless perennial averages 3–8" (7.5–20 cm) in height. The blue-violet five-petaled flowers are ¾" (18 mm) wide and the leaves are divided into narrow segments.

FLOWERING SEASON: May–June.

HABITAT: Grows in prairies and open woods.

RANGE: Occurs from Alberta east to Ontario, south to Ohio, and southwest to Arkansas, Oklahoma, and parts of the Southwest.

SIMILAR SPECIES: Bird's-foot Violet (*V. pedata*).

CONSERVATION: G5

Although globally secure, this species is critically imperiled in Mississippi, Ohio, and Wyoming in the U.S. and in Ontario in Canada. It is imperiled in Arkansas, Colorado, and Indiana. It is vulnerable in New Mexico as well as in Alberta and Saskatchewan.

■ NATIVE
■ RARE OR EXTIRPATED
■ INTRODUCED

ALTERNATE NAMES: Yellow Forest Violet

This species is native to eastern North America, where it prefers woodlands but can also be found growing in meadows. Downy Yellow Violet is named for its hairiness, a characteristic that helps to separate it from similar yellow-flowered species.

DESCRIPTION: This softly hairy plant grows 6–16" (15–40 cm) in height. It has yellow five-petaled flowers, ¾" (18 mm) wide, and heart-shaped, scallop-toothed leaves 2–5" (5–12.5 cm) wide.

FLOWERING SEASON: May–June.

HABITAT: Grows in woods and meadows.

RANGE: Occurs from southeastern Canada south to the Carolinas and west to Oklahoma.

SIMILAR SPECIES: Stream Violet (*V. glabella*), Prairie Yellow Violet (*V. nuttallii*), and Round-leaf Violet (*V. rotundifolia*).

CONSERVATION: G5
Globally secure, but critically imperiled in Delaware, Louisiana, Mississippi, and Rhode Island. It is imperiled in Prince Edward Island and Saskatchewan in Canada and Wyoming in the U.S.

■ NATIVE
■ RARE OR EXTIRPATED
■ INTRODUCED

ALTERNATE NAMES: Pine Violet

This yellow-flowered violet is a small plant native to western North America. The thick leaves are often tinted purple and are prominently veined. The shape of some of these leaves resembles that of a goose's foot, hence the common name.

DESCRIPTION: This is a small perennial violet growing 2–6" (5–15 cm) in height and featuring five-petaled yellow flowers, 1/4–1/2" (7–12 mm) long, often with a purplish or brownish tinge. The leaves are lanceolate to nearly round, thick and fleshy, and purplish-green.

FLOWERING SEASON: May–August.

HABITAT: Grows in open or partly shaded areas from lowlands to high mountains.

RANGE: Occurs from eastern British Columbia and eastern Washington south to southern California and east to Arizona, Colorado, Wyoming, and Montana.

SIMILAR SPECIES: Other yellow-flowered *Viola* species.

CONSERVATION: G5

Globally secure, but vulnerable in British Columbia and Wyoming.

- **NATIVE**
- **RARE OR EXTIRPATED**
- **INTRODUCED**

ALTERNATE NAMES: Hooded Blue Violet, Florida Violet, Sand Violet, Meadow Violet, Woolly Blue Violet, *Viola papilionacea*

This eastern North American native is known by many common names. It appears in a variety of habitats, including lawns and gardens, where it readily self-seeds. It has purple flowers that have a conspicuous white center. Common Blue Violet is a popular garden plant, and it has been used medicinally and for food; its leaves are high in vitamins A and C and are sometimes served in salads or cooked as greens.

DESCRIPTION: This stemless perennial plant grows 3–8" (7.5–20 cm) tall, featuring blue-violet, five-petaled flowers, ¾" (18 mm) wide, with white centers. The heart-shaped leaves grow on separate stalks and sport scalloped edges.

FLOWERING SEASON: March–June.

HABITAT: Grows in damp woods, moist meadows, lawns, and along roadsides.

RANGE: Occurs throughout eastern North America, except in Alberta.

SIMILAR SPECIES: The white centers of the flowers help to separate this species from Marsh Blue Violet (*V. cucullata*), which features flowers with dark blue centers. Missouri Violet (*V. missouriensis*) is similar but its flowers are often lavender to white. Western Dog Violet (*V. adunca*) has similar flowers but the plant is hairy.

CONSERVATION: G5

This species is globally secure.

■ NATIVE
■ RARE OR EXTIRPATED
■ INTRODUCED

ALTERNATE NAMES: Purple Passionflower, Maypop, Apricot-vine

This fast-growing plant is native to the southern United States, where it commonly grows in thickets, near riverbanks, along railroads, and various other open habitats. It is cultivated as a backyard garden plant and attracts a variety of bees and other pollinators. This species is a larval host plant for the Gulf Fritillary (*Dione vanillae*) and Variegated Fritillary (*Euptoieta claudia*) butterflies.

DESCRIPTION: This is a climbing or trailing vine featuring large, five-petaled flowers, 1½–2½" (4–6 cm) wide which are whitish or bluish and strikingly fringed. The leaves are palmately lobed and generally 3–5" (7.5–12.5 cm) wide. The green to yellow berries have edible pulp around each seed.

FLOWERING SEASON: June–September.

HABITAT: Grows in thickets, open woods, and fields.

RANGE: Occurs from Maryland south to Florida, west to Texas, and north to Nebraska and Missouri.

SIMILAR SPECIES: Yellow Passionflower (*P. lutea*).

CONSERVATION: (G5)

Although globally secure, this species is imperiled in Ohio and West Virginia, and vulnerable in Indiana.

■ NATIVE
■ RARE OR EXTIRPATED
■ INTRODUCED

ALTERNATE NAMES: Prairie Flax, Lewis Flax, *Linum perenne var. lewisii*

This wildflower is native to western North America, where it grows on ridges and dry slopes, in open woods, and in meadows and grasslands. It is sometimes planted in backyard gardens and is visited by various butterfly species including Checkered White (*Pontia protodice*), Rocky Mountain Duskywing (*Erynnis telemachus*), and Silvery Blue (*Glaucopsyche lygdamus*). This plant's species name honors the North American explorer Meriwether Lewis.

DESCRIPTION: A tufted perennial growing 1–2' (30–60 cm) in height with numerous leaves, each ½–¾" (1.5–2 cm) long. The blue flowers are 1–1½" (2.5–3.7 cm) wide and arranged in branched clusters.

FLOWERING SEASON: April–October.

HABITAT: Grows on dry slopes and in open woods and grasslands.

RANGE: Occurs throughout western North America from Alaska south to California, east to Texas and northeast to Ontario.

CAUTION: This species should not be ingested; animals have been poisoned, sometimes fatally, by eating plants in the *Linum* genus.

SIMILAR SPECIES: Common Flax (*L. usitatissimum*) is a closely related and widely cultivated plant introduced from the Old World. Meadow Flax (*L. pratense*) is a similar species native to the dry regions of the American Southwest; it is an annual.

CONSERVATION: Ⓖ5

Globally secure, critically imperiled in Kansas and Nebraska, and imperiled in West Virginia.

■ NATIVE
■ RARE OR EXTIRPATED
■ INTRODUCED

This slender plant is native to dry, open areas of southeastern Canada and the eastern United States. It produces small, yellow flowers from one or more stems growing from the root system. It flowers in midsummer and the petals frequently fall early.

DESCRIPTION: This perennial with smooth stems grows 6–36" (15–90 cm) in height. It features yellow five-petaled flowers about ½" (1.25 cm) across. The lance-shaped leaves are usually alternate and grow up to 1" (2.5 cm) long.

FLOWERING SEASON: July–August.

HABITAT: Grows in dry woods, fields, and other open places.

RANGE: Occurs from southern Wisconsin east to Ontario and Maine, south to Florida, and west to Texas.

SIMILAR SPECIES: Grooved Flax (*L. sulcatum*) and other yellow-flowered *Linum* species.

CONSERVATION:  G5

Although globally secure, this species is critically imperiled in Vermont and vulnerable in Ontario. It is possibly extirpated in Iowa.

■ NATIVE
■ RARE OR EXTIRPATED
■ INTRODUCED

ALTERNATE NAMES: Stiff-stem Flax

This common prairie plant is native to western and central North America. It features slender, wiry stems and yellow flowers with reddish centers. The petals usually drop after one or two days.

DESCRIPTION: This annual grows 6–20" (15–51 cm) in height and features loose clusters of yellow flowers, ¾–1" (18–24 mm) wide, with reddish centers. The leaves are narrowly lance-shaped, alternate, and inconspicuous.

FLOWERING SEASON: May–August.

HABITAT: Grows in upland prairies.

RANGE: Occurs from Alberta east to Manitoba, south to Texas, and west to New Mexico.

CAUTION: Animals have been poisoned, sometimes fatally, by eating plants in the *Linum* genus.

SIMILAR SPECIES: Grooved Yellow Flax (*L. sulcatum*) and Berlandier's Flax (*L. berlandieri*).

CONSERVATION: **G5**
Globally secure, but vulnerable in Manitoba, Iowa, and Wyoming.

■ NATIVE
■ RARE OR EXTIRPATED
■ INTRODUCED

ALTERNATE NAMES: Red-stem Storksbill, Red-stem Filaree, Clocks, Afilaria, Pin Clover

This Eurasian native was introduced to North America in the 18th century and now grows in fields and along roadsides across the continent, particularly common in desert areas of the Southwest. The long seed pod of this species resembles the shape of a stork's bill, inspiring the common name.

DESCRIPTION: An annual growing 6–12" (15–30 cm) in height with hairy stems and mostly basal leaves, pinnately divided into lobed and toothed leaflets, each up to 1" (2.5 cm) long. The small, five-petaled flowers are about 1/3" (10 mm) wide, pink or purplish-pink and arranged in clusters of six to nine.

FLOWERING SEASON: February–June.

HABITAT: Grows in disturbed areas, along roadsides, and in fields.

RANGE: Occurs throughout much of North America except in parts of the Far North.

SIMILAR SPECIES: Texas Storksbill (*E. texanum*) and Round-leaf Storksbill (*California macrophylla*) are related plants native to the Southwest. The latter is a vulnerable species found in California and a single county in southern Oregon.

CONSERVATION:
The conservation status of this introduced species is unranked.

■ NATIVE
■ RARE OR EXTIRPATED
■ INTRODUCED

ALTERNATE NAMES: Carolina Cranesbill, Carolina Geranium, Wild Geranium

This North American native is widespread across the continent and can be found in a variety of habitats. It is a member of the geranium family (Geraniaceae), leafy herbs known for showy clusters of white, pink, or purple flowers. This species features light pink flowers with shallowly notched petals.

DESCRIPTION: This perennial plant grows 1–2' (30–60 cm) in height. It has light pink or whitish flowers, ¼–⅓" (6–10 mm) wide, with five shallowly notched petals, arranged in loose clusters of two to five. The grayish-green leaves are deeply lobed and generally 4–5" (10–12.5 cm) wide.

FLOWERING SEASON: March–July.

HABITAT: Grows in woods, clearings, thickets, and fields.

RANGE: Occurs from Ontario and Maine south to Florida, west to Texas, and north to Minnesota, and in scattered areas of the West.

SIMILAR SPECIES: Dove's-foot Geranium (*G. molle*).

CONSERVATION: **G5**

Globally secure, but critically imperiled in Quebec, New Hampshire, and Utah. Imperiled in Alberta and Wyoming. Vulnerable in Manitoba and Saskatchewan. It is possibly extirpated in Vermont.

■ NATIVE
■ RARE OR EXTIRPATED
■ INTRODUCED

ALTERNATE NAMES: Spotted Cranesbill

Native to eastern North America, this flowering plant grows primarily in woodlands and can become quite abundant. It goes by many common and colloquial names, including Alum Root and Old Maid's Nightcap. The name Cranesbill, shared with a few other species, refers to the beak-like capsule of these plants.

DESCRIPTION: This perennial plant grows 1–2' (30–60 cm) in height. The five-petaled pinkish or whitish flowers are 1–1½" (2.5–3.8 cm) wide, arranged in loose clusters of two to five. The grayish-green leaves are about 4–5" (10–12.5 cm) wide and palmately divided into deeply toothed lobes.

FLOWERING SEASON: April–June.

HABITAT: Grows in woods, thickets, and mountain meadows.

RANGE: Occurs from Ontario and Maine south to Georgia, west to Arkansas, and north to Minnesota.

SIMILAR SPECIES: Carolina Geranium (*G. carolinianum*).

CONSERVATION: G5

Although globally secure, this species is critically imperiled in Manitoba in Canada and Louisiana, Nebraska, and South Dakota in the U.S. It is imperiled in Quebec and Kansas. It is possibly extirpated in North Dakota.

- ■ NATIVE
- ■ RARE OR EXTIRPATED
- ■ INTRODUCED

ALTERNATE NAMES: Water-willow

The sole representative of the genus *Decodon*, this shrubby plant grows in areas with shallow water throughout eastern North America. It can be recognized by its magenta flowers, narrow leaves, and typically arching stems.

DESCRIPTION: This perennial grows 2–8' (60–244 cm) in height. The magenta flowers are about ¼" (6 mm) long. They have four or five petals and are clustered in axils of leaves, which are either in opposite pairs or whorls of three or four, and up to 4" (10 cm) long.

FLOWERING SEASON: June–August.

HABITAT: Grows in wet meadows, along the edges of marshes and lakes, and in roadside ditches.

RANGE: Occurs throughout much of eastern North America.

SIMILAR SPECIES: Winged Loosestrife (*Lythrum alatum*).

CONSERVATION: G5

Globally secure, but critically imperiled in New Brunswick and Prince Edward Island in Canada and Iowa, Missouri, and West Virginia in the U.S. It is imperiled in Mississippi and vulnerable in Nova Scotia, Quebec, Kentucky, Minnesota, and Tennessee.

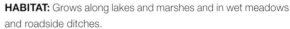

■ NATIVE
■ RARE OR EXTIRPATED
■ INTRODUCED

ALTERNATE NAMES: Spiked Loosestrife, Purple Lythrum

This colorful flowering plant is native to Europe, Asia, Africa, and Australia and extremely invasive in North America. It was introduced to North America as an ornamental in the 19th century and has since spread across the continent, crowding out native plant species wherever it goes. It can be found along lakes, marshes, streams, and many other wet habitats.

DESCRIPTION: This perennial plant grows 2–7' (60–210 cm) in height. The five-petaled purplish flowers are ½–¾" (12–18 mm) wide, arranged in crowded spikes. The leaves are narrow, opposite, and notched at the base, generally growing 1¼–4" (3–10 cm) long.

FLOWERING SEASON: August–September.

HABITAT: Grows along lakes and marshes and in wet meadows and roadside ditches.

RANGE: Occurs throughout much of North America, except parts of the Far North and parts of the southernmost United States.

SIMILAR SPECIES: Other *Lythrum* species.

CONSERVATION: G5

This species is globally secure. It is introduced and invasive in North America, and is often the target of removal or control efforts. Infestations can result in severe disruption to river flows and native wetland ecosystems. Purple Loosestrife can completely crowd out native cattails (*Typha* spp.), an important food and cover plant for native waterfowl and other wildlife.

■ NATIVE
■ RARE OR EXTIRPATED
■ INTRODUCED

ALTERNATE NAMES: Water Caltrop

Native to Eurasia and Africa, this is an extremely invasive plant growing in slow-moving waters in southeastern Canada and the northeastern United States. It often creates expansive mats across wide spans of water, restricting light penetration and thereby reducing oxygen levels for fish and other aquatic wildlife. The species has been cultivated in Asia for thousands of years for its edible seeds.

DESCRIPTION: This is an annual aquatic plant with long stems reaching lengths of 16' (5 m). It has four-petaled white flowers ½" (12 mm) wide and two types of leaves: The submerged leaves are arranged in pairs along the stem, finely divided into feathery segments, while the floating leaves grow in large rosettes and are toothed along the edges.

FLOWERING SEASON: July–August.

HABITAT: Grows in shallow lakes, ponds, and slow-moving streams and rivers.

RANGE: This invasive plant has been recorded from Ontario and Quebec south to Virginia.

SIMILAR SPECIES: *T. bicornis* is another invasive species also commonly known as Water Chestnut.

CONSERVATION:
The conservation status of this introduced species is unranked. If left uncontrolled, this highly invasive plant can create nearly impenetrable mats up to 1' (30 cm) thick atop vast areas of water, disrupting or destroying native aquatic ecosystems beneath.

■ NATIVE
■ RARE OR EXTIRPATED
■ INTRODUCED

ALTERNATE NAMES: Narrowleaf Fireweed, Great Willowherb, *Chamerion angustifolium, Epilobium angustifolium*

This showy wildflower is known for appearing in areas cleared by fire, hence the common name. Its seeds are dispersed over great distances by long, silky hairs. In Britain this species is locally known as Bombweed due to its rapid appearance at bomb sites during World War II. It is now often used to reestablish vegetation at various disturbed sites. Sources disagree on the proper name of the genus, with some preferring *Chamerion* to *Chamaenerion*.

DESCRIPTION: This perennial grows 2–6' (60–180 cm) in height. It features a tall stem with a terminal, spike-like cluster of deep pink, four-petaled flowers, about 1" (2.5 cm) wide, and alternate narrow leaves growing up to 8" (20 cm) long.

FLOWERING SEASON: June–September.

HABITAT: Grows in recently cleared woodlands and disturbed sites, such as following a fire or oil spill.

RANGE: Occurs throughout southern Canada and the western and northeastern United States.

SIMILAR SPECIES: River Beauty (*C. latifolium*).

CONSERVATION: G5

Globally secure, but critically imperiled in Indiana, Ohio, and Tennessee, and vulnerable in Iowa, Illinois, and Nebraska.

- ■ NATIVE
- ■ RARE OR EXTIRPATED
- ■ INTRODUCED

ALTERNATE NAMES: Small Enchanter's Nightshade

This wildflower is native mostly to forested areas throughout the Northern Hemisphere. In North America, the flowers are a favorite of the Mustard White butterfly (*Pieris oleracea*). The species is a complex that includes six subspecies that interbreed to create intergrading populations throughout its global range. The genus *Circaea* is named for Circe, the enchantress in Greek mythology known for using herbs for her magical purposes.

DESCRIPTION: This perennial herb grows 4–10" (10–25 cm) in height. It features slender stems and tiny white flowers arranged in sparsely-flowered racemes. The heart-shaped leaves are opposite, toothed, and 1–2½" (2.5–6.5 cm) long.

FLOWERING SEASON: May–July.

HABITAT: Grows in shady woods, swamps, and bogs.

RANGE: Occurs throughout much of North America except in the south-central and southeastern United States.

SIMILAR SPECIES: The similar Large Enchanter's Nightshade (*C. canadensis*) grows taller and has larger leaves that are more shallowly toothed.

CONSERVATION:  G5

Globally secure, but critically imperiled in Illinois, imperiled in New Mexico, and vulnerable in Yukon, Iowa, North Carolina, and North Dakota. It is presumed extirpated in Indiana.

■ NATIVE
■ RARE OR EXTIRPATED
■ INTRODUCED

ALTERNATE NAMES: Herald-of-summer

This showy plant's native range is limited to the coastal hills and mountains of western North America. As the common name suggests, the flowers of this species begin to bloom at the end of spring. It is commonly cultivated as an ornamental plant.

DESCRIPTION: This annual grows 1–3' (30–91 cm) in height. The cup-shaped, four-petaled flowers are 2–3" (5–7 cm) wide, pink to purple, rarely white, with reddish central blotches. The leaves are lanceolate and ¾–3" (2–7.5 cm) long.

FLOWERING SEASON: June–August.

HABITAT: Grows on grassy slopes and coastal bluffs.

RANGE: Occurs from British Columbia to central California.

SIMILAR SPECIES: This species resembles Speckled Clarkia (*C. cylindrica*) of south-central California and other *Clarkia* species.

CONSERVATION: G5
This species is globally secure.

- ■ NATIVE
- ■ RARE OR EXTIRPATED
- ■ INTRODUCED

ALTERNATE NAMES: Dense-flower Spike-primrose

This flowering plant is native to western North America, where it prefers moist soils but can be found in varied habitats. It features pinkish flowers with four deeply notched petals, and fuzzy green foliage.

DESCRIPTION: This annual ranges in height from 6–40" (15–100 cm). The four-petaled flowers are 1/4–2/3" (6–16 mm) wide, usually pinkish-purple, sometimes nearly white, with dark veining. The leaves are narrow and lance-shaped; the upper leaves are hairy.

FLOWERING SEASON: May–October.

HABITAT: Grows in moist or wet places.

RANGE: Occurs from British Columbia south to California and Arizona.

SIMILAR SPECIES: Smooth Willowherb (*E. pygmaeum*) is similar but has finely toothed leaves; it tends to be shorter and is more often limited to mudflats and vernal pools, whereas Dense-flower Willowherb grows in a wider variety of habitats.

CONSERVATION: **G5**

Globally secure, but critically imperiled in Utah and imperiled in British Columbia. It is possibly extirpated in Montana.

■ NATIVE
■ RARE OR EXTIRPATED
■ INTRODUCED

ALTERNATE NAMES: Linear-leaf Willowherb

This plant is native to wet habitats of northern and eastern North America. The genus name *Epilobium* is derived from the Greek words *epi*, meaning "upon," and *lobos*, meaning "pod" or "capsule," referring to the position of the petals above the ovary.

DESCRIPTION: This perennial herb grows up to 3' (1 m) tall. The small white to pink flowers are ½–¾" (12–18 mm) wide, arranged in a raceme, and the leaves are linear in shape and reach up to 3" (7.5 cm) in length.

FLOWERING SEASON: July–September.

HABITAT: Grows in wet meadows, marshes, bogs, and other moist places.

RANGE: Occurs throughout much of Canada and the northern United States scattered in the West south to Texas.

SIMILAR SPECIES: See Dense-flower Willowherb (*E. densiflorum*).

CONVERSATION: G5

Globally secure, but critically imperiled in Missouri, Oklahoma, and Tennessee. Imperiled in Kansas, Maryland, North Carolina, New Jersey, Virginia, and Wyoming. Vulnerable in Alberta and Newfoundland in Canada and West Virginia in the U.S.

Eulobus californicus **CALIFORNIA SUNCUP**

- NATIVE
- RARE OR EXTIRPATED
- INTRODUCED

ALTERNATE NAMES: Mustard-evening-primrose

This flowering plant is native to the southwestern United States and northwestern Mexico, where it grows in varied dry habitats and begins to bloom in late winter. It is an annual herb that produces a rosette of basal leaves on the ground and then bolts a long, slender stem that terminates in an inflorescence of bright yellow flowers.

DESCRIPTION: This lanky annual plant grows 6"–3' (15–90 cm) in height and features bright yellow flowers ⅔–¾" (15–18

mm) wide,, often with red spots at the bases of the petals. The stem leaves are narrow and jagged-toothed to sharply lobed, and the basal leaves are pinnately lobed.

FLOWERING SEASON: February–June.

HABITAT: Grows in dry, open areas.

RANGE: Found from southern California east to Arizona and south into Mexico.

SIMILAR SPECIES: This species bears a superficial resemblance to some members of the mustard family (Brassicaceae).

CONSERVATION: G4

Although apparently secure globally, this species is critically imperiled in Nevada.

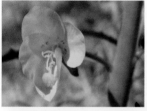

■ NATIVE
■ RARE OR EXTIRPATED
■ INTRODUCED

ALTERNATE NAMES: Squared-pod Water-primrose

This wildflower is native to wetlands and lakeshores of eastern North America. It is named for its squarish fruits, which appear like boxes and are filled with numerous seeds. These seeds rattle around in the boxes when dry.

DESCRIPTION: This perennial plant grows 2–3' (60–90 cm) in height and features a single, short-stalked, yellow flower, ½–¾" (12–18 mm) wide, in each upper leaf axil. The leaves are lanceolate and untoothed and generally grow 2–4" (5–10 cm) long. The distinctive fruits are square-shaped with many seeds.

FLOWERING SEASON: June–September.

HABITAT: Grows in swamps and other places with wet soil.

RANGE: Occurs from Iowa northeast to Ontario and New York, south to Florida, and west to Texas, Oklahoma, and Kansas.

SIMILAR SPECIES: Shrubby Water-primrose (*L. octovalvis*) and Wing-stem Water-primrose (*L. decurrens*).

CONSERVATION: G5

Globally secure, but critically imperiled in Ontario and Nebraska and vulnerable in Iowa and Michigan.

Oenothera biennis **COMMON EVENING PRIMROSE**

■ NATIVE
■ RARE OR EXTIRPATED
■ INTRODUCED

ALTERNATE NAMES: King's cure-all

This plant is native to eastern and central North America. The flowers open in the evening and close by noon the following day. It has been widely used as an ornamental plant, for medicinal purposes, and as a food source for centuries. Hummingbirds often visit these flowers in search of nectar and insect prey.

DESCRIPTION: This biennial grows 2–5' (60–150 cm) in height. The large, lemon-scented, yellow flowers, 1–2" (2.5–5 cm) wide, are situated atop a leafy stalk, with lanceolate, slightly toothed leaves 4–8" (10–20 cm) long.

FLOWERING SEASON: June–September.

HABITAT: Grows in fields, alongside roads, and in disturbed places.

RANGE: Occurs from Alberta east to Newfoundland, south to Florida, west to Texas, north to North Dakota, and in scattered other parts of the West.

SIMILAR SPECIES: The similar Northern Evening Primrose (*O. parviflora*) generally produces smaller flowers and has a small ridge or knob just below its sepal tips. Narrowleaf Evening Primrose (*O. fruticosa*) is similar but is a perennial found in the East; its overwintering basal rosette is evergreen and reddish-purple.

CONSERVATION: **G5**

This species is globally secure but vulnerable in Kansas.

TUFTED EVENING PRIMROSE *Oenothera caespitosa*

- ■ NATIVE
- ■ RARE OR EXTIRPATED
- ■ INTRODUCED

ALTERNATE NAMES: Gumbo Evening-primrose, Gumbo Lily

This short-stemmed or stemless flowering plant is native to western and central North America. As with other evening primroses, this wildflower blooms at night, depending on species such as the Five-spotted Hawkmoth (*Manduca quinquemaculata*) for pollination.

DESCRIPTION: This perennial plant grows up to 12" (30.5 cm) in height. The flowers are large and white, 3–4" (7.5–10 cm) wide, becoming pinkish with age, with yellow stamens. The lobed or toothed leaves are arranged in a basal cluster.

FLOWERING SEASON: April–August.

HABITAT: Grows on exposed hillsides, in canyons, clearings, open woods, and along roads.

RANGE: Occurs from British Columbia east to Manitoba, south to New Mexico, and west to California.

SIMILAR SPECIES: Dune Evening Primrose (*O. deltoides*).

CONSERVATION: G5

Although globally secure, this species is critically imperiled in Manitoba, imperiled in Washington, and vulnerable in Alberta and Saskatchewan.

■ NATIVE
■ RARE OR EXTIRPATED
■ INTRODUCED

ALTERNATE NAMES: Tall Evening Primrose

This tall wildflower grows in open places at a variety of elevations throughout western and central North America. It sports large, yellow flowers up to 4" (10 cm) across that often become pinkish or purplish with age.

DESCRIPTION: This flowering plant grows 1–5' (30–152 cm) tall. The large, yellow, four-petaled flowers are 2–4" (5–10 cm) wide and arranged in a raceme. The numerous narrow leaves range 6–12" (15–30 cm) long, becoming progressively smaller from the base to the top of the reddish and usually unbranched stem.

FLOWERING SEASON: June–September.

HABITAT: Grows on open slopes, roadsides, in grassy areas, and other open places.

RANGE: Occurs from eastern Washington to southern California, east to Texas, north to Colorado.

SIMILAR SPECIES: Common Evening Primrose (*O. biennis*) and Garden Evening Primrose (*O. erythrosepala*). Flowers of similar species have smaller petals.

CONSERVATION: **G5**

This species is globally secure.

NARROWLEAF EVENING PRIMROSE   *Oenothera fruticosa*

■ NATIVE
■ RARE OR EXTIRPATED
■ INTRODUCED

ALTERNATE NAMES: Common Sundrops

This plant is native to eastern North America, where it prefers dry open habitats. The yellow flowers, which typically first appear in late spring, are open on sunny days. In the southern part of this plant's range, the overwintering basal rosette is evergreen and reddish-purple in appearance.

DESCRIPTION: This perennial grows 15–30" (37.5–75 cm) in height. The flowers are 1–2" (2.5–5 cm) wide, yellow with darker veins, and the leaves are lanceolate and 2–3" (5–7.5 cm) long. The fruits are club-shaped.

FLOWERING SEASON: April–May.

HABITAT: Grows in open woods, fields, and disturbed places.

RANGE: Occurs from New England south to Florida, west to Louisiana, and north to Michigan.

SIMILAR SPECIES: See Common Evening Primrose (*O. biennis*).

CONSERVATION: G5

Globally secure, but imperiled in Nova Scotia and vulnerable in Missouri. It is presumed extirpated in Ontario and possibly extirpated in Connecticut.

■ NATIVE
■ RARE OR EXTIRPATED
■ INTRODUCED

ALTERNATE NAMES: Bottle Evening Primrose

This wildflower is native to the eastern United States and can be found in many other parts of the world as an introduced species. The flowers open at dusk and whither by morning. In some southern parts of its native range, this species blooms year-round. It is sometimes grown as an ornamental.

DESCRIPTION: This annual grows 6–30" (15–76 cm) in height. The pale yellow, cup-shaped flowers, ½–1½" (1.3–3.8 cm) wide, become reddish with age. The leaves are elliptical, irregularly toothed to pinnately lobed, and usually hairy.

FLOWERING SEASON: March–October.

HABITAT: Grows in fields and other open places.

RANGE: Occurs throughout much of the eastern and central United States.

SIMILAR SPECIES: Other *Oenothera* species.

CONSERVATION: G5
Although globally secure, this species is critically imperiled in New York and Wyoming, and vulnerable in North Dakota.

PALE EVENING PRIMROSE *Oenothera pallida*

◼ NATIVE
◼ RARE OR EXTIRPATED
◼ INTRODUCED

ALTERNATE NAMES: White-pole Evening-primrose, Sawtooth Evening-primrose

This attractive wildflower boasts large, fragrant white flowers and is often cultivated as an ornamental plant. It attracts a variety of bees, butterflies, and other pollinators. This species prefers sandy or rocky places in its native North American range.

DESCRIPTION: This perennial grows 6–24" (15–60 cm) in height. The flowers are large and white, ¾–1½" (2–4 cm) wide, with four petals. The leaves are linear with saw-toothed edges. The reddish stems become whitish and flaky with age.

FLOWERING SEASON: April–September.

HABITAT: Grows on dunes and sandy steppes.

RANGE: Occurs from Washington east to Montana, south to Texas, and west to Arizona.

SIMILAR SPECIES: See Tufted Evening Primrose (*O. caespitosa*).

CONSERVATION: **G5**
Globally secure, but critically imperiled in Montana and imperiled in British Columbia.

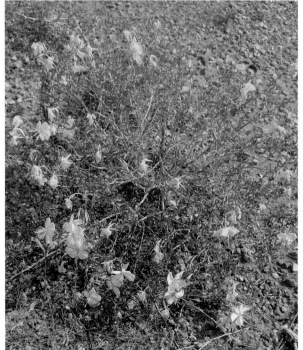

■ NATIVE
■ RARE OR EXTIRPATED
■ INTRODUCED

ALTERNATE NAMES: Yellow Evening-primrose

This leafy wildflower grows in dry open areas of central North America. It flowers for a long period from spring to fall, but the best blooming occurs in spring. The yellow flowers open in the morning and close in the afternoon.

DESCRIPTION: This leafy perennial ranges in height from 8–20" (20–50 cm). The small, four-petaled flowers are yellow, ½–1" (12–24 mm) wide, fading to orange or pink with age. The alternate toothed leaves are narrow and crowded along the branched stems.

FLOWERING SEASON: April–July.

HABITAT: Grows in prairies, on hills, and alongside roads.

RANGE: Occurs from Alberta east to Manitoba and Wisconsin, south to Texas, and west to New Mexico and parts of Arizona.

SIMILAR SPECIES: The similar Lavender-leaf Sundrops (*Calylophus lavandulifolius*) has grayish, untoothed leaves and has a smaller range in the southwestern United States.

CONSERVATION: G5

Although globally secure, this species is critically imperiled in Indiana. It is imperiled in Wisconsin. It is vulnerable in Alberta and Manitoba in Canada and in Montana, Texas, and Wyoming in the U.S. It is possibly extirpated in Arkansas and Kentucky.

■ NATIVE
■ RARE OR EXTIRPATED
■ INTRODUCED

ALTERNATE NAMES: Pink-ladies, Pink Evening-primrose

This attractive flowering plant is native to open habitats of the central United States, now naturalized throughout much of the country and in parts of Mexico. The species name is derived from Latin meaning "showy." This plant is frequently grown in backyard gardens and sometimes spreads, becoming invasive. The leaves and stems of this species can be cooked or served in salads.

DESCRIPTION: This perennial wildflower grows 8–24" (20–60 cm) in height. The flowers are usually large, 2–3½" (5–9 cm) wide, showy, and pinkish, often opening white, with four rounded or shallowly notched petals. The leaves are linear to lanceolate and generally 2–3" (5–7.5 cm) long.

FLOWERING SEASON: May–July.

HABITAT: Grows in prairies, meadows, open woods, and alongside roads.

RANGE: Occurs from Nebraska east to parts of Pennsylvania and Connecticut, south to Florida, and west to Texas and scattered areas of the Southwest.

SIMILAR SPECIES: See other *Oenothera* species.

CONSERVATION: G5
Globally secure, but vulnerable in Iowa.

■ NATIVE
■ RARE OR EXTIRPATED
■ INTRODUCED

ALTERNATE NAMES: Scarlet Beeblossom

This North American native grows in many habitat types, including urban areas. It is primarily pollinated by night-flying moths that are attracted to the white evening-opening flowers; these flowers become pink by the next morning and remain open for less than a day.

DESCRIPTION: A perennial growing 6–24" (15–60 cm) tall with ½–2½" (1.5–6.5 cm) long, crowded leaves. The four-petaled flowers are ½" (12 mm) wide, white in the evening and pink by midmorning.

FLOWERING SEASON: May–September.

HABITAT: Grows in grasslands and a wide variety of other places.

RANGE: Occurs from central Canada and Montana southwest to southern California and Mexico, and east to Minnesota, Missouri, and Texas.

SIMILAR SPECIES: See *Chamaenerion* species.

CONSERVATION: G5
This species is globally secure.

■ NATIVE
■ RARE OR EXTIRPATED
■ INTRODUCED

ALTERNATE NAMES: Tansy-leaf Suncup

This wildflower is native to the western United States, found in varied habitats from sagebrush plains to Ponderosa Pine forests. Lacking stems, this plant grows in the form of a rosette with bright yellow, broadly petaled flowers at the center.

DESCRIPTION: A perennial herb growing 1–4" (2.5–10 cm) tall with yellow, four-petaled flowers ¾–1½" (2–4 cm) wide. The narrow leaves range 2–8" (5–20 cm) in length and feature deep, irregular lobes.

FLOWERING SEASON: June–August.

HABITAT: Grows in meadows, alongside roads and rivers, in forests, and other habitat types.

RANGE: Occurs from Washington south through California and east to Idaho and Montana.

SIMILAR SPECIES: Diffuseflower Evening Primrose (*T. subacaulis*).

CONSERVATION: G5

This species is globally secure.

Rhexia virginica VIRGINIA MEADOW BEAUTY

■ NATIVE
■ RARE OR EXTIRPATED
■ INTRODUCED

ALTERNATE NAMES: Handsome-Harry

This familiar wildflower can be found throughout eastern North America, but is most common in the Southeast. It grows best in moist soils. Plants in this genus have distinctly shaped fruit that author Henry David Thoreau once compared to "little cream pitchers of graceful form."

DESCRIPTION: This plant grows 1–2' (30–60 cm) tall with showy, four-petaled, lavender flowers 1–1½" (2.5–3.7 cm) wide, arranged in terminal clusters. The three-veined, mostly lance-shaped leaves average ¾–2½" (2–6.5 cm) long.

FLOWERING SEASON: July–September.

HABITAT: Grows in open, wet places.

RANGE: Occurs from Ontario and Newfoundland south to Florida, west to Texas, and north to Wisconsin.

SIMILAR SPECIES: Awn-sepal Meadow Beauty (*R. aristosa*) and Maryland Meadow Beauty (*R. mariana*).

CONSERVATION: G5

Globally secure, but critically imperiled in Iowa and Vermont, and vulnerable in Nova Scotia, Michigan, Ohio, and Wisconsin.

- ■ NATIVE
- ■ RARE OR EXTIRPATED
- ■ INTRODUCED

ALTERNATE NAMES: Balloon Vine

This species is a climbing plant, which gets its common name from its balloon-like fruit. *C. halicacabum* is considered a weed and it's fast-growing, known to take over and kill existing vegetation. It is particularly damaging to soybean and sugarcane crops. *C. halicacabum* is native to Central and South America and introduced to the southern and southeastern regions of the United States.

DESCRIPTION: An annual, partially woody vine that's part of the shrub family. The plants climb with forked tendrils that can reach up to 10' (3 m) long. Broad, compound leaves can reach 4" (10 cm) long and 36' (13 m) tall. Leaflets are lance shaped and toothed. This plant's bright green fruit turns tan as it matures. The capsules grow up to 1.13" (3 cm) in diameter and contain three black seeds each, distinguished by a white, heart-shaped marking.

FLOWERING SEASON: Summer through fall.

HABITAT: Roadsides, grasslands, woodlands, forest margins, the margin of ponds, lakes, and swamps, floodplains, and rocky areas.

RANGE: From New York south to Florida, west to Texas and Kansas.

CONSERVATION: **G5**

Globally secure. *C. halicacabum* is considered an invasive species in the United States. Four states—Alabama, Arkansas, South Carolina, and Texas—have placed this plant on their noxious weed lists.

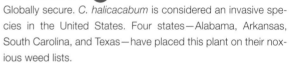

■ NATIVE
■ RARE OR EXTIRPATED
■ INTRODUCED

ALTERNATE NAMES: Dwarf Abutilon

This perennial herb has a multi-branched stem and often becomes tangled in itself. *A. parvulum* grows from a woody root and spreads or trails, reaching a maximum height of 8–24" (20–60 cm). The plant is native to the southwestern United States and northern Mexico.

DESCRIPTION: The shiny, glistening leaves are oval or heart-shaped, growing up to 2" (1-4 cm) wide. The stem and leaves are covered with star-shaped fuzz, while the fruit, nearly 1 cm long, is also hairy, containing five segments. The plant has small, solitary flowers with rounded petals that grow a few millimeters long in a pinkish-orange to red color.

FLOWERING SEASON: Throughout the year.

HABITAT: Slopes, hills, and plains.

RANGE: California east through Colorado and Texas.

CONSERVATION: G5
Globally secure but imperiled in California, Nevada, and Utah.

NATIVE
RARE OR EXTIRPATED
INTRODUCED

ALTERNATE NAMES: Pie-maker, Indian Mallow, Butterprint, Buttonweed

Velvetleaf is native to India. It was introduced to North America in the 18th century and has become a highly invasive and competitive plant, which can cause severe damage to agriculture–especially corn and soybean crops. Each plant produces 700–17,000 seeds, which can live 50–60 years, even after passing through animal digestive systems. Velvetleaf grows tall and can starve surrounding plants of sunlight. It also carries diseases, like maize pest and tobacco pest. Once killed, Velvetleaf further damages crops by releasing a chemical odor. Velvetleaf can be eradicated by mowing when it's young or digging it up or pulling it out manually. It should not be tilled or plowed, as this will allow seeds to germinate.

DESCRIPTION: An annual reaching 3–8' (1–2.5 m) tall and covered in hairs. Its heart-shaped and rounded leaves are velvety and can be 2–10" (5–25 cm) wide. *A. theophrasti* has solitary flowers with five yellow petals 1" (2.5 cm) wide with numerous stamens, which form a tube. The fruit is semi-rounded to cup-shaped, resembling the edges of a pie crust, hence the common name, Pie-maker. The flower has many carpels, each containing 2–9 seeds.

FLOWERING SEASON: July–October.

HABITAT: Found where the soil has been disturbed; also roadsides, fields, open areas, and barnyards.

RANGE: Native to India; it is introduced in midwestern and northeastern regions of the United States and Eastern Canada.

CAUTION: Toxic to livestock.

SIMILAR SPECIES: *Malva neglecta* (Common Mallow) has similar heart-shaped seedlings, but the plant is not covered in hairs. Common Mallow's leaves are also round instead of heart-shaped.

CONSERVATION:
A. theophrasti is on the quarantine list in Washington State, where it's prohibited to transport, buy, sell, or distribute plants, seeds, or plant parts. It is considered an invasive species and a weed.

■ NATIVE
■ RARE OR EXTIRPATED
■ INTRODUCED

ALTERNATE NAMES: Winecup, Buffalo Rose

A sprawling perennial native to Missouri. Purple Poppy-mallow has a long taproot, making it very drought-resistant. Its flowers resemble a winecup, hence one of its common names. The plant can be impacted by rust during the wet seasons in the Great Plains.

DESCRIPTION: Purple Poppy-mallow's stems sprawl along the ground, forming a thick mat that reaches up to 2' (60 cm) with support. The leaves are rounded, hairy, and divided into

5–7 lobes, like a finger. The solitary, cup-shaped flowers have five petals that close in the evening and open in the morning. The flowers also have a white spot at the base and are maroon-colored, about 2.5" (6 cm) wide. After pollination, the flowers close.

FLOWERING SEASON: March–June in the South; May–July in the North.

HABITAT: Open woods; plains; sandy or rocky hills.

RANGE: Michigan and Indiana to Eastern Wyoming, south to Texas and New Mexico.

SIMILAR SPECIES: *Callirhoe digitata* (Fringed Poppy-mallow) is another Missouri native, but by contrast it stands tall instead of sprawling. It grows 2–3' (0.5–1 m) tall.

CONSERVATION: ⓖ⑤
Globally secure, possibly extirpated in Louisiana.

■ NATIVE
■ RARE OR EXTIRPATED
■ INTRODUCED

ALTERNATE NAMES: Paleface

A small shrub found in deserts in the Southwest. The flower is small but showy and can be pale white, pale lavender or light pink, hence the common name Paleface. The petals can be very thin, and even almost translucent on some plants. The flower is bowl-shaped, while the petals are round and overlap, somewhat similar to rose petals.

DESCRIPTION: Paleface is covered in short, whitish hairs along the stem and leaf. The branches are thin and the leaves are a yellowish-green color ½–1" (1.3–2.5 cm) long, with shallow, rounded teeth. The plant reaches a height of 1–3' (30–90 cm). The petals are up to 1' (2.5 cm) long, with many short pinkish stamens joined at the base. Petals are bowl-shaped and dark purple at the center. They can be white, pink, or lavender.

FLOWERING SEASON: February to October.

HABITAT: Found in desert washes, rocky slopes, canyons, and mesas.

RANGE: Southern California to western Texas and northern Mexico.

SIMILAR SPECIES: *Trixis californica*, *Prosopis pubescens*, and *Sphaeralcea incana* are all similar, but more flamboyant than *H. denudatus*.

CONSERVATION: G5

Globally secure, but critically imperiled in Nevada.

- NATIVE
- RARE OR EXTIRPATED
- INTRODUCED

ALTERNATE NAMES: Smooth Rose Mallow or Halberd-leaved Rose

This plant's leaves are recognized by their resemblance to a halberd, a combination of a spear and a battleaxe used in medieval times. The showy, creamy-white or pink flowers are large and hard to miss. These flowers require exposure to sunlight to open up properly, and then last only a single day.

DESCRIPTION: A herbaceous perennial, 3–6' tall (1–2 m) and 2–3' wide (60–90 m), branching sparingly. The stems grow upright and produce large, pink or white flowers with a maroon-colored inside. Each flower is 5" (13 cm) across, with 5 rounded petals. The flowers can be solitary or clustered. The leaves are alternate, up to 6" (15 cm) long and 4" (10 cm) wide, divided into 3–5 pointed lobes with toothed margins. The middle leaf of three-lobed leaves is much larger than the other two. The seeds are large, with fine, white or brown hairs that are irregularly shaped. The fruit is an ovoid capsule and has many seeds. *H. laevis* spreads by reseeding itself. The stalks die down in the winter and grow back in the spring.

FLOWERING SEASON: Mid- to late summer, lasting only one month.

HABITAT: Habitats include marshes and wet areas, such as swamps, bogs, rivers and ponds. It is not often found in highly disturbed areas, and doesn't compete well against the invasive Sandbar Willow (*Salix interior*).

RANGE: New York south through Florida, west through Minnesota and Texas; Ontario.

SIMILAR SPECIES: *Hibiscus laevis* is most similar to *H. moscheutos,* but that species has more star-shaped hairs on the lower leaf. It is also similar to *H. syriacus,* which also produces large flowers, but it is a woody, branching shrub, its leaves are more egg-shaped with sharp teeth, and the flowers appear tightly clustered on short stalks.

CONSERVATION: G5
Globally secure.

CRIMSON-EYED ROSEMALLOW *Hibiscus moscheutos*

■ NATIVE
■ RARE OR EXTIRPATED
■ INTRODUCED

ALTERNATE NAMES: Marshmallow Hibiscus, *Hibiscus palustris*

This is a winter hardy plant with flowers the size of a dinner plate. The large flowers are similar to the garden plant Hollyhock *(Alcea rosea)*. There are many varieties of Crimson-eyed Rosemallow in nature and many hybrids have been created to sell in nurseries.

DESCRIPTION: A tall, shrubby perennial growing 3–8' (90–240 cm) tall with many stems arising from a single crown. The flowers are showy and appear white or pink and large, measuring 4–7" (10–17.5 cm) wide. The flowers are also musky-smelling and have five petals with a conspicuous red or burgundy color at the center, from which a tubular column of yellow stamens extends. The leaves are 4" (10 cm) long and heart-shaped, appearing grayish-green or yellow above and white-hairy beneath. The capsules contain many seeds.

FLOWERING SEASON: July–September.

HABITAT: Grows in undisturbed soils, and wet areas like swampy forests, tidal marshes and inland freshwater marshes.

RANGE: Southern and northeastern North America. Ontario; Massachusetts and New York south to Florida, west through Texas to New Mexico, and northeast to Kansas, Illinois, and Wisconsin; Utah and California.

CONSERVATION: **G5**
Secure, but considered vulnerable in Ontario and Michigan.

■ NATIVE
■ RARE OR EXTIRPATED
■ INTRODUCED

ALTERNATE NAMES: Bladder Hibiscus, Bladder Ketmia, Bladder Weed, Modesty, Puarangi, Shoofly, and Venice Mallow

This plant has just one flower, which closes in the shade, hence the common name. This central African native plant was introduced in the United States as an ornamental. It has since spread rapidly and is considered a noxious weed.

Flower-of-an-hour is a weedy annual, growing 1–2' (30–60 cm) tall, with palmate leaves. It has one white or pale yellow flower that grows about 2" (5 cm) wide, with a dark purplish color in the center. Its sepals form a bladder around the capsules, dividing into five sections. It has many leaves that grow ¾–1¼" (2–3 cm) wide. Some of its stems appear lying on the ground, while others are more erect. This plant can spread by both outcrossing and self-pollinating, which is how it thrives in many different environments. If there is no pollen available, the plant bends over and connects with its own anthers to self-pollinate.

FLOWERING SEASON: August–September.

HABITAT: Open land, disturbed areas.

RANGE: Introduced throughout most of the United States and in Canada from Saskatchewan eastward.

CONSERVATION:
This non-native species is considered a noxious weed in North America.

■ NATIVE
■ RARE OR EXTIRPATED
■ INTRODUCED

ALTERNATE NAMES: Streambank Wild Hollyhock, Mountain Globe-mallow

The several western globe-mallow species are recognizable by their maple-like leaves and pink or rose petals. The seeds of *I. rivularis* remain viable for a century but require heat for germination. The plant flowers in abundance after wildfires but is quickly replaced with other plants.

DESCRIPTION: A stout perennial plant with large, heart-shaped, maple-like leaves and showy pink or pinkish-lavender flowers in long, loose racemes at the top of the stem, and in shorter racemes in upper leaf axils. The plant reaches 3–7' (90–210 cm) tall and 1–2" (2.5-5 cm) wide. The leaves grow 2–8" (5–20 cm) wide and are nearly round, with five or seven triangular lobes.

The flowers have five petals with many stamens joined at the base, with each branch ending in a tiny knob. The plant also produces many segments in a ring, each containing three or four seeds. The smooth seeds have a hard coat.

FLOWERING SEASON: June–August.

HABITAT: Wet, moist areas, including springs, forests, and along mountain streams.

RANGE: British Columbia through eastern Washington to eastern Oregon; east to Montana; south to Utah and Colorado.

SIMILAR SPECIES: This species is distinguished from other members of the mallow family by having more than one seed in each ovary.

CONSERVATION:
Although not ranked globally, the species' population is apparently secure in Canada. However, it is critically imperiled in Alberta. It is also considered vulnerable in Colorado and Wyoming.

■ NATIVE
■ RARE OR EXTIRPATED
■ INTRODUCED

ALTERNATE NAMES: Virginia Fen-rose, Saltmarsh Mallow, *Kosteletzkya virginica*, Coastal Mallow, Virginia Saltmarsh Mallow

This mallow is especially abundant in southern Louisiana. The flowers open for just a day and then close.

DESCRIPTION: The plant grows 1–3' (30–90 cm) tall and 1½–2½" (4–6.5 cm) wide. It has pink flowers and yellow stamens, which grow in leaf axils or atop stems. The long, gray-green ovate leaves grow 2–5" (5–2.5 cm) long and are slightly hairy or rough, with divergent, basal lobes. The flower has five petals and numerous stamens, forming a tubular column around the style, with anthers outside. It contains a flat ring of one-seeded sections.

FLOWERING SEASON: May–October.

HABITAT: Moist areas, brackish to nearly freshwater marshes.

RANGE: New Jersey and Pennsylvania south to Florida and west to Texas.

SIMILAR SPECIES: It is distinguished from *Hibiscus* species by the flat ring of fruit segments. *K. altheaeifolia,* found on the east coast of Texas, is recognized by some botanists as a separate species; some consider it a variety of *K. virginica.*

CONSERVATION: **G5**
Globally secure but considered vulnerable in New Jersey.

■ NATIVE
■ RARE OR EXTIRPATED
■ INTRODUCED

ALTERNATE NAMES: Dwarf Mallow, Common Mallow, Cheesewood, Cheese Plant, Buttonweed

This Eurasian species, with its attractive foliage and flowers, can be found growing all year in much of North America, where it is considered an invasive weed. Mallows have thick seeds, which send out a taproot, allowing them to survive for a long period of time and making the seeds difficult to remove by hand.

DESCRIPTION: A low-trailing creeper plant growing up to 2' (60 cm) long. It has small, whitish-lavender flowers ½–¾" (1.5–2 cm) wide; with five petals notched at the tips and stamens that form a column around the style. The heavily-veined leaves are about 1½" (4 cm) wide, round, and scallop-edged. The fruit of this plant is unique—round, flattish, and forming a ring, which looks like a wheel of cheese, hence the common name.

FLOWERING SEASON: April–October.

HABITAT: Disturbed areas.

RANGE: Native to Eurasia. It is found throughout North America, except in the Far North and parts of the southeastern United States.

SIMILAR SPECIES: Low Mallow (*M. rotundifolia*) is similar to *M. neglecta*, but the flowers are smaller, about ¼" (6 mm) wide. This plant is also similar to the weedy native Carolina Geranium (*Geranium carolinianum*), but Carolina Geranium has more deeply-dissected leaves.

CONSERVATION:
This introduced species in North America is unranked.

- ■ NATIVE
- ■ RARE OR EXTIRPATED
- ■ INTRODUCED

ALTERNATE NAMES: Rocky Mountain Checker-mallow, Salt Spring Checkerbloom

An attractive wildflower. Similar to Hollyhock (*Alcea rosea*), New Mexico Checker-mallow has many large, pink flowers. It is often used as an ornamental in gardens.

DESCRIPTION: New Mexico Checker-mallow grows 1–3' (30–90 cm) tall and 1–1½" (2.5–3.8 cm) wide. This plant has many dark pink flowers that grow in a cluster at the top of leafy, mostly hairless stems. The flowers have five petals and many stamens joined at the base, forming a tube around the style. The lower leaves of this plant and the upper leaves look slightly different. The lower leaves grow 4" (10 cm) wide and are nearly round with five to seven shallow lobes and coarse teeth. The upper leaves are smaller and palmately divided into seven lobes.

FLOWERING SEASON: June–September.

HABITAT: Moist, often heavy soil, in mountain valleys and along streams and ponds at lower elevations.

RANGE: Eastern Oregon to southern California; east to New Mexico, Colorado, and Wyoming; also found in northern Mexico.

SIMILAR SPECIES: There are more than 20 species of checker-mallows. Many of them have pink flowers and survive in coastal marshes and mountains. All of them are native to western North America, making them very difficult to distinguish from each other.

CONVERSATION: G4
Apparently secure. Imperiled in California and Wyoming.

■ NATIVE
■ RARE OR EXTIRPATED
■ INTRODUCED

ALTERNATE NAMES: Copper Mallow, Scarlet Globemallow, Red False Globemallow

There are about 60 species of *Sphaeralcea* globe-mallows. The plants commonly appear in the West and are known to have five bright orange-red petals, making them difficult to differentiate.

DESCRIPTION: This leafy plant grows 20" (50 cm) tall and 1–1¼" (2.5–3.1 cm) wide. It's known for its red-orange or red flowers that bloom in narrow clusters. The flowers have five petals and many stamens, joined at the base and forming a tube around the style. The leaves grow ¾–2" (2–5 cm) wide and are covered in white hairs, making them appear silvery green. The leaves are also roundish and divided into three broad or narrow lobes, which may be further divided or toothed.

FLOWERING SEASON: April–August.

HABITAT: Semi-desert regions; sandy, dry areas; open ground; arid grassland; and among pinyon and juniper.

RANGE: Central Canada; south to western Montana, most of Utah, northeastern Arizona, and most of New Mexico; east to Texas and Iowa.

SIMILAR SPECIES: *S. coccinia* is especially similar to *S. parvifolia* and *S. grossulariifolia*. Distinguishing the plants often requires careful examination of the mature fruit, but there are some that are easier to identify. Scaly Globe-mallow (*S. leptophylla*) is covered in gray hairs and has narrow upper leaves that are not divided or toothed. Desert Globe-mallow (*S. ambigua*) has the typical red-orange flowers of this species in southwestern Utah, Arizona, and farther west, but Desert Globe-mallow's flowers appear lavender or whitish between Phoenix and Tucson and in northern California. Polychrome Globe-mallow (*S. polychroma*), similarly, has white, lavender, and red-orange plants in southern New Mexico and farther south.

CONSERVATION: 🅖5

Globally secure, but critically imperiled in Iowa.

- ■ NATIVE
- ■ RARE OR EXTIRPATED
- ■ INTRODUCED

ALTERNATE NAMES: Smallflower Globemallow, Nelson Globemallow

This plant is one of the most common of the globe-mallow species. Juices from Small-leaf Globe-mallow have been used to treat ailments like sores, cuts, and even broken bones, while the root has been used to treat constipation.

DESCRIPTION: Small-leaf Globe-mallow is a perennial with a large taproot, growing 2–3' (1 m) tall, with several stems. The stems are slender and very hairy, especially when the plant is young. The plant has whitish-gray foliage and orange-red flowers with five petals, appearing clustered. The leaves are alternate on the stems and are thick, round, and three-lobed. They have five noticeable veins on the undersides.

FLOWERING SEASON: May–July.

HABITAT: Dry mesas and slopes at elevations of 4000–7000' (1200–2100 m), semi-desert areas, foothills, and woodlands.

RANGE: West Colorado to New Mexico, Arizona, and Nevada.

SIMILAR SPECIES: There are many species of globe mallows that are difficult to differentiate. Small-leaf Globe-mallow is similar to *S. coccinea*, but *S. parvifolia* can be distinguished by its leaves, which are lobed with wavy edges.

CONSERVATION: **G5**
The population is secure.

■ NATIVE
■ RARE OR EXTIRPATED
■ INTRODUCED

ALTERNATE NAMES: Long-branch Frostweed, Canada Frostweed, Rock Frost, Frostplant, or Frostwort

This plant flowers only in the sunlight and the flower lasts only one day. Ice crystals are known to form from sap, which comes through cracks near the base of the stem in the late fall, hence the common names. This species' genus name comes from the Greek words *krokos*, for "saffron," and *anthemon*, which means "flower."

DESCRIPTION: A perennial growing 8–18" (20–45 cm) tall. The stems are hairy and each stem has one showy yellow flower, or sometimes (rarely) two flowers. The flowers are ¾–½" (2–4 cm) wide, with five petals shaped in a wedge and many stamens. Later in the season, clusters of inconspicuous, bud-like flowers without petals form. These flowers are ⅛" (3 mm) wide, with three to five stamens. The leaves are about 1" (2.5 cm) long and narrow, appearing dull green, with white hairs underneath.

FLOWERING SEASON: May–July.

HABITAT: Dry, sandy, or rocky, open woods, near thin tree covers.

RANGE: Ontario east to Nova Scotia, south to Georgia, west to Alabama, Kentucky, and Missouri, and north to Minnesota.

SIMILAR SPECIES: This species is similar to Hoary Frostweed (*C. bicknellii*), which blooms later in the year, but that species has more than two flowers and has branched star-shaped hairs on its surfaces. Bushy Frostweed (*C. dumosum*), another similar species, blooms earlier in the year and has widely diverging branches.

CONSERVATION: **G5**

Globally secure, but critically imperiled in Georgia, Kentucky, and Tennessee.

ALTERNATE NAMES: False Heather, Sand Golden Heather

The flowers of this species open only in the sunlight for just one day. This plant needs some disturbance, but it's susceptible to human trampling. This plant can grow in sand soils with poor nutrients in some areas, since it can take nutrients from green sands colonized with nitrogen-fixing blue-green algae, notably in Alberta.

DESCRIPTION: A low, matted, gray-green, somewhat woody evergreen reaching a height of 3–8" (7.5–20 cm). It has many small, sulfur-yellow flowers atop short branches. The flowers have five petals and are about ¼" wide. The leaves grow close to the stem and are tiny, scale-like and gray-woolly.

FLOWERING SEASON: May–July.

HABITAT: Sand dunes.

RANGE: Alberta east to Labrador, south to North Carolina, and northwest to Ohio, Illinois, Iowa, and North Dakota.

SIMILAR SPECIES: This species is similar to Golden Heather (*H. ericoides*), which has greenish foliage and outward-spreading leaves. It's found in dry pinelands or sands from Newfoundland and Nova Scotia south to Delaware and South Carolina.

CONSERVATION: G5
Globally secure, but critically imperiled in Illinois, Indiana, Iowa, and North Dakota, Vermont, and West Virginia. It is vulnerable in Virginia, Labrador, and Ontario.

NATIVE
RARE OR EXTIRPATED
INTRODUCED

ALTERNATE NAMES: Yellow Spider-Flower

Yellow Bee Plant often grows in large colonies. The genus includes plants that are evolutionarily similar to mustards in the family Brassicaceae. The flowers indeed resemble those of mustards, but the ovary on a jointed stalk and its compound leaves distinguish this plant as a member of the caper family.

DESCRIPTION: A branched plant growing 1½–5' (45–150 cm) tall with erect, hairless, green stems. The yellow flowers grow in elongated clusters at the tip of the stem and have four petals, about ¼" (6 mm) long, and six long stamens. The leaves are palmately compound and grow widely apart with three to seven leaflets, each ¾–2½" (2–6.3 cm) long and lanceolate. The seedpod is ½–1½" (1.3–3.8 cm) long, slender, on long arched stalks and jointed at the middle.

FLOWERING SEASON: May–September.

HABITAT: Desert plains and lower valleys in mountains, commonly near water or areas formerly filled with water, as well as in woodlands and openings.

RANGE: Eastern Washington to eastern California; east to southern Arizona, northern New Mexico, western Nebraska, and Montana.

SIMILAR SPECIES: Golden Spider-Flower (*P. platycarpa*), a similar species, has hairy stems and leaves.

CONSERVATION: G5
Globally secure, but critically imperiled in Montana.

Polanisia dodecandra RED-WHISKER CLAMMYWEED

This plant's leaves have an unpleasant sulfur-like odor when crushed. The name Clammyweed comes from the sticky or clammy leaves, which leave residue after the plant is touched.

DESCRIPTION: This annual has erect, branched, hairy stems, reaching 6"–3' (15–91 cm) tall. The stems develop a cluster of white, cream-colored, or pinkish flowers at the top, which are surrounded by leaves. Reddish-purple stamens extend well beyond the petals, which is where the name Red-whisker comes from. The light green leaves are hairy, sticky, and cold and clammy to the touch. They are divided into three lance-shaped leaflets, each 1½" (3.8 cm) long and ½" (1.2 cm) wide; some are smaller. Leaves on the upper part of the stem are smaller and singular. The seedpods are slender, 1–2" (2.5–5.1 cm) long, with many reddish-brown seeds, about 2 mm in length.

FLOWERING SEASON: May–October.

HABITAT: Streambanks, waste ground, roadsides, and open places.

RANGE: Alberta to California and Texas; Quebec to Virginia and Tennessee; Quebec to Alberta; Massachusetts to Washington; Virginia to Oregon; Louisiana to California, plus Georgia and Alabama; not in Delaware or Rhode Island.

SIMILAR SPECIES: Yellow Bee Plant (*Peritoma lutea*) is very similar, but the seedpods project outward or down.

CONSERVATION: **G5**
Although this widespread species is globally secure, it is of special concern in Connecticut and critically imperiled in Maryland.

GARLIC MUSTARD *Alliaria petiolata*

ALTERNATE NAMES: Jack-in-the-bush

This plant, native to Eurasia and northern Africa, is known for smelling and tasting like garlic. The genus name *Alliaria* means "resembling allium," and comes from the garlic smell of the crushed leaves. The leaves are used to flavor sauces and salads and are best when young.

DESCRIPTION: An erect biennial, 1–3' (30–90 cm) tall, that grows from a thin taproot. It is sometimes slightly branched and has either kidney-shaped or triangular leaves with white flowers clustered at the top of the stems. The leaves are 1–6" (2.5–15 cm) long, stalked, and toothed and have a garlic odor when crushed. In the first year of growth, plants grow in round clumps of green leaves, close to the ground; these rosettes remain green through the winter and develop into mature flowering plants the following spring. The flowers, about ¼" (6 mm) long and cross-shaped, appear in small clusters of four white petals. The fruit, a green, slender pod, about 1–2½" (2.5–6.5 cm) long, has two rows of many small, shiny black seeds.

FLOWERING SEASON: April–June.

HABITAT: Waste places and woods.

RANGE: This Old World native is exotic in North America. It is introduced in Ontario east to New Brunswick, south to Georgia, west through Tennessee to Oklahoma, and north to North Dakota; also in parts of the West.

CAUTION: Garlic mustard contains a high level of cyanide, which is toxic to many vertebrates.

CONSERVATION:

This plant spreads rapidly and is highly invasive in North American forests, where it can adversely affect native species.

■ NATIVE
■ RARE OR EXTIRPATED
■ INTRODUCED

ALTERNATE NAMES: Cream-flower Rock-cress

The scientific name of Hairy Rockcress was formerly *Arabis hirsuta*. This plant grows in a wide variety of habitats. Its leaves and stems appear larger and thicker in the shade and smaller in the sun. It has long, rough hairs on the stems and leaves, which differentiates it from other species in the mustard family.

DESCRIPTION: This erect perennial grows 1–2½' (30–75 cm) tall and is covered in stiff hairs on the stem and leaves. The stems are often singular, although multiple stems may arise from the base of the plant. The stem becomes progressively less hairy farther up. Basal leaves are 1–3" (2.5–7.5 cm) wide with short petioles, appearing oblong to ovate in shape and toothless or with a few sharp teeth, and rounded tips. The stem leaves are linear, alternate, and clasping. Flowers are about ¼" (0.63 cm) wide and mostly erect, with four rounded white petals and six cream-colored stamens in the center. The fruit is a slender pod up to 3" (7.3 cm) long, usually erect and close to the stem.

FLOWERING SEASON: May–July.

HABITAT: Open woods, ledges, meadows, streambanks.

RANGE: Throughout North America; absent in South Carolina and from Florida through Oklahoma and Texas.

SIMILAR SPECIES: There are two varieties in North America. *Arabis pycnocarpa* var. *adpressipilis* is native only to eastern North America, while *Arabis pycnocarpa* var. *pycnocarpa* is widespread across North America.

CONSERVATION: G5
Secure.

GARDEN YELLOWROCKET *Barbarea vulgaris*

■ NATIVE
■ RARE OR EXTIRPATED
■ INTRODUCED

ALTERNATE NAMES: Common Winter Cress, Bittercress, Rocketcress, Yellow Rocketcress, Winter Rocket, Yellow Rocket, and Wound Rocket

This plant was introduced from Eurasia. It was historically used to treat wounds caused by explosions, which is where the genus name *Barbarea* comes from. *Barbarea* is derived from Saint Barbara, the saint of artillerymen and miners. The young leaves and flower buds can be used in salads or cooked as greens.

DESCRIPTION: This plant grows in tufts, 1–2' (30–60 cm) tall. It has showy clusters of small, bright yellow flowers atop erect, leafy stems. The stems are branched and hairless, with many arising from the base. The lower leaves and upper leaves grow differently. The lower leaves grow 2–5" (5–12.5 cm) long, and are pinnately divided into five segments. The upper leaves are dark and glossy, are lobed, and grow close to the stem. Flowers grow in broccoli-like, rounded clusters ⅜" (8 mm) wide and have four yellow petals, forming a cross. The flowers also have six yellow-tipped stamens and four yellow or green sepals. Fruits appear in green, erect pods, about ¾–1½" (2–3.8 cm) long, and usually curve up or have a beak. The pods contain many brown seeds.

FLOWERING SEASON: April–August.

HABITAT: Moist fields, meadows, brooksides, and waste areas.

RANGE: Naturalized from Manitoba east to Newfoundland, south to Florida, west to Oklahoma, and north to North Dakota; also in much of the West.

CAUTION: Toxic to animals if ingested in large quantities.

SIMILAR SPECIES: Early Winter Cress (*B. verna*), looks similar, but it occurs farther to the east and differs from Yellow Rocket in having basal leaves with 4–10 lobes. The seedpods of Early Winter Cress are also longer. The native American Yellow Rocket (*B. orthoceras*) is another similar species, but some stem leaves of *B. orthoceras* have hairs around the auricles, while Garden Yellow Rocket's auricles are hairless.

CONSERVATION:
This exotic plant is naturalized and widespread in North America.

■ NATIVE
■ RARE OR EXTIRPATED
■ INTRODUCED

This short-lived perennial produces flowers and stems that are so small they can easily be overlooked. Drummond's Rockcress is distinguished from similar species by its erect fruit and mostly hairless stems and leaves.

DESCRIPTION: This slender plant reaches 1–3' (30–90 cm) tall. Flowers appear in a loose raceme at the top of the plant and are ⅓–½" (8–18 mm) across with four narrow, white or lavender petals and six yellow stamens. Four erect, hairless sepals, about half as long as the petals, surround the flowers. Lance-shaped leaves appear in a clump at the base, each 3" (8 cm) long with sparse hairs. Farther up the stem, the leaves are more widely spaced, hairless, and mostly toothless. The fruit is erect and slender, appearing in a two-sectioned pod, 2–4" (5–10 cm) long, and close to the stem. Seeds are up to 2.2 mm long and oval with papery wings around the edge.

FLOWERING SEASON: May–August.

HABITAT: Disturbed areas, rocky slopes, open conifer forests, and mountain meadows.

RANGE: Native to much of North America, including most of Canada and the western and northeastern United States; Maine south through Delaware, Michigan and Ohio west through Washington and California.

SIMILAR SPECIES: Drummond's Rockcress is distinguished by its erect fruit, and mostly hairless stems and leaves. This plant resembles Spreading-pod Rock Cress (*B. grahamii*), but has hairier basal leaves. The erect fruits are similar to Tower Mustard (*Turritis glabra*) and Hairy Rock Cress (*Arabis pycnocarpa*), but both have smaller flowers and hairier leaves.

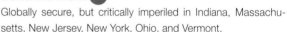

CONSERVATION: G5

Globally secure, but critically imperiled in Indiana, Massachusetts, New Jersey, New York, Ohio, and Vermont.

■ NATIVE
■ RARE OR EXTIRPATED
■ INTRODUCED

This edible plant is native to northern Africa and Eurasia. It's part of the genus that includes vegetables like cabbage, cauliflower, kale, broccoli, and Brussels sprouts. Black Mustard seeds can be ground and used as a spice, which is especially popular in Indian cuisine. The seeds are also used to make the condiment mustard and sometimes sold as bird food. Since the 1950s, however, the Asian native Brown Mustard (*B. juncea*) has become more popular for commercial growing, partly because it can be harvested more efficiently. There are several known medicinal uses of Black Mustard, including cold relief and treatment for respiratory infections. The seeds, mixed with honey, are used as a cough suppressant in eastern Europe.

DESCRIPTION: This erect annual grows 2–3' (60–90 cm) tall and about ½" (1.5 cm) wide. It's widely branched and has rough hairs at the base until it turns smooth farther up the stem. The plant has narrow clusters of small, yellow flowers near the top of the stem, containing four petals. The leaves are 1½–3" (4–7.5 cm) long. The lower leaves have a large terminal lobe and usually four lateral lobes. The upper leaves are not lobed and are typically lanceolate and toothed. A four-sided seedpod, ½" (1.5 cm) long, appears close to the stem. The brown or black seeds are just 1 mm in size.

FLOWERING SEASON: June–October.

HABITAT: Fields and waste places; floodplains; meadows and fields; and shores of rivers or lakes.

RANGE: This exotic plant is found throughout much of North America, except the Far North; it is rarely established in the southeastern United States.

CONSERVATION:

This Old World native is widely cultivated and has been introduced throughout much of North America.

■ NATIVE
■ RARE OR EXTIRPATED
■ INTRODUCED

This annual is the most common of the searocket species. Research has found this species treats unrelated neighboring plants with an aggressive root system to keep them away, but responds more kindly if the nearby plant is related. The young leaves are edible and taste somewhat like horseradish.

DESCRIPTION: A low branching plant growing 6–20" (15–50 cm) tall. The leaves are 3–5" (7½–12½ cm) long and ovate to lanceolate in shape, with wavy-toothed or lobed margins. The flowers appear mostly pale lavender, but can also be white and reddish, stretching ¼" (6 mm) wide, with four petals. The green fruit appears in a two-jointed pod, growing ¾" (2 cm) long. The top joint is longer than the lower joint. The fruit doesn't split open when it turns ripe and the fruit isn't released until the season is over.

FLOWERING SEASON: July–September.

HABITAT: Beaches, dunes.

RANGE: Greenland and Labrador south along the coast to Florida and west to Louisiana; local around the Great Lakes; also along the Pacific Coast.

SIMILAR SPECIES: European Searocket (*C. maritima*)—introduced from Europe and occurring along the East's Atlantic and Gulf beaches and on the Pacific Coast—is similar but has very deeply lobed leaves.

CONSERVATION: G5

Secure, but critically imperiled in Illinois.

■ NATIVE
■ RARE OR EXTIRPATED
■ INTRODUCED

Native to Eurasia, this plant is now found throughout much of the world. The genus name *Capsella* means "little box" and refers to its fruit. *Bursa-pastoris* means "purse of the shepherd." It's cultivated in Asia and used in many food dishes. Shepherd's Purse was used as a spice in colonial New England. It's also been used as a supplement to animal feed and used in medicine to stop bleeding.

DESCRIPTION: Shepherd's Purse grows from a rosette of lobed leaves at the base, 6–16" (15–40 cm) tall. A stem grows from the base with a few leaves topped with flowers. The lanceolate leaves are 1–2½" (2.5–6.5 cm) long and shallowly or deeply lobed. The small, white flowers, about ⅓" (3 mm) wide, have four petals and six stamens, and grow in loose racemes. The seedpod, about ¼" (6 mm) long, contains many seeds and is shallowly notched across the broad end.

FLOWERING SEASON: Throughout the year, earlier in the South than in the North.

HABITAT: Disturbed areas, cultivated land, lawns, and waste places.

RANGE: This Old World native is introduced throughout North America.

CAUTION: This plant can be toxic if consumed in high amounts. Signs of toxicity include sedation, pupil enlargement, and difficulty breathing.

CONSERVATION:

This species is widely cultivated and its population in the wild is unranked.

■ NATIVE
■ RARE OR EXTIRPATED
■ INTRODUCED

ALTERNATE NAMES: Bittercress, Bulbous Bittercress

Spring Cress is fairly common but not often encountered due to its muddy, swampy habitat. It was formerly named *C. rhomboidea*.

DESCRIPTION: This perennial wildflower reaches a height of 6–24" (15–60 cm) with an erect, mostly unbranched stem. It has both basal leaves and alternate leaves. The leaves at the base are 1–1½" (2.5–4 cm) long and oval or roundish, with long stalks. The leaves on the stem are slightly longer and are oblong to lanceolate, usually toothed, without stalks. A raceme of flowers appears at the top of the stem, spanning ½" (1.5 cm) wide when the flower is fully open. The flowers contain four white petals, four sepals, and six stamens, with the petals appearing significantly larger than the sepals. The flowers bloom for about three weeks and then are replaced by narrow seedpods that are 1" (2.5 cm) long. The seeds appear in a single row and are mostly flat and wingless.

FLOWERING SEASON: March–June.

HABITAT: Along springs and brooks, muddy areas, swamps, and cleared wetland.

RANGE: Manitoba east to Nova Scotia, south to Florida, west to Texas, and north to North Dakota.

SIMILAR SPECIES: Purple Cress (*C. douglassii*) is very similar to Spring Cress. Both species appear in similar habitats and both bloom during the spring, but Purple Cress blooms about two weeks earlier than Spring Cress. Purple Cress also has purple-pinkish flower petals, not white, and very dark purple sepals. The stem of Purple Cress is hairy at the base.

CONSERVATION:

Secure, but critically imperiled in Kansas, Maine, North Dakota, and Vermont.

NATIVE
RARE OR EXTIRPATED
INTRODUCED

ALTERNATE NAMES: Large Mountain Bittercress

The flowers and young leaves of Heartleaf Bittercress are edible and taste like horseradish. Some plants in this family have also been used to treat heart ailments.

DESCRIPTION: Heartleaf Bittercress has an extensive underground root system and reaches a height of 4–32" (10–80 cm). The leaves are simple and undivided, stretching from ¾–4" (2–10 cm) wide and growing in a heart shape at the base, hence the common name. The large, bright white flowers have four petals, growing ½–¾" (1.3–2 cm) long. The fruit is flat and slender, growing in pods ¾–1½" (2–3.8 cm) long.

FLOWERING SEASON: June–September.

HABITAT: Along mountain streambanks, in streams and alpine meadows.

RANGE: British Columbia; south to northern California; east to New Mexico, Wyoming and Idaho.

CAUTION: Although edible, ingesting this species from the wild should be avoided because it grows near stagnant water that can carry disease-causing parasites.

SIMILAR SPECIES: Heartleaf Bittercress is distinguished from other species of *Cardamine* by its large flowers and undivided leaves.

CONSERVATION: G5

Secure, but imperiled in Wyoming.

■ NATIVE
■ RARE OR EXTIRPATED
■ INTRODUCED

ALTERNATE NAMES: Two-leaf Toothwort, *Dentaria diphylla*

Crinkleroot is an important species to wildlife. The West Virginia White butterfly (*Pieris virginiensis*) and the Falcate Orangetip butterfly (*Anthocharis midea*) use the plant to lay eggs, and bees collect pollen from this species. The leaves and roots are edible and can be eaten raw or cooked.

DESCRIPTION: An upright perennial growing 8–16" (20–40 cm) tall. The paired leaves are sharply toothed and may even appear as compound; they are divided into three to five sections. A cluster of white to light pink flowers appear at the end of the stem. The fruit grows in a pod that splits open and releases seeds at maturity. It can take three or four years before seedlings bloom.

FLOWERING SEASON: March–June.

HABITAT: Wooded slopes; deep woods.

RANGE: Nova Scotia to southern Ontario and Wisconsin, south to South Carolina and Mississippi.

SIMILAR SPECIES: This plant is similar to Cut-leaved Toothwort (*C. concatenata*), but Crinkleroot's leaves appear in a whorl.

CONSERVATION: ⬤G5

Secure, but critically imperiled in Mississippi.

FLIXWEED *Descurainia sophia*

■ NATIVE
■ RARE OR EXTIRPATED
■ INTRODUCED

ALTERNATE NAMES: Herb-sophia, Tansy mustard

Flixweed, a winter annual, reproduces solely from seed. This Eurasian native is a common weed of roadsides and agricultural fields in North America.

DESCRIPTION: Flixweed grows 2' (60 cm) high or taller. Its hairy basal leaves grow 1–4" (3–10 cm) long, but wither when the plant flowers. The leaves, growing in an alternate pattern, are two to three times finely divided and become smaller and less lobed farther up the stem. The small, yellow flowers are up to ⅛" (0.3 cm) long, with four petals and four yellow-greenish sepals that are mostly hairless. Flixweed's upright stems are usually branched and hairy, appearing green to purplish. The fruit is about 1" (2.5 cm) long and grows in a pod, containing a single row of 12–25 seeds.

FLOWERING SEASON: May–July.

HABITAT: Disturbed habitats, fields, roadsides, logged-over forests, agricultural lands, vineyards, orchards, gardens, canyon bottoms, and disturbed desert areas.

RANGE: Native to Eurasia; it is introduced and naturalized in the United States and Canada. It has appeared in every state except Alabama and Florida.

CAUTION: All parts of the plant are poisonous if eaten and can cause livestock to go blind.

CONSERVATION:
This widespread introduced species has not been ranked.

■ NATIVE
■ RARE OR EXTIRPATED
■ INTRODUCED

ALTERNATE NAMES: Hoary Whitlow-grass

This is a perennial herb native to California. There have long been disagreements about the name and characteristics of this plant. The name *D. lanceolata* was given in the 1800s to a Eurasian species and mistakenly given to this plant in the 19th and 20th centuries. There are also wide disagreements about the plant's length, the number of stems it has, the length of the seedpods, and whether it grows in clumps or not. The most distinctive feature of *D. cana* is that it has hairs on all parts of the plant, which gives it an ashy gray or blue-greenish color.

DESCRIPTION: *D. cana* grows up to 4–12" (10–30 cm) tall from a rosette of leaves at the base. The basal leaves are about 0.8" (2 cm) long and covered in hairs. The plant also has 3–10 leaves scattered along the hairy stem, which are usually smaller than the basal leaves. There are 15–47 flowers that have four small, green sepals and four white petals in a compact raceme at the top of the stem, which may elongate in fruit. The fruits are flat, slender, hairy, and close to the stem. They grow about 0.4" (1 cm) long.

FLOWERING SEASON: May–August.

HABITAT: Exposed, dolomite cliffs, rock crevices, and small ledges.

RANGE: Throughout Canada and in the northernmost and Western United States.

SIMILAR SPECIES: This plant is similar to Rock Whitlow-grass (*D. arabisans*), but *D. cana* has hairs on the outside of the fruits that appear stellate.

CONSERVATION: G5

D. cana is critically imperiled in Maine, New Hampshire, Vermont, and Washington. This species is endangered in Minnesota and Wisconsin. It's also listed as threatened in Michigan.

SPRING DRABA *Draba verna*

NATIVE
RARE OR EXTIRPATED
INTRODUCED

ALTERNATE NAMES: Spring Whitlow-grass

This annual was introduced from Europe and has naturalized in much of North America. As the name suggests, it often blooms in the spring. The plant was also once known for treating finger sores called whitlows, hence its other common name.

DESCRIPTION: This small plant grows up to 8" (20 cm) tall. It has slender, leafless stalks and miniature flowers, just ⅛" (4 mm) wide, with four white petals that are yellow at the base and are very deeply notched, appearing as eight petals. The center of the flower has six yellow stamens. The leaves, in a basal rosette, are long and hairy, spanning ½–1" (1.5–2.5 cm). They can be oblong or spatulate in shape. Stems are green to purplish and unbranched. The lower stem is initially hairy, becoming hairless above. *D. verna* produces fruit containing 40 golden-brown seeds in a flattened elliptical pod.

FLOWERING SEASON: March–June.

HABITAT: Fields, roadsides, and open places.

RANGE: Introduced from Ontario east to Nova Scotia, south to Georgia, west to Mississippi, and north to Iowa and Wisconsin; also in parts of the West.

SIMILAR SPECIES: *D. verna* is similar to Carolina Whitlow-grass (*D. reptans*), but *D. reptans'* flower petals are not deeply cleft. *D. reptans'* lower stem also has leaves, while *D. verna's* leaves appear in a basal rosette.

CONSERVATION:

This species is exotic in North America and has naturalized in many areas.

■ NATIVE
■ RARE OR EXTIRPATED
■ INTRODUCED

ALTERNATE NAMES: Sanddune Wallflower, Prairie Rocket

Erysimum capitatum is an attractive perennial, cultivated for ornamental use. It is widespread throughout western North America, with smaller native populations occurring farther eastward in the United States. In Zuni culture, the entire plant is used externally to soothe muscle aches. The flower and the fruit can also be eaten to induce vomiting for stomachaches.

DESCRIPTION: *E. capitatum* reaches a height of 32" (80 cm). The plant has thin, upright stems containing dense clusters of mostly bright yellow-orange flowers, though the colors can vary and may appear as red, white, or purple in some regions. Each flower has four flat petals. The leaves are linear and finely toothed while the seedpods grow almost parallel to the stem.

FLOWERING SEASON: May–July.

HABITAT: Inland sand dunes, plains, foothills, and high elevation coniferous forests.

RANGE: Widespread in western North America; also occurs from Kansas and Oklahoma and to the northeast in scattered populations in the Midwest.

SIMILAR SPECIES: Western Wallflower widely varies in size and flower color. It is similar to the cultivated wallflower (*Cheiranthus* spp.).

CONSERVATION: G5

Globally secure but scattered easterly populations are critically imperiled in Indiana, Ohio, Tennessee, and West Virginia; and imperiled in Arkansas and Virginia. Contra Costa Wallflower, a subspecies endemic to Contra Costa County east of San Francisco, California, is endangered.

■ NATIVE
■ RARE OR EXTIRPATED
■ INTRODUCED

ALTERNATE NAMES: Mother-of-the-evening, Damask-violet, Dame's-violet, Dames-wort, Dame's Gilliflower, Night-scented Gilliflower, Queen's Gilliflower, Rogue's Gilliflower, Summer Lilac, Sweet Rocket, Good & Plenties, and Winter Gilliflower

This is a biennial or short-lived perennial, native to Eurasia and widely cultivated for its attractive flowers. The genus name *Hesperis* comes from the Greek word *hesperos* ("evening"), a reference to the scent of the flowers, which is strongest at night. The species name, meaning "matronly," refers to its many common names, including Mother-of-the-evening.

DESCRIPTION: *Hesperis matronalis* is an upright, multi-branched plant reaching a height of 1–4' (30–120 cm). It features terminal racemes of fragrant flowers appearing as white, pink, lavender, or purple. The flowers are about ¾" (2 cm) long, on stalks ½" (1.5 cm) long and contain four sepals, four petals, and six stamens. The leaves are toothed and lanceolate-shaped, growing 2–6" (5–15 cm) long. The upper leaves are stalkless or short-stalked, while the lower are short to long-stalked. The fruit appears as a long, slender pod, 2–4" (5–10 cm) long, with many seeds in a row.

FLOWERING SEASON: April–August.

HABITAT: Roadsides, open woods, woodland borders, thickets, and waste places.

RANGE: Eurasian native, widely introduced in much of the West and throughout the East, except the Gulf States and the Arctic.

CONSERVATION: **G4**
Globally the species is apparently secure. An exotic in North America, it is considered a weed in some states and prohibited in Massachusetts.

■ NATIVE
■ RARE OR EXTIRPATED
■ INTRODUCED

ALTERNATE NAMES: Asp of Jerusalem

During the Middle Ages in Europe, the indigo chemical Dyer's Woad produces was extracted to create a blue dye for cloth and body paint. Woad was also cultivated for its medicinal properties and was used to treat wounds, reduce fevers, and treat ulcers. Today, Dyer's Woad is considered an aggressive weed in certain parts of North America and its chemicals are known to inhibit growth of nearby plants.

DESCRIPTION: This species is a short-lived perennial, growing 1–4' (30–120 cm) tall. The leaves are bluish-green, 1–7" (3–18 cm) long, and covered with fine hairs. When the rosette grows, up to 20 stems can be produced. Usually seven or eight of these stems mature, producing flowering branches. The leaves on these branches are lance shaped, alternate and close to the stem. The flowers are small, wide-branching, and grow in flat-topped clusters. The small fruit appears as blackish pods, containing one seed each.

FLOWERING SEASON: June–July.

HABITAT: Roadsides, agricultural fields, and other disturbed areas.

RANGE: Native to western and central Asia; it is introduced in much of western North America and scattered areas in the East.

CONSERVATION:

This species is exotic in North America. It is listed as a noxious weed in several western states. A 20-year eradication effort in Montana removed the plant from 9 of the 13 counties in which it had been introduced.

■ NATIVE
■ RARE OR EXTIRPATED
■ INTRODUCED

ALTERNATE NAMES: Heart-pod Hoarycress, Whitetop, *Cardaria draba*

Native to Eurasia, this plant was accidentally introduced to North America in the early 1900s, where it is now considered a noxious weed. Hoarycress has an extensive root system and plants can resprout from small root fragments. The plant also spreads rapidly due to the vast number of seeds it produces.

DESCRIPTION: Hoarycress, a showy perennial, reaches a height of 8–20" (20–50 cm). Leaves and stems are grayish and have small hairs. Patches of leafy stems produce numerous tiny white flowers in racemes. There are four petals, about ⅛" (3 mm) long. The four sepals drop upon opening and are about half as long as the petals. The leaves are oblong and grow 1½–4" (3.8–10 cm) long. They are further pointed at the tip and have

edges with small teeth. The base of the blade is attached to the stem in a notch between two backward-projecting pointed lobes. The fruit is a flat pod about ¼" (6 mm) wide, with two lobes containing one seed each. The fruit is roundish, slightly flat, and smooth.

FLOWERING SEASON: April–August.

HABITAT: Roadsides, fields, old lots, agricultural sites, and disturbed land.

RANGE: Introduced throughout most of western North America; in the East south to Missouri, Kentucky, and Virginia.

SIMILAR SPECIES: *L. draba* is fairly unique. While most *Lepidium* species are annuals, *L. draba* is a perennial. Its four-petaled flowers with flattish clusters also don't elongate, which further separates it from other *Lepidium* species. The less common Globe-pod Hoary Cress (*C. pubescens*) has a downy seedpod.

CONSERVATION:
This exotic species is widely introduced and considered invasive in North America.

■ NATIVE
■ RARE OR EXTIRPATED
■ INTRODUCED

ALTERNATE NAMES: Poorman's Pepperwort, Poor-man's Pepper, Peppergrass

Virginia Pepperweed is one of the most common species of peppergrass. It is considered a weed among crop growers and readily grows in disturbed sites. The leaves contain high amounts of protein, vitamin A, and vitamin C, and can be cooked or eaten raw in salads. The fruit, which forms seedpods, can be used as a substitute for black pepper, hence the common name.

DESCRIPTION: The plant reaches a height of 6–24" (15–60 cm) and width of ⅟₁₆" (2 mm). It contains elongated clusters of small, white flowers with four petals and two (rarely four) stamens. The leaves at the base of the plant are about 2" (5 cm) long and toothed, with a large, terminal lobe and several small side lobes and stalks. The leaves on the stem are lanceolate-shaped without stalks. The fruit forms a dry, rounded, flattened pod, which is notched at the tip.

FLOWERING SEASON: June–November.

HABITAT: Waste areas, roadsides, and other disturbed sites.

RANGE: Native throughout much of North America, except the Prairie Provinces and parts of the Far North.

SIMILAR SPECIES: Field Peppergrass (*L. campestre*)—a common weed introduced from Europe—is a similar species; unlike *L. virginicum*, it has six stamens and a hairy stem. The fruit pods are longer than they are wide. Clasping Peppergrass (*L. perfoliatum*), also from Europe, has yellow flowers. Its lower leaves are finely cut and fern-like. The upper leaves are circular and wrap around the stem.

CONSERVATION: **G5**

Globally secure, but listed as vulnerable in Wyoming.

NATIVE
RARE OR EXTIRPATED
INTRODUCED

ALTERNATE NAMES: Watercress

Botanists disagree about this plant's genus. Some put it in the genus *Rorippa*, others put it in the genus *Nasturtium* and call it *N. officinale*. Nasturtium comes from the Latin words *nasi tortium* ("distortion of the nose"), which gives a nod to the plant's strong smell. Despite being mostly water, its leaves add a peppery taste to salads.

DESCRIPTION: A creeper plant reaching 10" (10–5 cm). It has leafy branched stems that mostly float in water or appear in mud, with upturned tips from which tiny white flowers grow in racemes. The flowers have four petals, each about ¼" (5 mm) long. It has six yellowish stamens. The leaves are 1½–5" (4–12.5 cm) long and pinnate with ovate leaflets. The terminal leaflet is the largest. The fruit is a slender pod, about ½–1" (1.5–2.5 cm) long, gently curved and pointing upward. The reddish-brown seeds grow in two rows.

FLOWERING SEASON: April–October.

HABITAT: Brooks, streams, and springs.

RANGE: Throughout much of temperate North America, except North Dakota, Manitoba, Quebec, and the Far North.

CAUTION: Although edible, the leaves should be thoroughly washed first to remove insects and parasites from the water it grows near.

SIMILAR SPECIES: When not fruiting, True Watercress is almost identical to *N. microphyllum*, but *N. microphyllum* has one row of seeds, not two. Nasturtium (*Tropaeolum majus*) also has a sharp flavor, but it is not related.

CONSERVATION:

This species is unranked. It's considered invasive in some areas, where it can spread rapidly out on the surface of the water and destroy native ecosystems.

■ NATIVE
■ RARE OR EXTIRPATED
■ INTRODUCED

ALTERNATE NAMES: Alpine Pennycress

This plant has gone through more than a dozen name changes since it was given the genus name *Noccaea* in the 1800s. It's sometimes called *Thlaspi montanum* or *N. montanum*. Wild Candytuft grows in large patches, with few other plants around it. Several short, unbranched stems with small, arrow-shaped leaves grow from a basal rosette of leaves that terminate in racemes of small white flowers.

DESCRIPTION: The plant reaches 1¼–16" (3.1–40 cm) tall. The flowers have four petals, each about ¼" (6 mm) long. The leaves at the base are ½–2½" (1.3–6.3 cm) long, with ovate blades at first, which become a slender petiole up the stem. The leaves on the stem are smaller and don't have petioles. The flat, ovate seedpods are ³⁄₁₆–½" (5–13 mm) long, pointed at the tip, and notched.

FLOWERING SEASON: February–August.

HABITAT: Open slopes in mountains from moderate to high elevations.

RANGE: British Columbia to northern California; east to the Rocky Mountains from Alberta to southern New Mexico and western Texas.

SIMILAR SPECIES: Candytufts may look like peppergrasses, but the chambers of Candytufts' pods have two seeds, while peppergrasses have just one seed.

CONSERVATION: G5
Secure.

■ NATIVE
■ RARE OR EXTIRPATED
■ INTRODUCED

ALTERNATE NAMES: Popweed, Lesquerella

This perennial, native to the southwestern United States and northern Mexico, is one of the earliest plants to flower in its area.

DESCRIPTION: This tufted plant grows to 1–16" (2.5–40 cm) high and the surface is covered with tiny, star-like scales, making it appear silvery-gray. The lanceolate leaves are up to 4" (10 cm) long, with the leaves at the base sometimes having teeth on the edges. The yellow flowers grow in loose racemes at the ends of stems and have four petals, each about ½" (1.5 cm) wide. They have a red color at the throat. The smooth fruit is a spherical pod, ¼–⅜" (6–9 mm) long. It has 6–25 seeds.

FLOWERING SEASON: March–June, often again after summer rains.

HABITAT: Rocky or sandy soil, especially that derived from limestone, in arid grasslands or deserts.

RANGE: Southern Utah east to western Kansas and south through eastern Arizona, New Mexico, and western Texas to northern Mexico.

SIMILAR SPECIES: Gordon's Bladderpod (*L. gordonii*) is similar, but the annual has slender stems that lie on the ground, turning up at the tips. It is not tufted.

CONSERVATION: **G5**
Secure, but imperiled in Utah.

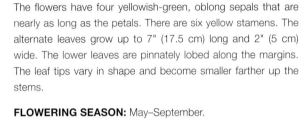

■ NATIVE
■ RARE OR EXTIRPATED
■ INTRODUCED

ALTERNATE NAMES: Bog Yellowcress, Marsh Yellowcress

Common Yellowcress is a widespread plant that has adapted to many soils. The species name *palustris* is Latin for "of the marsh" and is a nod to marshes and bogs where this plant typically grows.

DESCRIPTION: Common Yellowcress, a biennial, grows 1–3' (30–90 cm) tall. The stems are light to reddish-green, mostly smooth or hairy, and strongly furrowed or ribbed. The upper stems end in racemes of flowers about 4–12" (10–30 cm) long.

The flowers have four yellowish-green, oblong sepals that are nearly as long as the petals. There are six yellow stamens. The alternate leaves grow up to 7" (17.5 cm) long and 2" (5 cm) wide. The lower leaves are pinnately lobed along the margins. The leaf tips vary in shape and become smaller farther up the stems.

FLOWERING SEASON: May–September.

HABITAT: Muddy shores, marshy ground.

RANGE: Throughout North America.

CONSERVATION:  G5
Secure, but vulnerable in Wyoming and imperiled in Kansas.

- NATIVE
- RARE OR EXTIRPATED
- INTRODUCED

ALTERNATE NAMES: Jim Hill Mustard

Sisymbrium altissimum, an annual, is native to Europe and has become invasive throughout the United States. It germinates early in the season and grows quickly. When the fruit dries, it snaps off at the base of the plant and tumbles via the wind, through roadsides and open fields, dispersing its seeds, hence the common name. Each pod contains more than 100 seeds and each plant can produce hundreds of pods.

DESCRIPTION: This is a weedy, rounded plant that reaches 5' (1.5 m) tall. The lower leaves are large and pinnately divided, up to 8" (20 cm) long. The upper leaves are small and divided into threadlike segments. A tight cluster of yellow flowers appear twig-like. Each flower has 4 small, light yellow petals and 4 green sepals. As the flower matures, it becomes a seedpod, 2–4" (5–10 cm) long, containing a single row of up to 100 seeds.

FLOWERING SEASON: June–August.

HABITAT: Disturbed habitats, roadsides, fields, and sagebrush communities.

RANGE: Native of Eurasia; it is introduced in North America and reported in nearly every state and province.

CONSERVATION:

This species is not ranked, but it is widely naturalized throughout most of the world.

■ NATIVE
■ RARE OR EXTIRPATED
■ INTRODUCED

ALTERNATE NAMES: The Plant for Singers

Hedge Mustard is a native of Europe that has been widely introduced around the world, including throughout North America. Young shoots can be eaten raw or cooked and have a cabbage-like taste, while the seeds can be ground up to flavor foods. The Greeks believed Hedge Mustard was the cure for poisons, while in folk medicine, this plant has been called "The Plant for Singers" because it is believed to clear a hoarse throat.

DESCRIPTION: This is a hairy plant, 1–3' (30–90 cm) tall, with a wiry stem. Most leaves appear in the lower half of the plant. The lower leaves are oblong, up to 8" (20 cm) long, and pinnately cleft. The upper leaves are smaller, with a few narrow lobes. The stem is covered in hairs that stand out or point downward. Branches in the upper part of the plant end in racemes of small, yellow flowers with four petals, each about ¼" (5 mm) long. The fruit appears in an erect, slender pod, ⅜–½" (9–15 mm) long, and pressed tightly against the stem.

FLOWERING SEASON: March–September.

HABITAT: Fields, roadsides, and vacant lots.

RANGE: An exotic that's well-established throughout Canada and the United States; in the West, it occurs more frequently west of the Cascade Range and Sierra Nevada.

SIMILAR SPECIES: Hedge Mustard is similar to Tumbling Mustard *(S. altissimum)* and Loesel Tumble Mustard *(S. loeselii)*, but the seedpods are different. Hedge Mustard's fruit is close to the stem, unlike the other two species. The weedy invasive London Rocket *(S. irio)* is also similar but the flowers are yellower.

CONSERVATION:
This exotic species is unranked but established in North America and throughout the world.

NATIVE
RARE OR EXTIRPATED
INTRODUCED

ALTERNATE NAMES: Golden Prince's-plume

This is a conspicuous and showy plant. Its flowers are usually taller than in nearby plants that share its preferred habitats.

DESCRIPTION: This plant reaches 1–4' (30–120 cm) tall. Smooth, pale green leaves at the base are 2–8" (5–20 cm) long, pinnately divided, and broadly lanceolate. The leaves on the stem are also pinnately divided but smaller. It has yellow flowers, ⅜–1" (0.9–2.5 cm) long, with four yellow petals and four yellow sepals appearing at the top of tall, smooth, bluish-green stems with many leaves. The stamens and pistil extend beyond the petals. The flowers are very hairy on the inside. The seedpods are slender and about 1¼–3" (3.1–7.5 cm) long, curving downward and appearing on stalks measuring ½–¾" (1.3–2 cm) long.

FLOWERING SEASON: May–July.

HABITAT: Dry soil in deserts and plains to lower mountains, often with sagebrush.

RANGE: Southeastern Oregon south to southeastern California and east to western Great Plains from North Dakota to western Texas.

CAUTION: This species and others in the genus absorb selenium from the soil. This can be dangerous to livestock and humans who ingest the plant.

SIMILAR SPECIES: This plant is distinguished by its hairy flowers. Most related species have yellow flowers without hairs, except White Desert Plume *(S. albescens)*, which has white petals, not yellow, with hairs on the inside.

CONSERVATION: G5
Secure.

■ NATIVE
■ RARE OR EXTIRPATED
■ INTRODUCED

Douglas-fir Dwarf Mistletoe is native to North America. This parasitic plant grows inside trees and kills them slowly by consuming their nutrients. It causes the deformity known as witch's broom, where the branches form a vertical mass of shoots that look like a broom or a bird's nest. The deformity is most common on the upper branches.

DESCRIPTION: This plant is most commonly found on Douglas Fir trees (*Pseudotsuga menziesii*), but it is occasionally seen on Pacific Silver Fir (*Abies amabilis*), White Fir (*Abies concolor*), White Spruce (*Picea engelmannii*), Blue Spruce (*Picea pungens*), and others. Most of the plant grows inside the tree with a root-like structure that absorbs water and nutrients from the host, but its scaly stems are noticeable on the host tree's bark.

The small, branching stems are greenish, olive, or orangish, with bud-like tips appearing on tree limbs. Its flowers are red or purplish and staminate. The fruit, a berry, disperses seeds near the parent plant and host tree.

FLOWERING SEASON: March–June.

HABITAT: Conifer limbs, forests, woodlands.

RANGE: British Columbia and Alberta south throughout western United States to California and Texas.

CAUTION: All mistletoes may be poisonous if ingested.

CONSERVATION: G5

Globally secure, although possibly extirpated in Wyoming.

■ NATIVE
■ RARE OR EXTIRPATED
■ INTRODUCED

ALTERNATE NAMES: Umbellate Bastard Toadflax

This plant makes its own food through photosynthesis, but it is considered a parasite, feeding on other plants nearby to meet its nutritional needs through its rhizomes. The genus name comes from the Greek *come* ("hair") and *andros* ("a male") and refers to the way the anthers attach to the sepals by its hair. It's a host to Comandra Blister Rust, a fungus that kills pine trees. Blister Rust spreads from tree to tree through host plants.

DESCRIPTION: This is a parasitic plant, 6–16" (15–40 cm) tall, with compact, terminal clusters of small, greenish-white, funnel-shaped flowers. The flowers are ⅛" (4 mm) wide and petal-less, with five sepals, which are often connected to anthers by chunks of hair. It has five stamens and one pistil. The oblong leaves are ¾–1¼" (2–4 cm) long, with a pale underside. The fruit is a small nut that turns blue or brown as it matures and has one seed.

FLOWERING SEASON: April–June.

HABITAT: Dry fields and thickets.

RANGE: Throughout North America except Alaska, Florida, Louisiana, and the Arctic.

SIMILAR SPECIES: Northern Comandra (*Geocaulon lividum*) is a similar plant, but it is smaller, with purple flowers, and orange to red, juicy, berry-like fruit. Flowering Spurge (*Euphorbia corolla*) is also similar when young, but it's larger at maturity and blooms later in the year. Flowering Spurge also has whiter and thicker sap.

CONSERVATION: **G5**

Globally secure but vulnerable in Georgia, Illinois, Minnesota, North Carolina, and West Virginia.

■ NATIVE
■ RARE OR EXTIRPATED
■ INTRODUCED

ALTERNATE NAMES: Sea Thrift, Sea Pink, Western Thrift

This dense evergreen is one of the few plants that can tolerate salty conditions near the sea. It resembles a small onion plant, but it's unrelated.

DESCRIPTION: A low, mat-forming plant, 2–16" (5–40 cm) tall, with a basal cluster of many narrow leaves. The leaves are stiff and grass-like, 2–4" (5–10 cm) long. It has pale lilac or whitish flowers, which grow in clusters, atop the slender, leafless stalk.

The flowers have several purplish, papery bracts beneath. The flower head is ¾–1" (2–2.5 cm) wide and has a funnel-like calyx of pinkish parchment.

FLOWERING SEASON: March–August.

HABITAT: Beaches and coastal bluffs, or slightly inland on prairies.

RANGE: The Pacific Coast from British Columbia to southern California; in Arctic North America and Eurasia.

CONSERVATION: **G5**

Globally secure but vulnerable in Ontario and Yukon, and critically imperiled in Manitoba and Saskatchewan.

ALTERNATE NAMES: *Limonium nashii*, Marsh Rosemary

Carolina Sea Lavender is a slow-growing plant. It's showy in late summer on the tidal marshes and it is often collected to make wreaths and other flower arrangements. It has salt glands on its leaves, allowing it to survive in coastal environments.

DESCRIPTION: This is a smooth, saltmarsh perennial that reaches 1–2' (30–60 cm). Small, pale purple flowers grow in a cluster on the side of the stems. Flowers are about ⅛" (3 mm) wide and appear as a calyx with five teeth and 10 ribs. The corolla is funnel-like, with five spatulate lobes. Basal leaves are 2–10" (5–25 cm) long and lanceolate with smooth or somewhat wavy margins.

FLOWERING SEASON: July–October.

HABITAT: Saltmarshes.

RANGE: Newfoundland and Quebec south to Florida; west to Mississippi and Texas.

CONSERVATION: G5
Globally secure but vulnerable in Quebec and imperiled in Newfoundland.

■ NATIVE
■ RARE OR EXTIRPATED
■ INTRODUCED

ALTERNATE NAMES: American Bistort, Miner's Toes, *Polygonum bistortoides*

Western Bistort is one of the most common wildflowers in high mountains, sometimes covering meadows with thousands of clusters of white flowers so thickly they look like snow. Although the flowers are attractive, they are unpleasant to smell, hence the common name Miner's Toes. The roots were once an important food source for Native Americans. Young leaves may be cooked as greens. The seeds can be dried and ground into flour or roasted.

DESCRIPTION: This plant reaches a height of 8–28" (20–70 cm). Dense white or pale pink flower clusters appear atop slender, erect, reddish stems. The clusters are 1–2" (2.5–5 cm) long.

Each flower is less than ¼" (6 mm) long, with five petal-like segments. Lanceolate leaves are 4–8" (10–20 cm) long and mostly near the base of the stem, each with a brownish, papery sheath at the nodes.

FLOWERING SEASON: May–August.

HABITAT: Moist mountain meadows or along mountain streams.

RANGE: Western Canada southward to southern California, Arizona, and New Mexico.

CONSERVATION: G5

Globally secure but vulnerable in British Columbia and imperiled in Manitoba.

■ NATIVE
■ RARE OR EXTIRPATED
■ INTRODUCED

This annual is distinguished by being tall and erect, with one or just a few stems per plant. The plants grow in groups and the flowers look like lace covering the landscape. The flowers particularly stand out as they turn pink in the fall. Annual Wild-buckwheat has seen many uses over the years. The Lakota used the plant to treat sores in the mouths of children from teething while the Kiowa used the leaves to stain bison and deer hides.

DESCRIPTION: An annual, sometimes biennial that grows from a slender, tan taproot, 1½–4' (45–120 cm) tall. There is just one erect stem or a few, appearing silvery gray from its fine, wooly hairs. Each stem has clusters of small or pinkish flowers at the top of its branches. Each flower is 1–3 mm long, with six petal-like tepals. The basal leaves wither early while the stem leaves are hairy and white, 1–3½" (3–9 cm) long and less than ½"

(1.25 cm) wide. They are smooth and often curl under. The fruit is 2 mm long and brown.

FLOWERING SEASON: April–November.

HABITAT: Open, often grassy places.

RANGE: The Great Plains and central United States, Minnesota west to Montana, south to Texas; also a disjunct population found in Indiana and Illinois.

SIMILAR SPECIES: Annual Wild-buckwheat is similar to Sorrel Buckwheat (*E. polycladon*), but its stems are usually more branched, and it has more tightly clustered flowers all along the branches.

CONSERVATION: G5

This species is widespread and considered a weed in parts of the United States. It is critically imperiled in Arkansas and vulnerable in Wyoming.

■ NATIVE
■ RARE OR EXTIRPATED
■ INTRODUCED

ALTERNATE NAMES: Cushion Wild Buckwheat

There is much variation in this widespread species. Alpine plants are generally dwarf, while plants growing in sagebrush are usually taller. Flowers vary from cream to yellow when young and may become reddish or purple at maturity.

DESCRIPTION: This is a matted plant, often grayish-haired, reaching a height of 1–12" (2.5–30 cm). The small flowers grow from a cylindrical cup with five teeth and are red, purple, or cream-colored. They appear in round heads, about 1" (2.5 cm) wide, on long, erect, leafless stems. Each flower is about ⅛" (3 mm) long, with six petal-like segments. The basal leaves are ½–5" (1.5–12.5 cm) long, varying from short, spatula-shaped, and without stalks to long-stalked, with roundish blades.

FLOWERING SEASON: May–August.

HABITAT: In open areas from sagebrush plains to open coniferous woodlands, alpine ridges, and rocky mountain slopes.

RANGE: British Columbia south to northern California and east to the Rocky Mountains from New Mexico to Alberta.

CONVERSATION:  **CONSERVATION:** G5

Globally secure but vulnerable in Alberta.

■ NATIVE
■ RARE OR EXTIRPATED
■ INTRODUCED

ALTERNATE NAMES: Sulphur Flower, Wild Buckwheat

Species in the genus *Eriogonum* are known for tiny flowers, which grow from cups. Like other buckwheats, Sulphur Buckwheat is highly variable, with many varieties, making it difficult to identify.

DESCRIPTION: This is a perennial with long, erect stalks, 4–12" (10–30 cm) tall. The branches are composed of numerous little cups from which several flowers grow on very slender stalks. The flowers can be purple, yellow, or cream-colored, about ¼" (6 mm) long and 2–4" (5–10 cm) wide. They grow in ball-like clusters at the ends of woody branches. The flowers have six petal-like segments and are hairy on the outside. The ovate leaves grow ½–1½" (1.5–4 cm) long, on slender stalks. The leaves are two to three times as long as they are wide and are usually hairy on the lower side.

FLOWERING SEASON: June–August.

HABITAT: Dry areas from sagebrush deserts to foothills and alpine ridges.

RANGE: British Columbia to southern California; east to the eastern flank of the Rocky Mountains from Colorado to Montana.

CONSERVATION: ⑤

Globally secure but vulnerable in Alberta.

■ NATIVE
■ RARE OR EXTIRPATED
■ INTRODUCED

ALTERNATE NAMES: Climbing Buckwheat

This aggressive perennial has attractive fruits and flowers, but it climbs high and easily destroys other vegetation. Botanists disagree about the taxonomy of this species. Some divide *Fallopia scandens* into two or three separate species, while others recognize *Fallopia scandens* as a single species with a few different varieties.

DESCRIPTION: This twining vine grows up to 20' (6 m) long if near an object to climb, or else it sprawls. The red or green stems are hairless, round, and angular or slightly ridged. Stipular sheaths at the base of the stems are papery and smooth. The leaves are alternate and heart-shaped, 4" long and 2" wide. One or more racemes of green to white flowers (about 2–8") grow from the axils of the leaves. Each flower, about ⅙" (4 mm) long, has five light green tepals and eight stamens. The three outer tepals have noticeable wings, which can be smooth or wavy. The fruit is winged, about ⅓" (8 mm) long, and appears greenish-white before turning brown. The seeds have three angles and are dark brown or black and shiny. The fruit can float on water or spread through the wind to reseed itself.

FLOWERING SEASON: July–September.

HABITAT: Disturbed places, thickets, woods.

RANGE: Alberta east to Quebec, south to Florida, and west to Texas.

SIMILAR SPECIES: Crested Climbing Buckwheat (*Fallopia cristata*) is similar but the fruits of *F. scandens* are slightly longer. Black Bindweed (*F. convolvulus*) is also similar, but it grows shorter, about 6' (2 m) long. The flowers and fruits of Black Bindweed are also keeled, not winged, and the sides of its three-angled seeds are dull, not shiny.

CONSERVATION: G5
Globally secure but vulnerable in Manitoba, New Brunswick, Nova Scotia, Quebec, and Saskatchewan.

■ NATIVE
■ RARE OR EXTIRPATED
■ INTRODUCED

ALTERNATE NAMES: Wood Sorrel, Alpine Sorrel, Alpine Mountain-sorrel

Mountain Sorrel's genus, *Oxyria*, is derived from a Greek word meaning "sour". As the name suggests, the leaves are edible and high in vitamin C, but have a sour taste. Mountain Sorrel is an important food source for animals, favored by deer and elk in the wet, snowy places where it grows.

DESCRIPTION: This perennial grows 12" (30 cm) tall. It's easily recognized by its reddish stems and leaves. The stems are unbranched and hairless, while the leaves are kidney- or heart-shaped. The flowers are small and greenish, growing a dense, upright cluster from leaf rosettes at the base of the stem. The flowers swell and become reddish as they seed. The round fruit, a nut, is reddish and encircled by a papery wing.

FLOWERING SEASON: July–September.

HABITAT: Moist rocky areas at alpine and subalpine elevations.

RANGE: Circumboreal; in North America it is found from Alaska to Newfoundland, southward in the mountains of the West to California, Arizona, and New Mexico, and in the East to New Hampshire.

CONSERVATION: G5

Critically imperiled in New Hampshire and South Dakota.

■ NATIVE
■ RARE OR EXTIRPATED
■ INTRODUCED

ALTERNATE NAMES: Swamp Smartweed, Water Lady's-thumb

Water Smartweed is a common aquatic plant with short, pink, oval flower clusters. Its seeds provide food for waterfowl. This widespread species can be found throughout the Northern Hemisphere in many wet habitats.

DESCRIPTION: Water Smartweed is the subject of disagreement among botanists. Some sources treat the related land plant Scarlet Swampweed (*P. coccinea*) as a variety of this species. When treated this way, the terrestrial form is labeled var. *emersa*, while the aquatic form is called var. *stipulacea*. Scarlet Smartweed, also called Longroot Smartweed, has lanceolate leaves and short flower clusters. It spreads quickly and can become a weed. The plant reaches 1–3' (0.3–1 m) tall. The aquatic form has floating, hairless leaves. The terrestrial form has lance-shaped, hairy leaves. Both forms of the plant have pink, dense flower clusters that are spike-like. The fruit is seed-like, dark brown or black and lens-shaped.

FLOWERING SEASON: July–September.

HABITAT: Shallow water, wet shores, moist meadows.

RANGE: Throughout much of North America.

SIMILAR SPECIES: Water Smartweed's showy flowers distinguish it from other plants, although it is similar to Scarlet Smartweed (*P. coccinea*). The floral racemes of Water Smartweed are generally shorter than Scarlet Smartweed's.

CONSERVATION: **G5**
Secure, but critically imperiled in Georgia and North Carolina.

■ NATIVE
■ RARE OR EXTIRPATED
■ INTRODUCED

ALTERNATE NAMES: Mild Wildpepper

This native perennial is both an aquatic and wetland plant, commonly seen throughout North America. The genus name is from the Greek words *poly* ("many") and *gona* ("knee" or "joint"), and refers to leaf sheaths that surround the stem.

DESCRIPTION: A terrestrial or aquatic plant with tiny, dark pink flowers that grow in clusters. Terrestrial plants reach a height of 2–3' (60–90 cm), while aquatic plants with stems grow to 4' (1.2 m) long. Terrestrial plants have lanceolate leaves up to 8" (20 cm) long, tapering at both ends. The young shoots are usually very hairy. Aquatic plant leaves grow to 6" (15 cm) long, and float on the water; they are thin and lanceolate or ovate, with a rounded or heart-shaped base. The young shoots are smooth. Aquatic plants have flower clusters that grow 3" (7 cm) above water. The calyx is divided into five parts, with absent petals. The clusters are 1½"–7" (4 –17.5 cm) long. The fruit is seed-like, dark brown or black and lens shaped.

FLOWERING SEASON: April–October; year-round in southernmost areas.

HABITAT: Shorelines, wet prairies, swamps, ponds, and quiet streams.

RANGE: Throughout North America, except Georgia, Alabama, Florida, and the Arctic.

SIMILAR SPECIES: Swamp Smartweed can be difficult to distinguish. Bristly Smartweed (*P. setacea*) is very similar and some botanists consider it a variety of Swamp Smartweed. Others recognize it as a separate species. Bristly Smartweed is a rare plant with hairy leaves, while the more common Swamp Smartweed's leaves are mostly hairless. Swamp Smartweed is also similar to Dotted Smartweed (*P. punctata*) and Waterpepper (*P. hydropiper*). Dotted Smartweed has dots on its tepals and its mature leaves are more widely spaced along the stems. Similarly, Waterpepper's flowers are more widely spaced.

CONSERVATION: Ⓖ5

Globally secure and common. Possibly extirpated in Rhode Island. Critically imperiled in Indiana, Ohio, and New York.

■ NATIVE
■ RARE OR EXTIRPATED
■ INTRODUCED

ALTERNATE NAMES: Jesusplant, Spotted Lady's-thumb

An exotic species from Eurasia that is widely naturalized, this is a common weed throughout North America. The dark green mark in the center of the leaf was thought to resemble a lady's thumbprint, hence the common name.

DESCRIPTION: This annual grows 8–31" (20–80 cm) tall, with oblong or cylindrical spikes of small, pink or purplish flowers at the top of pinkish stems. The stems may appear unbranched or branching. The flowers are about ⅛" (4 mm) long. There are four to six sepals appearing in clusters, about ½–2" (1.5–5 cm) long. The leaves are 2–6" (5–15 cm) long and narrowly or broadly

lanceolate, with a dark green triangle in the center. The fruit is black and glossy, seed-like, and three-sided.

FLOWERING SEASON: June–October.

HABITAT: Roadsides, damp clearings, and cultivated sites.

RANGE: Throughout much of North America, except Northwest Territories.

SIMILAR SPECIES: Lady's Thumb's spike-like racemes make it somewhat distinct from other smartweeds. Smartweeds generally have racemes that are more slender, and less crowded with flowers.

CONSERVATION: G4
Apparently secure globally; this species is introduced throughout North America.

PENNSYLVANIA SMARTWEED *Persicaria pensylvanica*

■ NATIVE
■ RARE OR EXTIRPATED
■ INTRODUCED

ALTERNATE NAMES: Pinkweed, Pink Knotweed

Although a native North American wildflower, Pennsylvania Smartweed is considered an invasive species in some areas, and is considered a weed on agricultural land. In its native range, it is an important food for birds and insects. At least 50 species of birds have been observed eating its seeds, and dense stands provide excellent cover for young waterfowl and marsh birds.

DESCRIPTION: This plant grows to a height of 1–4' (30–120 cm). It has upright clusters of small, bright pink flowers on sticky-hairy stems. The narrow leaves are 4–6" (10–15 cm) long and lanceolate, with a round sheath at the base. Each flower is about ⅛" (3 mm) long, with six sepals that look like petals, although actual petals are absent. The flowers have three to nine stamens that grow in a cluster measuring ½"–2½" (1.5–6.5 cm) long. The smooth fruit is lens-shaped and concave, with two sides.

FLOWERING SEASON: May–October.

HABITAT: Moist waste places and fields.

RANGE: Throughout the East, except the Prairie Provinces; also in parts of the West.

SIMILAR SPECIES: Pale Smartweed (*P. lapathifolium*) is similar. It has white or pinkish flower spikes with mostly smooth stems.

CONSERVATION: Ⓖ5

Globally secure, but critically imperiled in Wyoming.

- ■ NATIVE
- ■ RARE OR EXTIRPATED
- ■ INTRODUCED

Formerly called *Polygonum punctatum*, this is one of the most common smartweeds in wetland areas. It is extremely variable. Dotted Smartweed has dots on its tepals, although seeing them may require microscopic examination.

DESCRIPTION: This plant's round, hairless stems reach 6"–3' (15–91 cm) high. The leaves are ovate and alternate, up to 6" (15 cm) long and ¾" (2 cm) wide. They become short petioles and are usually hairless. A sheath appears at the base of each leaf, wrapping around the stem. This sheath is mostly hairless and falls away at maturity. The flowers appear in branching clusters and are greenish with white edges, ⅛" long with five tepals that appear slightly open.

FLOWERING SEASON: July–September.

HABITAT: Swamps, shallow standing water, and along the borders of ponds and small streams.

RANGE: Maine south through Florida, west through Minnesota, Nebraska and Texas (not known in Mississippi), Washington and California; in Canada, Quebec west through British Columbia (not in Alberta).

SIMILAR SPECIES: There are 35 species of smartweed, which may be confused with each other. Dotted Smartweed is unique with its narrow leaves and upright flowers.

CONSERVATION: **G5**
Globally secure, but vulnerable in Montana and North Dakota and imperiled in Oregon.

PROSTRATE KNOTWEED *Polygonum aviculare*

■ NATIVE
■ RARE OR EXTIRPATED
■ INTRODUCED

ALTERNATE NAMES: Birdweed

An introduced annual from Eurasia, this plant has become a common weed throughout North America. Each plant can produce thousands of seeds that stay viable for years. Prostrate Knotweed is a problem for farmers because it is difficult to eradicate from crop fields. It also appears in cities, where it grows through cracks of pavement and along roads. It carries powdery mildew, which gives the leaves a whitish appearance.

DESCRIPTION: Prostrate Knotweed may grow 4–16" (10–40 cm) high. The leaves are hairless, with short stalks and round bases. There are few upper leaves, which are stalkless. The flowers are green with white or pink margins, with five to eight stamens. The fruit is a dark brown with three edges. The seeds can lay dormant for years, requiring light to germinate.

FLOWERING SEASON: June–October.

HABITAT: Waste places, fields, wetlands, roadsides, railway embankments.

RANGE: This exotic species has been noted in nearly all states and provinces.

CONSERVATION:
Unranked.

■ NATIVE
■ RARE OR EXTIRPATED
■ INTRODUCED

ALTERNATE NAMES: Mountain Knotweed

Douglas' Knotweed is widespread in North America, especially in the northern and western U.S. and throughout temperate Canada. *Polygonum douglasii* represents a species complex of variable plants with many closely related species or subspecies that are difficult to distinguish.

DESCRIPTION: This is an erect, hairless annual growing up to 4–20" (10–50 cm) tall. It has long, upright branches angled just below the joints. The leaves are alternate and narrow, ½–2" (1.3–5 cm) long, up to ⅓" (8.5 mm) wide and pointed at the tip. Clusters of two to four white to rose-purple flowers grow from widely-spaced leaf axils on the stems. Flowers are about ⅛" (3 mm) long and have five tepals that are all about the same size.

There are eight stamens in the center. The fruit droops as it matures while the seeds are black, three-sided, and smooth.

FLOWERING SEASON: July–August.

HABITAT: Open, often disturbed places.

RANGE: Maine west through Washington, south through Iowa, New Mexico and California; Maryland. In Canada, Quebec west through British Columbia.

SIMILAR SPECIES: Slender Knotweed (*P. tenue*) is similar, but has shorter leaves that form a "W" in the middle. The fruits are mostly erect, not drooping.

CONSERVATION: G5
Globally secure.

■ NATIVE
■ RARE OR EXTIRPATED
■ INTRODUCED

A native of East Asia, this plant looks superficially like bamboo but is not related. It was introduced as a garden plant but has become an invasive species, especially in the East. The root system can grow 10' (3 m) deep and can become so large that it can damage foundations. Japanese Knotweed can also quickly spread, even if just a few centimeters of it are left behind after eradication efforts. It's mostly seen along rivers and streams, where it can kill native plants. In many areas, Japanese Knotweed is prohibited. Its young shoots taste like rhubarb and can be cooked and eaten like asparagus, and songbirds readily eat the seeds.

DESCRIPTION: This is a large, bushy plant 3–7' (90–210 cm) tall. The stems are hollow and have nodes, with clusters of greenish-white flowers growing at the ends. Flowers are about ⅛" (3 mm) long. They have no petals, but do have five sepals. The flowers are 23" (5–7.5 cm) long. The rounded or ovate leaves are 4–6" (10–15 cm) long, tapering to a point and straight across at the base. The black fruit is three-sided and looks like a seed.

FLOWERING SEASON: August–September.

HABITAT: Waste places, streambeds, roadsides.

RANGE: Introduced and found from Manitoba east to Newfoundland, south to Georgia, west to Louisiana and Oklahoma, and north to Nebraska and Minnesota; also found in parts of the West.

SIMILAR SPECIES: Giant Knotweed (*P. sachalinense*), also from Asia, is similar but has heart-shaped leaves at the base.

CONSERVATION:

This exotic species is unranked and often considered invasive in North America.

■ NATIVE
■ RARE OR EXTIRPATED
■ INTRODUCED

ALTERNATE NAMES: Common Sheep Sorrel

Native to Europe and widely introduced across North America, this is a vigorous, sour-tasting weed that readily occupies disturbed sites. The flowers are attractive to bees and butterflies, birds eat the seeds, the leaves—or even whole plants—are favored by rabbits and deer. It's also used in food preparation, including in curdling for cheesemaking.

DESCRIPTION: This is a weedy perennial with a reddish, erect stem growing 6–12" (15–30 cm) tall. It has arrowhead-shaped basal leaves that curve out. The leaves are ¾–2" (2–5 cm) long. The flowers are small and grow in spike-like clusters. The male and female flowers grow on separate plants. Male flowers are yellowish-green while female flowers are reddish. Both male and female flowers have calyxes with six parts and are about ¹⁄₁₆" (2 mm) long, with absent petals. The male flowers appear in nodes on short-jointed stalks, while female flowers develop fruit about the same length as the calyx. The calyx eventually drops off. The fruit is golden-brown, three-sided and seed-like.

FLOWERING SEASON: June–October.

HABITAT: Open sites, especially in acidic soil.

RANGE: Throughout much of North America, except the Northwest Territories.

CAUTION: The leaves of *Rumex* species contain oxalic acid and can be toxic or even fatal to animals if consumed in high quantities. How toxic a plant is varies by the season and the plant's age.

SIMILAR SPECIES: Engelmann's Sorrel (*R. hastatulus*) is a similar species but it has unisexual flowers. It doesn't have rhizomes, and its calyx is longer than the fruit. Its leaves also have outward-curving, basal lobes. Green Sorrel (*R. acetosa*), also unisexual, has leaves 4–6" (10–15 cm) long, with basal lobes that do not curve outward.

CONSERVATION:
This plant is exotic in North America and unranked.

■ NATIVE
■ RARE OR EXTIRPATED
■ INTRODUCED

Native to Europe and West Asia, this introduced plant with a large root structure has become a weed in many parts of North America. The young leaves have a bitter, lemonish flavor and can be consumed in small quantities with other greens in salads, but the leaves should be washed thoroughly before consumption to get rid of toxins. The leaves become more bitter as the plant matures. The seeds are also edible once dried.

DESCRIPTION: This is a short plant, 2–4' (60–120 cm) tall, with small, reddish or greenish, bisexual flowers. The flowers appear in a cluster at the top of the stem and are about ⅛" (4 mm) long, with two sepals in a series of three. The petals are absent. The leaves have wavy edges and are 6–10" (15–25 cm) long and oblong to lanceolate in shape. The fruit is brown and seed-like, with three sides, encased in a three-winged calyx with smooth edges. The encasing allows the seeds to disperse efficiently by floating on water or attaching to animals.

FLOWERING SEASON: June–September.

HABITAT: Old fields and waste places.

RANGE: Introduced throughout much of North America, except the Northwest Territories.

CAUTION: The leaves of *Rumex* species contain oxalic acid and can be toxic or even fatal to animals if consumed in high quantities. How toxic a plant is also varies by the season and the plant's age.

SIMILAR SPECIES: Bitter Dock (*P. obtusifolius*) is a similar species, but it has heart-shaped leaves with reddish veins and calyx lobes.

CONSERVATION:
Unranked. It is exotic in North America and classified as a noxious weed in Arkansas and Iowa.

■ NATIVE
■ RARE OR EXTIRPATED
■ INTRODUCED

ALTERNATE NAMES: White Dock, White Willow Dock, Triangular-valved Dock

Mexican Dock is a common and widespread, highly variable native perennial. Some sources treat the species as a variety of Willow Dock (*R. salicifolius*).

DESCRIPTION: Mexican Dock is an erect plant that grows 1–3½' (30.5–107 cm) tall. The flowers appear at the top of branched, hairless stems and have triangular tepals, ⅛" (3 mm) long, which appear in whorls of 10–25 yellowish-green, white, or pink flowers. The outer tepals are smaller than the inner tepals. The leaves are flat, wavy, and alternate, 2½–8" (6.5–20 cm) long, up to 1½" (4 cm) wide, and hairless. The fruit is a brown or reddish capsule with three-sided seeds.

FLOWERING SEASON: June–July.

HABITAT: Ditches, fields, roadsides, waste areas, shores.

RANGE: Throughout most of North America, except the southeastern United States.

SIMILAR SPECIES: Pale Dock (*R. altissimus*) and Swamp Dock (*R. verticillatus*) are both similar. *R. altissimus* has wider leaves, and larger tepals, growing 6 mm long. *R. verticillatus* has less dense flower clusters with tepals up to 5 mm long.

CONSERVATION: G5
Globally secure.

■ NATIVE
■ RARE OR EXTIRPATED
■ INTRODUCED

Pink Sundew's long-stalked leaves are covered with sticky hairs that trap insects, which the plant consumes, allowing it to take in important nutrients it might not get from the soil.

DESCRIPTION: This carnivorous plant grows up to 12" (30 cm) tall, with a rosette of leaves at the base. The stem is curved and becomes leafless farther up, with a cluster of flowers at the top. The cluster is about ½" (12 mm) wide with five petals, appearing white or pink. The leaves are spoon-shaped, reddish and sticky, covered with long hairs and glands.

FLOWERING SEASON: April–August.

HABITAT: Pinelands, bogs, wet ditches.

RANGE: Maryland south to Florida and west to Texas.

CONSERVATION: G5
Globally secure but critically imperiled in Maryland and Tennessee. Historically present but presumed extirpated in Delaware.

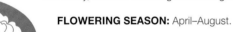

■ NATIVE
■ RARE OR EXTIRPATED
■ INTRODUCED

This plant, like other sundews, is carnivorous. The genus name comes from *droseros*, a Greek word meaning "dewy," which refers to the moist drops on the leaves. The drops are sticky and sugary, making them attractive to insects, which the plant entraps. The plant uses enzymes to consume the insects and nutrients it needs to survive in otherwise poor soil. The glands near the edges of the leaves also bend inward, allowing the plant to further take in the trapped insects and digest them.

DESCRIPTION: This plant grows in a slender stalk that bends over, with a basal rosette of leaves. It reaches up to 10" (25 cm) tall. White or pinkish flowers, about ⅜" (9 mm) wide with four to eight petals, appear near the top of the stems. The leaves are 1–4" (2.5–10 cm) long, with stout stalks and round blades, ¼–½" (6–13 mm) wide, covered with reddish stalked glands.

FLOWERING SEASON: June–September.

HABITAT: Bogs.

RANGE: Throughout North America south to Florida, Mississippi, Iowa, Colorado, Idaho, and California.

SIMILAR SPECIES: Narrow-leaved Sundew (*D. anglica*) is similar, but has erect leaves with long, narrow blades, not basal.

CONSERVATION: G5

Globally secure but vulnerable or imperiled in a number of states, including Alabama, Colorado, Delaware, Georgia, Idaho, Iowa, Maryland, Montana, North Carolina, North Dakota, Ohio, Tennessee, and West Virginia.

■ NATIVE
■ RARE OR EXTIRPATED
■ INTRODUCED

ALTERNATE NAMES: Common Corncockle

Corncockle's scientific name refers to the two toxic chemicals inside the plant—githagin and agrostemma acid—which can be fatal. The plant is native to Europe, where it was a problem weed in wheat fields because the seeds were often unintentionally harvested along with the crop. Before machine harvesting became commonplace in the 20th century, separating the poisonous seeds from the wheat was very time consuming.

DESCRIPTION: A slender, erect plant with few hairs and few branches forming narrow forks, 1–3' (30–90 cm) tall. Each branch has one dark pink to reddish-lavender flower, about 1" (2.5 cm) wide, with five petals and five sepals. The sepals are each ¾–1½" (2–4 cm) long and

the petals are slightly smaller, with a narrow tip, joined at the base to form a tube with 10 ribs. The capsule has five styles. The narrowly lanceolate leaves are 2–6" (5–15 cm) long and opposite, growing nearly erect against the stem.

FLOWERING SEASON: May–July.

HABITAT: Fields, roadsides, and waste places.

RANGE: Introduced throughout much of the United States and southern Canada.

CAUTION: The entire plant can be poisonous if consumed.

CONSERVATION:
This exotic, introduced species is unranked.

- NATIVE
- RARE OR EXTIRPATED
- INTRODUCED

ALTERNATE NAMES: Common Mouse-ear Chickweed

This widely introduced plant from Europe has become a problem weed in North America, especially in gardens. The plant easily spreads along the ground and takes root wherever fallen stems contact the soil. Its leaves can be boiled and eaten as greens. The genus name comes from the Greek word for "horned" and refers to the shape of the capsule.

DESCRIPTION: A low growing, matted perennial about 6–12" (15–30 cm) tall. It has hairy, sticky stems and fuzzy leaves. The small, white flowers, ¼" (6 mm) wide, grow in clusters at the top of slender stalks. The flowers have five petals that are deeply notched and five sepals, about the same length as the petals, with fine hairs. There are usually 10 stamens that are yellow, green, or reddish at the tip. The stalkless leaves are about ½"

(1.5 cm) long and paired. The bottom leaves are egg-shaped, while the top are more oblong. The small fruit capsule is somewhat cylindrical, with 10 teeth forming the shape of a crown at the tip. It has many reddish-brown seeds.

FLOWERING SEASON: May–September.

HABITAT: Waste places, fields, and roadsides.

RANGE: Introduced throughout much of temperate North America.

SIMILAR SPECIES: This plant is somewhat similar to Common Chickweed (*Stellaria media*), but Mouse-ear Chickweed's leaves and stems are fuzzier.

CONSERVATION:
This exotic plant is unranked and considered a weed in many areas.

■ NATIVE
■ RARE OR EXTIRPATED
■ INTRODUCED

ALTERNATE NAMES: American Field Chickweed, Field Mouse-ear Chickweed, *Cerastium arvense* ssp. *strictum*

This plant is the North American native representative of a species complex common in Europe. Field Chickweed has heavy stems that lean over and may form a tangled mat on the ground. This plant may look different depending on where it's growing. Field Chickweed is very hairy in coastal areas, but plants growing inland may be less hairy. The genus name comes from the Greek word *keras*, meaning "horn," and refers to the fruit, which may bend in slightly, resembling a cow's horn.

DESCRIPTION: This perennial has weak stems and grows to a height of 2–20" (5–50 cm). There are a few leaves, each about ½–1½" (1.25–4 cm) long, opposite, narrowly lanceolate. The flowers, 1" (2.5 cm) wide, sometimes grow in clusters at the top of the stems, but some plants have just one white flower atop each branch. Flowers have five green sepals and five petals, each about ½" (1.25 cm) long and deeply notched at the tip. The fruit is a cylindrical capsule, with 10 tiny teeth at the tip, containing brown seeds.

FLOWERING SEASON: April–August.

HABITAT: Open fields, grassy areas, rocky slopes, coastal beaches, and disturbed sites.

RANGE: Widespread throughout much of North America.

SIMILAR SPECIES: Mountain Chickweed (*C. beeringianum*) is a similar species, but it occurs at higher elevations. American Field Chickweed is very similar to the European subspecies of Field Chickweed and is difficult to distinguish, but the non-native plants are more likely to be found in sites disturbed by humans, such as along roadways.

CONSERVATION: G5
Secure, but vulnerable in West Virginia.

■ NATIVE
■ RARE OR EXTIRPATED
■ INTRODUCED

ALTERNATE NAMES: Mountain Pink

Introduced from Europe, this plant once grew abundantly in fields near Deptford, now an industrial area of London, hence the common name. It is often grown as an ornamental plant and has widely naturalized across most of the United States and southern Canada.

DESCRIPTION: This plant has one or several erect stems, growing 8–24" (20–60 cm) tall. The flowers are red or pink and grow in a tight, hairy cluster with five petals that are about ¾–1" (2–2.5 cm) long. The calyx is about ½" (1.5 cm) wide and narrow, with five pointed lobes. The leaves are 1½–4" (4–10 cm) long, opposite and very narrow. The leaves on the stem are erect. The fruit, a capsule, has two styles.

FLOWERING SEASON: June–August.

HABITAT: Dry fields and roadsides.

RANGE: Introduced throughout the East, except the Prairie Provinces and the Arctic; also found in much of the West.

CAUTION: This plant is toxic if eaten and can cause minor skin irritation.

SIMILAR SPECIES: The widely cultivated Carnation (*D. caryophyllus*) is similar.

CONSERVATION:
This exotic species is unranked.

■ NATIVE
■ RARE OR EXTIRPATED
■ INTRODUCED

ALTERNATE NAMES: Cuckoo Flower

This plant is native to Europe and is naturalized across much of North America. The colloquial name Cuckoo Flower references the Common Cuckoo (*Cuculus canorus*) in Europe, which is especially active in the spring when this plant blooms. The name Ragged Robin refers to the ragged appearance of the flower petals. Ragged Robin and other *Lychnis* species are placed in the closely related genus *Silene* in some sources.

DESCRIPTION: Ragged Robin grows up to 1–3' (30–90 cm) tall with thin branching stalks. The flowers are dark pink or white and grow in terminal clusters, ½" (1.5 cm) wide, with five petals that are deeply divided into four thin ragged-looking lobes. The stems are sticky and hairy and grow beyond the foliage. The opposite, lanceolate leaves are 2–3" (5–7.5 cm) long, are hairy and become smaller up the stem. The fruit is a capsule with many seeds.

FLOWERING SEASON: May–July.

HABITAT: Moist fields, meadows, and waste places.

RANGE: Naturalized from Ontario east to Newfoundland, south to Maryland, and west to Ohio.

SIMILAR SPECIES: Mullein Pink (*L. coronaria*) is a similar plant, but its stems and leaves are woollier.

CONSERVATION:
This species is introduced in North America and is unranked.

■ NATIVE
■ RARE OR EXTIRPATED
■ INTRODUCED

ALTERNATE NAMES: Bluntleaf Sandwort

This delicate wildflower is common throughout most of Canada. The width of the leaves and the hairiness vary by location. Grove Sandwort differs from other species by the fleshy, nutrient-rich appendage (elaiosome) attached to its seeds.

DESCRIPTION: A perennial, 2–8" (5–20 cm) tall, with erect narrow green stems that are mostly branched. The leaves are opposite, 1–1½" (3–4.25 cm) long and ⅛–½" (3 to 12 mm) wide and oval in shape. One to five flowers appear on top of Grove Sandwort's stems in a cluster. The flowers have five white petals and 10 stamens, with white at the tip. The flowers are ¼" (6 mm) across, and surround a greenish ovary with three spreading styles in the center. There are small, scale-like bracts at the base of the flower and five light green sepals with pointed tips. A fleshy food appendage is attached to the seeds.

FLOWERING SEASON: May–June.

HABITAT: Woods, meadows, gravelly shores.

RANGE: Found throughout Canada and in Alaska; from Washington south through Nevada and New Mexico, east through Maine and Virginia.

SIMILAR SPECIES: Large-leaved Sandwort (*M. macrophylla*), which is very similar, is rare. It is more erect and has larger leaves, up to 2" long. Grove Sandwort's flowers are similar to those of *Minuartia* spp. (also commonly called sandworts), but it has wider leaves and its seed capsules have six teeth, not three. Grove Sandwort is also similar to Thyme-Leaved Sandwort (*Arenaria serpyllifolia*), but Grove Sandwort's leaves are longer and narrower.

CONSERVATION: G5

Globally secure, but vulnerable in several midwestern and Great Plains states. It is presumed extirpated in West Virginia. Critically imperiled in Maryland and Virginia.

MOUNTAIN SANDWORT   *Mononeuria groenlandica*

■ NATIVE
■ RARE OR EXTIRPATED
■ INTRODUCED

ALTERNATE NAMES: Greenland Stitchwort, Mountain Daisy

As the common name suggests, this delicate plant is often found in high elevations, especially where bedrock is exposed. It is a rare perennial that grows in clumps on rocky ledges or in gravel, with small, white flowers that bloom above its thick foliage.

DESCRIPTION: This is a mat-forming plant growing 2–5" (5–12.5 cm) tall with many small, white or translucent flowers. The flowers appear at the tips of slender stalks, which rise up from basal leaves. The flowers are ½" (1.5 cm) wide, with five petals. The petals are surrounded by five green sepals. The leaves of each plant are seen in three to five pairs. The leaves are opposite, about ½" (1.5 cm) long, and narrow or needle-like. The fruit is a capsule with many seeds.

FLOWERING SEASON: June–August.

HABITAT: Granite crevices and gravelly sites, often at high elevations in mountains.

RANGE: In the East from Arctic Canada south to Newfoundland, New York, and Ontario; rare in the Carolinas and Tennessee.

CONSERVATION: G5

Although classified as globally secure, this rare species is vulnerable to human disturbance (trampling). It is listed as threatened or endangered in most states and provinces where it occurs. It is critically imperiled in Ontario and New Brunswick. Isolated populations are imperiled in North Carolina and critically imperiled in Tennessee, Vermont, Virginia, and West Virginia. It's vulnerable throughout New England.

■ NATIVE
■ RARE OR EXTIRPATED
■ INTRODUCED

ALTERNATE NAMES: Soapwort

This attractive perennial native to Eurasia looks like phlox (*Phlox* spp.). It has naturalized throughout much of North America and has become invasive in some native habitats. It is sometimes called Soapwort because the crushed foliage can be used as a soap to clean delicate fabrics. The genus name comes from the Latin *sapo*, meaning "soap." "Bouncing Bet" was an English term once used as a nickname for a woman who vigorously scrubbed laundry.

DESCRIPTION: This plant has leafy, unbranched stems growing 1–3' (30–90 cm) tall in patches. The white or pink flowers are sweet-smelling and about 1" (2.5 cm) wide, with five petals. The calyx is long and tubular, with five pointed red teeth. The flowers are most fragrant at night. They open in the evening and stay open for three days. The flowers release pollen on the second night of blooming, and the stigma fully develops on the third night. The leaves are 1½–5" (4–12.5 cm) long, opposite, and lanceolate.

FLOWERING SEASON: July–September.

HABITAT: Roadsides and disturbed areas.

RANGE: Introduced throughout North America, except the Far North.

CAUTION: The plant contains toxic saponins that can be harmful to humans or animals if ingested.

CONSERVATION:
This exotic species is unranked.

■ NATIVE
■ RARE OR EXTIRPATED
■ INTRODUCED

ALTERNATE NAMES: Wild Pink

Sticky Catchfly is a perennial native to the eastern United States. It is an important source of nectar for insects and butterflies in the early spring. There are several subspecies of this plant, each growing in different areas. The nominate subspecies predominates in the southeastern part of the range. 'Carolina Pink' (subsp. *pensylvanica*) occurs farther north and east. 'Wherry's Catchfly' (subsp. *wherryi*) is a vulnerable subspecies that grows in the western part of the species' range.

DESCRIPTION: Sticky Catchfly grows in a compact mound, 3–8" (7.5–20 cm). It has slender, sticky stems, from which showy rose-colored flowers with five petals appear in loose clusters. The calyx is hairy and sticky. Each plant can produce 50–100 flowers, which grow with narrow basal leaves, which are about 4" (10 cm) long. The leaves grow smaller as they ascend the stem. The fruit, a capsule up to ½" (1.3 cm) long, forms after the flowering season.

FLOWERING SEASON: April–June.

HABITAT: Dry, open woods.

RANGE: New Hampshire and Massachusetts to Ohio and Kentucky, south to Florida and Alabama, west to Missouri.

SIMILAR SPECIES: This plant resembles the subspecies of Cardinal Catchfly (*S. laciniata* subsp. *greggii*) that grows in Arizona, New Mexico, and Texas.

CONSERVATION: G5

Globally secure but critically imperiled in Delaware, Florida, and Tennessee; and imperiled in Virginia and Georgia.

■ NATIVE
■ RARE OR EXTIRPATED
■ INTRODUCED

ALTERNATE NAMES: Southern Indian Pink, Mexican Campion, Mexican Catchfly

Cardinal Catchfly is a popular flower for Western hummingbirds. Rabbits and other wildlife are also known to eat the leaves, which are covered in sticky hairs. The species is subdivided geographically. *S. laciniata* ssp. *major* is the form that occurs in California, with var. *angustifolia* occurring near the coast and var. *latifolia* inland. Subspecies *greggii* occurs in Arizona, New Mexico, and Texas.

DESCRIPTION: Cardinal Catchfly has one to several stems that are slender and weak, growing 1–2½' (30–75 cm) tall. This plant often leans against others for support. The opposite leaves have sticky hairs and are about 2" (5 cm) long, and linear. Each plant has just one flower or many, with five red petals, deeply divided. The sepals are red or green and have 10 veins. There are also 10 stamens.

FLOWERING SEASON: May–August.

HABITAT: Pine forests; grassy or brushy slopes.

RANGE: Texas to California, south to Mexico.

SIMILAR SPECIES: This plant resembles Sticky Catchfly (*S. caroliniana*) in the East.

CONSERVATION: **G5**
Secure.

■ NATIVE
■ RARE OR EXTIRPATED
■ INTRODUCED

ALTERNATE NAMES: Bladder Campion, *S. latifolia* subsp. *alba*

A widespread Eurasian native widely introduced in North America, White Campion was formerly called *Lychnis alba* but is now classified within *Silene latifolia*. The scented flowers open in the evening and close at noon or later on cloudy days. The alternate name Bladder Campion also is used commonly to identify *S. vulgaris*, but they are different species.

DESCRIPTION: This plant has several erect stems, growing 1½–4½' (45–135 cm) tall, covered with glandular hairs. Flowers on the end of each branch appear in a white cluster, about ¾" (2 cm) wide. The male and female flowers occur on separate plants. The female flowers are white, with a calyx about ¾" (2 cm) long when

the flower blooms; they have five petals, each about 1" (2.5 cm) long, divided by several veins near the middle. Male flowers have 10 pale yellow stamens and are more heavily veined; the upper part of the flower is deeply notched at the tip.

FLOWERING SEASON: June–August.

HABITAT: Fields, roadsides, and waste places.

RANGE: Introduced throughout much of North America, except the Gulf States and southwestern deserts.

SIMILAR SPECIES: White Campion is often confused with Night-flowering Catchfly (*S. noctiflora*), but Night-flowering Catchfly has both male and female parts on the same flower.

CONSERVATION:

This exotic species in North America is unranked. It tends to be weedy.

■ NATIVE
■ RARE OR EXTIRPATED
■ INTRODUCED

ALTERNATE NAMES: Widow's-frill

This native plant occurs in nearly every state in the eastern half of the U.S., and especially favors rich woodlands. It attracts a variety of butterflies and moths for pollination.

DESCRIPTION: This delicate wildflower grows 2–3' (60–90 cm) tall. It has slender pale green or reddish stalks that terminate in deeply fringed white flowers that grow individually or in clusters of two or three. The flowers are ¾" (2 cm) wide, and joined with bell-shaped sepals. The smooth, lanceolate leaves are 1½–4" (4–10 cm) long and appear on the stems, mostly in whorls of four. However, the uppermost and lowermost leaves are usually opposite. The fruit is a capsule with many seeds.

FLOWERING SEASON: June–September.

HABITAT: Open woods.

RANGE: Ontario southeast to New York and Massachusetts, south to Georgia, west to Texas, and north to South Dakota.

CONSERVATION: G5
Globally secure but imperiled in Connecticut, Louisiana, and Michigan. Vulnerable in Illinois.

■ NATIVE
■ RARE OR EXTIRPATED
■ INTRODUCED

ALTERNATE NAMES: Maiden's-tears, *Silene cucubalus*

This attractive perennial is native to Europe and widely naturalized in North America. It is edible and has a range of medicinal uses but has a tendency to escape cultivation. The plant attracts insects and birds. Once mature, the purplish calyx looks like a paper lantern.

DESCRIPTION: A usually smooth, hairless plant with pale green stems, reaching 3' (90 cm) high. Some of the stems terminate in a cluster of white flowers, each with a large, swollen, bladder-like calyx that is green or pinkish. The weight of the flowers may cause the plant to lean over to one side when in bloom. The flowers are about ½" (1.5 cm) wide, while the calyx is about ¾" (2 cm) long, containing many veins. The upper rim of the calyx has five triangular teeth. There are five petals deeply notched at the tip, with a long, stalk-like base and three styles. The opposite leaves are 1¼–3" (3–7.5 cm) long and lanceolate. The flowers become a three-celled capsule with brown seeds.

FLOWERING SEASON: June–August.

HABITAT: Fields and roadsides.

RANGE: Naturalized throughout much of North America, except the Far North; scattered in the southern United States.

SIMILAR SPECIES: Bladder Campion's large calyx distinguishes it from other species, as do the veins on the calyx. Night-flowering Catchfly (*S. noctiflora*) is similar but has a hairy calyx. White Campion (*S. latifolia*) is also similar to Bladder Campion and some sources use the names interchangeably. White Campion is short-lived, typically occurring as an annual or biennial, and reproduces only through seeds, whereas Bladder Campion is a perennial that can reproduce both through seeds and vegetatively.

CONSERVATION:
This species is introduced in North America and is unranked.

■ NATIVE
■ RARE OR EXTIRPATED
■ INTRODUCED

ALTERNATE NAMES: Lesser Sea-spurrey, *Spergularia salina*

These salt-tolerant plants are adapted to seawater and other saline environments. There has been much debate over the name of this species. In the 1700s, botanist and taxonomist Carl Linnaeus described the plant as a variety of the European native Red Sand-Spurry. It was subsequently elevated to a separate species using his original name for the variety (*marina*). Many sources designate this plant *Spergularia salina*, a name given in the early 1800s recognizing it as a newly described North American species, but others argue Linnaeus' name has nomenclatural priority.

DESCRIPTION: This is a sprawling annual or sometimes, but rarely, a perennial, with slender stems up to 14" (35 cm) long. It has five red or pink petals, appearing whte at the base, with sepals usually longer than the petals. Its green leaves are simple and opposite. The fruit is dry and splits open when ripe.

FLOWERING SEASON: April–September.

HABITAT: Marshes, disturbed areas, and sea beaches.

RANGE: Most of the United States and Canada.

CONSERVATION: G5

Globally secure, but critically imperiled in North Carolina and Ontario.

■ NATIVE
■ RARE OR EXTIRPATED
■ INTRODUCED

Native to Eurasia, Common Chickweed has been widely introduced in North America, where it sometimes can be invasive. It is common in lawns and gardens and can be eaten as a green. It is a favorite plant among birds and can host several species of butterflies and moths.

DESCRIPTION: A weak, branched plant growing 3–8" (7.5–20 cm) tall, with trailing stems up to 16" (40 cm) long. The stems are hairy on just one side. Its small, white flowers are ¼" (6 mm) wide, with five hairy sepals. There are five petals, but they are deeply cleft such that each flower looks like it has 10 petals. The flowers either grow in terminal clusters or solitarily in leaf axils. Smoothish, ovate leaves grow ½–1" (1.5–2.5 cm) long, and are opposite. The lower leaves have petioles; the upper leaves do not. The flowers become a capsule with many seeds.

FLOWERING SEASON: February–December.

HABITAT: Lawns and disturbed areas.

RANGE: Throughout most of North America, except the Arctic islands.

SIMILAR SPECIES: Common Chickweed can be confused with other chickweed species but is distinguished by having hairs on only one side of the stem and hairs on the sepal.

CONSERVATION:
This exotic plant in North America is unranked.

■ NATIVE
■ RARE OR EXTIRPATED
■ INTRODUCED

ALTERNATE NAMES: Green Amaranth, Red-root Amaranth

Native to tropical America, this widespread plant grows prolifically and is considered a weed in some areas. Its common name relates to its frequent appearance in pastures where hogs and other livestock graze. The leaves of this plant can be eaten like spinach and the seeds can be roasted or ground into meal.

DESCRIPTION: This plant has a rough, hairy stem with small, greenish flowers, about 2–4' (60–120 cm) tall. The leaves are oval or lanceolate, 3–6" (7.5–15 cm) long, stalked and untoothed. The flowers grow in loose clusters, about 2½" (6.5 cm) long. The

flowers intertwine with spiny, bristle-like bracts. Both female and male flowers are present on each plant. The fruit, a tiny capsule, is less than ⅟₁₆" (2 mm) long and has a single black seed inside.

FLOWERING SEASON: August–October.

HABITAT: Waste places, roadsides, and cultivated soil.

RANGE: Throughout much of North America, except the Arctic.

CAUTION: This plant can be toxic to animals if ingested in high quantities, but it can be useful feed for livestock in moderation.

CONSERVATION: G5
Secure.

■ NATIVE
■ RARE OR EXTIRPATED
■ INTRODUCED

ALTERNATE NAMES: Mexican-fireweed, Kochia

Burning Bush, a native of Eurasia, has been planted for erosion control and as an ornamental for its red fall foliage in gardens, but it has rapidly spread in the wild and become a weed in many areas. The seeds disperse by tumbling through wind and water in the fall, when the whole plant detaches from its stems, forming tumbleweeds. The plant contains toxins that inhibit the growth of nearby plants, and it readily develops resistance to herbicides used to control it.

DESCRIPTION: This is a bushy plant that grows up to 6½' (2 m) tall. The leaves are ¾–4" (2–10 cm) long and linear and usually have silky hairs on the lower part. The leaves are bright green in the summer and turn red in the fall. Greenish flowers grow in leaf axils. The fruit, ⅛" (2–3 mm) long, has a single brown seed.

FLOWERING SEASON: July–September.

HABITAT: Open areas, including neglected yards, roadsides, disturbed areas, fields, rangelands, and waste sites.

RANGE: Introduced into North America from Asia; now growing in the wild in most of the United States (except parts of the Southeast) and in Quebec and the Prairie Provinces in Canada.

CAUTION: Can be toxic to animals if eaten in large quantities.

CONSERVATION:
This introduced plant species in North America is unranked.

■ NATIVE
■ RARE OR EXTIRPATED
■ INTRODUCED

ALTERNATE NAMES: Pigweed, White Goosefeet

The nominate variety of this species (*C. album* var. *album*) is a widely naturalized European native that grows quickly and is considered a weed or invasive in many areas. It is cultivated in some regions around the world as a food crop for humans or livestock; the leaves taste like spinach. Missouri Lamb's-quarters (*C. album* var. *missouriense*) is native to North America and found mainly east of the Rocky Mountains and in California and Nevada.

DESCRIPTION: Reddish stems of this branching plant grow 1–6' (30–180 cm) tall. It grows upright at first, but its heavy foliage weighs it down. The unstalked greenish flowers, less than 1/16" (2 mm) wide, grow in clustered spikes. They have no petals. The leaves are triangular or diamond-shaped, 1–4" (2.5–10 cm) long and coarsely toothed, appearing white underneath.

FLOWERING SEASON: June–October.

HABITAT: Cultivated land, disturbed sites, and roadsides.

RANGE: The nominate variety (var. *album*) is introduced throughout North America, except the Arctic islands. Missouri Lamb's-quarter is native to the central and eastern United States, California, and Nevada.

CAUTION: This plant can kill livestock and other animals if consumed in high quantities.

SIMILAR SPECIES: Mexican Tea (*C. ambrosioides*) is a similar species, but it has oblong or lanceolate leaves with wavy-toothed edges. Another similar species, Jerusalem Oak (*C. botrys*), has oak-like leaves.

CONSERVATION: G5

Both the native and non-native varieties are globally secure. Missouri Lamb's-quarters is vulnerable in Illinois and West Virginia; it is presumed extirpated in New York.

■ NATIVE
■ RARE OR EXTIRPATED
■ INTRODUCED

ALTERNATE NAMES: Pickleweed, Saltwort, Woody Saltwort, *Sarcocornia ambigua*, *Sarcocornia perennis*

This plant has thick stems, allowing it to take in a lot of water, which helps it survive in conditions where other saltmarsh plants can't. The leaves and stems can be cooked or pickled, hence one of the common names.

DESCRIPTION: A creeping perennial that grows in extensive mats of 12" (30 cm). The succulent stems are smooth and very branched. The leaves are scale-like and hardly visible, but if they can be seen, they appear opposite and stalkless. The leaves join together at the base. The flowers are greenish and small. They grow in spikes at the ends of the branches. The flowers contain both male and female parts.

FLOWERING SEASON: June–November.

HABITAT: Saltmarshes, alkaline flats.

RANGE: On the Atlantic Coast from New England to Florida, west along the Gulf Coast to Texas. Also on the Pacific Coast from southern Alaska and British Columbia to Baja California.

SIMILAR SPECIES: *Salicornia depressa* is similar, but it is an annual, not a perennial. *S. depressa*'s center flower in a trio is also longer than the other two.

CONSERVATION:
Unranked. It is common in many areas but vulnerable to habitat destruction from human activity.

■ NATIVE
■ RARE OR EXTIRPATED
■ INTRODUCED

ALTERNATE NAMES: Red Saltwort, Red Swampfire

This annual succulent turns reddish in the fall where it grows in northern tidal marshes. It can be pickled or added raw to salads.

DESCRIPTION: A cylindrical creeping plant with opposite branches, reaching 6–18" (15–45 cm) tall. Its stem joints are longer than they are wide. It's mostly leafless, though it does have very small green leaves with opposite scales, appearing in groups of three, ⅛" (3 mm) wide.

FLOWERING SEASON: August–November.

HABITAT: Coastal saltmarshes, especially on bare peat, salt licks, and inland saltmarshes.

RANGE: New Brunswick and Nova Scotia south along the coast to Georgia; local in Michigan, Wisconsin, and Illinois.

SIMILAR SPECIES: Dwarf Glasswort (*S. bigelovii*) is similar, but usually unbranched, with joints wider than long. Woody Glasswort (*S. virginica*), another similar plant, has creeping stems that form extensive mats.

CONVERSATION:
Apparently secure, but imperiled in Washington and Oregon.

■ NATIVE
■ RARE OR EXTIRPATED
■ INTRODUCED

ALTERNATE NAMES: Baby Sun-rose, Ice Plant

Sea Fig is a non-native succulent that has become established in coastal habitats in California and Oregon. This species has been used as an ornamental but can become invasive. The foliage can be eaten, either cooked or used in salads. The fruit is also used medicinally as a laxative, and the juice from the leaves can be used as an antiseptic. The flowers open in the morning and close at night.

DESCRIPTION: This is a mat-forming creeper, with flowering branches reaching 5" (12.5 cm) tall and trailing stems reaching up to 6' (1.8 m) long. The erect, succulent leaves are 1½–3" (4–7.5 cm) long, mostly opposite and with three sides and a smooth outer angle. The large, deep reddish-lavender flowers are 1½–2½" (4–6.5 cm) wide with many petals. The sepals vary in length. The larger sepals are leaf-like. The edible fruit is green and plump, with 8–10 chambers.

FLOWERING SEASON: April–September.

HABITAT: Coastal sands and bluffs.

RANGE: Introduced from southern Oregon south to Mexico.

SIMILAR SPECIES: Sea Fig hybridizes with Sour Fig (*C. edulis*)—another troublesome non-native plant—on coastal dunes and along roads near the coast. Both species have flowers ranging in color from yellow to magenta and both are commonly called Ice Plants. Other similar cultivated species in different genera have flat or nearly cylindrical leaves.

CONSERVATION: G5

Globally secure, this non-native plant can be a problematic invasive species in coastal ecosystems.

■ NATIVE
■ RARE OR EXTIRPATED
■ INTRODUCED

An annual, succulent herb native to eastern North America, Slender Sea-Purslane is easy to overlook, especially in the northern part of its range where the plants tend to grow smaller and are only infrequently found.

DESCRIPTION: This is a leaning or somewhat erect plant, up to 12" (30 cm) high with many branches, forming a mat. The leaves are opposite and fleshy, 0.75" (2 cm) long and less than 0.25" (0.6 cm) wide. The flowers are solitary with very small or absent petals. Each flower has five pink or purple sepals and two to three styles, with horn-like appendages. There are 30–50 blackish seeds in each fruit capsule.

FLOWERING SEASON: June–November; may flower year-round in Southeast Texas.

HABITAT: Saltmarshes, beaches, dunes.

RANGE: Along the East and Gulf coasts of North America, from Rhode Island to Texas. Its distribution in select inland areas, including Kansas and Oklahoma, is not well known.

SIMILAR SPECIES: The species is sometimes confused with Western Sea-purslane (*Sesuvium sessile*) found in Texas. Shoreline Sea-purslane (*S. portulacastrum*) is also mat-forming and appears similar, but has five styles per flower, not two to three. Common purslane (*Portulaca oleracea*) is also similar, but the flowers are yellow.

CONSERVATION: G5
Globally secure but may be in decline due to development of coastal environments. It is critically imperiled in Kansas, Maryland, North Carolina, and New York; it is presumed extirpated in Oklahoma.

■ NATIVE
■ RARE OR EXTIRPATED
■ INTRODUCED

ALTERNATE NAMES: American Pokeweed, Common Poke-weed

The leaves and stems of Pokeweed are edible if the plant is young, before the pink color appears, but the plant is toxic to mammals, including humans, after it matures. Poke sallet (salad), a traditional southern Appalachian food, is made from young greens that are repeatedly boiled to remove the toxic chemicals. The toxins in the berries do not affect songbirds, which often eat them and help to distribute the seeds. Juice from the berries was also once used as a dye. Native Americans and early settlers used the plant for various medicinal purposes, but modern medical research has not found evidence of health benefits in humans.

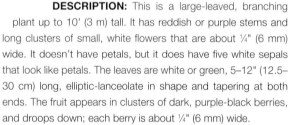

DESCRIPTION: This is a large-leaved, branching plant up to 10' (3 m) tall. It has reddish or purple stems and long clusters of small, white flowers that are about ¼" (6 mm) wide. It doesn't have petals, but it does have five white sepals that look like petals. The leaves are white or green, 5–12" (12.5–30 cm) long, elliptic-lanceolate in shape and tapering at both ends. The fruit appears in clusters of dark, purple-black berries, and droops down; each berry is about ¼" (6 mm) wide.

FLOWERING SEASON: July–September.

HABITAT: Open woods, damp thickets, clearings, and road-sides.

RANGE: Ontario to southern Quebec, New England, and New York; south to Florida; west to Texas and Mexico; north to Minnesota.

CAUTION: All parts of this plant can be toxic. Older leaves and the shoots, berries, and roots of this plant are potentially fatally poisonous to humans and livestock. The plant can be more or less toxic depending on the season and stage of growth.

CONSERVATION: G5
Secure.

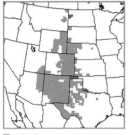

- ■ NATIVE
- ■ RARE OR EXTIRPATED
- ■ INTRODUCED

ALTERNATE NAMES: Sand Verbena, Fragrant Verbena, Snowball, Heart's-delight

Snowball Sand-verbena is a fragrant and elegant perennial, named for its ball-shaped clusters of (usually) white flowers. Like others in the four-o'clock family, the sweet-smelling flowers of this species open in the late afternoon and close in the morning.

DESCRIPTION: This plant can be erect or widely sprawling, ranging in height or width from 8–40" (20–25 cm). It has 25–70 long, funnel-shaped flowers that are usually white, but are sometimes green, lavender, or pink. The flowers have many blossoms appearing in ball-shaped clusters at the ends of the sticky-hairy, branched stems.

FLOWERING SEASON: April–September.

HABITAT: Sandy plains and rocky hillsides.

RANGE: Northern Arizona to West Texas and Oklahoma, north through the Rockies and western plains.

SIMILAR SPECIES: Dwarf Sand-verbena (*Abronia elliptica*), Largefruit Sand-verbena (*A. macrocarpa*) endemic to Texas, and Desert Sand-verbena (*A. villosa*) of the Desert Southwest are all very similar, and some botanists disagree about whether they should be separate species. The fruit and seed shape may be the only difference. The seed of *A. fragrans* has a thicker wall than *A. villosa*, with wings that do not fold together. The fruit of *A. fragrans* is also arrowhead-shaped, not heart-shaped like in *A. elliptica*.

CONSERVATION: ⬤G5
Globally secure but vulnerable in Wyoming and imperiled in Oklahoma.

- NATIVE
- RARE OR EXTIRPATED
- INTRODUCED

ALTERNATE NAMES: Trailing Windmills

Despite the common name, the flowers of this perennial remain open most of the day. Native Americans used this plant to treat swelling and added it to baths to combat fever. A decoction was also used to ease bowel and kidney ailments.

DESCRIPTION: This trailing creeper plant with showy flowers has stalks reaching up to 4" (10 cm) and stems to 3' (90 cm) long. The bright pink flowers appear in a cluster near the ground. Each cluster has three flowers and is ¼–1" (6–25 mm) wide, but it looks like one bilaterally symmetrical flower. The flowers are short-stalked and have three bracts. The leaves are reddish with wavy margins. Fruit appears beneath the cluster and is less than ¼" (6 mm) wide and convex on one side, with two rows of three or five curved teeth on the concave side.

FLOWERING SEASON: April–September.

HABITAT: Dry gravelly or sandy soils in the sun.

RANGE: Southeastern California to southern Utah and Colorado; south to Texas, Mexico, and beyond.

SIMILAR SPECIES: Smooth Trailing Four-o'clock (*A. choisyi*) is a similar species that grows from Arizona to Texas and southward. Its perianth is small, ³⁄₁₆" (5 mm) long or less. The curved edges of the fruit each have five to eight slender teeth that have glands at the tips.

CONSERVATION: G5
Secure. Imperiled in Colorado.

- ■ NATIVE
- ■ RARE OR EXTIRPATED
- ■ INTRODUCED

ALTERNATE NAMES: Red Boerhavia, Hogweed

This plant is widespread and highly variable. The native range is uncertain, partly because this plant is sometimes misidentified as Red Spiderling (*B. diffusa*).

DESCRIPTION: A sprawling perennial that grows in a mat with stems reaching over 3' (1 m) long. The stems are hairy and have sticky glands. The leaves are also sticky and oval to lance shaped, on short petioles. The leaf pairs tend to be unequal in size. The flowers are very tiny, about ⅛" (3 mm) long, and grow in very dense clusters of about five flowers. They can be red, pink, yellow, or white.

FLOWERING SEASON: Year-round in Arizona, California, and Texas. April–November elsewhere.

HABITAT: Open, often disturbed sites.

RANGE: Maryland south to Alabama and Florida, Louisiana west to California.

SIMILAR SPECIES: This plant is often confused with the closely related Red Spiderling (*B. diffusa*), especially in the Gulf States where both occur. The flowers of *B. diffusa* grow in less dense clusters than *B. coccinea*. *B. coccinea* is distinguished by shorter branches and more leaves.

CONSERVATION: **G5**
Secure.

WHITE FOUR-O'CLOCK *Mirabilis albida*

■ NATIVE
■ RARE OR EXTIRPATED
■ INTRODUCED

ALTERNATE NAMES: Hairy Four-o'clock

This perennial's leaves and stems are densely covered in hairs. This plant has showy flowers, but they only bloom at night—opening in the late afternoon and closing soon after sunrise the following morning.

DESCRIPTION: This plant grows 1–3' (0.3–1 m) tall. It has branching stems that are covered in thick, dark hairs at the top, making it sticky. The leaves are egg-shaped, about 4" (10 cm) long, and also very hairy, with wavy edges and a pointed tip. The flowers are white or pink and grow in terminal clusters. Each flower is about ½" (1.25 cm) across, with five sepals, at least three long, protruding pinkish stamens.

FLOWERING SEASON: June–October.

HABITAT: Dry areas, rocky soil, fields, prairies.

RANGE: California east to Arizona, Colorado, New Mexico and Texas, up to Montana, south to Mexico, as well as southern states.

SIMILAR SPECIES: White Four-o'clock flowers are similar to Wild Four-o'clock (*Mirabilis nyctaginea*), but the former are much hairier.

CONSERVATION: G5

Globally secure but critically imperiled in Georgia, Nebraska, and North Carolina. It's considered vulnerable in Iowa and Texas and imperiled in Alabama, Mississippi, and Tennessee.

■ NATIVE
■ RARE OR EXTIRPATED
■ INTRODUCED

ALTERNATE NAMES: Colorado Four-o'clock

The Latin name *Mirabilis multiflora* translates to "marvelous multi-flowered plant." This four-o'clock is often grown as an ornamental and requires little extra water once established. The flowers generally open only in the evening, but may open earlier on a cloudy day. Indigenous people used the large root to treat inflammation or various aches.

DESCRIPTION: A bushy plant up to 1½" (45 cm) tall with vibrant, showy pink flowers growing in leaf axils. Mature plants can have hundreds of flowers. The flowers are about 1" (2.5 cm) wide and petal-like, with five lobes and five stamens. Each flower grows from a cup, which has six to eight flowers that open on successive days. The opposite leaves are 1–4" (2.5–10 cm) long and ovate or heart-shaped, on short stalks. The round-ish fruit is brown or black, about ½" (1.5 cm) long, and seed-like.

FLOWERING SEASON: April–September.

HABITAT: Open sandy areas among juniper and pinyon trees, extending into deserts and grassland.

RANGE: Southern California to southern Colorado; south into northern Mexico.

SIMILAR SPECIES: Green's Four-o'clock (*M. greenei*) occurs on dry slopes in northern California, and MacFarlane's Four-o'clock (*M. macfarlanei*) grows in canyons in northeastern Oregon and adjacent Idaho.

CONSERVATION: ⓖ⑤
Secure.

■ NATIVE
■ RARE OR EXTIRPATED
■ INTRODUCED

ALTERNATE NAMES: Fringed Redmaids

This widespread and common plant is highly variable in its size but identifiable by its flowers, fruits, and seeds. It is a hardy species and can become weedy where it is introduced outside of its native range. This genus is named after Jean-Louis Calandrini, a Swiss botanist and professor. *Calandrinia menziesii* was formerly called *Calandrinia ciliata*, but genetic research has led to this plant's reclassification. All parts of *C. menziesii* are edible. The flowers open in the day and close at night.

DESCRIPTION: This small annual grows 2–16" (5–40 cm) tall with branching stems. It has bowl-shaped magenta flowers that bloom on short stalks, about ½" (1.3 cm) wide, with five petals and two sepals with many whitish stamens. The flowers grow in leafy racemes at the top of the stems. The leaves are ½–3" (1.3–7.5 cm) long and narrow, appearing sparsely on the stem. The upper leaves are much smaller than the lower leaves.

FLOWERING SEASON: April–May.

HABITAT: Disturbed areas, trailsides, often with weeds.

RANGE: California north to British Columbia; east to southwestern New Mexico.

CONSERVATION: G4

Apparently secure, but critically imperiled in Idaho. Vulnerable in New Mexico and British Columbia.

■ NATIVE
■ RARE OR EXTIRPATED
■ INTRODUCED

ALTERNATE NAMES: Pussypaws

This hardy perennial plant is able to establish itself in inhospitable climates, including dry sandy or gravelly soil in alpine environments. The common name comes from the flowers, which form a dense circle around a rosette of leaves, resembling the paws of a cat.

DESCRIPTION: This is a creeper with branching stalks extending 2–10" (5–25 cm). Its flowers appear in dense pink or white clusters at the ends of stems, which generally lean over to the ground from the weight of the flowers. There are two sepals that can be light pink or translucent, each about ½" (1.5 cm) long. There are three stamens with yellow or red anthers and four petals, about ¼" (6 mm) long, which quickly die. The petals and sepals are about the same size. The narrow leaves are shaped like a spoon and grow ¾–3" (2–7.5 cm) long in a dense rosette.

FLOWERING SEASON: May–August.

HABITAT: Alpine climates, loose soil in coniferous forests.

RANGE: British Columbia south to Baja California and east to Utah, Wyoming, and Montana.

SIMILAR SPECIES: One-seeded Pussypaws (*C. monosperma*) is a similar species, but generally has more than one flower cluster per stem.

CONSERVATION: G4

Apparently secure, but imperiled in Utah. Vulnerable in Wyoming and British Columbia.

■ NATIVE
■ RARE OR EXTIRPATED
■ INTRODUCED

ALTERNATE NAMES: Lanceleaf Springbeauty

As the name suggests, Western Springbeauty has pretty flowers that bloom in the spring, often near snowbanks. The entire plant is edible and the deeply buried underground stems can be eaten like potatoes. The genus *Claytonia*, once part of the purslane family, was reclassified to the Montiaceae family in 2009.

DESCRIPTION: A small, slender, delicate plant with a loose raceme of 3–15 white, pink, or rose bowl-shaped flowers. It reaches a height of 2–10" (5–25 cm). The succulent leaves are ½–3½" (1.5–9 cm) long, narrow, opposite, and lanceolate. It usually has one to two leaves near the base of the stem, which often die before the flowers bloom. The star-shaped flowers are ¼–¾" (6–20 mm) wide, with five petals and two sepals. If the flowers are white, they often have dark veins with five stamens and a yellowish color near the base of each petal. The fruit capsule contains two black seeds.

FLOWERING SEASON: April–July.

HABITAT: High mountains where it's snowy year-round.

RANGE: British Columbia south to southern California and east to Rocky Mountains from New Mexico to Alberta.

SIMILAR SPECIES: Rydberg's Springbeauty (*Claytonia multiscapa*) is similar with a smaller distribution. *C. multiscapa* is also a shorter plant, with smaller leaves and flowers.

CONSERVATION: (G5)
Secure.

■ NATIVE
■ RARE OR EXTIRPATED
■ INTRODUCED

ALTERNATE NAMES: Winter Purslane, *Montia perfoliata*

This plant was eaten by miners in the California gold rush to prevent scurvy, thus its common name. The leaves provide a good source of vitamin C and can be eaten in a salad. This annual used to be the namesake of the genus *Montia* before being reclassified.

DESCRIPTION: A succulent, trailing plant with slender stems that grows 1–14" (2.5–35 cm) tall. The leaves are narrow and grow in pairs that appear joined together as one circular leaf, about 2" (5 cm) wide around the stem. The mature leaves are succulent and edible. The small, white flowers grow in a raceme of 5–40 flowers above the leaves. The flowers are ⅛ –¼" (3–6 mm) wide, with two sepals, and five petals that are longer than the sepals.

FLOWERING SEASON: March–July.

HABITAT: Disturbed areas, moist soils.

RANGE: Throughout California. Alaska and British Columbia south to Baja California and east to Arizona, Colorado, Wyoming, and South Dakota.

CAUTION: This plant can contain high amounts of oxalates, which can lead to kidney stones if consumed in high quantities.

SIMILAR SPECIES: A pair of studies published in 2012 suggest *C. perfoliata*, Indian Lettuce (*C. parviflora*), and Redstem Springbeauty (*C. rubra*) interbreed, making up a polyploid pillar complex.

CONSERVATION: G5
Secure, but imperiled in Wyoming.

■ NATIVE
■ RARE OR EXTIRPATED
■ INTRODUCED

ALTERNATE NAMES: Virginia Springbeauty

This is an attractive spring perennial with small, potato-like underground tubers that taste like sweet chestnuts. Native Americans and European colonists gathered them for food. The plant's scientific name honors botanist John Clayton (1694–1773), who arrived in Virginia from England in the early 1700s.

DESCRIPTION: A low, trailing plant that grows 6–12" (15–30 cm) tall. The dark green leaves are 2–8" (5–20 cm) long, often growing in a single opposite pair halfway up the stem. The flowers are ½–¾" (1.3–2 cm) wide; with five pink, white, or sometimes yellow petals; two sepals; five stamens; and pink anthers. The flowers bloom for just three days, while the stamens only appear for one day. The fruit is a small capsule.

FLOWERING SEASON: March–May.

HABITAT: Moist deciduous forests, moist areas, lawns, city parks, wetlands, roadsides.

RANGE: Ontario to Quebec and southern New England; south to Georgia; west to Louisiana and Texas; north to Minnesota.

SIMILAR SPECIES: Carolina Springbeauty (*C. caroliniana*) is similar, but the leaves are longer and more oval.

CONSERVATION: G5

Secure, but imperiled in Vermont, Nebraska, and Massachusetts.

■ NATIVE
■ RARE OR EXTIRPATED
■ INTRODUCED

ALTERNATE NAMES: Pygmy Bitterroot, Alpine Lewisia, *Oreobroma pygmaeum, Talinum pygmaeum*

This is a small, highly variable plant found in the mountains in the West. It sometimes breeds with other *Lewisia* species, which makes it difficult to identify. The genus name *Lewisia* is in honor of American explorer Meriwether Lewis, who collected the related *Lewisia rediviva* in Montana in 1806.

DESCRIPTION: Pygmy Bitterroot is a small woody perennial, growing 2" (5 cm) tall. The leaves grow up to 3½" (9 cm) long, and appear in a basal rosette. The leaves are narrow, though fleshy and linear. They usually grow taller than the stems. A single pink, red, or white flower appears at the top of each stem. The flowers are ½" (12 mm) wide, with six to nine petals and two green to red sepals. The flowers may have a dark vein.

FLOWERING SEASON: June–August.

HABITAT: Dry alpine to subalpine meadows, moist areas, mountain ridges, sandy areas, sagebrush, coniferous woods, at elevations of 7500–13,500' (2300-4200 m).

RANGE: From Alaska and Alberta to California and New Mexico, to Utah, Nevada and Arizona.

SIMILAR SPECIES: Alpine Bitterroot is very similar to Nevada Bitterroot (*Lewisia nevadensis*). Some botanists consider them the same species, but there are some subtle differences. Alpine Bitterroot has small teeth at the top of its sepals and has glandular tips at the sepals, unlike *L. nevadensis*.

CONSERVATION: G5

Secure, but critically imperiled in South Dakota and imperiled in Alberta.

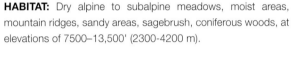

■ NATIVE
■ RARE OR EXTIRPATED
■ INTRODUCED

ALTERNATE NAMES: Prairie Flameflower, Small-flowered Flameflower

As the common name suggests, Sunbright grows in hot, sunny habitats. Its small, succulent leaves, thick stems, and flowers—which open just for a day—help this plant survive in dry climates. Native Americans used the roots to treat sores.

DESCRIPTION: This woody perennial grows 8" (20 cm) with branching, or erect stems. The leaves are slender and alternate. The leaves are slightly wider at the base, appearing about ½–2" (1.5–5 cm) long in a dense basal cluster. A single bloom of pink flower clusters appear at the top of slender stalks or in the cluster of leaves. The flowers are ⅓–½" (8–13 mm) wide and star-shaped with five petals. There are 5–10 stamens with yellow tips.

FLOWERING SEASON: June–July.

HABITAT: Open woods, shrublands, grasslands, slopes.

RANGE: The central United States from North Dakota and Minnesota south to Texas and Louisiana; east to Alabama and Illinois and west to Arizona.

SIMILAR SPECIES: Rough-seeded Fameflower (*Phemeranthus rugospermus*) is almost identical, but *P. rugospermus* has more stamens—about 12–25—while *P. parviflorus* has 5–10.

CONSERVATION: **G5**
Globally secure but threatened in several states in which it occurs. Critically imperiled in Alabama, Illinois, and Iowa. Imperiled in North Dakota and Wyoming. Vulnerable in Louisiana and Arizona.

■ NATIVE
■ RARE OR EXTIRPATED
■ INTRODUCED

ALTERNATE NAMES: Little-hogweed, Pusley, Verdolagas

This cultivated annual is widespread and variable. It's considered a weed in North America, especially in hot climates, where it grows quickly. Common Purslane has a sour or sweet taste and is dense in nutrients, including iron and omega-3 fatty acids. The stems, leaves, and flower buds can be eaten raw or cooked. The flowers open only in hot, sunny weather.

DESCRIPTION: This is a succulent, matted weed with smooth, red stems reaching about 2' (60 cm) long. The appearance of this plant varies by the amount of water available. Sometimes it is more erect, up to 6" (15 cm) tall. It has small, yellow flowers that grow singularly or in small clusters in leaf axils or at the tips of smooth, shiny stems. Flowers are about ¼" (6 mm) wide, with two sepals, four or five petals, and many yellow stamens. Just one flower opens in each cluster at a time. The alternate leaves are ½–1½" (1.5–4 cm) long, flat, and oval-shaped. They appear widest at the tip. Small black or brown seeds grow in a pod, which splits open.

FLOWERING SEASON: June–September.

HABITAT: Roadsides, waste sites.

RANGE: Introduced throughout much of North America, except the Far North.

CONSERVATION:
Exotic to North America and unranked.

■ NATIVE
■ RARE OR EXTIRPATED
■ INTRODUCED

ALTERNATE NAMES: Saguaro

Saguaro is the state flower of Arizona and is the largest cactus in the United States. Its Latin name is given in honor of philanthropist and steel industry magnate Andrew Carnegie. Saguaro National Park in southeastern Arizona was designated in 1994 to conserve this plant's habitat. Giant Saguaro grows very slowly and can live 150–200 years, although many plants are killed or injured by lightning in the desert. Saguaro can store great quantities of water, which it uses only when needed, making it highly drought tolerant. The pulp from the fruit can be eaten raw or preserved, while the sap can be made into an intoxicating drink, and the seeds can be ground into meal. Native Americans used the dead, woody ribs to build shelters. Various birds and insects feed on the flowers.

DESCRIPTION: A tall, tree-like cactus with thick, spiny stems. The plant generally reaches up to 50' (15 m) tall, but can reach as high as 80' (24 m). The branch-like stems are usually erect, but can be twisted, extending 8–24" (20–60 cm) wide, with 12–24 ribs. The spines grow up to 2" (5 cm) long, in clusters of 10–25 on ribs. The flowers are white, 2½–3" (6.5–7.5 cm) wide, and grow in crown-like clusters at the tips of branches with many petals. The green fruit is 2½–3½" (6.5-9 cm) long and oval with a red color on the inside.

FLOWERING SEASON: Late spring to late summer.

HABITAT: Rocky or gravelly soils of desert foothills, especially on south-facing slopes.

RANGE: The Sonoran Desert in Arizona south to Sonora, Mexico; very local in southeastern California; occurs at 700–3500' (213–1067 m).

CONSERVATION: G5
Globally secure, but critically imperiled in California.

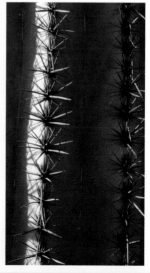

- ■ NATIVE
- ■ RARE OR EXTIRPATED
- ■ INTRODUCED

This species appears as spiny, branched trees or shrubs. The flower buds are edible, but the spines should be removed first.

DESCRIPTION: A spreading to erect shrub or tree-like cactus that spans 3½–13' (1–5 m) tall. The main trunk is short, branching open. The terminal joints are light green, 4¾–20" (10–50 cm) long and ¾–1½" (2–4.25 cm) wide, with 6–25 yellow or reddish spines that turn gray as they age. The flowers are yellow, or reddish to purple, 1¼–1¾" (3–4.5 cm) long and 1–2" (2.5–5 cm) wide, with light pink or red styles. The fruit is dry, shriveled and covered in spines.

FLOWERING SEASON: May–June.

HABITAT: Sandy or gravelly soil of hillsides, ledges, mesas, flats, and washes.

RANGE: In the Mojave and Sonoran deserts, at the southern tip of Nevada, extreme southwestern Utah, southern California, and southern and western Arizona.

SIMILAR SPECIES: This species is known to hybridize with other members of the genus *Cylindropuntia*. Buckhorn Cholla and Silver Cholla (*C. echinocarpa*) both grow in the same range. Silver Cholla's stems are usually shorter than those of Buckhorn Cholla. Silver Cholla also has one main trunk and a tubercle that is one or two times as long as it is wide, while Buckhorn Cholla usually has several small trunks and a tubercle length that's three to several times the width. Staghorn Cholla (*C. versicolor*) is also similar but has smoother fruit that stays on the plant while flowering. Walkingstick Cactus (*C. spinosior*) is another similar species but has bumpy, spineless fruit.

CONSERVATION: Ⓖ4
Apparently secure.

NATIVE
RARE OR EXTIRPATED
INTRODUCED

ALTERNATE NAMES: Walkingstick Cholla, Candelabra Cactus

Tree Cholla plants are sometimes used for decoration and the fruits are a popular food source for birds. Once the stems die, they become hollow with many holes.

DESCRIPTION: A spiny, leafless bush or small tree with cylindrical, jointed branches and deep pink to reddish-lavender flowers near ends reaching a height of 3–7' (90–210 cm). The flowers are 2-3" (5–7.5 cm) wide with many petals. The stems grow in joints, 5–16" (12.5–40 cm) long, ¾–¼" (2–3 cm) wide, and have very spiny knobs. The spines grow in clusters of 10–30 and are ½–1" (1.5–2.5 cm) long. The yellow fruits are 1–2" (2.5–5 cm) long and oval.

FLOWERING SEASON: May–July.

HABITAT: Plains, deserts, and among pinyon and juniper.

RANGE: Southern Colorado and Kansas to Arizona, New Mexico, Texas, and North Mexico.

CAUTION: This plant has sharp spines as well as tiny barbed hairs called glochids that can be difficult to remove from the skin.

CONSERVATION: G5
Secure, but imperiled in Kansas.

■ NATIVE
■ RARE OR EXTIRPATED
■ INTRODUCED

ALTERNATE NAMES: Christmas Cholla

This plant has the slenderest stems of all southwestern chollas. It grows red berries in winter that can be intoxicating if consumed. The fruits are crushed and mixed with a drink by some Native American groups for narcotic effects.

DESCRIPTION: A small bush with many spiny, intertangled, slender branches, up to 3' (90 cm) tall. Greenish or yellow flowers grow along the stem and open only in the late afternoon. They are about ½–1" (1.5–2.5 cm) wide. The stems are about ¼" (6 mm) thick and branched. The spines are tan or gray and 1–2½" (2.5–6.5 cm) long. Clusters of tiny, reddish bristles have one spine each. The fruit, ½" (1.5 cm) long, is fleshy and bright red, and appears around December and stays on the stem through most of winter.

FLOWERING SEASON: May–June.

HABITAT: Flats, slopes, and along washes in deserts and grasslands.

RANGE: Western Arizona to southern Oklahoma; south to North Mexico.

CAUTION: These cacti have sharp spines and barbed hairs called glochids that can penetrate skin. The fruit can have intoxicating effects if consumed.

CONSERVATION: G4
Apparently secure.

ENGELMANN'S HEDGEHOG CACTUS *Echinocereus engelmannii*

■ NATIVE
■ RARE OR EXTIRPATED
■ INTRODUCED

ALTERNATE NAMES: Saints Cactus, Strawberry Cactus

This is one of the most common hedgehog cacti in the south-western United States. It is sometimes used for decoration.

DESCRIPTION: Engelmann's Hedgehog Cactus stems are initially erect and grow in cylindrical clumps up to 2' (60 cm) high and 3' (90 cm) wide. There are 8–20 spines per areole and four to six central spines, 2–3" (5–7 cm) long, that can be whitish, grayish, yellow, or reddish-brown to nearly black. Its purple to magenta flowers are 2–3" long. The fruit is green initially and then turns pink and dries when ripe. The seeds are black.

FLOWERING SEASON: March–June.

HABITAT: Rocky, gravelly, or sandy hillsides and flats in deserts, desert grasslands, chaparral, pinyon-juniper woodlands, and montane forests.

RANGE: In the Sonoran, Mojave, and Great Basin deserts of southern California, southern Nevada, southern Utah, and Arizona.

CAUTION: This cactus has very sharp spines.

SIMILAR SPECIES: There are nine varieties that differ based on stem, flower size, and central spine characteristics. Some varieties are very rare. One variety, var. *nicholii*, has golden-yellow spines and tall, slender stems.

CONSERVATION: G5
Secure.

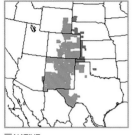

■ NATIVE
■ RARE OR EXTIRPATED
■ INTRODUCED

ALTERNATE NAMES: Nylon Hedgehog Cactus, Small-flowered Hedgehog Cactus

This is an attractive cactus with sweet-tasting fruit. This plant withdraws under the ground when water is scarce.

DESCRIPTION: This small cactus is mostly unbranched, growing 1–10" (2.5–25 cm) tall. It grows one cylindrical stem or several in a clump, and has yellowish-green or magenta flowers. The flowers are ¾–1" (2–2.5 cm) wide, near the top of the stem, with many petals. The stems are 4" (10 cm) wide with 6–14 ribs.

The spines are ½–1" (1.5–2.5 cm) long and appear red, brownish, white, gray, or greenish-yellow. The fruit is green or purplish.

FLOWERING SEASON: May–July.

HABITAT: Dry plains and hills.

RANGE: Southeastern Wyoming and western South Dakota south to eastern New Mexico and western Texas.

SIMILAR SPECIES: This species is very similar to Davis' Hedgehog Cactus (*E. davisii*), but *E. davisii* flowers earlier.

CONSERVATION: G5
Secure.

■ NATIVE
■ RARE OR EXTIRPATED
■ INTRODUCED

ALTERNATE NAMES: Beehive Cactus, *Coryphantha vivipara*

This species occurs throughout western North America in a wide range of habitats and at various elevations. This frost tolerant plant is one of the most widespread cacti in the U.S. It is one of just a handful of cacti that naturally occur in Canada.

DESCRIPTION: This is a small, round cactus growing to 1½–6" (4–15 cm). It is densely covered in star-shaped, straight white spines; with yellow, pink, red, or purple flowers; about 1–2" (2.5–5 cm) wide with many petals that are ¼–½" (6–13 mm) long. It has small, spherical stems, which typically grow in clusters, up to 3" (7.5 cm) wide and 6" tall. There are 3–10 straight central spines, each ½–¾" (1.5–2 cm) long, with pink, red, or black at the tip. These are surrounded by 12–40 slightly shorter white spines. Younger plants have only one radial spine. Older plants have grooves on the upper side of the tubercles. The fruit is about ½–1" (1.5–2.5 cm) long, green, and smooth, with brown seeds.

FLOWERING SEASON: April–June.

HABITAT: Rocky, desert slopes and rocky or sandy soil among pinyon, juniper, oaks, and Ponderosa Pine (*Pinus ponderosa*) trees.

RANGE: Central Canada south to southeastern Oregon and southeastern California, and east to the Great Plains from Manitoba and Minnesota to western Texas; also in northern Mexico.

SIMILAR SPECIES: Nipple Cactus (*C. missouriensis*) is very similar, but has only one central spine in each cluster, with greenish-white flowers, and reddish fruit with black seeds. Spinystar can easily be mistaken for Mountain Ball Cactus (*Pediocactus simpsonii*), which has smooth tubercles, while *E. vivipara*'s tubercles are grooved. *P. simpsonii* also has straighter spines.

CONSERVATION: **G5**

Secure, but critically imperiled in Minnesota and Manitoba and vulnerable in Wyoming and Alberta. Overcollecting may pressure local populations.

■ NATIVE
■ RARE OR EXTIRPATED
■ INTRODUCED

Eastern Prickly-pear is the only cactus native to New England, but is most common west of the Appalachian Mountains. It has long been overlooked by botanists and has been considered the same species as Low Prickly-pear (*O. humifusa*) until a recent revision of the taxonomy of the *O. humifusa* complex.

DESCRIPTION: This species is a short shrub with large, white spines that are smooth to the touch and turn gray as they age. There are one or two spines per areole and a lower layer of reddish-brown glochids. The flowers are yellow, 1–2.5" (3–6 cm) long, with 9–10 inner tepals and a red center. The stigmas are white and contain 6–10 lobes. It has flat, dark green pads. The fruit is green, ¾–2" (2–5 cm) long, with glochids on the surface. The seeds are smooth.

FLOWERING SEASON: May–July.

HABITAT: Sandy fields and plains, open oak forests, pastures, roadsides, disturbed areas.

RANGE: Kentucky and Indiana east to Massachusetts; it mainly occurs west of the Appalachian Mountains. The closely related Low Prickly-pear (*O. humifusa*), sometimes considered the same species, is more widespread from the Great Plains to the East Coast and most prominent on the coasts.

CAUTION: This plant, like most cacti of the *Opuntia* genus, has sharp spines and small barbed hairs called glochids that can be difficult to remove from the skin.

SIMILAR SPECIES: This species closely resembles Low Prickly-pear (*O. humifusa*) and was long considered the same species, but *O. humifusa* does not have spines. Eastern Prickly-pear is also similar to Twistspine Prickly-pear (*O. macrorhiza*) and Erect Prickly-pear (*O. mesacantha*) but its spines are not as barbed.

CONSERVATION:

This species is not yet formally ranked due to its inclusion within the *O. humifusa* complex until recently. Its population in Canada is endangered, limited to two small sites in Ontario on the northern shore of Lake Erie. It is critically imperiled in New York and vulnerable in Ohio.

FRAGILE PRICKLY-PEAR *Opuntia fragilis*

■ NATIVE
■ RARE OR EXTIRPATED
■ INTRODUCED

ALTERNATE NAMES: Pygmy Prickly-pear, Brittle Prickly-pear, Little Prickly-pear

This is one of the most common of the prickly-pears. As the common name suggests, it is very fragile and the joints and pads can easily become detached by rain, wind, or contact with humans or wildlife.

DESCRIPTION: A low matted clump that grows to a height of 8–10" (20–25 cm). The spiny, jointed stems extend 1–3' (30–90 cm) wide and are mostly flat. The joints are bright green and fragile, easily becoming detached. The areoles are whitish with white or yellow bristles. The white or grayish spines are ½–1" (1.5–2.5 cm) long and grow in a cluster of one to nine. The plant has few flowers, if any. The flowers appear at the top of the joints, about 1½–2" (4–5 cm) wide, with seven yellow petals and a green color at the center. The fruit is ½" (1.5 cm) long and oval or egg-shaped. The fruit appears green at first but turns brown at maturity with large seeds inside.

FLOWERING SEASON: May–June.

HABITAT: Dry, open areas.

RANGE: Western Washington and southern British Columbia; south on the east side of the Cascade Mountains to northern California and northern Arizona; east to northern Texas and southern Michigan.

CAUTION: This plant, like most cacti of the *Opuntia* genus, has sharp spines and small barbed hairs called glochids that can be difficult to remove from the skin.

SIMILAR SPECIES: Twistspine Prickly-pear (*Opuntia macrorhiza*) is similar, but has more flowers that are larger. The joints and pads are less fragile.

CONSERVATION:  G5

Globally secure but critically imperiled in California, Illinois, Iowa, and Michigan. Imperiled in Wyoming. Vulnerable in Wisconsin and Ontario.

■ NATIVE
■ RARE OR EXTIRPATED
■ INTRODUCED

ALTERNATE NAMES: Purple-fruited Prickly-pear, Tulip Prickly-pear

Desert Prickly-pear is common in the deserts of Texas and the cool, moist forests of the Rocky Mountains. The edible fruits taste like watermelon or pear and are an important food source for wildlife in the desert.

DESCRIPTION: This plant grows 3' (1 m) tall and 5' (1.5 m) wide. It can be sprawling or erect. The flowers, 2–3" (5–7.5 cm) long, are bright yellow with a pale green center, or sometimes a reddish center. The stem grows up to 18" (45 cm) and appears bluish-green or yellow in dry areas. The pads are protected by clusters of spines. Each cluster has one to four brown, reddish-brown, or gray spines, which are often over 1¼" (3 cm) in length.

FLOWERING SEASON: April–June.

HABITAT: Sandy or gravelly soils of hillsides, flats, canyon rims, and mesas in grasslands, deserts, oak woodlands, chaparral, pinyon-juniper woodlands, and montane forests.

RANGE: Found in the Great Basin, Mojave, and Chihuahuan deserts in southern Nevada, Utah, Colorado, southern California, Arizona, New Mexico, western Kansas, Oklahoma, and the western two-thirds of Texas.

CAUTION: This plant, like most cacti of the *Opuntia* genus, has sharp spines and small barbed hairs called glochids that can be difficult to remove from the skin.

SIMILAR SPECIES: This species has up to 10 or more varieties, making identification difficult. The varieties are distinguished by pad size, spine distribution on the pad, spine color and size, and fruit length.

CONSERVATION: G5
Secure.

SIMPSON'S HEDGEHOG CACTUS *Pediocactus simpsonii*

■ NATIVE
■ RARE OR EXTIRPATED
■ INTRODUCED

ALTERNATE NAMES: Snowball Cactus

Unlike most other cacti, Simpson's Hedgehog Cactus grows in cool areas. This is one of the most cold-hardy species and can be found in elevations as high as 4500–11,500' (1400–3500 m). This cactus is frequently camouflaged under other plants. It can even shrink below the ground in the winter. The flowers are generally pink in the East and yellowish in the West.

DESCRIPTION: This plant grows 2–8" (5–20 cm) tall. It has several bell-shaped flowers that are white, rose, or yellow at the top of one or a few stems. The flowers usually remain closed on cloudy days. Stems are shiny and nearly spherical, 2–3" (5–7.5 cm) wide. Spines are ⅜–¾" (9–20 mm) long, straight, and brownish.

FLOWERING SEASON: May–July.

HABITAT: Powdery soils among sagebrush and juniper.

RANGE: Found in high elevations, throughout many of the mountain states in the western U.S., from eastern Washington to west-central Nevada and northern Arizona; east to northern New Mexico, western Colorado, western South Dakota, and western Montana.

CAUTION: This cactus has very sharp spines.

CONSERVATION: G5

Globally secure but critically imperiled in Arizona. Vulnerable in Idaho, Montana, New Mexico, and Wyoming. The plant is widespread and locally abundant, but faces localized threats from horticultural collecting, mining, and changes in land use.

■ NATIVE
■ RARE OR EXTIRPATED
■ INTRODUCED

ALTERNATE NAMES: Climbing Hydrangea

This widespread hydrangea is native to the southeastern United States, where it thrives in wet bottomland forests. Woodvamp is sometimes grown as an ornamental.

DESCRIPTION: A high-climbing woody vine (attached by rootlets) that grows 30' (9 m) long through shrubs and up trees. It only flowers while climbing. The white and creamy-colored flowers smell sweet and grow in clusters, appearing 1–2' from the climbing surface. The dark green leaves grow all along the vine and are 1–4" (3–10 cm) long, shiny, smooth, ovate, and pointed. The fruit, a brown capsule, about ¼", appears in the late summer.

FLOWERING SEASON: May–July.

HABITAT: Wet woods, swamps.

RANGE: Southeastern United States from Virginia south to Florida, west to Arkansas and Louisiana; also found in southern Delaware and on Long Island in New York.

CONSERVATION: G5
Globally secure but critically imperiled in Arkansas and Delaware.

■ NATIVE
■ RARE OR EXTIRPATED
■ INTRODUCED

ALTERNATE NAMES: Smooth Hydrangea, Sevenbark

This attractive plant resembles a tree and is often grown as an ornamental. Wild Hydrangea root was also used by Native Americans to treat kidney and bladder stones. The stem bark peels off in layers, revealing multiple colors, hence the common name "Sevenbark."

DESCRIPTION: Wild Hydrangea is often wider than it is tall. It grows 3–6' (1–2 m) high in a small mound shape. The many stems are grayish-brown and have sharply toothed, dark green leaves, 2–6" (5–15 cm) long, that are light green underneath. The lower leaves have fine hairs. Very small greenish or white flowers, ⅓" (1 cm) long, are sometimes absent; when present, the flowers can be heavy enough to weigh the stem to the ground. Fruit appears in a brown capsule, about 2 mm long.

FLOWERING SEASON: May–September.

HABITAT: Moist or rocky slopes, ravines, streambanks.

RANGE: New York to Florida west to Illinois, Missouri, Oklahoma and Louisiana.

CONSERVATION: G5

Globally secure but critically imperiled in Florida and imperiled in New York.

■ NATIVE
■ RARE OR EXTIRPATED
■ INTRODUCED

ALTERNATE NAMES: Stinging Serpent

This native perennial shrub attracts bees, butterflies, and birds. Its thick leaves distinguish it from other species but it should not be touched because its stinging hairs can cause a rash.

DESCRIPTION: This plant reaches 2' (60 cm) tall. The lower leaves are thick and lobed with wavy edges and stinging hairs that vary in length. The leaves also have yellow dots of glands. The upper leaves have no stems and are smaller and more lanceolate in shape. The clustered flowers are hairy and mostly only bloom at night. They have five yellow sepals and five petals. The flowers are subtended by red-orange bracts.

FLOWERING SEASON: March–October.

HABITAT: Open areas.

RANGE: Arizona, New Mexico, and Texas.

CAUTION: This plant should not be touched. It has sharp, barbed hairs, some needle-like and stinging. Skin contact can result in a rash.

CONSERVATION:

The conservation status of this species has not been formally ranked; its range and estimated number of occurrences in the wild are fairly limited. It is believed to be extirpated in Oklahoma.

■ NATIVE
■ RARE OR EXTIRPATED
■ INTRODUCED

As its name suggests, Ten-petal Blazingstar is known for its large, bright flowers with 10 petals. It is a short-lived perennial native to dry parts of central and western North America.

DESCRIPTION: Ten-petal Blazing Star grows 1–3' (30–91 cm). There can be one stem or several exfoliating near the base. Each stem is whitish, erect, and stout. Large cream-colored to pale yellowish flowers appear in clusters at the top of the stems. The flowers have 10 petals, 1½–2¾" (4–7 cm) long, with about 100–200 stamens. The flowers only open in the late afternoon and close around midnight. When open, the flowers often overlap. The leaves are alternate, simple, and lanceolate, spanning 2–6" (5–15 cm) long and ½–1½" (1.3–4 cm) wide. They are covered in clingy hairs. The fruit, a capsule, is 1¼–2" (3–5 cm) long and ½–¾" (1.5–2 cm) wide.

FLOWERING SEASON: August–October.

HABITAT: Plains, slopes, and roadsides.

RANGE: Alberta to Manitoba south to Nevada, New Mexico, and Texas, with a limited range in extreme western Iowa.

SIMILAR SPECIES: Bractless Blazingstar (*Mentzelia nuda*) is similar, but has shorter petals. The flowers of *M. nuda* also close near sunset and do not overlap when open.

CONSERVATION: G5

Globally secure, but possibly extirpated in Manitoba and critically imperiled in Iowa. Imperiled in New Mexico. Vulnerable in Montana, Wyoming, Alberta, and Saskatchewan.

- ■ NATIVE
- ■ RARE OR EXTIRPATED
- ■ INTRODUCED

ALTERNATE NAMES: Northern Blazingstar, Evening Star, Stickleaf, Smoothstem Blazingstar, *Nuttallia laevicaulis*

This species is colloquially called Stickleaf or "nature's Velcro" because of the leaves' barbed hairs, which stick easily to clothing and other fabric.

DESCRIPTION: This biennial or short-lived perennial grows 1–3' (30–90 cm) tall. The sticky leaves are 4–12" (10–30 cm) long and narrowly lanceolate. The edges have large, irregular teeth. The yellowish, star-like flowers grow 2–5" (2.5–12.5 cm) wide with five petals at the top of a white stem. It has numerous stamens. The flowers open at dusk and remain open through the night and the following morning before closing in the afternoon.

FLOWERING SEASON: June–September.

HABITAT: Gravelly or sandy slopes and plains, mostly in arid regions.

RANGE: Southeastern British Columbia south to southern California and east to Colorado, Wyoming, and Montana.

SIMILAR SPECIES: This species' large flowers make it distinct from others in its genus, which includes 60 species in the western United States. Most other plants in the genus also have at least eight petals, while Giant Blazingstar has just five.

CONSERVATION: G4

Apparently secure. Vulnerable in Montana and Wyoming.

■ NATIVE
■ RARE OR EXTIRPATED
■ INTRODUCED

ALTERNATE NAMES: Ground Dogberry, Canadian Bunchberry, Canada Dwarf-dogwood, *Cornus canadensis*

This species, sometimes classified as *Cornus canadensis*, is sometimes grown as an ornamental. It's a low-growing, mostly herbaceous plant, among the smallest dogwoods, which are mainly shrubs and trees. The flowers have an appendage that catapults pollen when touched; the explosive release of pollen is one of the fastest known plant motions. In the wild, these plants use this catapulting action to lodge pollen into the hairs of insect pollinators that come into contact with the flowers.

DESCRIPTION: This slow-growing plant appears in an extensive carpet-like mat, reaching 2–8' (5–20 cm) tall. It has six clustered leaves that appear in a whorl, growing ¾–3" (2–7.5 cm) long and narrowly ovate. Flowers are greenish and grow in a cluster, resembling one large flower on a short stalk above the leaves.

Bracts, just below the flower, are white or pinkish, growing 4" (10 cm) wide. The fruit grows in a cluster of bright red, round berries. The seeds are hard and crunchy.

FLOWERING SEASON: June–August.

HABITAT: Cool, moist woods, disturbed areas, and damp openings.

RANGE: Across Canada to Labrador and southern Greenland; across much of the northern United States, including Alaska; isolated populations in Colorado and New Mexico.

CONSERVATION: **G5**

Globally secure. Small or isolated populations are critically imperiled in Indiana, Illinois, Iowa, Maryland, New Jersey, New Mexico, Ohio, and Virginia. Imperiled in California, Rhode Island, West Virginia, and Wyoming.

■ NATIVE
■ RARE OR EXTIRPATED
■ INTRODUCED

ALTERNATE NAMES: Spotted Touch-me-not

This native annual has bright orange flowers, which attract pollinators such as hummingbirds and bees. Orange Jewelweed is highly competitive with other plants and is considered a weed in some areas. Native Americans have used its sap and leaves to relieve itching from poison ivy and other ailments. The sap was also used as an antifungal remedy.

DESCRIPTION: This is a tall, leafy plant that grows 2–5' (60–150 cm) high with weak, succulent stems. The leaves are 1½–3½" (4–9 cm) long, oval shaped and alternate, with toothed margins. The flowers are orange or golden, about 1" (2.5 cm) long and form a sac, about ¼" (6 mm) long. The flowers have three sepals—two of which are green and one is the same color as the petals. Some of the flowers grow on nodes on the stem and never open. The fruit is a capsule that opens at maturity to release its seeds. This action can be triggered by a light touch.

FLOWERING SEASON: July–October.

HABITAT: Wetlands, moist areas, and woods.

RANGE: Alberta east to Newfoundland, south to Florida, west to Texas, and north to North Dakota; also in northwestern United States and Canada.

SIMILAR SPECIES: Pale Touch-me-not (*I. pallida*) is a similar species but has yellow flowers, not orange.

CONSERVATION: G5

Globally secure. This species is considered a weed in some areas.

■ NATIVE
■ RARE OR EXTIRPATED
■ INTRODUCED

ALTERNATE NAMES: Collomia

This phlox-like annual is widespread and considered a weed in some areas outside of its native range in northern and western North America. Native Americans mashed the leaves to treat injuries and bruises. *Collomia* comes from the Greek word *colla*, meaning "glue," and refers to its glue-like seeds.

DESCRIPTION: This plant has a velvety stem, reaching 4–16" (10–41 cm). Atop the stem is a cluster of about 20 white, pink, or lilac flowers, about 1 cm across, with five small petals. Some of the flowers appear in a cluster among dense upper leaves. The long, green leaves are generally narrow, lance-shaped, and alternate.

FLOWERING SEASON: May–August.

HABITAT: Dry, open woods; yards, fields, waste areas.

RANGE: New Brunswick and Gaspé, Quebec, west to British Columbia, southwest to Wisconsin, Nebraska, New Mexico, and California; widely naturalized elsewhere.

SIMILAR SPECIES: Large-flower Mountain-trumpet (*Collomia grandiflora*) has a similar shape, but its petals are yellow or orange, not pink.

CONSERVATION: G5
Secure.

■ NATIVE
■ RARE OR EXTIRPATED
■ INTRODUCED

ALTERNATE NAMES: Blue-head Gily-flower, Blue Field-gilia, Bluehead Gilia

This attractive flower, part of the phlox family, contains many subspecies and can vary in appearance. It's often grown in gardens and is often part of wildflower seed mixes purchased at a store. The blue flowers grow in tight clusters that resemble a globe or a ball, hence the common name.

DESCRIPTION: This is a tall annual growing up to 20–24" (50–60 cm) high. The stems are branching and sometimes hairy, topped with a dense spherical cluster of about 50 to 100 tiny pale blue-violet to white flowers. The basal leaves are pinnately divided and the upper leaves are smaller than the lower leaves.

FLOWERING SEASON: April–July.

HABITAT: Rocky or sandy areas, dry slopes below 6000' (1829 m).

RANGE: Most of the northern two-thirds of California, north to British Columbia and Idaho; naturalized locally in the Santa Cruz Mountains and in several northeastern states.

CONSERVATION: **G5**
Secure.

■ NATIVE
■ RARE OR EXTIRPATED
■ INTRODUCED

ALTERNATE NAMES: Bird's-eyes

This attractive wildflower is endemic to California, but is widely cultivated around the world and is frequently included in wildflower seed mixes. The white and lavender flowers contain plenty of nectar and are very attractive to butterflies, bees, and hummingbirds. The flowers have a sweet or musky scent, similar to chocolate.

DESCRIPTION: This annual reaches 4–12" (10–30 cm) high, with light blue or purplish tubular flowers with five sepals, five petals, and five stamens, appearing in clusters. The flower has a dark purple ring at the top of its tubular throat, which is yellow on the inside, giving *Gilia tricolor* its name. The basal leaves are slender and grow like a fern or a feather.

FLOWERING SEASON: Year-round.

HABITAT: Grassy plains, mountain ranges, foothills, and slopes.

RANGE: Throughout California, with many populations in the Bay Area and the Central Valley. There are also some wild-growing populations in Massachusetts and Texas, likely descended from escaped cultivated plants thanks to this species' popularity in garden seed mixes.

SIMILAR SPECIES: Globe Gilia (*Gilia capitata*) is a similar species but is more variable and has a much wider range, found throughout much of the West. Its throat opening may be white, pink, lavender, or light blue.

CONSERVATION: G5
Secure, although its status may need review because it is endemic to a single state.

■ NATIVE
■ RARE OR EXTIRPATED
■ INTRODUCED

ALTERNATE NAMES: Scarlet Gilia, Scarlet Skyrocket, Skunk Flower, *Gilia aggregata*

Skyrocket is very common in the West and is easily distinguished by its red flowers, with pointed lobes that look like rockets. The flowers have a skunk-like scent, hence the common name Skunk Flower. This plant is similar to morning glories, and the genus name *Ipomopsis* is Latin for "similar to *Ipomoea*," which is a main genus of morning glories. This plant was once considered to be part of the genus *Gilia*.

DESCRIPTION: Skyrocket can widely vary in size and color depending on elevation. It grows to a height of 6–84" (15–210 cm) and is topped with clusters of attractive trumpet-shaped flowers that are usually red. The flowers can also appear pink in higher elevations and may even be yellow or orangish in some areas. The flowers are ¾–1¼" (2–3.1 cm) long, with five pointed lobes. The leaves are 1–2" (2.5–5 cm) long and fern-like, appearing densest at the base of the plant.

FLOWERING SEASON: May–September.

HABITAT: Dry, sandy areas.

RANGE: Eastern Oregon to southern California; east to western Texas; north through the Rocky Mountains to western Montana; also northern Mexico.

CONSERVATION: Ⓖ5
Secure.

■ NATIVE
■ RARE OR EXTIRPATED
■ INTRODUCED

ALTERNATE NAMES: White-flower Skyrocket

This plant appears thin and delicate, but it grows vigorously. The flowers are attractive, but Pale Trumpet can be difficult to identify if it's not flowering because its thin, wispy stems blend in with undergrowth. Pale Trumpet's nectar is favored by moths.

DESCRIPTION: This slender, branching plant grows to 2' (60 cm) tall, with hairy stems and leaves. The leaves are sparse. The lower leaves are about 1½" (4 cm) long, and larger than the upper leaves, and pinnately divided into a few narrow lobes. The flowers are trumpet-shaped, about 1–1½" (2.5–4 cm) long, appearing bluish, purple, or white, and may grow in pairs or singly. The flowers are quite pointed and appear like a five-pointed star.

FLOWERING SEASON: March–October.

HABITAT: Dry, sandy deserts, low elevations, and grasslands.

RANGE: Southern Utah northeast to western Nebraska and south to western Texas, New Mexico, Arizona, and northern Mexico.

SIMILAR SPECIES: Pale Trumpets are often compared with Scarlet Gilia (*I. aggregata*), but have fewer flowers and basal leaves that wilt over time.

CONSERVATION: G5
Secure, but imperiled in South Dakota and Wyoming.

■ NATIVE
■ RARE OR EXTIRPATED
■ INTRODUCED

ALTERNATE NAMES: Bristly Langloisia, Bristly-calico

Great Basin Langloisia is a western native annual with funnel-shaped flowers that are mainly white but may be tinged blue, lavender, or magenta. This is the only species in the genus *Langloisia*.

DESCRIPTION: This is a leafy annual, growing to 1½–8" (4–20 cm) tall and covered in bristles. The leaves are 2–3 cm long, arranged in a spiral, and covered in branched hairs, or bristles, with a toothed or lobed margin. The flowers are ½–2 cm wide and tipped with bristles. The calyx is funnel-like, appearing white to lavender, with five equal-sized lobes. The corolla may show faint patterns of dots or stripes.

FLOWERING SEASON: April–June.

HABITAT: Deserts, rocky slopes.

RANGE: Western North America from Oregon and Idaho south to eastern California and Arizona.

SIMILAR SPECIES: Desert Calico (*Loeseliastrum matthewsii*) is a close relative that's relatively common in the Mojave and Sonoran deserts. The flowers have a distinct maroon arch over a white patch on each of the upper lobes. Schott's Calico (*Loeseliastrum schottii*) is another similar species common in the Desert Southwest; its spotted or mottled flowers are speckled with purple or orange and arranged with three lobes on the upper lip and two lobes on the lower.

CONSERVATION: **G4**
Apparently secure, but imperiled in Utah.

■ NATIVE
■ RARE OR EXTIRPATED
■ INTRODUCED

Many species formerly classified in the genus *Linanthus*— including this attractive annual herb—are endemic to California. Most of them have prickly leaves and large flowers that are clustered at the top of the plant. The drought-tolerant species is cultivated as an ornamental plant for its small, colorful flowers.

DESCRIPTION: This is a spindly plant that grows 2–12" (5–30 cm) high with a narrow stem. The leaves are prickly and needle-like, about ½–1¼" (1.5–3 cm) long and divided into five to nine narrow, pointed lobes. Most of the leaves are clustered at the top. The flowers are trumpet-shaped with five pointed lobes. The corolla is about ½–¾" (1.5–2 cm) wide, and may appear pink, purple, white, or yellow.

FLOWERING SEASON: April–June.

HABITAT: Grassy slopes and open areas.

RANGE: Most of California west of the Sierra Nevada.

SIMILAR SPECIES: The related True Babystars (*Leptosiphon bicolor*) is similar, but False Babystars is distinguished by its larger flowers.

CONSERVATION: **G5**
Secure.

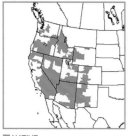

■ NATIVE
■ RARE OR EXTIRPATED
■ INTRODUCED

ALTERNATE NAMES: Nuttall's Desert-trumpets

This western native plant has a sweet scent and often grows near other plants in a variety of habitat types. It's typically found in patches at high elevations.

DESCRIPTION: This is a clump-forming perennial that grows up to 1' (30 cm) with small, hairy, erect stems. It has white or cream-colored flowers that appear in whorls of two to five flowers at the top of the stems. The leaves are ¾" (2 cm) long, opposite, and divided into five to nine needle-like lobes. The corolla is about ½" (1.5 cm) wide, with a narrow tube and five broad, white lobes.

FLOWERING SEASON: June–September.

HABITAT: Open or sparsely wooded, often rocky slopes in mountains.

RANGE: Western North America from British Columbia to California to New Mexico and Montana, and northwestern Mexico.

SIMILAR SPECIES: Other *Leptosiphon* species.

CONSERVATION: G5

Secure. It is imperiled in Wyoming and vulnerable in Montana.

■ NATIVE
■ RARE OR EXTIRPATED
■ INTRODUCED

ALTERNATE NAMES: Granite Gilia

Granite Prickly Phlox is a drought-tolerant plant that can grow in just about any environment, from mountain slopes to deserts. Its ability to grow in dry environments with thin or poor soils makes it a dominant species in some areas. It may appear as a mat in higher elevations.

DESCRIPTION: This perennial grows up to 2.5' (80 cm), with several branching stems covered in narrow, needle-like leaves. The flowers have a strong scent and are funnel-shaped, about ½–1" (1–2.5 cm) long, appearing white, yellowish, or pinkish. The fruit is a capsule with three valves that hold 5–10 seeds each.

FLOWERING SEASON: April–June.

HABITAT: Woodlands, pine forests, grassy areas, montane areas, and alpine climates. It can tolerate a variety of soil conditions, including dry, shallow, or salty soils.

RANGE: Western North America, from British Columbia to California, east to Montana and New Mexico.

SIMILAR SPECIES: See the related *Leptosiphon* species.

CONSERVATION: G5
Secure.

■ NATIVE
■ RARE OR EXTIRPATED
■ INTRODUCED

ALTERNATE NAMES: *Phlox gracilis*

This plant is native to most of the United States and Canada, but was introduced in Alaska. It is an annual herb whose flowers include a tubular, yellowish throat. It may vary in shape depending on where it's growing. This is the only species in the genus *Microsteris*, and some sources place it in the genus *Phlox* instead.

DESCRIPTION: This annual grows 3–10" (7.5–25 cm) tall with hairy, sometimes branched, erect stems that tend to be brown

and sticky. The narrow leaves are reddish and lance shaped. The upper leaves are alternate, while the lower leaves are opposite. A small flower appears at the top of the stem, which may be pink, white, or yellow. The calyx is hairy and the corolla is flat, with five lobes about 2 mm long.

FLOWERING SEASON: March–August.

HABITAT: Found in a variety of dry to moist areas, including open areas, woodlands, and shrublands.

RANGE: Western North America, east to Alberta, Montana, South Dakota, and Nebraska and south into Mexico.

CONSERVATION: **G5**
Secure.

■ NATIVE
■ RARE OR EXTIRPATED
■ INTRODUCED

This annual is native to California. It thrives in a variety of habitats, including moist areas and drying pools.

DESCRIPTION: This plant grows 2–8" (5–20 cm) tall. It has slender, brown stems, which may be erect or spreading, with white hairs. The leaves, up to 1" (3 cm) long, are divided into three narrow, needle-like lobes of unequal length. The flowers grow in clusters at the top of the stem and are tubular-shaped and white or blue. The flowers are usually covered in dense hairs.

FLOWERING SEASON: April–June.

HABITAT: Wetlands, meadows, and slopes.

RANGE: British Columbia to California, east to North Dakota and south to Nebraska and New Mexico. Disjointed populations have been recorded in Massachusetts, Ohio, and Tennessee.

CONSERVATION: G5
Secure.

■ NATIVE
■ RARE OR EXTIRPATED
■ INTRODUCED

ALTERNATE NAMES: California Stinkweed

This spreading annual herb is native to the West Coast. As the common name suggests, it smells like a skunk.

DESCRIPTION: This hairy plant grows 4–24" (10–60 cm) tall. The flowers grow in dense heads at the ends of the stems and are tubular. They appear lilac-pink to deep blue, up to ½" (12 mm) wide. The flowers have five lobes and the heads are encircled by spiny sepals and bracts. The leaves are pinnately lobed and spiny, about 2½" (6 cm) long.

FLOWERING SEASON: June–September.

HABITAT: Low elevations, open areas, wet soils, and slopes.

RANGE: The West Coast from British Columbia to California. It is naturalized elsewhere.

CONSERVATION: **G5**
Secure.

■ NATIVE
■ RARE OR EXTIRPATED
■ INTRODUCED

ALTERNATE NAMES: Mat Phlox

Spreading Phlox is a widely distributed western native perennial. It's an important source of nectar for butterflies and bees in the spring. The genus name comes from the Greek word *phlox*, for "flame," and refers to the colorful flowers of some varieties.

DESCRIPTION: This perennial grows in a woody mat close to the ground, up to 8" (20 cm) tall. The stem is densely covered in yellowish, narrow, needle-like leaves that are opposite. The leaves are only about 1.5 cm long. The flowers are small but showy and can be white to lavender or pinkish. There are often many blossoms. The flowers have five petals with a white or blue tubular corolla.

FLOWERING SEASON: May–August.

HABITAT: Dry, rocky mountain areas.

RANGE: Southern British Columbia and northwestern Montana south to California and New Mexico.

CONSERVATION: **G5**
Secure.

■ NATIVE
■ RARE OR EXTIRPATED
■ INTRODUCED

ALTERNATE NAMES: Annual Garden Phlox, Drummond's Phlox

This plant is native to Texas but is commonly found throughout the southeastern United States. It is often grown as an ornamental and can withstand most weather conditions. Unlike most *Phlox* species, the leaves are alternate, not opposite. The species is named after Thomas Drummond, a Scottish botanist who sent seeds of this and other plants collected on his expedition in Texas to England in 1835.

DESCRIPTION: This branching plant grows 8–18" (20–45 cm) high. The leaves are 1–3" (2.5–7.5 cm) long and ovate to lanceolate and alternate. It has bright red, pink, or white trumpet-shaped flowers, about 1" (2.5 cm) wide, that grow in clusters at the ends of stems. The corolla has five spreading lobes and five short stamens.

FLOWERING SEASON: April–July.

HABITAT: Waste places, fields.

RANGE: North and South Carolina south to Florida, west to Texas.

CONSERVATION: G5
Secure.

■ NATIVE
■ RARE OR EXTIRPATED
■ INTRODUCED

ALTERNATE NAMES: Carolina Phlox, Thickleaf Phlox

This phlox continues to bloom in cold temperatures, until frost. As the common name suggests, this phlox species is usually hairless. Its sweet-smelling flowers attract several native butterflies and insects, including Parenthesis Lady Beetle (*Hippodamia parenthesis*), Eastern Giant Swallowtail (*Papilio cresphontes*), and various sulphur butterflies (*Colias* spp.).

DESCRIPTION: This is an erect clump-forming perennial, growing 24—30" (60–76 cm) tall. The stems are slender and streaked with red. The tubular-shaped flowers are lavender or pink and have a sweet scent. The flowers also have five lobes and are typically seen in loose, terminal clusters, though flowers can also grow singly. The leaves are bright green, thin, opposite, and leathery, up to 4" long.

FLOWERING SEASON: May–July.

HABITAT: Woodland edges and openings.

RANGE: Georgia to Texas, north to Maryland, Missouri, Illinois, and Indiana.

CONSERVATION: G5

Globally secure, but critically imperiled in Maryland and West Virginia and vulnerable in Ohio and North Carolina.

■ NATIVE
■ RARE OR EXTIRPATED
■ INTRODUCED

As the common name suggests, this western native plant has many, relatively long leaves, unlike other phlox species that have short, often prickly leaves. This plant also produces looser flower clusters than most other phlox species. The flowers smell sweet and attract butterflies, including Becker's White (*Pontia beckerii*).

DESCRIPTION: This plant grows up to 4–16" (10–40 cm) tall with woody stems at the base. The stems and leaves are covered in short hairs. The leaves are very narrow and opposite,

about 3" (7.5 cm) long, and especially hairy on the upper sides. The flowers are white, purple, or pink, and numerous, but less dense than other mat-forming phlox species. The corolla has five rounded lobes and is about 1" (2.5 cm) wide, with a slender tube about ½– ¾" (1.3–2 cm) long. It has a five-pointed calyx.

FLOWERING SEASON: April–July.

HABITAT: Dry habitats, hillsides, rocky areas in low elevations.

RANGE: Southern British Columbia to Oregon and Washington, eastern areas of California and the Great Plains.

CONSERVATION: **G5**
Secure, but vulnerable in Wyoming.

- ■ NATIVE
- ■ RARE OR EXTIRPATED
- ■ INTRODUCED

ALTERNATE NAMES: Fall Phlox

As the common name suggests, this phlox is largely cultivated as an ornamental plant for its colorful flower display, which extends into the fall. It has likely escaped cultivation and become well adapted in the wild. This perennial plant is also widely used as a medicinal herb. The leaves are boiled and used for treating boils and other ailments.

DESCRIPTION: This clump-forming plant has a pyramidal cluster of white, pink, or lavender flowers atop its erect stem. Garden Phlox grows to 2–6' (60–180 cm) tall. The narrow leaves are 3–5" (7.5–12.5 cm) long, opposite, and ovate, with prominent side veins. There are up to 40 pairs of leaves below the flowers. The trumpet-shaped corolla is 1" (2.5 cm) wide, with five spreading lobes. It has a hairy tube with five stamens, one pistil, and three stigmas. The fruit is a capsule.

FLOWERING SEASON: July–October.

HABITAT: Open woods and thickets.

RANGE: Ontario east to Nova Scotia, south to Georgia, west to Louisiana and Oklahoma, and north to Nebraska and Minnesota.

SIMILAR SPECIES: Large-leaved Phlox (*P. amplifolia*) is very similar, but the leaves are longer and the stem is hairy, not smooth. *P. amplifolia* has fewer leaves, only about 6–15 leaf pairs below the flower cluster.

CONSERVATION: G5
Secure.

■ NATIVE
■ RARE OR EXTIRPATED
■ INTRODUCED

This plant's scientific name *pilosa* means "hairy" and refers to this plant's numerous hairs on the stems and flowers. This plant is favored by pollinators, especially butterflies. It was also used by Native Americans in tea to purify the blood. An infusion of Downy Phlox's leaves were used to treat eczema.

DESCRIPTION: This perennial grows in a clump and has erect stems, up to 1–2' (30–60 cm) tall. The unbranched stems and sepals are covered with fine hairs, making the plant feel sticky. The leaves are opposite and widely spaced, with pointed tips,

about 3½" (8.75 cm) long and ½" (1.25 cm) wide. The fragrant flowers grow in clusters of pink, lavender, or purple and have a long corolla, with five petal-like lobes, about ½–¾" (13–19 mm) wide.

FLOWERING SEASON: April–May in the South, May–July in the North.

HABITAT: Open prairies, woodlands, rocky areas.

RANGE: New York south to Florida, west through North Dakota and Texas.

CONSERVATION: G5

Globally secure, but critically imperiled in Maryland, North Dakota, Pennsylvania, and Virginia. It is vulnerable in North Carolina.

ALTERNATE NAMES: Pink Phlox, Desert Mountain Phlox

This perennial herb is native to the Desert Southwest. As the common name suggests, it prefers dry habitats. It often grows in desert and plateau scrub or woodlands, where it may grow up through other shrubs.

DESCRIPTION: This plant has an upright, subwoody stem that forms a clump, up to 4–8" high. The leaves are gray or green and covered in hairs, about 1" (3 cm) long, and oppositely arranged. The flowers are white or pink and the corolla is flat (sometimes curled in dry environments) with five lobes. The plant may be shorter in drier areas.

FLOWERING SEASON: April–June.

HABITAT: Woodland areas, sagebrush or pinyon-juniper slopes, deserts.

RANGE: California to Utah, New Mexico, and Texas.

CONSERVATION: G5
Secure.

■ NATIVE
■ RARE OR EXTIRPATED
■ INTRODUCED

ALTERNATE NAMES: Moss Phlox, Rock Pink

This phlox species grows in a carpet close to the ground. These plants are favored as ornamentals in rock gardens, where they bloom profusely for a few weeks in the spring. Their dense flowers are attractive, but may produce an odor. They have escaped gardens in some areas and become well established outside their native habitat.

DESCRIPTION: This creeping plant forms moss-like mats, up to 2–5" (5–12.5 cm). The leaves are ½" (1.5 cm) long, opposite, and needle-like. The flowers, about ¾" (2 cm) wide, grow in terminal clusters of pink or lavender colors, sometimes white. The corolla is tubular with five notched lobes and five stamens.

FLOWERING SEASON: April–May.

HABITAT: Dry sandy places, rocky slopes, and pine or oak barrens.

RANGE: Ontario east to Nova Scotia, south to North Carolina, Tennessee, and Louisiana, and north to Minnesota. It is introduced and naturalized in some of this range, especially farther north.

SIMILAR SPECIES: Trailing Phlox (*P. nivalis*) and Sand Phlox (*P. bifida*) are similar species, but are found in different habitats and have slightly different flowers. Trailing Phlox has un-notched petals and grows from Virginia south to Florida and west to Texas. Sand Phlox grows in sandy areas in the Midwest and has petals with notches at least ⅛" (3 mm) deep.

CONSERVATION: G5
Globally secure but critically imperiled in Ontario, North Carolina, and Tennessee.

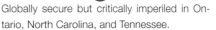

■ NATIVE
■ RARE OR EXTIRPATED
■ INTRODUCED

ALTERNATE NAMES: Western Jacob's-ladder

This western native perennial can be found in many moist habitat types. As the common name Jacob's-ladder suggests, the plant has long leaves divided into leaflets that resemble ladders growing from a long, erect stem.

DESCRIPTION: This leafy perennial grows up to 1–3' (30–90 cm) tall. The erect stems have narrow, pinnately compound leaves, with 19–27 lanceolate leaflets, each ½–1½" (1.5–4 cm) long. The flowers are funnel-shaped and purple to blue with a white throat, forming a branched cluster near the top of the stem. The corolla is ½–¾" (1.5-2 cm) wide, with five round lobes and sticky hairs.

FLOWERING SEASON: June–August.

HABITAT: Wet places at moderate elevations.

RANGE: Alaska, south to southern California, Nevada, Utah, and Colorado. Small, disjunct populations have been recorded in northern Minnesota and Wisconsin.

SIMILAR SPECIES: Leafy Polemonium (*P. foliosissimum*) is similar but is leafier and grows from Idaho and Wyoming south to Arizona and New Mexico.

CONSERVATION: (G5)

Secure, but critically imperiled in Minnesota and imperiled in Wyoming.

■ NATIVE
■ RARE OR EXTIRPATED
■ INTRODUCED

ALTERNATE NAMES: American Greek Valerian, Jacob's-ladder, Creeping Jacob's-ladder

This attractive wildflower native to eastern North America has delicate leaves and weak stems and may appear floppy. It's sometimes grown as an ornamental. The flowers are sources of pollen and nectar for a variety of native bees, beetles, syrphid flies, butterflies, and moths.

DESCRIPTION: This sprawling perennial has smooth, weak stems that appear green or red, growing up to 1–1½' (30–45 cm) tall. The leaves are pinnately compound, like a feather, with 5–21 ovate to lanceolate leaflets, each about 1½" (4 cm) long. The light blue-violet flowers are ½" (1.5 cm) wide. The corolla has five spreading lobes as long as the tube and five stamens, about equal with the corolla. The stigmas have three lobes.

FLOWERING SEASON: April–June.

HABITAT: Woodlands.

RANGE: Ontario, south to Georgia, west to Mississippi and Kansas, and north to Iowa, South Dakota, and Minnesota.

SIMILAR SPECIES: This plant resembles Western Jacob's-ladder (*P. occidentale*), but Western Jacob's-ladder isn't found in the East.

CONVERSATION: G5

Globally secure, but critically imperiled in Delaware, Georgia, Nebraska, and North Carolina. It is imperiled in Michigan, Mississippi, and Kansas.

■ NATIVE
■ RARE OR EXTIRPATED
■ INTRODUCED

ALTERNATE NAMES: Water Cabbage, Water Rose

This native perennial has a global distribution and can tolerate a variety of weather conditions. It grows both in water and on land. The flowers appear throughout the growing season, from spring to fall. The leaves taste bitter but are edible in salads and have been consumed in Europe and Asia as a laxative and as a vitamin C source.

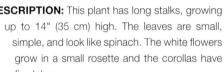

DESCRIPTION: This plant has long stalks, growing up to 14" (35 cm) high. The leaves are small, simple, and look like spinach. The white flowers grow in a small rosette and the corollas have five lobes.

FLOWERING SEASON: April–October.

HABITAT: Wetlands, riverbanks, marshes.

RANGE: Widely distributed throughout North America, except in the Rocky Mountains and northern Great Plains.

CONSERVATION: G5
Secure.

■ NATIVE
■ RARE OR EXTIRPATED
■ INTRODUCED

ALTERNATE NAMES: Shepherd's Weatherglass, Poor Man's Weatherglass, Birds Eye

This annual, introduced from Europe and widely naturalized, is often grown in flower gardens. The flowers open only in the sun and close in late afternoon and in overcast weather, earning the common name Shepherd's Weatherglass. This plant's leaves were once used medicinally, but they are toxic and can cause severe skin irritation.

DESCRIPTION: A sprawling, low-branched plant with weak stems, reaching 4–12" (10–30 cm). The egg-shaped leaves are green and soft, about ¼–1¼" (6–31 mm) long. The leaves are opposite, ovate, and stalkless. This plant has one star-like flower with small hairs rising from each leaf axil. The flowers are usually orange or reddish, sometimes white or blue, about ¼" (6 mm) wide, with five petals and small teeth. The fruit, a capsule, is so heavy that it bends the stem, and the seeds are transported by the wind or rain.

FLOWERING SEASON: June–August.

HABITAT: Sandy soil, waste places, and roadsides.

RANGE: Introduced and widely naturalized throughout North America.

CAUTION: Although at one time used to treat melancholy, this plant should not be consumed. The leaves may cause severe dermatitis.

CONSERVATION:
This plant is introduced in North America and considered a weed in some areas.

■ NATIVE
■ RARE OR EXTIRPATED
■ INTRODUCED

ALTERNATE NAMES: Maystar, Northern Starflower, *Trientalis borealis*, *Trientalis americana*

This woodland perennial is native to the East, especially farther north. As the common name suggests, this plant has small, pure white flowers in the shape of a star.

DESCRIPTION: This creeping plant has fragile stems and flowers, reaching about 4–8" (10–20 cm) tall. Each stalk has a whorl of five to nine leaves that are 1¾–4" (4.5–10 cm) long, lanceolate. One or two white flowers grow about ½" (1.5 cm) wide on smaller stalks from the whorl of leaves. The flowers are star-shaped, with seven petals, seven stamens and golden anthers. The fruit, a capsule, splits open along seams into five parts.

FLOWERING SEASON: May–August.

HABITAT: Woodlands and peaty slopes, ascending to subalpine regions.

RANGE: Alberta east to Newfoundland and Labrador, and across the northern tier of the U.S. from Minnesota to New England; limited populations occur south to Illinois, Tennessee, and Georgia.

SIMILAR SPECIES: Broadleaf Starflower (*Trientalis latifolia*) is the western equivalent; the two plants are considered subspecies of *Trientalis borealis* in some sources.

CONSERVATION: Ⓖ➎
Globally secure. Critically imperiled in Delaware, Georgia, Illinois, Kentucky, North Carolina, and Tennessee.

■ NATIVE
■ RARE OR EXTIRPATED
■ INTRODUCED

ALTERNATE NAMES: Fringed Yellow-loosestrife

This widespread native perennial occurs in moist woodlands and meadows. As the common name suggests, its leaf stalks and the base of the flowers are covered in small, stiff hairs, making them look fringed. It is cultivated as an ornamental plant, but it can spread aggressively by suckers if left uncontrolled.

DESCRIPTION: This plant has yellow flowers on long, slender stalks, which grow from leaf axils to a height of 1–4' (30–120 cm). The main stem of Fringed Loosestrife is hairless, but the leaf stalks are covered with stiff hairs, giving the plant a fringed appearance. The leaves are 1–2½" (2.5–6.5 cm) long, opposite, ovate, evenly distributed on the stem. They are round at the base and then taper out to a point. The flowers are about ¾" (2 cm) wide, with five round lobes. The base of the flowers are covered in tiny hairs. The fruit is a capsule.

FLOWERING SEASON: June–August.

HABITAT: Disturbed habitats, floodplains, riverbanks, forests, and meadows in full sun or shady areas.

RANGE: Widespread throughout North America except in parts of the Southwest.

SIMILAR SPECIES: Lowland Yellow Loosestrife (*L. hybrida*) is a similar species, but its leaf blades are rounded at the base, while *L. ciliata*'s leaf blades usually taper at the base. Southern Loosestrife (*L. tonsa*) is also similar, but it has smooth leafstalks with no hairs. Lance-leaved Loosestrife (*L. lanceolata*) occurs in a range similar to that of Fringed Loosestrife but not in Canada or the northern Plains states. Trailing Loosestrife (*L. radicans*) is also similar but has weaker stems that almost trail on the ground.

CONSERVATION: **G5**

Globally secure but imperiled in Utah and vulnerable in Wyoming.

NATIVE
RARE OR EXTIRPATED
INTRODUCED

ALTERNATE NAMES: Sea Milkweed, Black Saltwort

Sea-milkwort is found around the world in the Northern Hemisphere. This plant favors saltmarshes. It differs from all other species in its family by having petalless flowers. It was formerly placed alone in the genus *Glaux*.

DESCRIPTION: This low-growing plant has small stalkless flowers that rise from the leaf axis, growing to 2–12' (5–30 cm) tall. The leaves are fleshy, opposite, and oval, about ½" (1.5 cm) long. The flowers are tiny and white, pink, lavender, or crimson, about ⅛–¼" (3–5 mm) wide, with five sepals that join into a tube. The petals are absent. The fruit is a capsule with few seeds.

FLOWERING SEASON: June–July.

HABITAT: Seashores, saltmarshes, and alkaline meadows.

RANGE: Alberta east to Newfoundland and south to Massachusetts; also in Virginia, Nebraska, North Dakota, Minnesota, and throughout the West.

CONSERVATION: G5
Globally secure but critically imperiled in Nebraska and Minnesota. It is imperiled in Yukon and vulnerable in New Hampshire and Wyoming.

■ NATIVE
■ RARE OR EXTIRPATED
■ INTRODUCED

ALTERNATE NAMES: Creeping Jenny

This perennial gets its common name from its round leaves, which look like coins. This vigorous plant is native to Europe and spreads rapidly; it is invasive in some areas in North America and can be difficult to eradicate.

DESCRIPTION: This creeping plant grows 6–20" (15–50 cm) long and spreads quickly by stem rooting. It has opposite, nearly round leaves and showy, yellow flowers on slender stalks rising from leaf axils. The leaves are ½–1" (1.5–2.5 cm) long. The flowers are 1" (2.5 cm) wide and cup shaped, with five stamens and five petals that have a dark red dot. The fruit is a capsule.

FLOWERING SEASON: June–August.

HABITAT: Damp areas, roadsides, shorelines, and grasslands, often in shade.

RANGE: Widely introduced in the Midwest and East, and also along the West Coast.

CONSERVATION:
This introduced species is unranked and considered invasive in some areas.

■ NATIVE
■ RARE OR EXTIRPATED
■ INTRODUCED

ALTERNATE NAMES: Whorled Yellow-loosestrife, Prairie Loosestrife

Whorled Loosestrife is a native perennial found throughout the East. The genus name comes from the legend of Lysimachus, a king in ancient Sicily who purportedly used loosestrife to calm an angry bull. European colonists are said to have fed the plant to oxen to calm them down before working with them. Native Americans have used this plant to treat kidney and urinary issues.

DESCRIPTION: This plant has erect, hairy stems, with delicate, yellow, star-like flowers with long stalks rising from axils of whorled leaves, reaching a height of 1–3' (30–90 cm). The leaves are 2–4" (5–10 cm) long and widely spaced on the stems. They are light green in color and often occur in whorls of 3–7 (usually 4–5). The leaves are oval to lanceolate and pointed at the tips. The flowers are about ½" (1.5 cm) wide, with five yellow petals, each with red at the base and with streaks extending to the tip. There are five yellow stamens, tipped with red, and one pistil protruding beyond stamens. The fruit is a capsule.

FLOWERING SEASON: June–August.

HABITAT: Dry or moist areas, open woods, open fields.

RANGE: Ontario east to New Brunswick, south to Georgia, west to Alabama, and north to Illinois and Minnesota.

CONSERVATION: G5
Globally secure but critically imperiled in New Brunswick and Nova Scotia. It is vulnerable in Minnesota and Quebec.

■ NATIVE
■ RARE OR EXTIRPATED
■ INTRODUCED

Swamp Candles is a weedy wetland perennial, mainly found in the East but with a small, disjunct population in the Pacific Northwest. This plant was believed to be able to soothe animals. Some tied a branch of the plant to oxen to make them easier to work with.

DESCRIPTION: This plant has erect stems that grow up to 1–3' (30–90 cm) tall, with opposite leaves and terminal racemes of star-like, yellow flowers. The leaves are 2–6" (5–15 cm) long. The flowers are spike-like, about ½" wide, with five petals, with a reddish line down the center, and five yellow stamens. Sometimes, in place of flowers, you may see reddish-brown bulblets in the summer, which look like caterpillars. The fruit is a capsule.

FLOWERING SEASON: June–August.

HABITAT: Wet areas, marshes, moist thickets.

RANGE: Manitoba south to Georgia and east to the Atlantic Coast; also in Oklahoma and from British Columbia south to Oregon and Idaho.

SIMILAR SPECIES: Loomis's Loosestrife (*L. loomisii*) is a similar species but has smaller leaves, 1" (2.5 cm) long.

CONVSERVATION: **CONSERVATION:** G5

Globally secure, but critically imperiled in Georgia, Kentucky, Missouri, South Carolina, Tennessee. It is imperiled in Manitoba and vulnerable in Labrador and North Carolina.

- NATIVE
- RARE OR EXTIRPATED
- INTRODUCED

ALTERNATE NAMES: Mountain Saxifrage

Western Rockjasmine is native to open habitats in the Midwest and western North America. This annual has several varieties, which look similar but differ slightly in technical features.

DESCRIPTION: This plant forms a small clump of hairy basal leaves, growing up to 2–12" (5–30 cm) tall from an erect reddish stem. The leaves are small—about 2½" (6.5 cm) long—ovate and toothed. The flowers grow in a cluster of 5–10 tiny white flowers, each about ¾–2" (2–5 cm) wide. There are five white or pink petals, about ⅛" (3 mm) long, sometimes with two yellow spots at the base, forming a cup shape. The fruit is greenish or red and pod-like, about ¼" (6 mm) long.

FLOWERING SEASON: April–August.

HABITAT: Moist areas, meadows, rocky areas.

RANGE: British Columbia south to Arizona, east to Manitoba and Ohio.

CONSERVATION: G5

Secure, but critically imperiled in Arkansas, Nevada, Ohio, and Utah. It is imperiled in Indiana and vulnerable in Alberta, Manitoba, and Wyoming.

■ NATIVE
■ RARE OR EXTIRPATED
■ INTRODUCED

ALTERNATE NAMES: Jeffrey's Shootingstar, Sierra Shooting Star

Tall Mountain Shootingstar is a native perennial found in western mountain meadows and streambanks. The species is named after John Jeffrey, a Scottish botanist who explored western America for four years before he disappeared in 1854.

DESCRIPTION: This perennial has basal rosettes of long, rounded leaves with a leafless flowering stem, up to 2' (60 cm) tall. The flowers are reddish or purple and cyclamen-like, growing in clusters from the top of the stem. It has four to five showy petals that sweep backward. The stamens and stigma shoot forward and downward.

FLOWERING SEASON: June–August.

HABITAT: Wet mountain sites, from 2300–10,000' (700–3000 m)

RANGE: Mountains from North and central California to southern Alaska and Montana.

CONSERVATION: **G5**
Secure, but critically imperiled in Wyoming.

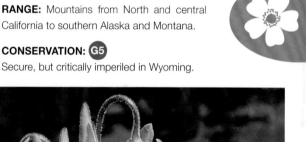

- ■ NATIVE
- ■ RARE OR EXTIRPATED
- ■ INTRODUCED

ALTERNATE NAMES: Pride-of-Ohio, Common Shootingstar

This native perennial is popular in gardens. As the common name suggests, the flowers resemble a shooting star. The flower stamens are united, making it difficult for bees and other pollinators to reach the nectar. The genus name comes from the Latin word *primus*, meaning "first," and *ulus*, meaning tiny, alluding to the tendency of this plant to bloom in early spring.

DESCRIPTION: This plant grows up to 8–20" with flowers that appear at the top of the stem from a basal rosette of leaves. The smooth leaves are up to 6" (15 cm) long, green, and lanceolate with reddish bases. The flowers grow in an umbel of 8–20 flowers, each about 1" (2.5 cm) long, with five petals that are pink, lilac, or white.

There are five yellow, protruding stamens. The fruit is a capsule with many seeds, opening lengthwise.

FLOWERING SEASON: April–June.

HABITAT: Open woods, meadows, prairies.

RANGE: Pennsylvania to Georgia; west to eastern Texas; north to Wisconsin.

SIMILAR SPECIES: Amethyst Shooting Star (*D. amethystinum*) is a similar plant, but has a green leaf base, not red.

CONSERVATION: G5

Globally secure, but critically imperiled in Florida, Michigan, Minnesota, Pennsylvania. It is imperiled in Louisiana, Mississippi, North Carolina, and South Carolina.

■ NATIVE
■ RARE OR EXTIRPATED
■ INTRODUCED

ALTERNATE NAMES: Brook Primrose

This is the largest of the native primroses in North America. It has attractive flowers, but the entire plant smells like a skunk. The species is heterostylous, with some of the plants having long styles and anthers attached low in the tube, and others having short styles and anthers attached high in the tube, which helps attract different pollinators and ensures genetic variation.

DESCRIPTION: This perennial grows up to 3–16" (7.5–40 cm) tall from a basal rosette of leaves. The stem is leafless. The leaves are egg-shaped and leathery and 2–12" (5–30 cm) long.

There are 3–25 pink flowers that grow in a loose umbel, covered in small hairs. The corolla is ⅗–1¼" (1.5–3.1 cm) wide, with five round lobes with a slender tube.

FLOWERING SEASON: June–August.

HABITAT: Wet ground, often along streams, at high elevations, in moist areas.

RANGE: Western United States from Idaho and Montana to Nevada, northern Arizona, and northern New Mexico.

CONSERVATION:
Apparently secure. Vulnerable in Wyoming and Montana.

■ NATIVE
■ RARE OR EXTIRPATED
■ INTRODUCED

ALTERNATE NAMES: Dark-throat Shootingstar

This is a common species, although it's highly variable. It may have broader leaves or hairs depending on its growing environment. It usually has up to six pinkish flowers that point backward and look like a dart.

DESCRIPTION: This plant has bright pink, dart-like flowers that grow atop an erect stalk, reaching 4–24" (10–60 cm) tall. The leaves grow in a basal cluster, while the stem is leafless. The leaves are 2–16" (5–40 cm) long and lanceolate. The margins are mostly smooth, or may have small teeth. The buds and flowers point downward and grow in clusters. The tube-like flower grows ¾–1" (2–2.5 cm) long. The corolla has four or five narrow lobes that point backward from a yellow, pink, or purple ring. The stamens are yellowish or purplish and project forward, like a dart.

FLOWERING SEASON: April–August.

HABITAT: Prairies, mountain meadows, streamsides, damp woods.

RANGE: Throughout much of the West.

CONSERVATION: G5

Globally secure but vulnerable in California, Manitoba, and Yukon.

■ NATIVE
■ RARE OR EXTIRPATED
■ INTRODUCED

ALTERNATE NAMES: *Dodecatheon alpinum*

This is a partially aquatic native wildflower of the West. As the common name suggests, this showy perennial is often found at high elevations, where it thrives in wet areas. The flowers resemble a shooting rocket.

DESCRIPTION: This plant has a smooth stalk growing from a basal rosette of leaves, with one to nine pinkish-purple flowers at the top. Alpine Shootingstar grows up to 4–12" (10–30 cm) tall. The leaves are 1¼–4" (3–10 cm) long, and up to ½" (1.5 cm) wide. Each flower looks like a small rocket and is ¾–1" (2–2.5 cm) long. The corolla has four reddish-lavender, narrow lobes that bend back. The base of the flower is bright yellow. The stamens are purple and form the nose of the rocket, while the stigma looks like a knob.

FLOWERING SEASON: June–July.

HABITAT: Mountain meadows and along mountain streams.

RANGE: Eastern Oregon south to southern California and east to Arizona and Utah.

SIMILAR SPECIES: Tall Mountain Shootingstar *(P. jefferyi)* is a similar species but has a different geographical range, found from the Alaskan mountains south to the southern Sierra Nevada and east to Idaho and Montana. Tall Mountain Shooting-star also has small hairs on the leaves and flowers. Another similar species, Scented Shootingstar *(P. fragrans)*, is densely covered with hairs.

CONSERVATION: G3
This species is globally vulnerable and classi-fied as imperiled in Utah.

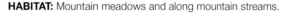

■ NATIVE
■ RARE OR EXTIRPATED
■ INTRODUCED

ALTERNATE NAMES: Wandflower, Wandplant

This native plant found mainly in the Appalachian Mountains has evergreen leaves and a long flower stalk. It is often planted in gardens, where it has escaped from cultivation. It has also been used to treat cuts and kidney issues. The genus name is from the Greek for "milk," and refers to the white flower color.

DESCRIPTION: This evergreen perennial has small milk-colored flowers that grow in a spike-like cluster on a long leafless flower stalk, growing up to 1–2½" (30–75 cm). The flowers are about ⅛" (4 mm) wide, with five petals united at the base. The leaves are leathery, 2–5" (5–12.5 cm) wide, and only grow at the base of the plant in clusters. They usually have rounded teeth.

FLOWERING SEASON: May–July.

HABITAT: Open woods.

RANGE: Found mainly in the southeastern U.S. It occurs from southern New York and Massachusetts south to Georgia and west to Alabama, Tennessee, and Kentucky.

SIMILAR SPECIES: Oconee Bells (*Shortia galacifolia*) is a similar species, but far more rare, and found only in the southern Appalachian Mountains.

CONSERVATION: G5

Globally secure but vulnerable in West Virginia. It is presumed extirpated in Maryland.

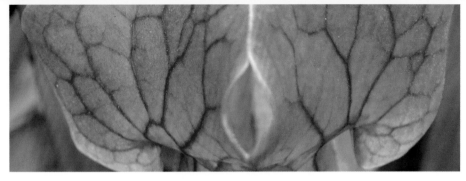

- ■ NATIVE
- ■ RARE OR EXTIRPATED
- ■ INTRODUCED

ALTERNATE NAMES: Trumpets

This plant is carnivorous, consuming insects through its hollow leaves, which fill with water and drown its prey. It is common in much of its native range in the southeastern United States and is also commonly cultivated as a garden plant.

DESCRIPTION: This plant grows 1½–3½' (45–105 cm), with hollow, inflated leaves that hold water. The leaves are 1–3' (30–90 cm) long, with a purple base. It has bright yellow, trumpet-shaped flowers that droop down. The flowers smell musty and are 3–5" (7.5–12.5 cm) wide, with five sepals and petals, many stamens and a large, round style. The fruit is a capsule.

FLOWERING SEASON: April–May.

HABITAT: Wet pinelands and bogs.

RANGE: Virginia south to Florida and west to Alabama.

SIMILAR SPECIES: Trumpet Pitcher-plant (*S. alata*) is a similar species, but its leaves are not purple at the base and it has a different range, from Alabama west to Texas. Green Pitcher-plant (*S. oreophila*) is also similar, but is rarer, found in North Carolina, Georgia, and Alabama.

CONSERVATION: **G5**
Globally secure but critically imperiled in Virginia and vulnerable in Georgia, North Carolina, and South Carolina.

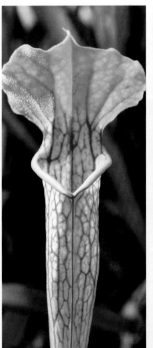

NATIVE
RARE OR EXTIRPATED
INTRODUCED

ALTERNATE NAMES: Purple Pitcher-plant

This carnivorous plant consumes insects by trapping them in its hollow, pitcher-like leaves. The insects are attracted into the colorful leaves in search of nectar and can't climb out due to downward-pointing hairs inside. The plant then releases enzymes to help digest the insects with the help of bacteria. Absorbing nutrients from prey insects helps the plant thrive where soil nutrients are otherwise scarce. Native Americans have used the roots of this plant for several medicinal purposes, including as a diuretic and an anti-diabetic.

DESCRIPTION: This plant has just one large, purplish-red flower on a leafless stalk growing 8–24" (20–60 cm) high. The leaves are hollow and inflated, forming a bronzy, reddish-green rosette. The leaves are 4–12" (10–30 cm) long and have a flaring, terminal lip covered with stiff, downward-pointing hairs. The flower is 2" (5 cm) wide, with five petals, numerous stamens, and a style that expands into an umbrella-like shape. The fruit is a capsule.

FLOWERING SEASON: May–August.

HABITAT: Peat bogs and other wetland areas.

RANGE: Throughout Canada from British Columbia to Newfoundland, the Great Lakes states, and along the Atlantic coast.

SIMILAR SPECIES: Parrot Pitcher Plant *(S. psittacina)* is a similar species found mostly in the South. It has many prostrate "pitchers" with hooked lips like a parrot's bill.

CONSERVATION: G5
Globally secure but less common at the western and southern margins of its native range. It is critically imperiled in Georgia and Illinois. It is imperiled in Delaware, Maryland, Ohio, and Virginia in the U.S. and British Columbia in Canada. It is vulnerable in Alberta, Indiana, New York, and North Carolina.

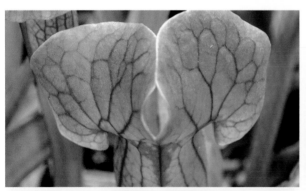

■ NATIVE
■ RARE OR EXTIRPATED
■ INTRODUCED

ALTERNATE NAMES: Sugarstick

Candystick lacks chlorophyll and instead absorbs nutrients from fungi growing on nearby plants. The flower stalk is unique in that it has white and red stripes, like peppermint sticks, hence the common name.

DESCRIPTION: This perennial has erect, scaly stems with red and white stripes and grows 4–12" (10–30 cm) tall. The leaves are reduced to scales. The flowers are white or pink and hang in a densely clustered raceme at the top of the stems. The flowers are usually turned outward or upward, forming what looks like an inverted bowl. There are five petals, each about ¼" (6 mm) long. The stamens project well beyond the petals. The anthers are maroon-colored.

FLOWERING SEASON: May–August.

HABITAT: Humus soil of coniferous forests.

RANGE: British Columbia south to California, as well as Idaho, Montana, and Nevada.

CONSERVATION: G4
Apparently secure, but vulnerable in Idaho and Montana.

ALTERNATE NAMES: Striped Prince's Pine, Spotted Pipsissewa, Striped Wintergreen

This is a conspicuous plant because of its white-and-green-spotted leaves. The common name "Wintergreen" refers to the evergreen leaves, but the leaves do have a mint-like fragrance. The flowers are very fragrant and only bloom in the summer, but the fruit, a brown capsule, remains through the winter.

DESCRIPTION: Spotted Wintergreen has a red or brown glabrous central stem and grows 3–9" (7.5–22.5 cm) tall. Waxy evergreen leaves appear mottled with white and green. The leaves are leathery, lanceolate, and striped with white along the midvein. The base leaves are whorled, while the upper leaves are opposite.

Each leaf is about ¾–2 ¾" (2–7 cm) long. The flowers are white or pinkish, appearing in an umbel-like cyme of usually two to five atop the stem. Each flower is ⅝" (16 mm) wide, with five white petals, five green sepals, 10 stamens, and one knobby pistil. The fruit is a brown capsule that lasts all winter.

FLOWERING SEASON: June–August.

HABITAT: Dry woods.

RANGE: Ontario, Quebec to Maine, south to Florida, west to Mississippi, and north to Illinois and Michigan. It also occurs in southern Arizona and into Central America.

SIMILAR SPECIES: Pipsissewa (*C. umbellata*) is a taller plant with shiny, dark green leaves that do not have spots.

CONSERVATION: G5

Globally secure but critically imperiled in Illinois. It is imperiled in Arizona, Maine, Mississippi, and Ontario; and vulnerable in Indiana and Vermont. It is presumed extirpated in Quebec.

■ NATIVE
■ RARE OR EXTIRPATED
■ INTRODUCED

ALTERNATE NAMES: Prince's Pine, Umbellate Wintergreen, Common Wintergreen

This native plant with shiny, evergreen leaves and conspicuous flowers has been used for a variety of purposes. It's been used to flavor candies and root beer and other soft drinks. The roots and leaves have been used in teas. Native Americans used this plant to treat tuberculosis and urinary disorders, and infusions made from the aboveground parts of the plant have been used to treat various infections and as a diuretic.

DESCRIPTION: A low shrub or evergreen perennial with slender, woody stems growing 4–12" (10–30 cm) tall. The leaves are shiny and leathery, with shallow teeth and fine hairs at the edges. The leaves grow in opposite pairs or whorls of three or four along the stem. Delicate flowers grow at the top of the stems. The flowers are pink or white, about ½" (12 mm) wide, and grow in a small umbel, with four or five petals spread flat around a fat, knob-like pistil.

FLOWERING SEASON: June–August.

HABITAT: Shady conifer and mixed forests.

RANGE: Circumboreal. In North America it occurs from Alaska to Newfoundland; south in the West to California, New Mexico, and South Dakota; and in the East in the Great Lakes states, the Northeast, and in the mountains to Georgia.

SIMILAR SPECIES: See Spotted Wintergreen (*C. maculata*).

CONSERVATION: G5

Secure, but critically imperiled in Illinois, Iowa, Ohio, and Yukon; and imperiled in Delaware and Newfoundland. It is vulnerable in Maryland, North Carolina, West Virginia, Wyoming, and Saskatchewan.

■ NATIVE
■ RARE OR EXTIRPATED
■ INTRODUCED

ALTERNATE NAMES: Mayflower

This eastern native plant is often found under fallen trees in the early spring. The genus name means "upon the earth" and refers to this species' sprawling growth. Native Americans have used this plant for a variety of ailments, including labor pains, kidney disorders, abdominal pain, diarrhea, and indigestion.

DESCRIPTION: This low-growing evergreen reaches 16" (40 cm) long with hairy stems. The flowers are pink or white, grow at the ends of the branches, and are very fragrant. The leaves are alternate, ¾–3" (2–7.5 cm) long, leathery, oval, with rust-colored hairy edges. The flowers are about ½" (1.5 cm) wide and hairy within a tubelike structure, with the corolla flaring into five equal lobes.

There are five stamens. The fruit is a fleshy capsule with a white interior.

FLOWERING SEASON: February–May.

HABITAT: Oak forests, sandy or rocky woods, especially in acidic soil.

RANGE: Manitoba east to Newfoundland south to Mississippi and Florida.

CONSERVATION: G5

Globally secure but imperiled in Florida and Labrador, and vulnerable in Indiana, Mississippi, Manitoba, and Newfoundland.

■ NATIVE
■ RARE OR EXTIRPATED
■ INTRODUCED

ALTERNATE NAMES: Checkerberry, Teaberry, Eastern Teaberry

This plant is not a mint, but the fruit and leaves are edible and have a peppermint or spearmint flavor. Extract from Wintergreen is used to flavor teas, candies, medicines, and chewing gum. The red fruit often stays on the plant through the winter.

DESCRIPTION: A low, evergreen shrub with creeping, underground stems that spread 2–6" (5–15 cm). The branches are erect with white or pinkish, bell-shaped, nodding flowers, appearing singly or in groups of two or three per stem. The leaves are 1–2" (2.5–5 cm) long, oval, slightly toothed, and leathery with a wintergreen scent when crushed. The corolla is about ⅜" (8 mm) long with five lobes. The fruit is a bright red capsule that looks like a berry but is actually dry.

FLOWERING SEASON: April–May.

HABITAT: Pine or hardwood forests, oak woods, or under evergreens, especially in sandy sites.

RANGE: From Manitoba east to Newfoundland and south to Georgia and Alabama.

CONSERVATION:  G5

Globally secure but critically imperiled in Newfoundland. It is presumed extirpated in Illinois, and vulnerable in Manitoba and South Carolina.

■ NATIVE
■ RARE OR EXTIRPATED
■ INTRODUCED

ALTERNATE NAMES: American Pinesap, Many-flower Indian-pipe, *Monotropa hypopitys, Monotropa uniflora*

This plant is a saprophyte; its leaves do not contain chlorophyll and it instead obtains nourishment from fungi, often found on the roots of oak or pine trees. Pinesap was formerly placed in the genus *Monotropa*, but is now part of *Hypopitys* as a result of DNA testing.

DESCRIPTION: All parts of this fleshy perennial appear yellowish, lavender, or reddish and grow 4–14" (10–35 cm) tall, with one to several erect and unbranched stems. The stems and leaves are usually the same color. The leaves are reduced to scales, clasping the stem. Each leaf is about ½" (1.5 cm) long, appearing denser at the base. The flowers are vase-like and grow in clusters of up to 11 on a downy, scaly stem. Each flower is about ½" (1.5 cm) long, with four or five petals. The flowers can be different colors at different times of the year; those blooming in the summer are yellow and somewhat hairy, while those blooming later in the year are red and hairier. The fruit is an erect capsule.

FLOWERING SEASON: June–November.

HABITAT: Pine woods, upland woods, usually in acidic soil.

RANGE: British Columbia east to Newfoundland, and widely distributed throughout the U.S. except for in the Great Plains and Desert Southwest.

SIMILAR SPECIES: Sweet Pinesap (*Monotropsis odorata*) is a similar plant, but its petals are united. Sweet Pinesap is a rare species found only in the eastern U.S.

CONSERVATION: ⬤G5

Globally secure but critically imperiled in Florida, Iowa, Kansas, Nebraska, and Oklahoma in the U.S. and Manitoba and Saskatchewan in Canada. It is imperiled in Alabama, Colorado, Delaware, Louisiana, and Wyoming. It is vulnerable in Indiana, New Jersey, Alberta, Newfoundland, and Prince Edward Island.

■ NATIVE
■ RARE OR EXTIRPATED
■ INTRODUCED

ALTERNATE NAMES: Single-delight, One-flowered Wintergreen

Wood Nymph is native to the wet coniferous forests of northern and western North America. The genus name is from the Greek for "single delight," and refers to this plant's attractive single flower. The flower is very fragrant and attractive to bees. Seeds are spread through buzz pollination — bees visit the flower and shake the anthers through the vibration of their wings. Wood Nymphs can produce up to 1,000 seeds per fruit with the help of bees. Native Americans have used Wood Nymph for colds and skin problems. Stem and leaf extracts from this plant may also be an effective antibiotic against the bacteria that cause tuberculosis.

DESCRIPTION: This small plant grows 2–6" (5–15 cm) high, with one nodding, white or pale pink, saucer-shaped flower, ¾" (2 cm) wide, with five thick petals, 10 bright yellow stamens, a bright green style, and five-lobed stigma. The leaves appear in

whorls of three or four at the base of the stem, while leaves on the upper part of the stem are oppositely arranged. Each leaf is ½–1" (1.5–2.5 cm) long, with lobed margins.

FLOWERING SEASON: May–August.

HABITAT: Coniferous forests, damp areas.

RANGE: Alaska and throughout Canada and south to New Mexico, Arizona, and California; also in the northern Great Lakes region and New England.

CONSERVATION: G5

Globally secure but critically imperiled in Connecticut and Rhode Island; imperiled in Arizona, California, Maine, and South Dakota; and vulnerable on Prince Edward Island.

■ NATIVE
■ RARE OR EXTIRPATED
■ INTRODUCED

ALTERNATE NAMES: Indianpipe, One-flower Indian Pipe, Ghost Plant

This waxy native plant has just one flower atop the stem. The entire plant is white or translucent with black patches, and turns darker as the fruit matures or when it's picked or dried. It feeds on decayed organic material. Sightings of aboveground stems and flowers are fairly rare; they are often found in small clusters.

The so-called "Indian Pipe" or "Ghost Plant" was Emily Dickinson's favorite flower, and she mentioned it extensively in her early work.

DESCRIPTION: This saprophytic, white or reddish plant grows 3–9" (7.5–22.5 cm). The leaves are reduced to scales. It has a thick, translucent or ghostly white stem covered with scaly bracts and terminated by one nodding flower that points downward. Each flower is ½–1" (1.5–2.5 cm) long and white or pink, with four or five petals, 10–12 stamens, and one pistil. The fruit is a capsule, becoming enlarged and erect as the seeds mature.

FLOWERING SEASON: June–September.

HABITAT: Woodland humus, usually in deep, shady woods.

RANGE: British Columbia east to Newfoundland, throughout the U.S. east of the Great Plains and in the Northwest from western Montana to northernmost California.

CONSERVATION: **G5**

Globally secure but critically imperiled in Alaska, Nebraska, Oklahoma, and South Dakota; imperiled in California; vulnerable in Florida, North Dakota, Alberta, and Labrador.

■ NATIVE
■ RARE OR EXTIRPATED
■ INTRODUCED

ALTERNATE NAMES: Pink Wintergreen, Liverleaf Wintergreen, Heartleaf Pyrola, Pink Pyrola

This widespread native *Pyrola* species has pink flowers, which distinguishes it from any other member of its genus. It is a small, evergreen, creeping perennial that produces long, nearly leaf-less flower stalks. It is most common in cool, moist climates throughout Canada and in the northern tier of the United States.

DESCRIPTION: This woodland plant has shiny leaves with racemes of 5–25 pinkish flowers, growing 6–16" (15–40 cm)

tall, on long stalks. The leaves are basal and kidney- or heart-shaped, about 3" (7.5 cm) long. The upper leaf surface is very shiny and leathery. The corolla is about ½" (1.3 cm) wide, with five roundish petals that form a bowl, with edges that curve down. The style is light green and tube-like, curving down.

FLOWERING SEASON: June–September.

HABITAT: Moist ground, generally in woods.

RANGE: Found in Alaska and throughout Canada; south in the West to southern California, Nevada, Utah, New Mexico, and South Dakota; across the northern United States in the East.

CONSERVATION: **G5**

Globally secure but critically imperiled in Indiana and Iowa. It is possibly extirpated in Massachusetts. Imperiled in New York, Vermont, and Prince Edward Island. Vulnerable in Labrador and Nova Scotia.

■ NATIVE
■ RARE OR EXTIRPATED
■ INTRODUCED

ALTERNATE NAMES: Waxflower Shinleaf, White Wintergreen, Elliptic Shinleaf, Large-leaved Shinleaf

This is one of the most common native *Pyrola* species. This perennial's elliptical leaves are evergreen. The leaves contain an aspirin-like substance and have been used by Native Americans to relieve pain.

DESCRIPTION: This evergreen has fragrant greenish-white flowers in an elongated cluster rising 5–10" (12.5–25 cm) tall. The dark green basal leaves grow on a red stalk and are about 2¾" (7 cm) long and broadly elliptical. The flowers are waxy and grow in clusters of about 21. Each flower is about ⅝" (16 mm) wide, with five oval petals and 10 orange-tipped stamens, with yellow anthers, and one pistil. There is one protruding style. The fruit is a capsule with five chambers.

FLOWERING SEASON: June–August.

HABITAT: Dry to moist woods.

RANGE: Southern Canada from British Columbia to Newfoundland, south to Virginia, and west to Illinois and South Dakota; also found in parts of the West.

SIMILAR SPECIES: Round-leaved Pyrola (*P. americana*) is a similar species, but has leathery, roundish leaves. One-sided Pyrola (*Orthilia secunda*) has flowers only along one side of the stem, and Wood Nymph (*Moneses uniflora*) is shorter, with a single flower on each stem.

CONSERVATION: Ⓖ5
Globally secure but critically imperiled in Iowa, Nebraska, North Carolina, and Virginia; imperiled in Delaware, Newfoundland, and Wyoming; vulnerable in Indiana and British Columbia. Land-use conversion and habitat fragmentation can impact this species.

■ NATIVE
■ RARE OR EXTIRPATED
■ INTRODUCED

ALTERNATE NAMES: Bog Cranberry, Swamp Cranberry

This woody, evergreen perennial grows near bogs or moist areas and is often growing on hummocks of peat moss. The edible berries have been used both medicinally and as a food by Native Americans.

DESCRIPTION: This plant has vine-like stems, growing 1–4" (2.5–10 cm) long. The branches are smooth or covered in fine hairs. The older branches are dark red or brown and usually smooth, spreading horizontally. The new branches are brownish and erect. The leaves are alternate, small and leathery, with dark green tops and a gray underside. It has four nodding, pinkish flowers that rise from leaf axils, about ¼–⅜" (6–9.5 mm) long, with four petals that tightly curl back and reddish stamens. The stalks are reddish, with fine hairs. The fruit is a red berry, ¼–½" (6–12 mm) wide, that often remains on the plant through the winter.

FLOWERING SEASON: June.

HABITAT: Bogs, especially in cold areas.

RANGE: Alaska and throughout Canada, south to Oregon and Idaho in the West and to Minnesota, Illinois, and Virginia in the East.

SIMILAR SPECIES: The flowers of Small Cranberry are very similar to Large Cranberry (*Oxycoccus macrocarpon*) but Large Cranberry has larger leaves. Creeping Snowberry (*Gaultheria hispidula*) is also similar but has rounder leaves with hairier stems and leaves.

CONSERVATION: G5

Globally secure but critically imperiled in Illinois and New Jersey. Imperiled in Indiana, Maryland, Ohio and Rhode Island. Vulnerable in Idaho and West Virginia.

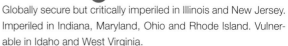

NATIVE
RARE OR EXTIRPATED
INTRODUCED

ALTERNATE NAMES: Virginia Buttonweed

This native perennial can spread quickly and is considered a nuisance weed in some areas. Buttonweed is susceptible to a virus that attacks the foliage, often making the leaves appear spotted or mottled with yellow.

DESCRIPTION: The small, sprawling plant has star-like flowers that grow from leaf axils to variable heights, depending on its habitat. The stems are squarish and floppy while the leaves are opposite and lanceolate to elliptical. The flowers are ½" (12 mm) and usually white, but sometimes pink. It has four hairy petal lobes that grow on a long, narrow tube.

FLOWERING SEASON: Year-round.

HABITAT: Ponds, wet pinelands, woods, swamps.

RANGE: Throughout the south-central and southeastern United States, northeast to New Jersey and Connecticut and west to Kansas and Texas.

SIMILAR SPECIES: Rough Buttonwood (*Diodia teres*) is a similar plant, but *D. virginiana* is more upright-growing, hairier, and has narrower leaves.

CONSERVATION: G5

Secure, but critically imperiled in Kansas and vulnerable in Indiana.

■ NATIVE
■ RARE OR EXTIRPATED
■ INTRODUCED

ALTERNATE NAMES: Sticky-willy, Bedstraw, Goosegrass, Catchweed

This widespread annual has several colloquial names that describe it. The plant is covered in bristles on the leaves, stem, and fruit, causing it to stick to clothing and animal fur, giving rise to the common names Cleavers, Catchweed, and Sticky-willy. Geese are also known to frequent this plant, hence the name Goosegrass. The leaves and fruits of this plant have often been consumed for various reasons, including as a coffee substitute. It is considered a noxious weed in some areas.

DESCRIPTION: This is a weak-stemmed, sprawling plant, reaching 8–36" (20–90 cm) long. The stem and leaves are Velcro-like and covered with backward-hooked bristles that cling to fur and clothing. The leaves grow in whorls of six to eight, with each leaf about 1–3" (2.5–7.5 cm) long and lanceolate to linear. The flowers are white or greenish and grow in clusters of two or three on stalks rising from axils of whorled leaves. Each flower is about ⅛" (3 mm) wide with a four-lobed corolla and four stamens. The fruit is dry, seed-like, and covered with hooked bristles that often attach to animal fur to help distribute seeds.

FLOWERING SEASON: May–July.

HABITAT: Woods, thickets, and waste places.

RANGE: Throughout North America, except the Northwest Territories and Newfoundland.

CAUTION: *G. aparine* may cause a rash or skin irritation if touched.

CONSERVATION: **G5**
Globally secure but critically imperiled on Prince Edward Island. It is imperiled in Nova Scotia and vulnerable in Manitoba and Wyoming.

NORTHERN BEDSTRAW *Galium boreale*

■ NATIVE
■ RARE OR EXTIRPATED
■ INTRODUCED

ALTERNATE NAMES: *Galium septentrionale*

Northern Bedstraw is a widespread perennial native to most temperate and subarctic parts of the Northern Hemisphere, including most of North America. This plant spreads both by seed and underground rhizomes, making it aggressive in some areas.

DESCRIPTION: This leafy perennial has square stems growing 8–31" (20–80 cm) tall. The leaves have fine hairs on the edges and are spaced evenly on the stem, about 2" (5 cm) long, and narrow, with three veins. The flowers are small but showy with a pleasant scent, growing in white clusters at the ends of branches. There are four petals, each less than ⅛" (3 mm) long, and pointed at the tip, spreading from the top of the ovary, with four stamens. The seeds have short, dark hairs.

FLOWERING SEASON: June–August.

HABITAT: Rocky soil, shorelines, and streambanks.

RANGE: Alaska and most of Canada, from Yukon and British Columbia east to Nova Scotia, south to Virginia, Tennessee, Missouri, and Nebraska; also found in much of the western U.S.

SIMILAR SPECIES: Northern Bedstraw is distinguished from its relatives by the whorls of four leaves and its smooth stem.

CONSERVATION: G5

Secure, but critically imperiled in Maryland and Massachusetts in the U.S. and Prince Edward Island in Canada. It is imperiled in Missouri, New Jersey, and Nova Scotia. It is vulnerable in Pennsylvania, Vermont, Virginia, and West Virginia and in New Brunswick.

■ NATIVE
■ RARE OR EXTIRPATED
■ INTRODUCED

ALTERNATE NAMES: Sweet-scented Bedstraw

As the common name suggests, Fragrant Bedstraw's leaves have a strong vanilla scent, especially when dried. The species name *triflorum* refers to this plant's flowers, which grow in clusters of three.

DESCRIPTION: Fragrant Bedstraw is a trailing perennial that grows on the forest floor, or over other plants, to 4' (1.2 m) long. The leaves are broad and grow in whorls of six, with small hairs on the leaf edges and on the underside. The flowers grow in a cluster of usually three flowers, and are about ⅛" (3 mm) across with four greenish-white petals sharply pointed at the tip, and four greenish-white stamens. It has a hairy stem that is slightly rough and sticks to clothing. The fruit is a round pod covered in hairs.

FLOWERING SEASON: May–September.

HABITAT: Moist woods and thickets; fields.

RANGE: Throughout North America. It is most common in Canada and the northern U.S. and occurs more sparingly in the southern states and into Mexico.

SIMILAR SPECIES: Rough Bedstraw (*G. asprellum*) is similar but is very sticky and much more branched, with flower clusters having more than three flowers.

CONSERVATION: G5
Secure but vulnerable in Wyoming and Yukon.

NATIVE
RARE OR EXTIRPATED
INTRODUCED

ALTERNATE NAMES: Tiny Bluets

This plant has tiny blue flowers and is easy to overlook. It is one of the earliest plants to bloom in the spring within its range.

DESCRIPTION: This tiny plant grows up to 2–6" (5–15 cm). The leaves are basal, opposite, and spoon-shaped, 2.5–10 mm long. The leaves are usually smooth but sometimes hairy and sometimes stalked. The flowers are blue and grow from a single branch from the leaf axil. The flowers have a yellow center, appearing about ¼" wide, with four lobes.

FLOWERING SEASON: April–June.

HABITAT: Woods, meadows, disturbed places.

RANGE: South Dakota south to Texas, east to Maryland and south to Florida; also found in southern Arizona.

CONSERVATION: G5
Globally secure but critically imperiled in Nebraska.

■ NATIVE
■ RARE OR EXTIRPATED
■ INTRODUCED

ALTERNATE NAMES: Two-eyed berry, Running Fox

This widespread Eastern native plant is an evergreen, non-climbing vine that has red, berry-like fruit. The species name is from the Latin for "creeping" and references the low-growing nature of the plant. Native Americans have made a tea from the leaves to alleviate childbirth pains.

DESCRIPTION: This attractive, trailing evergreen grows 4–12" (10–30 cm) long. The leaves are dark green and shiny, about ½–¾" (1.5–2 cm) long, opposite, roundish, with white veins and a yellowish midrib. The flowers are white and trumped-shaped, about ½–⅝" (13–16 mm) long, growing in pairs. Each flower rises from a calyx covered with fine hairs and has four spreading lobes, with a hairy inside. The flowers are united by ovaries. Each flower has one pistil and four stamens, but the flowers can have different styles. Some flowers have long styles and short stamens, others have short styles and long stamens. The fruit is red (rarely white), and berry-like.

FLOWERING SEASON: June–July.

HABITAT: Dry to moist woods.

RANGE: Throughout the East from Ontario to Newfoundland and south to the Gulf Coast.

SIMILAR SPECIES: Mountain Cranberry (*Vaccinium vitis-idaea*), which may share the common name Partridgeberry in some sources, is a different species. It is a small evergreen with prominent red berries, found farther north and west, restricted to Alaska, Canada, and northernmost New England.

CONSERVATION: **G5**
Globally secure but critically imperiled in Iowa and Prince Edward Island. Imperiled in Newfoundland.

■ NATIVE
■ RARE OR EXTIRPATED
■ INTRODUCED

ALTERNATE NAMES: Bluet, Narrow-leaved Bluet, Fine Leaf Bluet

As the common name suggests, this plant's flowers are tiny and diamond-shaped. This woody native perennial has a long taproot and can live in a variety of soils. The species in genus *Stenaria* were formerly assigned to the genus *Houstonia* or the more inclusive *Hedyotis*.

DESCRIPTION: This plant has a few to numerous slender, delicate stems that are upright or sprawling, about 2–20" (5–50 cm) high. The leaves are linear, very narrow and thread-like with rolled margins, about 1¼" long and ⅛" wide. Each leaf has a single center vein and a pointed tip. The diamond-shaped flowers are lilac or white and grow in crowded clusters, about ¼" wide. The flowers have a green calyx, four stamens, and a single protruding style.

FLOWERING SEASON: April–November in the South; April–July in the North.

HABITAT: Dry, rocky prairies and hillsides; rocky, open woods; roadsides.

RANGE: Florida to New Mexico, north to Ohio, Iowa, and Nebraska. It is common in the Ozarks in the central U.S.

CONSERVATION: G5
Globally secure but critically imperiled in Colorado and Virginia and presumed extirpated from Michigan. It is vulnerable in Georgia, Indiana, and Ohio.

■ NATIVE
■ RARE OR EXTIRPATED
■ INTRODUCED

ALTERNATE NAMES: Slender Centaury, Lesser Centaury

This European native has tiny, red or pinkish flowers that only open in the daylight and close in the afternoon. The flowers usually grow in clusters, but sometimes the plant has just one stem and one flower.

DESCRIPTION: This erect annual is only about 4" (10 cm) tall with tiny flowers. The base of the stem has whorled leaves that wither by the time the plant produces flowers. The stem leaves are opposite, narrow, and oval. The pink flowers are very small and star-like, usually growing in clusters with a short stalk and four or five narrow petals, about 1 cm across. The anthers are yellow. The fruit is a cylindrical capsule.

FLOWERING SEASON: June–September.

HABITAT: Roadsides, wetland areas, waste areas, and disturbed soils.

RANGE: Introduced in the Northeast and Midwest as well as Texas up to South Dakota.

SIMILAR SPECIES: Common Centaury (*C. erythraea*) is similar but much larger; it is also introduced in North America, naturalized in many of the same areas as *C. pulchellum* but also on the West Coast.

CONSERVATION:

This introduced species is unranked. It may be a weed in some areas.

■ NATIVE
■ RARE OR EXTIRPATED
■ INTRODUCED

ALTERNATE NAMES: Catchfly Gentian, Seaside Gentian

This annual can produce flowers throughout the year, depending on its environment. The genus name comes from the Greek words *eu*, meaning "good" or "beautiful," and *stoma*, meaning "mouth"—a nod to the attractive, cup-shaped flower.

DESCRIPTION: This plant has purple or white, cup-shaped flowers and erect stalks, growing 1–3' (30–90 cm) tall. The leaves are grayish, about 3" (7.5 cm) long, with a clasping stem. The leaves are also opposite and oblong. The flowers grow solitary or in terminal clusters, about 1½" (4 cm) wide. The petals are united at the base into a short tube, flaring out into five long, wide lobes.

FLOWERING SEASON: May–October.

HABITAT: Wet areas, sandy, coastal areas, and saline or freshwater marshes.

RANGE: The Great Plains to Montana and South Dakota, south to Mexico and the Gulf Coast, east to Florida, west to California.

CONSERVATION: G5

Secure, but critically imperiled in Nevada and vulnerable in Arizona.

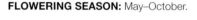

■ NATIVE
■ RARE OR EXTIRPATED
■ INTRODUCED

ALTERNATE NAMES: Elkweed, Green Gentian, Deer's Ears, *Swertia radiata*

This plant has wide leaves that look like the ears of deer. The plant usually grows separately from other species and sometimes doesn't flower when it grows in groups. The leaves are favored by elk and deer. This species was previously known as *Swertia radiata*.

DESCRIPTION: This perennial, cone-shaped plant grows from a woody base, surrounded by rosettes of leaves, to a height of 4–7' (1.2–2.1 m). It has one tall, erect stem. The leaves at the base are larger than the leaves on the stem. The basal leaves are about 10–20" (25–50 cm) long, lanceolate, and form a whorl with three or four leaves. The leaves on the stem are equally spaced and get progressively smaller. The flowers are star-like, 1–1½" (2.5–3.8 cm) wide, and grow in clusters. The yellowish-green, four-lobed corolla is joined at the base, with pointed lobes and purple spots, and with two oblong, fringed glands at the base. There are four stamens.

FLOWERING SEASON: May–August.

HABITAT: Rich soil in woodland openings, from moderate to high elevations.

RANGE: Eastern Washington to central California; east to western Texas, eastern Wyoming, and Montana; also northern Mexico.

SIMILAR SPECIES: Mullein (*Verbascum thapsus*) and Corn Lily (*Veratrum californicum*) are both similar species. Mullein has soft leaves with numerous hairs and yellow flowers. Corn Lily has numerous small, white flowers.

CONSERVATION: G4
Apparently secure.

■ NATIVE
■ RARE OR EXTIRPATED
■ INTRODUCED

This is one of the most common gentian species. As the common name suggests, the deep blue, bottle-like flowers of this plant usually remain closed, or have a small opening at the tip.

DESCRIPTION: This plant grows to a height of 1–2' (30–60 cm). The flowers grow in tight clusters and don't open or only open slightly, appearing at the top of the stem or in the upper leaf axils. The flowers are 1–1½" (2.5–4 cm) long. The corolla has a white base, with five lobes and fringed white bands between the lobes. The bands are slightly longer than petals. The leaves are ovate or lanceolate, with parallel veins, and grow to 4" (10 cm) long. The leaves appear in a whorl below the flower cluster.

FLOWERING SEASON: August–October.

HABITAT: Moist thickets and meadows, near streams and ponds.

RANGE: Saskatchewan east to Quebec, south to Nebraska, Missouri, Kentucky, and Virginia.

SIMILAR SPECIES: Blind Gentian (*G. clausa*) is a similar species, but the bands are smaller, usually not longer than the petals. Narrow-leaved Gentian (*G. linearis*) is also similar, but has very narrow leaves and open flowers, not closed. Stiff Gentian (*Gentianella quinquefolia*) also has open, light blue flowers.

CONVERSATION: G5

Globally secure but critically imperiled in Delaware and Massachusetts, and imperield in Maryland and Vermont in the U.S. and Saskatchewan in Canada.

■ NATIVE
■ RARE OR EXTIRPATED
■ INTRODUCED

ALTERNATE NAMES: Rainier Pleated Gentian

This species, like others in its genera, is among the most attractive alpine plants. This western species is often grown as a native garden flower and for ethnobotanical uses. The genus name comes from King Gentius of Illyria, who is credited with discovering a wide range of health benefits from drinking the roots of gentians in a tonic. Gentians are still used today to treat digestive problems, fever, high blood pressure, and muscle spasms, among other ailments.

DESCRIPTION: This perennial grows to a height of 2–12" (5–30 cm) with leafy stems and one to three blue funnel-shaped flowers. The leaves are thick and hardy, about ½–1¼" (1.3–3.1 cm) long. The leaves are also opposite, with those at the bases of the flowers joined together, forming a sheath around the stem.

The flowers grow from thin, red stems and are about 1–1½" (2.5–3.8 cm) long. The bell-shaped flowers have five lobes. The corolla is blue to yellowish-green, usually with greenish streaks and five pointed, nearly erect lobes.

FLOWERING SEASON: July–October.

HABITAT: Mountain meadows, alpine areas, and streambanks.

RANGE: British Columbia south to the Sierra Nevada of California, east to the Rocky Mountains.

CONSERVATION:  G4

Globally secure but imperiled in Alberta, British Columbia, and Wyoming.

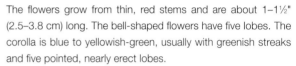

- ■ NATIVE
- ■ RARE OR EXTIRPATED
- ■ INTRODUCED

ALTERNATE NAMES: Soapwort Gentian

The vivid blue or purple flowers of this native perennial make it an attractive garden plant. The flowers are bottle-shaped and only ever partly open. The flowers are a favorite among bumblebees (*Bombus* spp.), which are a principal pollinator of the plant. It is somewhat rare in the wild.

DESCRIPTION: This perennial grows 6–24" (15–60 cm) with slender, unbranched stems and terminal clusters of blue-violet flowers. The central stem is reddish or green, while the leaves are dark green to shiny, up to 3½" (9 cm) long and 1½" (4 cm) across, lanceolate, opposite, with smooth margins. The flowers are 1½" long and bottle-shaped, or tubular, and nearly closed at the top. The flowers have five lobes. The calyx lobes are erect. The flowers often have vertical streaks of purple, green, or white.

FLOWERING SEASON: August–October.

HABITAT: Moist areas, Black Oak forests, margins of woodland creeks and ponds.

RANGE: New York to Michigan, south to Florida and Texas.

SIMILAR SPECIES: Bottle Gentian (*G. andrewsii*) is a similar species, but the flowers are often blue, not purple. The calyx lobes in flowers of Harvestbells are fairly straight and upright, while the calyx lobes of Bottle Gentian often curl outward. American Gentian (*G. catesbaei*), another common species, has brighter green leaves that are especially wide at the base.

CONSERVATION: G5

Globally secure but rare in many sites. It is critically imperiled in Ohio, Oklahoma, Pennsylvania, and New York; and vulnerable in Alabama, Arkansas, Delaware, Mississippi, New Jersey, and North Carolina. It is presumed extirpated in Michigan.

■ NATIVE
■ RARE OR EXTIRPATED
■ INTRODUCED

ALTERNATE NAMES: Autumn Dwarf Gentian

This leafy plant is found across northern and western North America as well as in Eurasia. It's one of the smaller, less showy species in this family. The flowers of this late-blooming biennial emerge from light yellowish buds but may be pink, purple, or blue.

DESCRIPTION: This plant has erect stems growing to 2–16" (5–40 cm). The flowers are small and trumpet-shaped, appearing purplish, bluish, or pinkish. The corolla is ½–¾" (1.5–2 cm) wide with a hairy fringe, appearing yellow, blue or lavender, with five lobes, flaring slightly. There are five stamens. The leaves are stalkless and opposite, growing ¼–½" (6–38 mm) long. The leaves sometimes have a purple tinge and five lengthwise veins.

FLOWERING SEASON: June–September.

HABITAT: Open areas, meadows and moist areas, mostly in the mountains.

RANGE: Alaska and throughout Canada, to the Atlantic Coast and northeastern United States, as well as south in the West to California, Arizona, New Mexico, and North Dakota.

CONSERVATION: G5

Globally secure but critically imperiled in Maine and Nova Scotia. It is imperiled in Labrador and vulnerable in Minnesota, Nevada, New Brunswick, and Newfoundland. It is possibly extirpated in Vermont.

■ NATIVE
■ RARE OR EXTIRPATED
■ INTRODUCED

ALTERNATE NAMES: Greater Fringed Gentian

This biennial is perhaps the most beautiful of the gentians, with its bright blue flowers and fringed petals. The flower opens during the day and closes at night. This flower is considered rare in some areas due to habitat loss and should not be picked or transplanted.

DESCRIPTION: This branching plant grows 1' (30 cm), sometimes to 3' (90 cm), with a single attractive blue flower at the end of an erect yellowish or green stem. The flower is fringed, about 2" (5 cm) long and tubular. The calyx has four unequal, pointed lobes, united below, and four green or reddish petals, each tipped with fringed segments 1/16–1/4" (2–5 mm) long. The leaves of first-year plants are basal, while the leaves of second-year plants appear opposite on the stem. Each leaf is 1–2" (2.5–5 cm) long and 3/8–3/4" (1–2 cm) wide, ovate to lanceolate, with a pointed tip. The leaves are rounded at the base.

FLOWERING SEASON: Late August–November.

HABITAT: Wet thickets and meadows and seepage banks.

RANGE: Manitoba east to Quebec, south to Georgia, and northwest to Tennessee, Illinois, Iowa, and North Dakota.

SIMILAR SPECIES: Lesser Fringed Gentian (*G. procera*) is similar but has narrower leaves and a shorter fringe.

CONSERVATION: G5

Globally secure but critically imperiled in Georgia, Maryland, New Hampshire, North Carolina, Virginia, and West Virginia; imperiled in North Dakota, Rhode Island, and Quebec; vulnerable in Maine, Iowa, Ohio, and Vermont. Due to its rarity in many sites, wild specimens should not be collected or transplanted.

■ NATIVE
■ RARE OR EXTIRPATED
■ INTRODUCED

ALTERNATE NAMES: Virginia Pennywort

Pennywort is a low-growing perennial that may be difficult to see because it grows under leaves on the forest floor. This is the only plant in genus *Obolaria*, which comes from the Greek word *obolos,* meaning a small coin—a reference to the roundish leaves. The common name is also a nod to the coin-like, round leaves.

DESCRIPTION: This low-growing fleshy plant reaches 3–8" (8–20 cm) tall. The leaves are oppositely arranged and grow under the flowers in a thick, round or wedge shape. Each leaf grows ½" (1.5 cm) long. The lower leaves are reduced to small scales. The upper leaves are purplish and bract-like. The flowers are dull white or purplish, usually growing in groups of three in the upper leaves and atop the stem. Each flower is about ½" (1.5 cm) long, with two sepals that are spatulate in shape and four petals, united at the middle.

FLOWERING SEASON: March–May.

HABITAT: Moist hardwoods and thickets.

RANGE: New Jersey and Pennsylvania south to Florida, west to easternmost Texas, Arkansas, and Missouri.

CONSERVATION: **G5**
Globally secure but imperiled in Arkansas, Missouri, and New Jersey; it is vulnerable in Delaware and Louisiana.

ROSE-PINK *Sabatia angularis*

■ NATIVE
■ RARE OR EXTIRPATED
■ INTRODUCED

ALTERNATE NAMES: Bitterbloom

This species' name refers to its rose- and pink-colored flowers. This plant's alternate name, "Bitterbloom," refers to its bitter taste. It has been made into a tonic and consumed for indigestion and fever, as well as other ailments.

DESCRIPTION: This biennial has square stems, reaching about 1–3' (30–90 cm) high, with dense basal leaves appearing the first year of the plant's life and flowering stalks in the second year. The flowering stems are multi-branched, with fragrant, rose or pinkish flowers, about 1" wide, appearing in a terminal, flat-topped cluster. Each flower has five lobes and a star-shaped, greenish-yellow center. The leaves are oval or heart-shaped, about 1½" long, and stalkless.

FLOWERING SEASON: June–September.

HABITAT: Moist, open woods; sand or peat thickets; acidic soils.

RANGE: Massachusetts west to southern Michigan, south to northern Florida, Texas, and extreme southeastern Kansas.

CONSERVATION: G5

Secure, but critically imperiled in New York and Kansas; imperiled in Michigan; vulnerable in Delaware. It is presumed extirpated in southern Ontario.

■ NATIVE
■ RARE OR EXTIRPATED
■ INTRODUCED

ALTERNATE NAMES: Large Marsh Pink

As its common names suggest, this native marsh plant has large, attractive, usually bright pink flowers. Two varieties are described, with var. *dodecandra* occurring along the Atlantic Coast in brackish areas and var. *foliosa* occurring farther inland from Texas to South Carolina and more frequently in freshwater habitats.

DESCRIPTION: A multi-branched, slender perennial growing 2½' (76 cm) high with opposite, lance-shaped leaves. The flowers often grow singly, appearing blue, purple, pink, red or white, about 3" wide. There are 8–13 petals and four to seven stamens.

FLOWERING SEASON: June–September.

HABITAT: Wetlands; saline, brackish, or rarely freshwater marshes and meadows.

RANGE: The Atlantic coastal plain, from southeastern Texas to Florida, north to Connecticut.

SIMILAR SPECIES: Plymouth Rose-gentian (*S. kennedyana*) is similar, but the flowers are about half the size of Marsh Rose-gentian and the plants mostly grow in sandy and peaty pond shore areas of the coastal plain.

CONSERVATION: G5

Globally secure but critically imperiled in Delaware; imperiled in Louisiana; and vulnerable in Maryland, North Carolina, and Virginia.

SALTMARSH SABATIA *Sabatia stellaris*

■ NATIVE
■ RARE OR EXTIRPATED
■ INTRODUCED

ALTERNATE NAMES: Rose-of-Plymouth, Marsh Pink, Sea Pink

Saltwater Sabatia is an inconspicuous native plant that's easy to overlook until it blooms. It has showy pink (sometimes white) flowers, and grows in saltwater or brackish waters.

DESCRIPTION: An annual with alternate branches reaching 6–18" (15–45 cm) tall. The leaves are linear to narrowly lanceolate in shape, about ¾–1½" (2–4 cm) long each, light green, and opposite. The flowers are pink to white and wheel-shaped with yellowish-red edges and star-shaped centers. Each flower is about ¾–1½" (2–4 cm) wide. The sepals are long and narrow, edged with red, shorter than petals. The corolla has four to seven lobes. The stamens are yellow, with a twisted style, divided below the middle.

FLOWERING SEASON: July–October.

HABITAT: Saline or brackish marshes and meadows.

RANGE: Along the Atlantic and Gulf coasts from New York and Massachusetts south to Florida and west to Louisiana.

SIMILAR SPECIES: Slender Marsh Pink (*S. campanulata*) is similar, but has sepals as long as the petals. Large Marsh Pink (*S. dodecandra*) has similar flowers, but the corollas are more divided, appearing with 8–13 lobes. Rose Pink (*S. angularis*) has a four-angled stem and flowers on opposite branches; it is mainly a southern and midwestern species.

CONSERVATION: ⓖ5
Globally secure but critically imperiled in Connecticut, Rhode Island and Massachusetts; imperiled in New York; and vulnerable in Delaware and North Carolina.

■ NATIVE
■ RARE OR EXTIRPATED
■ INTRODUCED

ALTERNATE NAMES: Felwort

This western perennial is one of the most attractive wildflowers that grows in mountains. Its star-shaped flowers are sometimes difficult to spot among other foliage. The genus name *Swertia* honors Emanuel Sweert, a 16th-century Dutch botanist, herbalist, and author.

DESCRIPTION: Star Gentian has several erect stems, growing 2–20" (5–50 cm) tall. The leaves appear denser at the base. The base leaves are spoon-shaped, about 2–8" (5–20 cm) long, while one or two pairs of smaller stem leaves, lanceolate in shape, are present. Star-like, pale bluish-purple flowers with greenish or white spots appear at the top of the stem in an open, branched, narrow cluster. Each flower is about ¾" (2 cm)

wide. The corolla has five pointed lobes joined at the base, with two fringed glands at the base of each lobe and a short, thick style.

FLOWERING SEASON: July–September.

HABITAT: Meadows, bogs, and moist places at high mountain elevations.

RANGE: Alaska to the southern Sierra Nevada of California; east to New Mexico; north in the Rocky Mountains to Montana and British Columbia.

CONSERVATION: G5
Globally secure but critically imperiled in Washington and Yukon, imperiled in Oregon, and vulnerable in Wyoming.

EASTERN BLUESTAR *Amsonia tabernaemontana*

■ NATIVE
■ RARE OR EXTIRPATED
■ INTRODUCED

ALTERNATE NAMES: Blue Dogbane

Eastern Bluestar is a clump-forming perennial native to the eastern and central U.S. It produces clusters of blue, star-like flowers in the spring, and the foliage turns yellow in the fall. The species name honors influential 16th-century German physician and herbalist Jakobus Theodorus Tabernaemontanus, who pioneered German botany.

DESCRIPTION: This erect, clump-forming plant grows to a height of 1–3' (30–90 cm). The leaves are lanceolate to elliptical in shape and short-stalked. The leaves are alternately arranged, but often grow very close and appear opposite each other. The flowers are sky blue and star-shaped, appearing in much-branched clusters.

The corolla has a downy outside. There are five narrow petals, each ¼–⅜" (6–10 mm) long, rising from a funnel-like tube. The slender fruit pods grow in pairs, each 3–4½" (9–12.5 cm) long, opening along one side.

FLOWERING SEASON: April–July.

HABITAT: Moist or wet woods and streambanks.

RANGE: Massachusetts south to Florida, west to Kansas and Texas.

CAUTION: The foliage of this species produces a toxic white latex sap when damaged that makes it unappetizing to deer and other browsers. This sap can cause mild skin irritation in humans.

CONSERVATION: G5

Secure, but critically imperiled in Kansas; vulnerable in Tennessee and Virginia.

■ NATIVE
■ RARE OR EXTIRPATED
■ INTRODUCED

ALTERNATE NAMES: Hemp Dogbane, Clasping Leaf Dogbane, *Apocynum sibiricum*

The genus name *Apocynum* means "poisonous to dogs," while the specific epithet *cannabinum* and the common names refer to its similarity to cannabis. Native Americans used this species extensively as a source of sturdy plant fibers to make items such as bows, nets, fishing lines, and clothing. Washes and weak teas made from this plant have also been used in traditional medicine to treat a wide range of ailments and as a sedative, but all parts of the plant are toxic if consumed and should not be taken internally.

DESCRIPTION: This plant has a single strong, erect, purplish or red stem, growing 3–4' (90–120 cm) tall. The branches ascend from the upper part. Long oval leaves often have a white coating or bloom. Small, cream-colored flowers are clustered at the ends of branches or on stalks from the leaf axils. Tufted seeds form in spindle-shaped pods.

FLOWERING SEASON: May–August.

HABITAT: Roadsides, thickets, fields, lakeshores, waterways, and disturbed areas.

RANGE: Widespread throughout the U.S. and Canada.

CAUTION: This plant is highly toxic and can cause nausea or cardiac symptoms, which can be fatal, if consumed.

CONSERVATION: G5

Globally secure but vulnerable in Wyoming in the U.S. and Alberta and Newfoundland in Canada. It may be an aggressive weed in some areas and can be difficult to control.

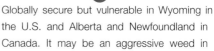

■ NATIVE
■ RARE OR EXTIRPATED
■ INTRODUCED

This is one of the most beautiful milkweeds, distinguished by its single terminal flower cluster and stalkless leaves that clasp the stem. The genus name honors Asklepios, the Greek god of medicine and gives a nod to the milkweeds' many uses in traditional medicine, although we now know most of these plants contain toxins that can be harmful if consumed and may irritate the skin of sensitive individuals.

DESCRIPTION: This plant has erect, waxy stems growing to a height of 1½–3' (45–90 cm). The stems are green or pinkish, with two to five pairs of widely spaced leaves. The leaves are oval and have wavy edges, 1½–5" (4–12.5 cm) long and ¾–3" (2–7.5 cm) wide, with a waxy appearance, and a whitish to pink midrib, usu-

ally without a stalk. A single cluster of 15–80 flowers grows at the end of each stem. Each flower has five pinkish-green petals that pull away from the pinkish crown. Usually there is just one terminal flower cluster, but sometimes a second cluster grows at the base of the plant.

FLOWERING SEASON: May–August.

HABITAT: Prairies, roadsides, open woods, fields.

RANGE: Throughout the eastern United States, from southern Minnesota to East Texas east to the Atlantic Coast.

CAUTION: All plants in the genus *Asclepias* are likely somewhat toxic to humans and animals. The sap may cause skin irritation.

SIMILAR SPECIES: Green Milkweed (*A. viridiflora*) is a similar species with broad leaves with wavy edges, but its leaves have short stalks.

CONSERVATION: G5

Globally secure but critically imperiled in Nebraska and Vermont; imperiled in Minnesota, New Hampshire, Rhode Island, and West Virginia; and vulnerable in Delaware and Ohio.

■ NATIVE
■ RARE OR EXTIRPATED
■ INTRODUCED

ALTERNATE NAMES: Mexican Whorled Milkweed

As the common names suggest, Narrowleaf Milkweed has long, narrow leaves that often grow in a whorl around the stem. Milkweeds in general are often covered in insects. This species is one of the most important native plants for Monarch butterflies (*Danaus plexippus*) in California and for various native bees.

DESCRIPTION: This perennial grows 24–36" (61–91.5 cm) with many thin, erect stems and long, very narrow leaves that often grow in whorls around the stem. The leaves are up to 6" (15 cm) long and ¾" (2 cm) wide. The flowers are pink-lavender or white and appear in erect clusters of 20 or more on each stem.

Each flower is ³⁄₁₆" (4–5 mm) with five recurved lobes that extend down. The fruits are pods that split open to release the seeds.

FLOWERING SEASON: June–September.

HABITAT: Dry climates, plains, roadsides, and disturbed grounds.

RANGE: Washington and Idaho south to southern Utah and California.

CAUTION: All plants in the genus *Asclepias* are likely somewhat toxic to humans and animals. The sap may cause skin irritation.

CONSERVATION: G5

Secure, but critically imperiled in Utah. This plant may be considered a weed in some areas.

The small flowers are very attractive to butterflies and are an important food source for Monarchs (*Danaus plexippus*). The stems of this plant exude a milky sap that is toxic to humans and animals, although Swamp Milkweed's sap is less thick than others in this genera. Although most milkweeds are poisonous, Native Americans have used milkweed sap in salves to remove warts, reduce swelling, and treat rashes, coughs, fevers, asthma, and other ailments.

DESCRIPTION: This is an erect, clump-forming plant, growing 1–4' (30–120 cm) tall with branching stems. The deep pink or white flowers appear in tight clusters atop the stems. Each flower is very fragrant and about ¼" (6 mm) wide, with five recurved petals and an ele-

vated central crown divided into five hoods. There are numerous narrow leaves, about 4" (10 cm) long, opposite, and lanceolate in shape. The seed is a pod, about 2–4" (5–10 cm) long, which splits open when ripe.

FLOWERING SEASON: June–August.

HABITAT: Swamps, shorelines, and thickets.

RANGE: Manitoba east to Nova Scotia, south to the Gulf of Mexico, west to the Rocky Mountain states.

CAUTION: All plants in the genus *Asclepias* are likely somewhat toxic to humans and animals. The sap may cause skin irritation.

CONSERVATION: G5

Secure, but critically imperiled in Arizona, Montana, and Prince Edward Island; imperiled in Arkansas, Idaho, Louisiana, and Wyoming; vulnerable in Georgia and Manitoba.

■ NATIVE
■ RARE OR EXTIRPATED
■ INTRODUCED

ALTERNATE NAMES: Corn Kernel Milkweed

The large flowers of this native perennial resemble clusters of corn kernels. This plant is sometimes sold commercially for use as a garden plant that attracts butterflies.

DESCRIPTION: A tall perennial herb with erect stems 2–3' (60–90 cm) tall. The round leaves are so thick and large that the green or yellowish flowers are often hidden within them. The leaves are dark green and coarse, about 3–4" (7–10 cm) long, 2" (5 cm) wide, attached directly to the main stem and growing in a loose whorl.

FLOWERING SEASON: June–September.

HABITAT: Roadsides, plains, prairies, rocky slopes, mesas.

RANGE: South Dakota to Texas west to Utah, Arizona, and California.

CAUTION: These plants are poisonous to livestock, including sheep, cattle, and goats.

CONSERVATION:  G5

Secure, but vulnerable in Nebraska.

SHOWY MILKWEED *Asclepias speciosa*

■ NATIVE
■ RARE OR EXTIRPATED
■ INTRODUCED

This attractive plant has pale fragrant blooms that are often described as looking like small crowns. Native Americans used fiber in this plant's stems to make ropes, baskets, and nets.

DESCRIPTION: A grayish, velvety plant with erect, leafy stems growing 1–4' (30–120 cm) tall. The leaves are 4–8" (10–20 cm) long, opposite, and lanceolate or ovate, with distinct veins from the midrib to the edge. The pinkish flowers are star-like and grow in clusters of 10 or more at the top of the stems. Each flower is about ¾" (2 cm) wide, with five reddish petals, each about ½" (1.5 cm) long, that bend back after blooming. It has five pink erect hoods with horns. The seed is a coarse pod, about 2–4" (5–10 cm) long, covered with velvet and small, soft spines.

FLOWERING SEASON: May–August.

HABITAT: Dry gravelly slopes, sandy areas, in brush and open forests.

RANGE: British Columbia to California, east of the Cascade Mountains; eastward to the central United States.

CAUTION: All plants in the genus *Asclepias* are likely toxic, and some are fatal to both humans and animals if consumed. The sap of some species causes skin irritation in humans.

SIMILAR SPECIES: Showy Milkweed is similar to Common Milkweed (*A. syriaca*), but Showy Milkweed has slightly longer petals.

CONSERVATION: G5
Secure, but vulnerable in Iowa and Alberta.

■ NATIVE
■ RARE OR EXTIRPATED
■ INTRODUCED

ALTERNATE NAMES: Horsetail Milkweed

All *Asclepias* should be considered toxic, but as the common name suggests, this species is particularly poisonous, containing cardenolides that are toxic and even fatal to livestock. This plant is especially toxic during the active growing season.

DESCRIPTION: This perennial grows in feathery clumps or patches with small white, star-like flowers, about 4' (1.2 m) tall. The flowers grow in umbrella-like clusters, with three to five very narrow leaves in whorls at nodes. The leaves are ¾–5" (2–12.5 cm) long, with dwarf branches. Each flower is ½" (1.5 cm) wide, with five tiny sepals, five petals that bend back and five roundish hoods with long horns arching toward the center. The fruit and flowers appear at the same time. The fruit is a wide, smooth pod, 2–4" (5–10 cm) long, containing many seeds with long silky hairs.

FLOWERING SEASON: May–September.

HABITAT: Sandy or rocky plains and desert flats and slopes; common along roadsides.

RANGE: Idaho, Nevada, and Arizona east to Missouri and south to Texas and into Mexico.

CAUTION: Farm animals may eat Poison Milkweed when other forage is unavailable. It can cause a variety of effects in animals, including spasms, bloating, and fevers, and is often fatal. The sap of some milkweeds causes skin irritation in humans.

CONSERVATION: G4
Apparently secure. It is possibly extirpated in Wyoming.

■ NATIVE
■ RARE OR EXTIRPATED
■ INTRODUCED

The widespread and sometimes weedy plant contains cardiac glycosides, which are used in humans to treat heart diseases, but can be toxic and even fatal, especially to animals. Plants in this genera are important food for Monarch butterflies (*Danaus plexippus*), and the glycosides consumed by its larvae make the adult butterflies toxic to birds and other predators. Swedish botanist Carl Linnaeus gave this species the name *syriaca* because he mistakenly thought it came from Syria.

DESCRIPTION: A tall downy plant with drooping, purplish to pink flower clusters, growing 2–6' (60–180 cm) tall. The light green to gray leaves are 4–10" (10–25 cm) long, opposite, and broad-oblong. The leaves exude a milky sap when cut or bruised. The flower clusters are about 2" (5 cm) wide. Each flower is ½" (1.5 cm) wide, with five reflexed petals, and a conspicuous central crown divided into five hoods. The fruit is a rough-textured pod, opening along one side.

FLOWERING SEASON: June–August.

HABITAT: Old fields, roadsides, and waste places.

RANGE: Saskatchewan east to Nova Scotia, south to Georgia, west to Texas, and north to North Dakota.

CAUTION: All plants in the genus *Asclepias* are likely somewhat toxic to humans and animals. The sap may cause skin irritation.

SIMILAR SPECIES: Showy Milkweed (*A. speciosa*) is similar, but it has longer flower petals.

CONSERVATION: G5
Globally secure but critically imperiled in Saskatchewan and vulnerable in Manitoba.

■ NATIVE
■ RARE OR EXTIRPATED
■ INTRODUCED

ALTERNATE NAMES: Pleurisy Root, Butterfly Milkweed

This perennial is native to eastern North America and the south-western U.S. The plant's copious supply of nectar and bright flowers attract various types of butterflies, including Queens (*Danaus gilippus*) and Monarchs (*D. plexippus*), although its principal pollinators are bees and wasps. Although classified alongside milkweeds, this plant does not produce the milky sap characteristic of other members of the genus.

DESCRIPTION: This plant has leafy, hairy stems and grows up to 3' (90 cm) tall. The leaves are 1¼–4½" (3–11 cm) long, almost opposite on the stem and narrowly or broadly lanceolate. Some of the leaves are indented at the base and the underside is covered in rough hairs. Umbels of star-like, orange, yellow, or red flowers appear in the upper leaf axils. Each umbel is 3" (7.5 cm) wide, while each flower is about ½" (1.5 cm) wide. There are five sepals, five petals bent backward, and five erect, scoop-shaped hoods with slender horns arching toward the center. The fruit, a narrow pod, is 3–6" (7.5-15 cm) long, containing many seeds with silky hairs.

FLOWERING SEASON: June–September.

HABITAT: Dry open soil, roadsides, and fields.

RANGE: Ontario and Quebec south to Florida, west to Texas, and north to South Dakota and Minnesota; also in the southwestern states.

CAUTION: All plants in the *Asclepias* genus contain toxic gly-cosides, which are harmful and perhaps fatal to humans and other animals.

SIMILAR SPECIES: This plant is distinguished from other milkweeds (*Asclepias* spp.) in having alternate leaves and non-milky sap. Lanceolate Milkweed (*A. lanceolata*) is similar, but it has fewer flowers and the stems become milky when broken.

CONSERVATION: G5

Globally secure. It is presumed extirpated in Maine and possibly extirpated in Vermont.

- ■ NATIVE
- ■ RARE OR EXTIRPATED
- ■ INTRODUCED

ALTERNATE NAMES: Sand Vine, Honeyvine Milkweed, Bluevine Milkweed, Climbing Milkweed, and Smooth Wallow-wort

Honeyvine can be a problem weed in crop fields. It has an extensive root system that grows rapidly and is difficult to eradicate. It is an important food plant for Monarch butterflies (*Danaus plexippus*) and Milkweed Tiger Moths (*Euchaetes egle*).

DESCRIPTION: This perennial vine has a twining stem, growing up to 15' (4.5 m) long. These stems are round and shiny, appearing light green or reddish-green, with fine white hairs. The leaves are opposite, growing to 4" (10 cm) long. Each leaf is entire with a wide cleft at the base. The flowers grow in small clusters, about ¾–1½" (2–4 cm) wide. Each flower is white or pinkish-white and up to ¼" long.

FLOWERING SEASON: June–September.

HABITAT: Woods, fields, disturbed places, and streambanks.

RANGE: Maryland south to Florida, west to central Kansas and Texas; it is also recorded in Idaho.

CONSERVATION: G5

Globally secure but critically imperiled in North Carolina; vulnerable in Georgia and Iowa.

■ NATIVE
■ RARE OR EXTIRPATED
■ INTRODUCED

ALTERNATE NAMES: Angular-fruit Milkvine

This plant is important for butterflies, especially female Queens (*Danaus gilippus*) and Monarchs (*D. plexippus*) that are laying eggs.

DESCRIPTION: This vine grows to approximately 10' (3 m), with opposite heart-shaped leaves. Each leaf is nearly 1" (2.5 cm) wide. The flowers are star-shaped, with five petals appearing in clusters in the leaf axils. The blooms are reddish in the center and greenish-yellow on the edge. The petals are thick and fleshy.

FLOWERING SEASON: June–August.

HABITAT: Woods and thickets.

RANGE: Maryland south to Florida, west to Kansas and Texas.

CONSERVATION: G5

Globally secure but critically imperiled in Kansas and vulnerable in Indiana.

■ NATIVE
■ RARE OR EXTIRPATED
■ INTRODUCED

ALTERNATE NAMES: Running Myrtle, Vinca, Lesser Periwinkle, Creeping Myrtle, Dwarf Periwinkle

This European plant, used in gardens for groundcovers, has escaped cultivation and frequently forms extensive patches in the wild. The stems have been used to make wreaths. The common name Periwinkle comes from the Latin word *pervinca*, meaning to bind or wind, and references the long stems.

DESCRIPTION: A low, evergreen, trailing plant with lavender flowers that look like phlox, growing 6–8" (15–20 cm) tall. The leaves are 1¼–2" (3–5 cm) long, shiny, dark green, and opposite. The flowers are about 1" (2.5 cm) wide, with a funnel-shaped corolla that has five lobes and a whitish star in the center. The fruits, paired cylindrical pods, are about ½–1¼" (1.5–3 cm) long.

FLOWERING SEASON: April–May.

HABITAT: Woodland borders, roadsides, abandoned sites, and cemeteries.

RANGE: Introduced from Ontario east to Nova Scotia, and from Nebraska south to Texas and east to the Atlantic Coast; also introduced in parts of the West.

SIMILAR SPECIES: Greater Periwinkle (*V. major*), also introduced in North America, is similar but its leaves are larger.

CONSERVATION:
The status of this exotic species is unranked. It is considered invasive in many areas.

■ NATIVE
■ RARE OR EXTIRPATED
■ INTRODUCED

ALTERNATE NAMES: Black Dog-strangling Vine

The flowers of this European vine are very dark, often appearing black, hence the common name. The flowers last just for one day. This plant has escaped cultivation and is an invasive species in many areas in North America.

DESCRIPTION: This perennial vine grows 6.5' (2 m) tall. It has ovate leaves that grow in pairs along the stem. Each leaf is about 3–4" (7.6–10.2 cm) long. The flowers are dark purple or black and shaped like a star. Each flower is 1½–2" (4–5 cm) wide, with five petals and white hairs. The seeds have white tufts of hair that allow them to disperse in the wind.

FLOWERING SEASON: June–July.

HABITAT: Roadsides, disturbed areas, and open woods.

RANGE: Introduced in much of the Midwest, Northeast, and southeastern Canada. It has also been reported in California.

CONSERVATION:

This exotic species is unranked. The plant spreads rapidly through rhizomes and is considered an invasive weed in many areas, often crowding out native species.

■ NATIVE
■ RARE OR EXTIRPATED
■ INTRODUCED

ALTERNATE NAMES: Small-flower Fiddleneck

Menzies' Fiddleneck produces clusters of yellowish flowers shaped like a violin or a fiddle. This plant's seeds are favored among Lawrence's Goldfinch (*Spinus lawrencei*) and other birds. It can be a problem plant in agricultural land, and some fiddlenecks are toxic to livestock.

DESCRIPTION: A bristly perennial with a terminal flowering whorl that curls like the neck of a violin or fiddle, growing to 6–30" (15–76 cm). The flowers are yellowish or orange and funnel-shaped with five lobes. The stem rolls or arches at the end and the flowers grow from the bottom up. The leaves are linear to narrowly lanceolate or oblong.

FLOWERING SEASON: Spring and early summer.

HABITAT: Open, disturbed, dry habitats, such as along roadsides, grasslands, croplands, and pastures.

RANGE: Most common in western North America, from Alaska south to California and Arizona, east through Manitoba, North Dakota, Nebraska and Texas. Also found in Missouri, Illinois, and parts of the Northeast.

CAUTION: The seeds of some fiddlenecks contain compounds toxic to humans and other mammals.

CONSERVATION: G5

Secure, but imperiled in Wyoming. This plant may be considered a weed in some areas.

■ NATIVE
■ RARE OR EXTIRPATED
■ INTRODUCED

ALTERNATE NAMES: Obscure Cat's-eye

This western native plant is covered in stiff, thick hairs. It grows in many types of habitats, from sagebrush scrub to open conifer forests.

DESCRIPTION: This annual has a branching stem with stiff hairs, reaching 4–14" (10–35 cm) tall. The leaves are hairy and bristly, up to 1½" (4 cm) long, and oblong to linear in shape. It has a dense clump of white flowers covered in bristles that appear at the end of the stem. Each flower is about 3–4 mm wide, with five lobes. The fruit is four small, coarse nutlets 2 mm wide.

FLOWERING SEASON: May–August.

HABITAT: Forests and open areas in dry, sandy, or gravelly sites.

RANGE: Western North America, from British Columbia to California east to Colorado.

SIMILAR SPECIES: Torrey Pine (*C. torreyana*) is similar, but has smoother nutlets.

CONSERVATION:  G4
Apparently secure, but imperiled in Utah and vulnerable in Wyoming.

A European plant, Houndstongue spreads quickly and is considered a weed in many areas in North America. The nutlets are coarse and prickly, often becoming embedded in the wool or hair of livestock, potentially causing eye damage if they become embedded in the eyes. The genus name comes from the Greek words *kynos*, meaning "dog," and *glossa*, meaning "tongue," referring to the shape and texture of the basal leaves.

DESCRIPTION: This weedy biennial has small red or purple flowers and erect stems densely covered with hairs, growing to 4' (1.2 m) tall. The basal leaves are gray or green and covered with soft hairs, while the stem leaves are lanceolate and also densely hairy. Each leaf is 6–8" (15–20 cm) long and ¾–2" (2–5 cm) wide. The flowers are funnel-shaped and grow in small clusters, with five petals and five sepals. Each flower is ¼" (6 mm) long. The tiny nutlets are grayish and covered in small barbs.

FLOWERING SEASON: May–September.

HABITAT: Disturbed sites such as pastures, roadsides, and fields.

RANGE: Throughout North America, except in parts of the South.

CAUTION: Houndstongue can be fatally poisonous to horses and cattle if eaten.

CONSERVATION:
The status of this exotic plant is unranked.

Echium vulgare COMMON VIPER'S-BUGLOSS

■ NATIVE
■ RARE OR EXTIRPATED
■ INTRODUCED

ALTERNATE NAMES: Blueweed, Blue Devil

This Eurasian plant has showy flowers, but is introduced in North America and is considered a weed in some areas. The nutlets are covered in bristles and are said to resemble snake heads, hence the common name. This plant was also once used as a treatment for snakebites, although it is now considered toxic. *Bugloss* is a Greek word for an ox tongue, and references the shape and rough hairs on the leaves.

DESCRIPTION: A hairy biennial with showy flowers, growing 1–2½" (30–75 cm) tall. The flowers are tubular and grow in clusters on one side of a branched spike. The clusters unravel as the flowers bloom. The flowers are blue, each about ¾" (2 cm) long, with five lobes and five protruding red stamens. The leaves are 2–6" (5–15 cm) long, with rough hairs, and oblong to lanceolate in shape. The coarse nutlets of this plant appear after the flowers.

FLOWERING SEASON: May–October.

HABITAT: Fields, roadsides, and waste places; often in limestone soil.

RANGE: Alberta east to Newfoundland, south to Georgia, west to Tennessee, Arkansas, and Texas, and north to South Dakota and Minnesota; also in parts of the West.

CAUTION: The hairs on this plant are very dense and can cause skin irritation. The plant is also toxic to livestock, especially cattle and horses, if consumed.

CONSERVATION:
The status of this exotic species is unranked.

SPOTTED HIDESEED *Eucrypta chrysanthemifolia*

NATIVE
RARE OR EXTIRPATED
INTRODUCED

ALTERNATE NAMES: Common Eucrypta

This native annual of the Southwest is small and inconspicuous, with foliage that resembles that of chrysanthemums, and with tiny white flowers. The inflorescence smells like vinegar. The genus name *Eucrypta* means "secret" or "well-hidden" and refers to the plant's hidden seeds.

DESCRIPTION: This species has an erect to leaning stem, about 2' (60 cm) tall. The leaves are oval-shaped, divided into many lobes and subdivided further into smaller lobes, like lace. The flowers are small, white, and bell-shaped, drooping as the fruit develops. Each flower is about ⅓" (1 cm) wide, appearing at the ends of the stems. The fruit is a capsule covered in bristles, about 3 mm wide.

FLOWERING SEASON: February–June.

HABITAT: Disturbed areas, recently burned areas.

RANGE: Arizona, California, and Nevada.

SIMILAR SPECIES: Dainty Desert Hideseed (*E. micrantha*), another native to the Desert Southwest, is the only other species in this genus. Its tiny flowers are white or purple with a yellow throat.

CONSERVATION: G4
Apparently secure.

■ NATIVE
■ RARE OR EXTIRPATED
■ INTRODUCED

ALTERNATE NAMES: Davis Mountain Stickseed, Large-flowered Stickseed

This western native wildflower is a favorite among many native bees and butterflies, including Indra Swallowtail (*Papilio indra*) and Pearl Crescent (*Phyciodes tharos*).

DESCRIPTION: This plant has a few leafy stems and grows about 3' (91 cm) tall, with coiled branches at the tip. The leaves are 1½–8" (3.8–20 cm) long and lanceolate. The lower leaves have petioles and become progressively smaller up the stem. The flowers are small, blue, and funnel-shaped, appearing in an open, branched cluster. Each corolla is about ¼" (6 mm) wide, with five petal-like lobes and yellow pads around the opening of the narrow tube. The fruit (hard nutlets) has barbed prickles on the edges, but not on the back.

FLOWERING SEASON: June–August.

HABITAT: Moist thickets and meadows, generally in coniferous forests

RANGE: From Washington to northern California and southern Nevada; east to northern New Mexico, Colorado, and Montana.

SIMILAR SPECIES: Jessica's Stickseed (*H. micrantha*) is very similar but has many stems and small prickles on the back of each nutlet, not on the edges. Forget-me-nots (*Myosotis* spp.) are also similar but lack the prickles on the nutlets.

CONSERVATION: G5
Secure, but critically imperiled in Texas and vulnerable in Nebraska.

■ NATIVE
■ RARE OR EXTIRPATED
■ INTRODUCED

ALTERNATE NAMES: Beggar's-lice, Sticktight

The plant is native to eastern North America. The fruits are covered in prickles and Velcro-like, often sticking to and becoming difficult to remove from clothing and animal fur. It is sometimes considered a nuisance plant because of these burs.

DESCRIPTION: This annual or biennial has stems with white hairs and grows 1–4' (30–120 cm) tall. The leaves are soft to the touch with short hairs and alternately arranged. The basal and lower leaves are up to 6" (15 cm) long, 3.5" (9 cm) wide, and triangular or ovate in shape. The upper leaves become progressively smaller. The upper stems terminate in racemes of tiny blue flowers that are tubular or saucer-shaped. Each flower is about ⅛" (3 mm) across, with five round petals. The fruits are round nutlets with dense, Velcro-like prickles covering the outer surfaces.

FLOWERING SEASON: June–August.

HABITAT: Woodlands, woodland borders, thickets, and areas disturbed by fire or grazing.

RANGE: Quebec and Ontario south to Georgia, west to the Great Plains.

SIMILAR SPECIES: Nodding Stickseed (*H. deflexa*) is similar but the leaves are smaller and have rough hairs, not soft hairs.

CONSERVATION: G5

Secure, but critically imperiled in South Carolina; imperiled in North Carolina and New Hampshire; vulnerable in Quebec.

■ NATIVE
■ RARE OR EXTIRPATED
■ INTRODUCED

ALTERNATE NAMES: Dwarf Monkey-fiddle

This western native plant grows in patches low to the ground. It has tiny white or blue flowers. The genus includes only one other species, California Hesperochiron (*H. californicus*), which occupies a similar but slightly smaller range.

DESCRIPTION: This low-growing perennial has one slender stalk growing from a basal rosette of leaves, reaching 1–2" (2.5–5 cm) tall. The leaves are 1–3" (2.5–7.5 cm) long and lanceolate in shape. Between one and five white or blue, saucer-shaped flowers grow at the end of the stem. The corolla is ½–1¼" (1.5–3 cm) wide, with five rounded lobes that appear different sizes, with fine hairs and purple or light blue veins. The center of the flower is yellow. Five bluish stamens grow from the center of the flower with a hairy style, topped by two stigmas.

FLOWERING SEASON: April–July.

HABITAT: Woodlands, moist meadows, flats, and slopes.

RANGE: British Columbia and in the western United States from Washington to Montana south to California and Arizona.

SIMILAR SPECIES: California Hesperochiron (*H. californicus*) is similar but has larger leaves and a funnel- or bell-shaped corolla.

CONSERVATION: G4

Apparently secure, but imperiled in Arizona, Wyoming, and British Columbia; vulnerable in Nevada. It is possibly extirpated in Colorado.

■ NATIVE
■ RARE OR EXTIRPATED
■ INTRODUCED

This species is called "hoary" due to the downy appearance of the leaves. The native Powhatan people used this perennial's roots to make a red dye used for pottery, basketry, and decoration in ceremonies. The common name "Puccoon" is from the Powhatan word *poughkone*, for "red dye."

DESCRIPTION: This hairy, grayish plant grows 6–18" (15–45 cm) tall. The leaves and stems are covered with fine, soft hairs, giving the plant a hoary look. Each leaf is ½–1½" (1.5–4 cm) long, narrow and mostly stalkless. It has terminal clusters of tubular yellowish-orange flowers that grow ½" (1.5 cm) wide. The corolla has five lobes that flare out. Each flower has up to two white or brownish nutlets.

FLOWERING SEASON: March–June.

HABITAT: Dry, rocky or sandy sites and borders of grasslands.

RANGE: Saskatchewan east to Ontario and Pennsylvania, south to Georgia, west to Texas and the edge of the Great Plains.

SIMILAR SPECIES: Hairy Puccoon (*L. caroliniense*) has harsher, longer hairs and is a larger plant. Corn Gromwell (*L. arvense*), a European native now found throughout the United States, is an annual with inconspicuous white flowers, not yellow, in its upper leaf axils.

CONSERVATION: Ⓖ5
Secure, but imperiled in Pennsylvania, North Carolina; vulnerable in West Virginia and Ontario.

■ NATIVE
■ RARE OR EXTIRPATED
■ INTRODUCED

ALTERNATE NAMES: Western Marbleseed

The flowers of this native plant grow in a crowded coil on one side of the stem, resembling a scorpion's tail. The classification of this species has changed multiple times; it was formerly placed in the genus *Onosmodium* before it was moved to *Lithospermum* based on a molecular study in 2009.

DESCRIPTION: This erect perennial is covered in white, spreading hairs on all parts of the plant, growing to 1–4' (30–120 cm). The leaves are gray-green and stalkless, those at the base are about 8" (20 cm) long, while those farther up the stem are smaller. The leaves are lance or oval shaped, with pointed tips and conspicuous veins. The flowers are greenish or white and tubular, growing in a coiled cluster at the ends of the branches, which uncoils as the flowers mature. Each flower is about 0.4" (1 cm) long, with five narrow, erect petals and a long stamen. The fruit is a nutlet, about 0.1" (3–3.5 mm) long.

FLOWERING SEASON: May–June.

HABITAT: Prairies, open woods, roadsides.

RANGE: Alberta east to Quebec, south to Georgia, and east to Arizona and New Mexico.

SIMILAR SPECIES: Some sources treat this plant as a subspecies of Softhair Marbleseed (*Lithospermum parviflorum*), formerly classified as *Onosmodium molle* or *O. bejariense*; related plants in this complex are generally taller, up to 48" (120 cm) tall, with larger leaves. Virginia Marbleseed (*L. virginianum*) is a similar species, but has pale yellow flowers, not green.

CONRSERVATION:

Apparently secure but critically imperiled in Georgia and Ontario, and vulnerable in Alberta, Manitoba, Saskatchewan, and Wyoming. It is rare and declining in many parts of its native range, threatened by habitat loss due to development.

■ NATIVE
■ RARE OR EXTIRPATED
■ INTRODUCED

ALTERNATE NAMES: Tall Fringed Bluebells, Streamside Bluebells

The sweet-smelling flowers of the western native plant start blue and become pinkish-red at maturity. The plant may carpet moist meadows and hillsides under ideal conditions. Species of *Mertensia* are also called Lungwort, after a European species that was believed to aid in treating lung disease. All parts of this plant are edible raw when young but should not be consumed in high quantities because they contain alkaloids.

DESCRIPTION: This perennial grows in clumps of leafy stems growing 6–60" (15–50 cm) tall. The leaves are veiny, about 1¼–6" (3.1–15 cm) long, and tapered at the base. The lower leaves are on long petioles. The flowers are bell-shaped and blue, turning pink with age. The flowers hang like bells from a stem that arches at the top and expands into a wider mouth. The corolla is five-lobed, about ½–¾" (1.3–2 cm) long. The tubular part of the flower is about the same length as the bell-like end.

FLOWERING SEASON: May–August.

HABITAT: Streambanks, seeps, and wet meadows.

RANGE: Central Idaho to central Oregon and the Sierra Nevada; east to western Montana, western Colorado, and northern New Mexico.

CAUTION: The plant is edible but should not be eaten in high quantities because it contains toxic alkaloids.

SIMILAR SPECIES: Virginia Bluebells (*Mertensia virginica*) is similar but occurs in the East and doesn't grow as tall.

CONSERVATION: **G5**
Secure, but critically imperiled in South Dakota and imperiled in New Mexico.

■ NATIVE
■ RARE OR EXTIRPATED
■ INTRODUCED

ALTERNATE NAMES: Virginia Cowslip

Virginia Bluebells are native to eastern and central North America. The flowers of this attractive spring perennial start off pink and become light blue as they mature. The flowers hang like bells from an arched stem.

DESCRIPTION: This is an erect plant with smooth, gray-green leaves with prominent veins growing to a height of 8–24" (20–60 cm). The bell-shaped flowers grow in clusters at the ends of arched stems. They appear as pink buds initially, eventually opening and becoming light blue at maturity. Each flower is about 1" (2.5 cm) long, with a five-lobed corolla, and five stamens. The basal leaves are 2–8" (5–20 cm) long, while those on the stems are smaller, oval, and untoothed.

FLOWERING SEASON: March–June.

HABITAT: Floodplains and moist woods, rarely meadows.

RANGE: Ontario east to Quebec and Massachusetts, south through the East to Arkansas and Georgia

SIMILAR SPECIES: Sea Bluebells (*M. maritima*) is similar but smaller, found farther north in Canada to coastal New England, and usually restricted to seashores. Tall Lungwort (*M. paniculata*) is also similar, but has hairy stems.

CONSERVATION: G5

Secure, but critically imperiled in Michigan, Mississippi, and Kansas; imperiled in Georgia, New Jersey, and North Carolina; and vulnerable in Delaware and Ontario.

■ NATIVE
■ RARE OR EXTIRPATED
■ INTRODUCED

ALTERNATE NAMES: Water Forget-me-not

The flowers of this introduced Eurasian plant coil up and resemble the coiled tail of a scorpion, hence the Latin name. True Forget-me-not is considered a weed in some midwestern states. The genus name comes from the Greek words meaning "mouse ear," and refers to the short, pointed leaves of some plants in this genus.

DESCRIPTION: This sprawling plant has tiny, light blue, tubular flowers growing to a height of 6–24" (15–60 cm). The leaves are hairy, shiny, semi-evergreen, and about 1–2" (2.5–5 cm) long. The leaves are oblong in shape and mostly stalkless. The flowers have a golden or yellow center and grow on curving branches that uncoil as the flowers mature. Each flower is ¼" (6 mm) wide with five lobes.

FLOWERING SEASON: May–October.

HABITAT: Streamsides, bogs, riversides, wet places.

RANGE: Widely introduced throughout eastern and western North America.

SIMILAR SPECIES: Small Forget-me-not (*M. laxa*) is a similar species, but the flowers are much smaller.

CONSERVATION: **G5**

Globally secure. This introduced species is considered a noxious weed in the Midwest.

NATIVE
RARE OR EXTIRPATED
INTRODUCED

ALTERNATE NAMES: Spring Scorpion Grass, White Forget-me-not

Spring Forget-me-not is a widespread native annual. As its name suggests, it is one of the first plants to bloom in the spring. Its hairy fruits often attach to animal fur and clothing.

DESCRIPTION: This plant has light green stems growing ½–1½' (15–45 cm) tall. The central stem and flower stalks are densely covered in hairs. The leaves are alternate, with a prominent central vein. Each leaf is about 2" (5 cm) long, becoming progressively smaller up the stem. The flowers grow in a raceme of 8–24 white, bell-shaped blooms at the tips of branching stems. The cluster is initially tightly curled at the tip with the flowers gradually opening. Each flower is about ⅛" (3 mm) wide, with five spreading, egg-shaped to oblong lobes, five sepals, and five stamens. The fruits are seedpods covered in hairs that readily stick to animal fur and clothing.

FLOWERING SEASON: April–July.

HABITAT: Woodlands, sand prairies, fields, and roadsides.

RANGE: Throughout much of eastern North America, from Minnesota and Ontario south and east; also found in the West from British Columbia south to California and sparsely through the northwestern U.S.

SIMILAR SPECIES: Big-Seeded Scorpion Grass (*M. macrosperma*) is a similar species with larger seeds.

CONSERVATION: G5
Secure, but critically imperiled in Wyoming, Vermont, Nebraska and Quebec; vulnerable in New Jersey and Maryland.

■ NATIVE
■ RARE OR EXTIRPATED
■ INTRODUCED

ALTERNATE NAMES: Oak-leaved Nemophila, Wood's Nemophila

This western native has very small flowers and is easy to overlook in the wild. It's typically found in wooded, shady areas.

DESCRIPTION: This delicate annual has slender, branched, sprawling stems growing to 6–20" (15–50 cm). The plant's stems are covered in a few stiff hairs. The leaves are hairy and opposite, divided into five toothed, unequal lobes. Each leaf is ⅖–1⅖" (1–3.5 cm) long and ⅓–1" (0.8–2.5 cm) wide. The flowers are white or lavender and bowl-shaped, about 4.5 mm wide, and solitary. The flowers grow in leaf axils, with two pairs of lateral lobes.

FLOWERING SEASON: April–July.

HABITAT: Meadows, forests, and disturbed areas.

RANGE: British Columbia to California west to Idaho and Utah.

CONSERVATION: G5
Secure.

■ NATIVE
■ RARE OR EXTIRPATED
■ INTRODUCED

ALTERNATE NAMES: Sleeping Popcorn Flower, *Plagiobothrys hispidus*

This southwestern native plant has very small flowers that are easy to overlook. The species name comes from the Latin *hispid*, meaning "hairy" or "bristly," and refers to the bristle-like hairs on the leaves.

DESCRIPTION: The stems of this plant grow upright or angled, up to 2–16" (5–40 cm) tall. The leaves are about 2" (5 cm) long, with a rough undersurface covered in bristly hairs. The basal leaves are linear in shape, while those higher up the stems are more oblanceolate. The flowers are white with a yellow center and somewhat funnel-shaped. The flowers appear smaller than the leaves, forming a green, five-lobed calyx. The calyx is about 1" (2.5 cm) long and curves out.

FLOWERING SEASON: April–August.

HABITAT: Mainly deserts.

RANGE: Southwestern United States in dry areas of Oregon, northeastern California, and northwestern Nevada.

SIMILAR SPECIES: Some *Myosotis* species (forget-me-nots) are similar, but have alternate leaves, and typically blue flowers with rounded petal lobes.

CONSERVATION:
Apparently secure.

■ NATIVE
■ RARE OR EXTIRPATED
■ INTRODUCED

ALTERNATE NAMES: Woody Tiquilia

Woody Crinklemat is a native of low-elevation deserts in the southwestern U.S. This is a small, woody plant that forms mats on the ground.

DESCRIPTION: This low-growing woody perennial grows up to 2' (60 cm) tall in mats. The leaves are fleshy, ovate in shape and densely covered in white hairs. Each leaf is about ½–¾" (13–19 mm) long. The flowers are small and pinkish or white and tubular, with five lobes. The calyx is 4–6 mm long.

FLOWERING SEASON: March–August.

HABITAT: Deserts, semi-desert grasslands.

RANGE: Southwestern United States, in California, Nevada, Arizona, Utah, New Mexico, Texas.

SIMILAR SPECIES: Fanleaf Crinklemat (*T. plicata*) and Pacific Gaper (*T. nuttallii*) are similar, but have conspicuously veined leaves. Hairy Crinklemat (*T. hispidissima*) has narrower leaves and bristle-like hairs.

CONSERVATION: G5

Secure, but critically imperiled in Utah.

■ NATIVE
■ RARE OR EXTIRPATED
■ INTRODUCED

The leaves of this western native herb are edible and were a staple food for several Native American groups. The flowers are covered in downy white hairs.

DESCRIPTION: This perennial grows from creeping rootstalks, up to 6–24" (15–60 cm) tall. The flowers are bright white or lavender and grow in a large ball, densely packed together. Each flower is about ⅖" (1 cm) wide, with a long, protruding style and five stamens. Both the stalks and flowers are covered in downy white hairs, growing above a canopy of leaves. The stalks are reddish-green and leafless. The dark green leaves are fuzzy and grow in pairs along the stem, arranged like a feather. The fruit is a spherical capsule with two seeds.

FLOWERING SEASON: April–June.

HABITAT: Mountain woods and thickets.

RANGE: Idaho and Oregon south to Arizona and California.

CONSERVATION: G4
Apparently secure, but vulnerable in Idaho.

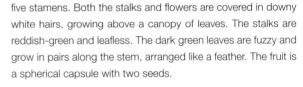

NATIVE
RARE OR EXTIRPATED
INTRODUCED

ALTERNATE NAMES: Shawnee-salad, John's-cabbage

Virginia Waterleaf is native to eastern and central North America. The common name Waterleaf refers to the water-stained appearance of the leaves. The young leaves are edible and can be used in salads. This woodland perennial has been used by Native Americans for medicinal purposes. Tea made from the roots was used to stop bleeding and to treat diarrhea and mouth sores.

DESCRIPTION: This clumping plant has stems with short hairs, growing 1–2½' (30–75 cm) tall. The leaves have blotches that resemble water stains and are pinnately divided, with five to seven lanceolate or ovate, sharply toothed leaflets. Each leaf is 2–5" (5–12.5 cm) long. The flowers are bell-shaped and grow in clusters

of white or dark violet on long stalks that extend beyond the leaves. The flowers have a five-lobed corolla, and five stamens, with hairy filaments extending beyond the petals. Each flower is ¼–½" (6–13 mm) long.

FLOWERING SEASON: May–August.

HABITAT: Moist woods and clearings.

RANGE: Ontario, Quebec, and New England south to North Carolina, west to the Great Plains.

SIMILAR SPECIES: Large-leaved Waterleaf (*H. macrophyllum*) is a similar species, but has larger leaves. Broadleaved Waterleaf (*H. canadense*) has shorter flower stalks, with maple-like leaves. Appendaged Waterleaf (*H. appendiculatum*), also with palmately lobed leaves, has lavender flowers with small appendages alternating with the sepals.

CONSERVATION: **G5**

Secure, but imperiled in Kentucky and New Hampshire; vulnerable in Connecticut, Delaware, and Tennessee.

■ NATIVE
■ RARE OR EXTIRPATED
■ INTRODUCED

ALTERNATE NAMES: Silverleaf Scorpion-weed, *Phacelia leucophylla*

Whiteleaf Phacelia is a variable western native species with deeply veined leaves, which distinguishes it from other *Phacelia* species. The leaves are densely covered in white or silver hairs.

DESCRIPTION: This perennial is covered in bristly white hairs, growing 8–20" (20–50 cm) tall. The leaves are grayish-green with prominent, almost parallel veins. Most of the leaves grow at the base of the plant and are lancolate in shape, with silvery hairs. Each leaf is 1¼–2½" (3.1–6.3 cm) long and may be divided into two small lobes. Small, very hairy, bell-shaped, white or pale purple flowers grow in coils. The corolla is ¼" (6 mm) long, with five round lobes at the end, and five protruding stamens. The fruit, a capsule, is also very hairy.

FLOWERING SEASON: May–July.

HABITAT: Dry, rocky places among sagebrush and in coniferous forests.

RANGE: Southern British Columbia to northern California; east to western Nebraska, Colorado, and Alberta.

CAUTION: Some *Phacelia* species can cause skin irritation when touched, similar to poison ivy.

CONSERVATION: G5

Secure, but vulnerable in Alberta.

■ NATIVE
■ RARE OR EXTIRPATED
■ INTRODUCED

ALTERNATE NAMES: Threadleaf Scorpion-weed, Linearleaf Phacelia

All parts of this showy annual are covered in white hairs. It has a wide corolla and narrow leaves, distinguishing it from other *Phacelia* species.

DESCRIPTION: This plant has slender, erect stems, reaching 4–20" (10–50 cm) tall. The leaves are narrowly lanceolate and hairy, sometimes with one to four pairs of small lobes appearing in the lower half of the plant. The flowers are reddish-lavender and broadly bell-shaped, growing in loose coils at the top of the stems. The corolla is ⅜–¾" (9–20 mm) wide, with five round lobes and five stamens. Each leaf is ½–4" (1.3–10 cm) long. The fruit, a small capsule, has 6–15 seeds with pitted surfaces.

FLOWERING SEASON: April–June.

HABITAT: Among brush and in open grassy areas in foothills and on plains, woodlands.

RANGE: Southern British Columbia and Alberta to northern California; east to Utah, Wyoming, and western South Dakota.

CAUTION: Some *Phacelia* species can cause skin irritation when touched, similar to poison ivy.

CONSERVATION: G5
Secure, but vulnerable in Alberta.

■ NATIVE
■ RARE OR EXTIRPATED
■ INTRODUCED

ALTERNATE NAMES: Purple Fringe, Silky Scorpionweed

This native western perennial's Latin name comes from *sericeus,* meaning "silky," referring to the fine hairs on the leaves, stems, and flowers. The flowers grow in a cylindrical cluster, which distinguishes this plant from other phacelias. It is an important plant for several species of native bumblebees (*Bombus* spp.).

DESCRIPTION: All parts of this plant are covered in silky silver-gray hairs. It has erect stems that grow from a woody base, reaching 16" (40 cm) tall. Most of the leaves grow at the base of the plant. The leaves are 1–4" (2.5–10 cm) long, broadly lanceolate, and feather-like. Purple or dark blue bell-shaped flowers grow in short coils in a tight cylindrical cluster, which appear fuzzy or fringed because of the long, protruding stamens. The corolla is covered in hairs, growing about ¼" (6 mm) long, with five round lobes.

FLOWERING SEASON: June–August.

HABITAT: Open or wooded rocky places in the mountains, often at high elevations.

RANGE: Alaska to southern British Columbia and Alberta, south to northeastern California, east to the mountains in Wyoming and Colorado.

CAUTION: Some *Phacelia* species can cause skin irritation when touched, similar to poison ivy.

CONSERVATION: **G5**
Secure, but imperiled in Alaska.

NATIVE
RARE OR EXTIRPATED
INTRODUCED

ALTERNATE NAMES: Desert Fiesta-flower

This native plant of the Southwest has bluish flowers, covered with backward-pointing bristles that act like Velcro, sticking to fabric and animal fur. In the time of Spanish ranchos, women would place these flowers on their dresses for decoration. These plants grow especially well after wildfires.

DESCRIPTION: This sprawling annual reaches 2' (61 cm) long. It has weak, sprawling stems covered in curved hooks or hairs. The leaves are opposite toward the bottom of the stem and alternate toward the top. Each leaf is deeply lobed, ½–3" (1.5–8 cm) long. The flowers are blue or purple, bell-shaped, and covered in hairs. Each flower is up to ½" (1.5 cm) long and 1" (3 cm) wide.

FLOWERING SEASON: February–April.

HABITAT: Hills, rocky slopes, woods, plains, streambanks.

RANGE: Canyons of the Colorado River in Arizona, California, and Nevada.

SIMILAR SPECIES: Distant Phacelia (*Phacelia distans*) is a similar sprawling species, but grows slightly larger.

CONSERVATION: G5
Secure.

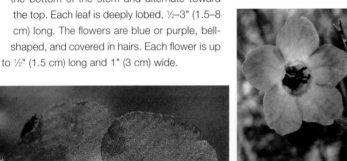

■ NATIVE
■ RARE OR EXTIRPATED
■ INTRODUCED

ALTERNATE NAMES: Quail Plant, Seaside Heliotrope

This widespread native perennial has attractive flowers that grow in a coil. It gets the colloquial name Quail Plant from the fact that quails often eat its fruit. The plant grows well in salty soils.

DESCRIPTION: This fleshy, bluish-green creeping plant stretches 4' (1.2 m) long with leafy stems that lie mostly on the ground. The flowers are white or purple and grow in paired coils, with flowering branches reaching about 16" (40 cm) long. The corolla is funnel-shaped, with five round lobes, about ¼–⅜" (5–9 mm) wide. The leaves are ½–1½" (1.5–4 cm) long, spatula-shaped,

and sometimes broader toward the tip. The fruit is divided into four small nutlets.

FLOWERING SEASON: March–October.

HABITAT: Open, alkaline or saline soil, often in sand or clay, or dried ponds.

RANGE: Western United States, east to the Great Plains and across the southern United States.

CAUTION: Consuming medicinal teas made from the leaves of this plant may cause liver disease.

CONSERVATION: G5

Globally secure but critically imperiled in Virginia and Manitoba, imperiled in Oregon, and vulnerable in Alberta.

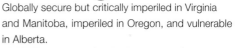

■ NATIVE
■ RARE OR EXTIRPATED
■ INTRODUCED

ALTERNATE NAMES: Hedge False Bindweed

This widespread North American native perennial can be weedy. It grows in a twine and often engulfs or suffocates nearby plants. It is difficult to eradicate without removing its roots. The genus name is from the Greek words *calyc* for "calyx" and *steg* for "cover," referring to the modified leaves that surround and cover the calyx. The species name, *sepium*, means "of hedges," and refers to this plant's natural habitat.

DESCRIPTION: This smooth, twining vine grows 3–10' (90–300 cm) long. It has white or pink funnel-shaped flowers with white stripes. Before the flowers bloom and open, the buds are hidden by large, green bracts. Each flower is 2–3" (5–7.5 cm) long when open, with a five-lobed calyx, five united petals, and two rounded stigmas. The leaves grow in a spiral. Each leaf is 2–4" (5–10 cm) long and arrow-shaped or triangular. The fruit is a spherical capsule.

FLOWERING SEASON: May–September.

HABITAT: Hedges, open woods, moist soil, thickets, roadsides, and waste places.

RANGE: Throughout North America, except the Far North.

SIMILAR SPECIES: Hedge Bindweed's flowers are very similar to morning glories (*Ipomoea* spp.) but have two stigmas, not one. Erect Bindweed (*C. spithamaeus*) is also similar, but its leaves are oval and the plant does not twine. Field Bindweed (*Convolvulus arvensis)* is a similar relative, but its calyx is only partially hidden by bracts.

CONSERVATION: Ⓖ5

Secure, but imperiled in Wyoming and vulnerable in North Carolina and Alberta.

■ NATIVE
■ RARE OR EXTIRPATED
■ INTRODUCED

ALTERNATE NAMES: Seashore False Bindweed

Beach Morning Glory grows on sandy beaches in temperate climates around the world, including on the West Coast in North America. It is a perennial vine that grows in beach sand and produces attractive flowers that resemble those of true morning glories (*Ipomoea* spp.).

DESCRIPTION: This creeper plant has trailing stems that grow from deep rootstalks. The branches grow up to 3" (7.5 cm) high, and stems to 20" (50 cm) long. The leaves are thick and kidney-shaped, 1–2" (2.5–5 cm) wide. The flowers grow from leaf axils in a funnel shape and appear pink or rose in color on short stalks. The corolla is 1½–2½" (3.8–6.3 cm) wide, with five short points. The calyx is partially hidden beneath two bracts.

FLOWERING SEASON: April–September.

HABITAT: Beach sands.

RANGE: Along the Pacific Coast from British Columbia to southern California.

SIMILAR SPECIES: The flowers of *C. soldanella* are very similar to those of morning glories (*Ipomoea* spp.), but unlike morning glories, the flowers stay open most of the day.

CONSERVATION: G5
Secure, but vulnerable in British Columbia.

- ■ NATIVE
- ■ RARE OR EXTIRPATED
- ■ INTRODUCED

ALTERNATE NAMES: Field Bindweed

This Eurasian native is a noxious weed in North America. It has very deep roots that make it difficult to eradicate. *Convolvulus arvensis* can be confused with a number of similar weeds and native plants, but the flowers are generally smaller than others and grow tighter together.

DESCRIPTION: This perennial has long-trailing and twining stems, growing 1–3' (30–90 cm) long. The young stems exude a milky sap when broken. The leaves are generally triangular, but sometimes arrow-shaped or ovate, about ¾–1½" (2–3.8 cm) long each. The flowers are white or pinkish and funnel-shaped. Each flower is about 1" (2.5 cm) wide. The corolla has five veins, leading to low lobes on the edge, with two narrow bracts about ¼" (6 mm) long on stalks well below the calyx.

FLOWERING SEASON: May–October.

HABITAT: Fields, lots, gardens, and roadsides.

RANGE: Introduced throughout much of North America.

SIMILAR SPECIES: Hedge Bindweed (*Calystegia sepium*) is a similar, widespread North American native, but Bindweed's flowers are smaller. Bindweed's flowers are also smaller than those of Wild Sweet Potato (*Ipomoea pandurata*), a native plant of the central and eastern U.S.

CONSERVATION:
The status of this introduced species is unranked.

■ NATIVE
■ RARE OR EXTIRPATED
■ INTRODUCED

ALTERNATE NAMES: Gronovius Dodder, Swamp Dodder, Common Dodder

This widespread native plant grows near swamps. All dodders are parasites, attaching to nearby plants for nutrients. Cultivating this plant may be prohibited in some states.

DESCRIPTION: This annual grows up to 4' (1.2 m) tall with wiry yellow or orange twisting stems. The leaves are scale-like or absent. Its white flowers grow in dense clusters; each is about ⅛" (3 mm) long, with five triangular-shaped lobes that fuse together. The calyx is triangular or roundish, about half as long as the floral tube, with tips. The flowers have five stamens that appear yellow at the tips.

FLOWERING SEASON: July–August.

HABITAT: Moist areas, swamps.

RANGE: Widespread, mostly in northern North America.

CONSERVATION: G5
Globally secure but critically imperiled in Alberta and Newfoundland; imperiled in Kansas, Nebraska, Prince Edward Island, and Saskatchewan; vulnerable in Manitoba.

■ NATIVE
■ RARE OR EXTIRPATED
■ INTRODUCED

The genus name comes from the Latin word *evolvo,* meaning "to untwist or unravel," referring to the non-vining habit of this species. Most true morning glories are vines with a twining growth habit. This species instead grows as a subshrub.

DESCRIPTION: This morning glory-like perennial does not vine and instead grows as a densely hairy subshrub, about 20" (51 cm) tall. The stems have hairy, silvery-green leaves, about 1" (2.5 cm) long. The flowers are very small and bell-shaped, either blue or lavender. The flowers open just in the morning and close at night and on cloudy days. Each flower is about ½" (1.3 cm) wide.

FLOWERING SEASON: June–October.

HABITAT: Dry, open sandy or rocky prairies.

RANGE: Montana and North Dakota south to Texas and Arizona.

CONSERVATION: G5

Secure, but vulnerable in Arkansas, Montana, Wyoming, and Tennessee.

■ NATIVE
■ RARE OR EXTIRPATED
■ INTRODUCED

ALTERNATE NAMES: Redstar

This showy vine is introduced in North America and has become a weed in many areas. The genus name derives from the Greek word for "worm-like" and refers to the plant's twining habit. The species name is Latin for "scarlet," and refers to the red color of the flowers.

DESCRIPTION: This twining, annual vine grows 3–9' (90–270 cm) long. The leaves are heart-shaped, and occasionally lobed, about 1½–4" (4–10 cm) long. It has small, scarlet flowers, each with a flaring tube that flattens into five shallow lobes. Each flower is about ¾" (2 cm) wide, with five sepals and bristle-like tips. The stamens and stigma extend beyond the flower.

FLOWERING SEASON: July–October.

HABITAT: Thickets, disturbed areas, and roadsides.

RANGE: This tropical plant has been introduced throughout much of the central and eastern U.S. and has naturalized as far north as Iowa, Michigan, and New York.

SIMILAR SPECIES: Scarlet Cypress Vine (*I. quamoclit*) is a related species, also introduced, but its leaves are divided into very narrow segments, like teeth on a comb.

CONSERVATION:
The status of this introduced species is unranked.

This twining plant is an aggressive weed in some areas. The species gets its name from English Ivy (*Hedera helix*), which has similar leaves. Ivy-leaf Morning Glory's foliage and hairiness of the stems may vary.

DESCRIPTION: A leafy, hairy, annual vine that grows 3–6' (90–180 cm) long. The flowers are bluish and funnel shaped, with a white inside, turning rose-purple in the afternoon. Each flower is 1½" (4 cm) wide and has five sepals with long tips, and a narrow hairy base. The corolla has five united petals. The leaves are 2–5" (5–12.5 cm) long and wide, with three deep lobes, tapering to points.

FLOWERING SEASON: July–October.

HABITAT: Fields and disturbed areas.

RANGE: Ontario, east to Maine, south to Florida, west to Texas, and north to North Dakota.

SIMILAR SPECIES: Ivy-leaf Morning Glory has three-lobed leaves, which distinguishes it from other morning glories (*Ipomoea* spp.) and bindweeds (*Calystegia* spp.).

CONSERVATION: G5
Secure.

■ NATIVE
■ RARE OR EXTIRPATED
■ INTRODUCED

ALTERNATE NAMES: Whitestar

This native annual grows mainly in the eastern United States. It is a climber that grows from a taproot and produces a potato-like tuber that Native Americans consumed. The small, mostly white flowers open mainly in the morning and have no scent.

DESCRIPTION: Small White Morning Glory has twining, some-what hairy vines that can grow up to 7' (2.1 m) long. The leaves are alternate, with a somewhat hairy top and a hairy underside. Each leaf is 2–4" (5–10 cm) long, on delicate and somewhat hairy stalks. The larger leaves are usually heart-shaped, while the smaller leaves are egg-shaped. Some three-lobed leaves may also be present. Each stalk has one to three funnel-shaped flowers that are mostly white, but occasionally light purple or pink. Each flower is about 1" (2.5 cm) long and ¾" (2 cm) wide. There are five petals, one white knobby stigma, one white style, five stamens, and five light green sepals. The seed capsules are spherical and hairy, surrounded by five small bracts.

FLOWERING SEASON: July–October.

HABITAT: Moist fields, roadsides, disturbed areas.

RANGE: Massachusetts to Florida, west to Kansas, Oklahoma, and Texas.

CAUTION: This plant can have a high alkaloid content, making it toxic to small animals.

SIMILAR SPECIES: Field Bindweed (*Convolvulus arvensis*) has similar flowers, but they are wider than those of *I. lacunosa* when fully open.

CONSERVATION: G5

Secure, but critically imperiled in Iowa and imperiled in West Virginia.

COMMON MORNING GLORY *Ipomoea purpurea*

ALTERNATE NAMES: Tall Morning Glory

Like most morning glories, the attractive flowers of this intro-duced annual open only in the morning. This species is native to Mexico and Central America and was formerly called *Convolvulus purpureus*. Its specific epithet *purpurea* means "purple," a reference to its purplish flowers, although they may be blue, pink, or white, especially in some cultivated varieties.

DESCRIPTION: This vine has large, funnel-shaped purple, blue, red, or white corollas and grows 3–10' (90–300 cm) long. The leaves are heart-shaped or three-lobed. Each leaf is 1¼–6" (3–15 cm) long. The corolla is 1½–2½" (4–6.5 cm) wide, with five sepals, each about ½" (1.5 cm) long. The corolla has stiff hairs on the back and narrow, green tips.

FLOWERING SEASON: July–October.

HABITAT: Fields, roadsides, and disturbed areas.

RANGE: Introduced throughout most of the United States and in Ontario.

SIMILAR SPECIES: Ivy-leaf Morning Glory (*I. hederacea*) is a similar plant with narrower sepal tips that curve outward and are longer than the body of the sepal. Japanese Morning Glory (*I. nil*), also has similar flowers, but the sepals are straighter.

CONSERVATION:

This introduced species is unranked.

- ■ NATIVE
- ■ RARE OR EXTIRPATED
- ■ INTRODUCED

ALTERNATE NAMES: Smallflower Morning Glory

This native annual is mainly found in the Gulf Coast states, with a few sites as far north as Missouri and Pennsylvania. The flowers open at night and close when the sun rises. Both the leaves and sap have been used for a variety of medicinal purposes; the leaves are used to clean wounds, and the scent they produce when crushed is used to alleviate headaches. An herbaceous annual vine with hairy, twining stems growing up to 6' (2 m) tall.

The leaves are hairy and alternate. Each leaf is 1–5" (3–12.7 cm) long and ⅜–3½" (1–9 cm) wide and ovate to elliptical in shape. The flowers are bright blue and have five sepals and long hairs. The corolla is funnel-shaped with a white center.

FLOWERING SEASON: May–October.

HABITAT: Roadsides, fields, and disturbed areas.

RANGE: The Gulf Coast states; also found sparingly north to Missouri and Pennsylvania.

CONSERVATION:
This species is unranked. It may be considered a noxious weed in some areas where it is not native.

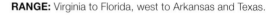

■ NATIVE
■ RARE OR EXTIRPATED
■ INTRODUCED

ALTERNATE NAMES: Southern Morning Glory

The white flowers of this native perennial of the Southeast close quickly in hot, direct sunlight. The species name comes from the Latin *humi*, meaning "ground" or "earth," and *strat*, meaning "a covering," referring to the plant's natural habitat.

DESCRIPTION: This vine has a hairy stem with attractive white or pink flowers that are funnel-shaped, about ⅓–1" (8–24 mm) long. The flowers have a two-parted style that unites at the base. The leaves are oblong to elliptical and alternate.

FLOWERING SEASON: May–July.

HABITAT: Pine oak woodlands, sandhills, dry hammocks, roadsides, and waste ground.

RANGE: Virginia to Florida, west to Arkansas and Texas.

CONSERVATION: G4
Apparently secure, but critically imperiled in Tennessee and vulnerable in North Carolina and Virginia.

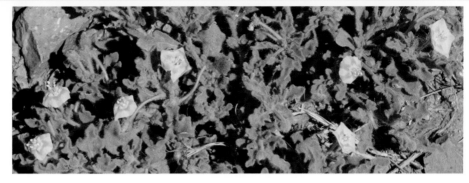

■ NATIVE
■ RARE OR EXTIRPATED
■ INTRODUCED

ALTERNATE NAMES: Hairy False Nightshade, Dingy Chamaesaracha

This herb native to southwestern plains and deserts is fairly common but often inconspicuous. It is a low-growing, dull green plant. The common name is a nod to this plant's hairy, pentagonal-shaped flower.

DESCRIPTION: This low-growing perennial is covered with fine, sticky hairs. The stems mostly hug the ground and grow 1' (30 cm) long. The leaves are dull green, pointed at the tip and tapering to the base, with wavy edges and sometimes low lobes. Each leaf is about 1½" (4 cm) long. A dingy whitish-green pentagonal flower grows in each upper leaf axil. The corolla is hairy and about ½" (1.5 cm) wide, with five lobes and five yellowish bands that grow from the center to the tips of the corolla. It has five slender stamens that alternate with greenish-yellow pads. The fruit, a berry, is ⅛–⅜" (4–8 mm) wide and tightly enclosed by the calyx.

FLOWERING SEASON: May–September.

HABITAT: Plains and deserts.

RANGE: Southern Arizona, southern New Mexico, and Texas.

CONSERVATION: G4
Apparently secure.

■ NATIVE
■ RARE OR EXTIRPATED
■ INTRODUCED

This tropical annual has prickly fruit and large, trumpet-shaped flowers that open at night. All parts of Jimsonweed are poisonous and psychoactive if consumed. The common name is derived from Jamestown, Virginia, where British soldiers suppressing Bacon's Rebellion in the late 1600s consumed the plant and spent nearly two weeks experiencing hallucinogenic effects.

DESCRIPTION: This tall plant has greenish or purplish stems and trumpet-shaped, white or violet flowers. It grows up to 1–5' (30–150 cm) tall. The leaves are ovate and irregularly lobed. Each leaf is about 8" (20 cm) long. The flowers are about 3–4" (7.5–10 cm) wide, with a green, tubular, five-lobed calyx and a five-lobed corolla that's funnel-shaped, about twice as long as the calyx. The fruit, an egg-shaped capsule, is covered in prickles, about 2" (5 cm) wide.

FLOWERING SEASON: July–October.

HABITAT: Waste places, fields, and barnyards.

RANGE: This plant's native range is uncertain, but it is likely native to Central America. It is an aggressive and often invasive weed, and has naturalized throughout temperate North America.

CAUTION: All parts of this plant, including the fruits, leaves, flowers, and nectar, are highly poisonous and may be fatal to humans and animals if consumed. Touching the leaves or flowers may also cause skin irritation.

SIMILAR SPECIES: Angel Trumpet (*D. wrightii*) is a similar species, but its flowers are about twice the size of those of Jimsonweed. Entire-leaved Thorn Apple (*D. metel*) is also similar, but has double flowers and grows slightly taller, up to about 7' (2.1 m).

CONSERVATION:
This tropical native species is considered introduced in North America and unranked. It is an aggressive weed in many areas.

■ NATIVE
■ RARE OR EXTIRPATED
■ INTRODUCED

The flowers of the western native annual normally open at dusk and remain open through the night to attract nocturnal hawkmoth (*Manduca* spp.) pollinators. The plants also host hawkmoth larvae and appear to adjust to predation by adjusting their schedules to open their flowers during the day to attract hummingbirds instead. The plant has been used for a variety of medicinal purposes by Native Americans. Smoke and tea made from this tobacco have been used to treat earaches and deafness, stomachaches, fevers, and swelling. It has also been smoked ceremonially.

DESCRIPTION: Coyote Tobacco is a sticky plant, covered sparsely in white hairs with erect stems that grow 2–5' (60–150 cm) tall. The lower leaves are oval while the upper leaves are more narrow in shape. Each leaf is about 4" (10 cm) long. The flowers are pink or green and tubular in shape. Each flower is

about 1¼" (3 cm) long, with five white lobes. The sepals are triangular. The fruit, a capsule, is about ⅖" (1 cm) long.

FLOWERING SEASON: May–October.

HABITAT: Well-drained slopes, disturbed areas, and after fires.

RANGE: From California north to British Columbia and east to New Mexico, Colorado, and northern Montana.

CAUTION: Like other tobaccos, Coyote Tobacco contains nicotine, which can be addictive.

CONSERVATION: G4
Apparently secure, but imperiled in Washington and Wyoming.

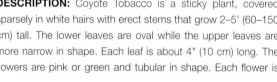

■ NATIVE
■ RARE OR EXTIRPATED
■ INTRODUCED

This sticky perennial has up to a dozen bell-like flowers on branching stems with colored centers and inflated bladders around the fruit. It is a widespread North American native plant sometimes divided into four subspecies based on variations in the thickness and margins of the leaves.

DESCRIPTION: The stems of this plant are covered in soft hairs, growing 1–3' (30–90 cm) tall. It has bell-shaped, greenish-yellow flowers with a purplish-brown center in each leaf axil. The leaves are heart-shaped with a few teeth, about 1–4" (2.5–10 cm) long. The flowers are ¾" (2 cm) wide, with a five-lobed corolla, and five stamens with yellow anthers. The fruit is a tomato-like, yellow berry.

FLOWERING SEASON: June–September.

HABITAT: Dry woods and clearings.

RANGE: Ontario east to Nova Scotia, south to Florida, west to Texas, north to North Dakota; also in much of the interior western U.S.

CAUTION: The leaves and the unripe fruit of this plant are poisonous, but the ripe fruit is edible and often used in jams or pies. Animals have been poisoned by feeding on the plants but generally avoid them unless other forage is scarce.

SIMILAR SPECIES: Virginia Groundcherry (*P. virginiana*) has very similar flowers, but the leaves are narrower and mostly toothless. Long-leaf Groundcherry (*P. longifolia*) also has similar flowers but is mostly hairless. Chinese Lantern Plant (*P. alkekengi*) is another similar plant with white flowers and a showy, inflated, orange calyx.

CONSERVATION: G5

Secure, but imperiled in Wyoming, and vulnerable in North Carolina and Quebec.

■ NATIVE
■ RARE OR EXTIRPATED
■ INTRODUCED

ALTERNATE NAMES: Horsenettle, Bull Nettle

This plant is a nightshade rather than a true nettle, but it is similarly covered in sharp prickles. This native perennial has attractive flowers but has a tendency to become invasive. It is often considered a weed, spreading both by seed and underground rhizomes, and its deep roots make it difficult to eradicate. Ground birds in the family *Phasianidae* (such as pheasants, quail, and turkeys) eat the fruits.

DESCRIPTION: The stems of this plant are erect and prickly, growing 1–3' (30–90 cm) tall. Scattered white leaves cover the spine and the underside of leaves. The leaves are rough and covered in prickles. Each leaf is elliptical-oblong, and coarsely lobed, about 3–5" (7.5–12.5 cm) long each. The flowers are star-like and white or pale. Each flower is about ¾–1¼" (2–3 cm) wide, with five petals, and five stamens with yellow, elongated anthers forming a central cone. The fruit is yellow and resembles a tomato, about ¾" (2 cm) wide.

FLOWERING SEASON: May–October.

HABITAT: Fields, waste places, and cultivated sites.

RANGE: Probably native to the eastern U.S., but widely naturalized throughout much of temperate North America.

CAUTION: The berries of this nightshade resemble small, yellow tomatoes but contain toxins that may be fatal to children if ingested. This plant is also poisonous to other mammals if ingested, but livestock tend to avoid the plant because of its prickles.

SIMILAR SPECIES: White Horsenettle (*S. elaeagnifolium*) is a similar plant but has silvery foliage. Buffalo Bur (*S. rostratum*) is also similar, but it's an annual, not a perennial, and has bright yellow flowers.

CONSERVATION: **G5**

Secure. It is considered a noxious weed in several U.S. states.

■ NATIVE
■ RARE OR EXTIRPATED
■ INTRODUCED

ALTERNATE NAMES: Bittersweet Nightshade, Deadly Night-shade

This climbing vine is native to Eurasia and Northern Africa, and is widely introduced in North America. The leaves and unripe fruit are poisonous and should not be eaten. The plant has been used in herbalism since the Ancient Greeks, and in the Middle Ages it was hung around homes in discrete places to protect against witchcraft. Modern medicine has identified compounds in the plant that may combat ringworm and other skin conditions.

DESCRIPTION: This climbing woody perennial vine grows 2–8' (60–240 cm) long. It has flat, loose clusters of drooping, blue or violet, star-shaped flowers. Each flower is ½" (1.5 cm)

wide, with a five-lobed corolla, and five stamens, with yellow anthers that form a cone in the center. The leaves are up to 3½" (9 cm) long, and halberd-shaped, with two basal lobes. The fruit looks like a tomato; it starts green before turning bright red.

FLOWERING SEASON: May–September.

HABITAT: Thickets and clearings.

RANGE: Introduced across North America, widely naturalized in southern Canada and the northern U.S.

CAUTION: The leaves and berries contain a poisonous alkaloid, solanine, and should not be ingested. Although the plant is sometimes called Deadly Nightshade, its toxin is not usually fatal. However, the berries are attractive to children and can cause poisoning if eaten in quantity.

CONSERVATION:
This introduced species is unranked.

■ NATIVE
■ RARE OR EXTIRPATED
■ INTRODUCED

ALTERNATE NAMES: Eastern Black Nightshade, *Solanum ptychanthum*

This mostly eastern native is commonly found mixed among crops of related nightshades that are grown agriculturally like potatoes and tomatoes. *Solanum emulans* is part of a complex of related black nightshade species. It was formerly called *S. ptychanthum* and still appears under that name in many sources; *ptychanthum* is from Greek words meaning "a fold" and "flower," referring to this plant's curving flowers.

DESCRIPTION: Common Nightshade is a smooth plant that grows to a height of 1–2½' (30–75 cm). It has a few white star-like flowers that grow in umbels and droop. Each flower is ⅜" (8 mm) wide, with five petals that curve backward, and five stamens with yellow anthers that form a cone in the center. The leaves are 2–4" (5–10 cm) long, thin, ovate, pointed, and wavy-toothed. The fruit is a berry that appears green before turning black.

FLOWERING SEASON: June–November.

HABITAT: Cultivated and disturbed areas and open woods.

RANGE: Alberta east to Newfoundland, south to Florida, west to Texas, and north to North Dakota; also in parts of the Interior West.

CAUTION: The leaves and berries of this plant contain the poisonous alkaloid solanine, which can be toxic if consumed in high quantities.

SIMILAR SPECIES: This plant is mislabeled in many collections as *S. americanum* or *S. nigrum*, which are different species. American Black Nightshade (*S. americanum*) is a more southern species and European Black Nightshade (*S. nigrum*) is the Eurasian equivalent. Cut-leaved Nightshade (*S. triflorum*) is a similar species, but has deeply dissected leaves. Hairy Nightshade (*S. villosum*) is also similar, but has hairy stems and yellow or red berries.

CONSERVATION: G5
Secure, but imperiled in Wyoming.

ALTERNATE NAMES: Buffalobur Nightshade, Horned Night-shade

This annual herb is a leafy weed indigenous to the central United States and adventive elsewhere throughout North America. It is densely covered in golden-yellow prickles that protect the plant from most browsers, and is highly toxic if consumed. This plant tends to spread into spoiled sites and can survive with virtually no water and poor soil.

DESCRIPTION: Buffalo Bur has dense golden-yellow prickles on the stems and calyx, growing 16–32" (40–80 cm) tall. The leaves are also covered in prickles. Each leaf is 2–6" (5–15 cm) long, with blades that are deeply parted. The flowers are yellow and star-like. Each corolla is ¾–1" (2–2.5 cm) wide, with five points and a slender, cone-like center formed from its unequal anthers. The fruit, a berry, is enclosed in the calyx.

FLOWERING SEASON: May–September.

HABITAT: Roadsides, edges of fields, old lots, and other heavily disturbed sites.

RANGE: Nnative to the central United States and naturalized throughout much of southern Canada and the U.S.; also in northern Mexico.

CAUTION: The leaves and roots can be highly toxic to livestock. It is also covered with sharp prickles and should not be touched.

SIMILAR SPECIES: Melonleaf Nightshade (*S. citrullifolium*) is a similar species, but has blue-violet corollas.

CONSERVATION: G5
Secure, but vulnerable in Wyoming.

■ NATIVE
■ RARE OR EXTIRPATED
■ INTRODUCED

ALTERNATE NAMES: One-flower Hydrolea

This native perennial herb is covered in sharp thorns and almost always occurs in wetlands. Many small mammals and native birds use this thorny plant for both food and cover.

DESCRIPTION: One-flower False Fiddleleaf has blue flowers that grow in clusters in thorny leaf axils, up to 4–30" (10–76 cm) tall. The leaves are slender, alternate, simple and untoothed. The corolla is deeply lobed in five sections, about ⅖" (11 mm) long.

FLOWERING SEASON: June–October.

HABITAT: Wet woods, ditches, and disturbed wetlands.

RANGE: Found primarily in the Lower Mississippi River region; it occurs from Indiana and Illinois south to Alabama and East Texas.

CAUTION: One-flower False Fiddleleaf has very sharp thorns and should not be touched.

SIMILAR SPECIES: Blue Waterleaf (*H. ovata*) is a similar plant but with smaller flowers.

CONSERVATION: G5
Secure, but critically imperiled in Illinois and Kentucky.

■ NATIVE
■ RARE OR EXTIRPATED
■ INTRODUCED

ALTERNATE NAMES: Broom Menodora, Tenfinger Menodora

This native southwestern shrub has rough leaves and multi-branched, broom-like stems that produce loose clusters of yellow flowers at the ends.

DESCRIPTION: This perennial plant has many broom-like stems and grows to a height of 5–14" (12.5–35 cm). The upper leaves are alternate and smaller than the lower leaves, while the lower leaves are opposite. Each leaf is ½–1½" (1.5–4 cm) long and broadly lanceolate. Pale yellow flowers grow in loose terminal clusters. Each flower is ½–¾" (1.5–2 cm) wide, with a short narrow corolla tube and five spreading lobes, with two hidden stamens. The fruit is a spherical capsule, nearly ¼" (6 mm) wide.

FLOWERING SEASON: March–September.

HABITAT: Grassy slopes, brushy deserts, dry mesas, canyons, slopes, rocky hillsides.

RANGE: Southeastern California and southern Utah east to western Texas, south to northern Mexico.

CONSERVATION: G5
Secure, but critically imperiled in Utah and imperiled in Colorado.

This native perennial forms dense tufts of short stems with white flowers that appear at the end of the growing season. It is an important food plant to Sachem butterflies (*Atalopedes campestris*) in the South. This is the only species in the genus, which has at times been placed in several different families.

DESCRIPTION: This multi-branched, spreading plant grows 8–12" (20–30 cm) tall. It has very small white flowers that typically grow in dense terminal clusters, although solitary flowers are possible. The corollas are 2–3 mm long, with four lobes and four stamens. The throat is covered in short hairs. The leaves are small, about ½–1" (1.5–2.5 cm) and very narrow, opposite and stalkless. The foliage turns red or brown in the fall.

FLOWERING SEASON: June–October.

HABITAT: Open areas, pond margins, disturbed areas.

RANGE: Eastern U.S. from New York to Florida, west to Texas.

SIMILAR SPECIES: This plant may look similar to some bluets like Longleaf Bluet (*Houstonia longifolia*) and Narrowleaf Summer Bluet (*Stenaria nigricans*), but Juniper-leaf's stamens don't protrude as in these species.

CONSERVATION: ⬤G5
Secure, but critically imperiled in New Jersey.

■ NATIVE
■ RARE OR EXTIRPATED
■ INTRODUCED

ALTERNATE NAMES: Herb-of-grace, Coastal Water-hyssop

This native perennial herb is found in wetlands across the southern tier of the United States and on every continent except Antarctica. Water Hyssop has been used in Ayurvedic traditional medicine for centuries to improve memory, reduce anxiety, and treat epilepsy, although the health benefits of *Bacopa monnieri* are unproven.

DESCRIPTION: This plant is a creeping, mat-forming, aquatic or semi-aquatic perennial. The small, white flowers are bell-shaped and bloom in the spring until frost and are often tinged with blue or pink, with four or five petals. The small, succulent leaves are evergreen, oval and shiny, about 4–6 mm thick.

FLOWERING SEASON: April–October.

HABITAT: Wet, tropical areas, wet sands, mudflats, and ponds or stream margins.

RANGE: The coastal plain from Maryland to Texas; also in southern California and Arizona.

CAUTION: *Bacopa monnieri* may cause nausea and upset stomach if consumed.

CONSERVATION: G5

Secure, but critically imperiled in Oklahoma; imperiled in Arizona; and vulnerable in North Carolina and Virginia. It is possibly extirpated in Maryland.

■ NATIVE
■ RARE OR EXTIRPATED
■ INTRODUCED

This native Eastern perennial is an important plant for butterflies, notably the Baltimore Checkerspot (*Euphydryas phaeton*), which uses White Turtlehead as its primary larval host plant. The common name comes from the appearance of its flower petals, which look like a tortoise head. The genus is named for Chelone, a nymph in Greek mythology who was transformed into a tortoise as a form of punishment.

DESCRIPTION: This plant grows 1–3' (30–90 cm), with white or lavender-tinged flowers with two lips that grow in terminal clusters and look like turtle heads. Each flower is 1–1½" (2.5–4 cm) long, with the upper corolla lip arching over the hairy lower lip, and five stamens, one of which is short and sterile. The leaves are 3–6" (8–15 cm) long, opposite, lanceolate, sharply toothed. The fruit is a capsule.

FLOWERING SEASON: July–September.

HABITAT: Wet thickets, streambanks, and low ground.

RANGE: Manitoba east to Newfoundland and south to Arkansas, central Mississippi and Alabama, and Georgia.

CONSERVATION: G5
Globally secure but critically imperiled in Arkansas; imperiled in Manitoba; vulnerable in Iowa, Mississippi, North Carolina, and Rhode Island.

■ NATIVE
■ RARE OR EXTIRPATED
■ INTRODUCED

ALTERNATE NAMES: Small-flower Blue-eyed Mary

This native annual is a fairly common plant in the western and northern parts of North America and often grows in masses. It has tiny white-and-blue flowers.

DESCRIPTION: This small, multi-branched annual has reddish stems and grows 2–16" (5–40 cm) tall. The leaves are narrowly lanceolate, appearing opposite near the base of the plant and growing in whorls of four in the flower cluster. Each leaf is about 2" (5 cm) long. The tiny, blue-and-white flowers grow in open clusters. The corolla is about ¼" (6 mm) wide with five lobes. The upper corolla is white and tinged with violet at the tip, with two lobes that bend upward. The lower corolla lip has three lobes, with two blue-violet lobes projecting forward and a middle lobe folded between them.

FLOWERING SEASON: April–July.

HABITAT: Open, gravelly flats and banks, often in sparse grass.

RANGE: Alaska south to California and east to Colorado and the Great Plains, and in the north to Michigan and Ontario.

SIMILAR SPECIES: Little Tonella (*Tonella tenella*) is a similar plant but its corolla is about half the size.

CONSERVATION: Ⓖ5
Secure, but critically imperiled in Manitoba; imperiled in Michigan and North Dakota; vulnerable in Saskatchewan and Yukon.

■ NATIVE
■ RARE OR EXTIRPATED
■ INTRODUCED

ALTERNATE NAMES: Purple Foxglove

This species is found across temperate Europe and has been introduced and naturalized in parts of North America. Foxglove is a biennial often grown as an ornamental for its showy purplish flowers that dangle in clusters on one side of the stem.

DESCRIPTION: Foxglove is a tall plant with large, spiraled leaves and attractive flowers that hang to one side, growing 2–7' (60–210 cm) tall. The leaves are ovate with toothed margins. Each leaf is up to 1' (30 cm) long, with the largest leaves growing at the base. The flowers are pinkish or purple and grow in elongated clusters atop the leafy stem. Each flower is tubular. The corolla is bilaterally symmetrical, about 1½–2½" (4–6.5 cm) long, and ¾" (2 cm) wide. The flower tube is heavily spotted with red or maroon dots on the lower inside.

FLOWERING SEASON: June–July.

HABITAT: Along roadsides and other disturbed, open sites.

RANGE: Introduced in North America. It is naturalized from British Columbia to California west of the Cascade Mountains and the Sierra Nevada, and spreading eastward into Montana and Colorado. It is also found in the East from Maine to Maryland.

CAUTION: The heart medication digitalis is derived from a relative of this plant, but the compounds found in Foxglove can be fatal if not properly administered. People have died from consuming teas and salads made from the leaves. Livestock tend to avoid browsing these plants, but they can be fatally poisonous to them if mixed with hay.

CONSERVATION:
This introduced species in North America is unranked.

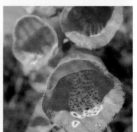

■ NATIVE
■ RARE OR EXTIRPATED
■ INTRODUCED

ALTERNATE NAMES: Gaping Penstemon, Gaping Beardtongue, Rothrock's Penstemon

The common name of this plant comes from the flowers, which form a gaping opening that resembles an open mouth.

DESCRIPTION: This low subshrub grows up to 3' (90 cm) tall with many thin stems. The leaves are shiny and evergreen, about 4 cm long, with toothed margins. The flowers grow in loose, hairy spikes. Each flower is up to 2 cm wide, with five whitish lobes and long, shiny hairs on the surface. The three lower lobes curl outward and under, while the two upper lobes are joined into a lip that curves forward, forming an open "mouth" with purple lines.

FLOWERING SEASON: May–July.

HABITAT: Dry, rocky slopes, below 8000' (2400 m).

RANGE: California Coast Ranges and Sierra Nevada, occasionally into western Nevada.

CONSERVATION: **G5**
Secure.

■ NATIVE
■ RARE OR EXTIRPATED
■ INTRODUCED

ALTERNATE NAMES: Old-field Toadflax, *Linaria canadensis*

These plants native to eastern North America resemble snapdragons. This is a larval host plant for Common Buckeye butterfly (*Junonia coenia*) and is frequently visited by birds, moths, and butterflies. The name Toadflax refers to the opening of the corolla, which looks like the mouth of a toad, and to the leaves, which resemble those of flaxes (genus *Linum*).

DESCRIPTION: This lanky plant with multiple stalks grows 6–24" (15–60 cm) tall. The small leaves at the base of the plant are oppositely arranged and form a rosette, with trailing stems. The leaves on the stem are shiny, smooth and linear. Each leaf is about 1½" (4 cm) long, and alternately arranged. The flowers are light blue, with 2 lips and spurred, appearing in a scattered elongated cluster on a slender stem. Each flower is ¼–½" (6–13 mm) long, with five sepals. The upper corolla lip has two lobes, while the lower corolla lip has three lobes, with two small white ridges and a long, thread-like spur projecting at the base. The fruit is a capsule.

FLOWERING SEASON: April–September.

HABITAT: Open, dry, shady or rocky sites and usually sandy, abandoned fields.

RANGE: Ontario east to New Brunswick, south to Florida, west to Texas, and north to North Dakota and Minnesota.

SIMILAR SPECIES: Texas Toadflax (*N. texanus*) is similar but has larger flowers and seeds.

CONSERVATION: **G5**

Secure, but critically imperiled in Ontario and Ohio; imperiled in Quebec and West Virginia; vulnerable in Iowa and Minnesota.

■ NATIVE
■ RARE OR EXTIRPATED
■ INTRODUCED

ALTERNATE NAMES: *Linaria genistifolia* ssp. *dalmatica*

This attractive Eurasian plant is considered an aggressive weed in much of North America, often invading native grasslands. It has long roots and can withstand cold weather.

DESCRIPTION: This introduced perennial has smooth, bluish waxy stems and leaves and grows up to 28" (70 cm) tall. Bright yellow flowers grow in dense, spike-like clusters with two lips and an orange hairy throat with a long, straight spur. Each flower is about 1¾" (4.5 cm) long. The leaves are ovate or lanceolate and clasp the stem.

FLOWERING SEASON: June–September.

HABITAT: Rangelands, roadsides, fields.

RANGE: This introduced plant is often invasive and is widely established throughout much of North America.

SIMILAR SPECIES: Butter-and-eggs (*L. vulgaris*), another introduced species, has similar flowers but has long, narrow leaves.

CONSERVATION: G5
Globally secure but non-native in North America.

■ NATIVE
■ RARE OR EXTIRPATED
■ INTRODUCED

ALTERNATE NAMES: Greater Butter-and-eggs, Common Toadflax, Yellow Toadflax

This introduced weedy plant thrives in dry areas. The common name Butter-and-eggs alludes to the yellow and orangish colors of the flowers. This plant has been used in folk medicine to treat a number of ailments, including stomachaches, jaundice, and fever. However, it is toxic to animals and should not be consumed in high quantities.

DESCRIPTION: This plant has yellow, two-lipped, spurred flowers that grow in a terminal cluster on leafy stems, about 1–3' (30–90 cm) tall. Each flower is about 1" (2.5 cm) long, with five sepals. The upper corolla lip has two lobes, while the lower corolla has three lobes, with orange ridges and a prominent spur at the base. The flower has four stamens and one pistil with a green style. The leaves are gray-green. The upper leaves are alternate and linear or grass-like, while the lower leaves are opposite or whorled. Each leaf is about 1–2½" (2.5–6.5 cm) long. The fruit is a capsule.

FLOWERING SEASON: May–October.

HABITAT: Dry fields, waste places, and roadsides.

RANGE: Alberta east to Newfoundland, south to Florida, west to Texas, and north to North Dakota; also throughout the West.

CAUTION: Can be toxic and in some cases fatal to animals if eaten in high quantities. Humans should generally avoid ingesting plants that are toxic to animals.

SIMILAR SPECIES: Dalmatian Toadflax *(L. dalmatica)* is a similar plant, but is larger than Butter-and-eggs, and has ovate or lanceolate leaves.

CONSERVATION:
This introduced species is unranked and is widely considered a weed in North America.

■ NATIVE
■ RARE OR EXTIRPATED
■ INTRODUCED

ALTERNATE NAMES: Roving-sailor, Snapdragon Vine, *Maurandella antirrhiniflora*

This native perennial is somewhat rare in the wild but is often grown as an ornamental for its attractive flowers. The central stem as well as the leaf and flower stalks twine and curve, as the common name suggests.

DESCRIPTION: Twining Snapdragon has twining stems that grow through other vegetation, reaching about 7' (2.1 m) long. Each leaf is about 1" (2.5 cm) long and shaped like an arrowhead. The tubular flowers are pale blue-violet or reddish-lavender, with curved stalks. The corolla is bilaterally symmetrical, about ¾–1" (2–2.5 cm) long, with a hairy cream patch at the base of the lower lip near the opening. There are five lobes. The two lobes of the upper lip are bent upward, while the three lobes of the lower lip are bent downward.

FLOWERING SEASON: June–September.

HABITAT: Sandy or gravelly soil, in deserts, sometimes on rock walls, and among pinyon and juniper.

RANGE: Southeastern California to western Texas; south into Mexico.

CONSERVATION: G4

Apparently secure, but critically imperiled in Nevada and imperiled in California. It is possibly extirpated in Utah.

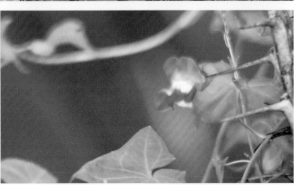

Neogaerrhinum filipes **YELLOW TWINING SNAPDRAGON**

■ NATIVE
■ RARE OR EXTIRPATED
■ INTRODUCED

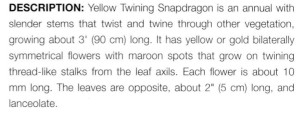

ALTERNATE NAMES: Twining Snapdragon, *Antirrhinum filipes*

This inconspicuous climber is native to desert washes in the Southwest. It is a twining plant with bright yellow flowers and bright green leaves and stems but is often hidden under other vegetation. Some sources place this species in the large genus *Antirrhinum*, which is subdivided into sections; others treat this group as a tribe composed of distinct genera.

DESCRIPTION: Yellow Twining Snapdragon is an annual with slender stems that twist and twine through other vegetation, growing about 3' (90 cm) long. It has yellow or gold bilaterally symmetrical flowers with maroon spots that grow on twining thread-like stalks from the leaf axils. Each flower is about 10 mm long. The leaves are opposite, about 2" (5 cm) long, and lanceolate.

FLOWERING SEASON: March–May.

HABITAT: Sandy deserts.

RANGE: Arizona, California, Nevada, and Utah.

CONSERVATION: Ⓖ③
Vulnerable. It is critically imperiled in Utah.

NATIVE
RARE OR EXTIRPATED
INTRODUCED

ALTERNATE NAMES: White-flower Beardtongue, Redline Beardtongue

This is a perennial herb native to central North America that is of special value to native bees. Its white flowers are also visited by a number of native butterflies, including Arogos Skipper (*Atrytone arogos*), Uncas Skipper (*Hesperia uncas*), and Riding's Satyr (*Neominois ridingsii*). All *Penstemon* species, commonly called beardtongues, were formerly in the Scrophulariaceae (figwort) family, but have been reclassified into the Plantaginaceae (plantain) family.

DESCRIPTION: This stout, erect perennial has hairy stems and grows 8–12" (20–30 cm) tall. The flowers are white and appear in dense whorl-like clusters of tubular blossoms. Each flower is about ¾" (2 cm) long with five lobes. The lower lip has three downward lobes, while the upper lip has two lobes. The inside of the tube has reddish or purplish lines that act as nectar guides for insects. The four stamens have black tips. The basal leaves are mostly hairless or somewhat hairy and oblong, about 2½" (6 cm) long. Leaves become stalkless and more lance-like up the stem, often with small teeth around the edges, oppositely attached.

FLOWERING SEASON: April–June.

HABITAT: Gravelly or sandy grasslands.

RANGE: Southern Manitoba to Alberta, south through western Minnesota and northwestern Iowa, to Oklahoma, Texas, and New Mexico.

SIMILAR SPECIES: Foxglove Beardtongue (*Penstemon digitalis*) has similar flowers, but is a much larger plant.

CONSERVATION: G5

Globally secure. The plant is imperiled in Iowa and vulnerable in Alberta and Manitoba.

■ NATIVE
■ RARE OR EXTIRPATED
■ INTRODUCED

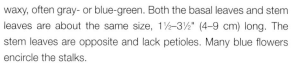

ALTERNATE NAMES: Whorled Penstemon

This drought-tolerant native herb of the Great Plains is a favorite among hummingbirds and butterflies, which visit its plentiful, bright blue flowers. Native Americans traditionally crushed the flowers of this plant to produce blue dye.

DESCRIPTION: This perennial typically has one erect stem, growing up to 2' (60 cm) tall. The leaves and stems are thick and waxy, often gray- or blue-green. Both the basal leaves and stem leaves are about the same size, 1½–3½" (4–9 cm) long. The stem leaves are opposite and lack petioles. Many blue flowers encircle the stalks.

FLOWERING SEASON: May–June.

HABITAT: Prairies, sand hills, semi-desert areas, openings, and shrublands.

RANGE: North Dakota to eastern Montana and central Wyoming, south to Oklahoma, New Mexico, Colorado, Utah, and Arizona.

CONSERVATION: G5
Secure but critically imperiled in Montana.

■ NATIVE
■ RARE OR EXTIRPATED
■ INTRODUCED

ALTERNATE NAMES: Cobaea Penstemon, Wild Foxglove

This clump-forming perennial native to the southern and central Great Plains is popular among various moths. The name *Penstemon* comes from the Greek words *penta* and *stemon*, meaning "five stamens." The flowers do indeed have five stamens, including one sterile stamen, which is easily distinguished from the others because it is covered in beard-like hairs.

DESCRIPTION: Cobaea Beardtongue is a clump-forming plant that grows 1–2½' (30–75 cm) tall. It has three stout stems that grow from a woody rhizome. Thick, lance-shaped leaves grow in pairs and clasp the stem, becoming progressively smaller up the stem. The lower leaves often die or wither by the time the flowers start blooming. Its large tubular flowers grow in loose panicles on the upper half of the stems. The flowers are white to pink with dark purple lines inside the floral tube. Each flower has five stamens.

FLOWERING SEASON: April–May in the South, May–June in the North.

HABITAT: Sandy or rocky, open hillsides; limestone outcrops.

RANGE: Most common from Nebraska to Texas; found east to Missouri and Arkansas and west to Arizona.

CONSERVATION: G4
Apparently secure but critically imperiled in Colorado and Iowa; vulnerable in Arkansas and Nebraska.

- NATIVE
- RARE OR EXTIRPATED
- INTRODUCED

Foxglove Beardtongue is native to the eastern U.S. The species name *digitalis* comes from the Latin word *digitus*, meaning "finger," a nod to the flowers that look like the finger of a glove.

DESCRIPTION: This perennial grows in a clump with erect stems, reaching 2–5' (60–150 cm) tall. It may be semi-evergreen in the South. The stems are topped with stalked clusters of white, two-lipped tubular flowers, which grow from the upper leaf axils. Each flower is 1¼" (3 cm) long. The stem leaves are lance-shaped while the basal leaves are elliptical.

FLOWERING SEASON: June–July.

HABITAT: Low, moist areas; prairies; open woodlands.

RANGE: Maine west to Minnesota and South Dakota, south to the Carolinas, central Alabama, and East Texas.

CONSERVATION: G5
Secure, but critically imperiled in Nebraska and Rhode Island.

◼ NATIVE
◼ RARE OR EXTIRPATED
◼ INTRODUCED

ALTERNATE NAMES: Slender Penstemon

This native perennial sports clusters of small, pale violet to lavender flowers that are popular among hummingbirds and various native bees.

DESCRIPTION: This is a slender, delicate plant that grows 8–24" (20–60 cm) tall. It has loose, spike-like clusters of small, lavender, snapdragon-like flowers. The flowers are tubular and about ¾" (2 cm) long. Each flower has a hairy, beard-like orange "flap" down the center, with five lobes and five yellow-tipped stamens. The leaves are dark green and narrow, about 3" (7.5 cm) long and ⅓" (1 cm) wide at the base, becoming progressively smaller up the stem, with small teeth at the edges.

FLOWERING SEASON: May–June.

HABITAT: Prairies; open woods at lower elevations.

RANGE: Ontario to British Columbia, south to Indiana, Illinois, and Nebraska, and along the eastern slope of the Rockies from Montana to New Mexico.

CONSERVATION: G5

Secure, but critically imperiled in Iowa and Michigan; vulnerable in Wyoming, British Columbia, and Manitoba.

■ NATIVE
■ RARE OR EXTIRPATED
■ INTRODUCED

ALTERNATE NAMES: Large-flower Penstemon, Showy Beard-tongue, Wild Foxglove

This plant native to the central U.S. is the showiest of the *Penstemon* species. It is endangered in many areas and is rare to see in the wild. It has large, tubular flowers that bloom for just a few weeks. Native Americans chewed the root of this plant for toothaches.

DESCRIPTION: This perennial reaches 2–4' (60–120 cm) tall with large, lavender tubular flowers that are horizontally arranged. Each flower is about 2" (5 cm) long. The corolla tube flares out abruptly into five lobes above the calyx. It has five sta-

mens, one of which is sterile and hairy (bearded). The leaves are opposite and blue-green, clasping the stem. Each leaf is 1–2½" (2.5–6.5 cm) long and broadly ovate, usually with a whitish bloom. The fruit is a capsule.

FLOWERING SEASON: May–June.

HABITAT: Woods and thickets.

RANGE: North Dakota east to Michigan and Ohio, south to Missouri, Oklahoma, and Texas; also scattered in parts of the West.

CONSERVATION:  G5

Globally secure but critically imperiled in Colorado, Illinois, Missouri, Montana, and Oklahoma; imperiled in Wyoming; and vulnerable in Iowa.

■ NATIVE
■ RARE OR EXTIRPATED
■ INTRODUCED

ALTERNATE NAMES: Northeastern Beardtongue

This perennial is native to northeastern North America. Its tubular flowers are visited by numerous native butterflies, bees, and other insects. Hairy Beardtongue is one of the earliest of the beardtongues to bloom. Its hairy stems distinguish it from other beardtongues.

DESCRIPTION: This upright perennial grows 1–3' (30–90 cm) tall. Its stems are covered with long, whitish hairs and a rosette of leaves grows at the base. The leaves are oppositely arranged and oblong to lanceolate in shape. Each leaf is about 2–5" (5–13 cm) long, stalkless, and toothed. The flowers are purple, blue, or white, each about 1" (2.5 cm) long. The upper corolla lip is erect and has two lobes, while the lower corolla lip has three lobes that form the mouth of the tube. Five stamens project forward, the fifth of which is sterile and hairy. The fruit is a capsule.

FLOWERING SEASON: June–July.

HABITAT: Dry, rocky ground in woods, fields, and on hillsides.

RANGE: Northeastern North America from Quebec and Maine to Wisconsin, and south to Virginia and Kentucky.

CONSERVATION: G4
Apparently secure, but critically imperiled in Massachusetts and Wisconsin. Imperiled in Quebec. Vulnerable in Vermont and Virginia.

■ NATIVE
■ RARE OR EXTIRPATED
■ INTRODUCED

ALTERNATE NAMES: Toadflax Beardtongue

Toadflax Penstemon is a common native penstemon in dry, open woodlands in the southwestern U.S. It has attractive, spike-like, blue-violet flowers that tend to grow on one side of the stem.

DESCRIPTION: This plant grows 6–16" (15–40 cm) tall with narrow, grayish-green leaves that are crowded at the base. Each leaf is slightly hairy and about ½–1" (1.3–2.5 cm) long. The flowers are blue-violet and bilaterally symmetrical, turned up at a 45-degree angle. The flowers tend to grow only on one side of the stem. The corolla is about ¾" (2 cm) long, with a long, narrow tubular base expanding abruptly into a throat. There are five lobes—two lobes of the upper lip are bent upward and three lobes of the lower lip are bent downward. There are five stamens; the fifth is bearded.

FLOWERING SEASON: June–August.

HABITAT: Open, often rocky soil at moderate elevations.

RANGE: Southern Nevada through Utah to Colorado, and in much of New Mexico and Arizona.

CONSERVATION: G5
Secure, but vulnerable in Arizona.

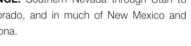

■ NATIVE
■ RARE OR EXTIRPATED
■ INTRODUCED

This plant is native in much of the East but has also been introduced elsewhere in the East outside of its natural range. The name *pallidus* comes from the Latin word meaning "pale," referencing the pale flowers. Penstemons are commonly called beardtongues because one of their stamens is sterile and has a tuft of small, beard-like hairs.

DESCRIPTION: This slender perennial grows 8–24" (20–60 cm) tall. It has white, snapdragon-like flowers with fine purple lines within the throat. There are five stamens and five lobes—two lobes on the upper corolla lip and three on the lower lip. Its basal leaves form a rosette. The stem leaves are narrow, partially clasp the stem, lanceolate in shape and about 2" (5 cm) long and ¾" (2 cm) wide.

FLOWERING SEASON: May–June.

HABITAT: Dry woods; fields.

RANGE: New England to Minnesota, south to Arkansas, central Alabama, and North Carolina.

CONSERVATION: G5

Globally secure but critically imperiled in Wisconsin, Georgia, and Illinois; imperiled in Kansas and Ohio; and vulnerable in West Virginia. It is presumed extirpated in Michigan.

■ NATIVE
■ RARE OR EXTIRPATED
■ INTRODUCED

ALTERNATE NAMES: Small-flower Beardtongue

This western native grows as a low, evergreen mat of foliage with tiny lavender-blue flowers. The species name *procerus* means "tall" in Latin, a reference to the plant's erect stems.

DESCRIPTION: This perennial grows in mats with some erect stems reaching up to 24" (60 cm) tall. The leaves are lance-shaped to oval, growing densely at the base. The stem leaves are smaller and arranged in opposite pairs. The flowers are purple or blue and tube-shaped, with a white throat and a lobed mouth. Each flower is about 1 cm long with yellowish or white hairs on the inside.

FLOWERING SEASON: June–August.

HABITAT: Alpine climate, meadows.

RANGE: Western North America, from Alaska to California to Colorado, east to Manitoba.

CONSERVATION: G5
Secure but critically imperiled in North Dakota and Manitoba.

■ NATIVE
■ RARE OR EXTIRPATED
■ INTRODUCED

ALTERNATE NAMES: Bridge's Penstemon, Mountain Scarlet Penstemon, *Penstemon bridgesii*

This southwestern native perennial is frequented by hummingbirds as well as bees, butterflies, and moths in search of nectar. The plant grows best in areas with cooler temperatures and some water during the summer.

DESCRIPTION: A low, multi-branched, sub-woody plant that grows 2–3' (60–90 cm) tall. The leaves are narrowly spoon-shaped and yellow-green, about 7 cm long. The flowers are red or orange and grow in clusters. The mouth of the flower has a hooded upper lip and a three-lobed lower lip. Each flower is 2–3 cm long. The lower lip is reflexed and the anthers are shaped like a horseshoe.

FLOWERING SEASON: June–August.

HABITAT: Dry slopes in pinyon and Ponderosa Pine forests; 4500–10,000' (1370–3050 m).

RANGE: Southwestern Colorado and western New Mexico, west to the Sierra Nevada in California.

CONSERVATION: G4
Apparently secure.

■ NATIVE
■ RARE OR EXTIRPATED
■ INTRODUCED

ALTERNATE NAMES: Meadow Beardtongue

The dark blue-violet whorls of small flowers help distinguish this common penstemon from most other species. It is also called Meadow Beardtongue for its tendency to grow in damp, grassy western meadows. The species is named for American botanist Per Axel Rydberg, the first curator of the New York Botanical Garden Herbarium.

DESCRIPTION: This plant has whorls of purplish flowers on erect stems, growing 8–24" (20–60 cm) tall. The corolla is narrow and funnel-shaped, about ½–¾" (1.5–2 cm) long. The upper corolla lip has two lobes projecting forward. The lower corolla lip has three lobes that spread downward and appear

hairy near the opening of the tube. There are five stamens, the fifth of which is sterile with beard-like hairs. The leaves are lanceolate and opposite, about 1½–3" (4–7.5 cm) long each. The leaves at the middle of the stem don't have stalks.

FLOWERING SEASON: June–July.

HABITAT: Open mountain slopes.

RANGE: Washington State south to California and east to northern Arizona and New Mexico, central Colorado and Wyoming, and southwestern Montana.

CONSERVATION: G4
Apparently secure, but imperiled in Arizona.

ROYAL BEARDTONGUE *Penstemon speciosus*

■ NATIVE
■ RARE OR EXTIRPATED
■ INTRODUCED

ALTERNATE NAMES: Royal Penstemon

Royal Beardtongue is a western native that inhabits a range of habitats and elevations, but is most common in mountainous scrubby areas and subalpine forests. This attractive plant is often grown as an ornamental.

DESCRIPTION: The branches of this clumped, shrub-like perennial grow up to 2' (0.6 m). It has narrow, spoon-shaped leaves that are opposite and oblanceolate. Whorls of bright blue tubular flowers grow 1" (2.5 cm) long. The flowers have white throats and five lobes.

FLOWERING SEASON: May–July.

HABITAT: Open juniper or pine woodlands, sagebrush steppes, dry plains, wooded or shrubby slopes; most common at elevations of 3500–8500' (1067–2591 m).

RANGE: Eastern Washington to California east to Idaho and Utah.

CONSERVATION: G5
Secure, but critically imperiled in Utah.

■ NATIVE
■ RARE OR EXTIRPATED
■ INTRODUCED

ALTERNATE NAMES: Rocky Mountain Penstemon

This drought-tolerant native perennial has showy flowers and is often grown as an ornamental. Native butterflies including Common Checkered Skipper (*Burnsius communis*) and Nevada Skipper (*Hesperia nevada*) frequent the flowers.

DESCRIPTION: Rocky Mountain Beardtongue is an evergreen that grows in low mats of foliage, about 1–3' (30–90 cm) tall. The leaves are paired and may be either narrow and grass-like or broad and lance-shaped, depending on the variety. The stem leaves are narrower than the basal leaves. The flowers are deep blue or purple. Each flower stalk has a whorl of one to four flowers at each node. The corolla is 1–1½" (2.5–4 cm) long. The tube is white at the opening and lightly bearded (fuzzy) with reddish lines inside the throat. The fruit is a capsule.

FLOWERING SEASON: May–June.

HABITAT: Subalpine to valley sagebrush and conifer forests.

RANGE: Wyoming to New Mexico and northeast Arizona.

CONSERVATION: G5

Globally secure but vulnerable in Arizona, Wyoming, and the Navajo Nation.

■ NATIVE
■ RARE OR EXTIRPATED
■ INTRODUCED

English Plantain is distinguished from other plantains by its narrow leaves. This widespread introduced plant is a frequent weed in North America, often growing on lawns and other open, disturbed areas. The seeds are often eaten by songbirds and the leaves are a favorite food of rabbits.

DESCRIPTION: This perennial has a leafless, slightly hairy stalk and grows 6–20" (15–50 cm) tall. Tiny whitish-green flowers grow in a round head, arranged in a spiral. Each flower is about ⅛" (3 mm) long, with a four-lobed papery corolla, four white protruding stamens, and papery bracts beneath. Narrow leaves grow in a basal rosette, with three or four conspicuous veins. Each leaf is about 4–16" (10–40 cm)

long and lightly covered in hairs. The fruit, a capsule with two seeds, opens in the middle.

FLOWERING SEASON: April–August.

HABITAT: Lawns, roadsides, pastures, and waste places, cracks in pavement.

RANGE: Introduced throughout subarctic North America.

CONSERVATION: G5

Globally secure. This species is exotic and widespread in North America.

■ NATIVE
■ RARE OR EXTIRPATED
■ INTRODUCED

ALTERNATE NAMES: Great Plantain, Broadleaf Plantain, Nipple Seed Plantain

This plant is native to Eurasia but widely naturalized around the world, including throughout North America. It is considered a weed in many areas. Like other plantains, this plant has broad leaves and inconspicuous flowers.

DESCRIPTION: This plant grows 6–18" (15–45 cm) tall from basal leaves that are broad and conspicuously veined. Each leaf is 6" (15 cm) long, 4" (10 cm) wide and ovate to elliptical. The small flowers are greenish-white and grow in a spike-like cylindrical mass. Each flower is ¹⁄₁₆" (2 mm) long, with a four-lobed papery corolla, four stamens, one pistil, and bracts beneath the flowers. The fruit is a capsule, with 12–18 seeds, splitting open around the middle.

FLOWERING SEASON: June–October.

HABITAT: Waste places, fields, roadsides, lawns, and disturbed areas.

RANGE: Introduced and widely naturalized throughout North America, except in parts of the Far North.

SIMILAR SPECIES: Pale Plantain (*P. rugelii*) is a very similar native species, but has a longer fruit capsule with fewer seeds that splits open at the base. Heart-leaved Plantain (*P. cordata*), also known as Water Plantain, is rarer, mostly occurring in the East. The flower stalks of *P. cordata* are hollow.

CONSERVATION: G5
Globally secure. This species is widely considered a weed in North America, especially in vineyards and orchards.

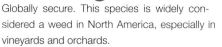

■ NATIVE
■ RARE OR EXTIRPATED
■ INTRODUCED

ALTERNATE NAMES: American Speedwell

This widespread native perennial herb has been used both medicinally and as a nutrient-rich source of food. The common name "brooklime" comes from this plant's preference for wet, muddy areas or lime soils.

DESCRIPTION: American Brooklime has erect or leaning, leafy, rounded stems, growing 4–40" (10–100 cm) tall. The leaves are opposite with short stalks, growing ½–3" (1.3–7.5 cm) long, oppositely arranged and broadly lanceolate. Small blue flowers grow in several open racemes, one in each upper leaf axil. Each flower is ¼–½" (6–13 mm) wide, with four lobes joined at the base, and two stamens.

FLOWERING SEASON: May–July.

HABITAT: Wet areas such as streams, wetlands, and bottomlands.

RANGE: Throughout much of North America, except in the Southeast. It is present but much less common in the Midwest.

SIMILAR SPECIES: Some members of the mint family (Lamiaceae) are somewhat similar but have square stems instead of rounded stems.

CONSERVATION: G5

Globally secure but critically imperiled in Delaware, Illinois, Iowa, Tennessee, and Texas; imperiled in Kansas and North Carolina; and vulnerable in Yukon. It is possibly extirpated in Missouri, Kentucky, and Indiana. This plant is highly threatened by land-use conversion, habitat fragmentation, and forest management practices and is vulnerable to sedimentation.

■ NATIVE
■ RARE OR EXTIRPATED
■ INTRODUCED

This plant introduced from Africa and Eurasia is a common and widespread weed of disturbed sites in North America. The species name *arvensis* is Latin for "living in open fields" and references this plant's tendency to invade lawns, gardens, and agricultural fields.

DESCRIPTION: This low, hairy plant with many branches grows 2–16" (5–40 cm) tall. The lower leaves are opposite, rounded or oval, and toothed. The upper leaves are alternate, unstalked, and lanceolate. Each leaf is about ½" (1.5 cm) long. It has one very tiny, nearly stalkless, deep blue flower in each upper leaf axil. Each flower is up to ¼" (5 mm) wide, with four sepals of unequal size, four united petals, and two stamens. The fruit is a capsule.

FLOWERING SEASON: March–August.

HABITAT: Waste places, pastures, open woods, and cultivated sites.

RANGE: Introduced throughout most of the United States, in British Columbia, and from Ontario east to Newfoundland.

SIMILAR SPECIES: Purslane Speedwell (*V. peregrina*) is a similar, widespread native plant but has white flowers, not blue.

CONSERVATION:

This exotic species in North America is unranked. It is largely considered a weed.

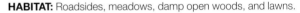

NATIVE
RARE OR EXTIRPATED
INTRODUCED

This small perennial is native to North America but also is frequently found outside its natural range. It is often considered a weed and is difficult to eradicate from lawns. It is so named because its leaves resemble those of thyme.

DESCRIPTION: This mat-forming plant reaches 2–10" (5–25 cm) tall. The leaves are oppositely arranged and ovate. Each leaf is ½–⅝" (13–16 mm) long and toothed. It has short, erect, narrow clusters of white or pale blue flowers with darker stripes. Each flower is about ⅛" (4 mm) wide, with four united petals, three of which are rounded and one (the lowest) is narrower. There are two stamens. The fruit is a capsule.

FLOWERING SEASON: April–July.

HABITAT: Roadsides, meadows, damp open woods, and lawns.

RANGE: Found throughout North America except in the Far North and less common in the central third of the continent.

SIMILAR SPECIES: A subspecies, Brightblue Speedwell (ssp. *humifusa*), is similar but its flowers are brighter blue.

CONSERVATION: G5

Secure, but critically imperiled in Manitoba and Saskatchewan; vulnerable in Alberta and Labrador.

■ NATIVE
■ RARE OR EXTIRPATED
■ INTRODUCED

Culver's Root has attractive flowers densely packed together. This perennial herb native to eastern North America has long been used in medicine as a laxative. Its common name refers to a Dr. Culver (or Coulvert), an 18th-century physician who is said to have used the purgative properties of the root of this plant to treat his patients; however, there is no record of this individual. The genus name refers to its resemblance to plants in the genus *Veronica*.

DESCRIPTION: This plant grows 3–7' (90–210 cm) tall. The leaves are lanceolate to narrowly ovate in shape, appearing mostly in whorls of three to seven, with sharply toothed, short stalks. Small white or purple flowers appear in dense spikes. Each flower is about ¼" (5 mm) wide, with four petals, four sepals, and two projecting stamens. Each spike is 2–6" (5–15 cm) long. The fruit is an ovoid capsule, opening at the tip.

FLOWERING SEASON: June–September.

HABITAT: Rich woods, thickets, roadsides, and prairies.

RANGE: Manitoba and Ontario; Maine south to Florida, west to Texas, and north to South Dakota.

CONSERVATION: (G4)
Apparently secure. It is critically imperiled in Delaware, Massachusetts, Nebraska, North Carolina, Oklahoma, South Carolina, Vermont, and Washington D.C. in the U.S. and in Manitoba and Ontario in Canada. It is imperiled in Georgia, Mississippi, New Jersey, and New York. It is vulnerable in Kentucky, Iowa, Virginia, and West Virginia. It is possibly extirpated in Louisiana and North Dakota.

■ NATIVE
■ RARE OR EXTIRPATED
■ INTRODUCED

ALTERNATE NAMES: American Figwort, Hare Figwort, Early Figwort

The flowers of Lanceleaf Figwort are tiny, but they are rich in nectar, known to attract Ruby-throated Hummingbirds (*Archilochus colubris*) among other birds, as well as butterflies and moths. This widespread native plant has also been used in medicine as a tea to treat swollen glands.

DESCRIPTION: This perennial grows in clusters of erect or spreading stems, up to 6' (1.8 m) tall. The leaves are toothed and oppositely arranged. Each leaf is about 5½" (14 cm) long and lanceolate in shape, on short petioles. It has many hairy, glandular branches with loose panicles of reddish-green flowers. The flowers have a tubular-shaped corolla, with hooded lobes. Each corolla is about ½" (1.3 cm) long and greenish-brown. The fruit is a brown capsule, about 1 cm long, with many black seeds.

FLOWERING SEASON: May–July.

HABITAT: Open woods, roadsides, disturbed areas, fields.

RANGE: Widespread across the U.S. except in the southernmost states, with limited distribution in southern Canada.

SIMILAR SPECIES: Late Figwort (*S. marilandica*) is similar but blooms later, generally July–October.

CONSERVATION: G5

Secure, but critically imperiled in North Carolina and Nova Scotia; imperiled in Kansas, Rhode Island, and New Brunswick; vulnerable in Maryland, New Mexico, Vermont, Wyoming, British Columbia, and Quebec.

■ NATIVE
■ RARE OR EXTIRPATED
■ INTRODUCED

ALTERNATE NAMES: White Moth Mullein

Moth Mullein is introduced from Eurasia and is now widely naturalized in North America, where it is often invasive. It is named for its fuzzy stamens that look like moth antennae.

DESCRIPTION: This biannual has erect stems and grows 2–4' (60–120 cm) tall. The leaves are variously lobed or toothed—the largest appear in a rosette at the base, while those on the stem become progressively smaller up the stem. Each leaf is 2–6" (5–15 cm) long. The flowers are yellow or white with brownish-purple marks on their backs, and they grow in open, spike-like clusters. The corolla is almost radially symmetrical, with five round lobes of about equal length, and five stamens with red-purple hairs on the filaments. The corolla is about 1" (2.5 cm) wide.

FLOWERING SEASON: June–September.

HABITAT: Old fields and roadsides.

RANGE: Introduced throughout much of temperate North America.

SIMILAR SPECIES: White Mullein (*V. lychnitis*) is a similar introduced species that has naturalized in the East from Ontario to New Hampshire, south to Virginia and Iowa. Wand Mullein (*V. virgatum*)—another introduced relative—is also sometimes called Moth Mullein; its flower stalks are shorter than the capsule and it is less widely naturalized in North America.

CONSERVATION:

This exotic species is not ranked in North America. It is often invasive and is considered a noxious weed in some states.

■ NATIVE
■ RARE OR EXTIRPATED
■ INTRODUCED

ALTERNATE NAMES: Great Mullein, Woolly Mullein

Introduced from Europe, this biennial has become widely naturalized and weedy in many areas of North America. The velvety leaves and spike-like flowers have been used for a variety of purposes. The plant was used to poison fish in Virginia in the 1700s. Roman soldiers supposedly used the flower spikes as torches after dipping them in grease. The leaves are still sometimes used as wicks. Native Americans and colonists lined their footwear with the leaves in cold weather. A tea made from the leaves was used to treat colds, and the flowers and roots were used medicinally to treat earaches, croup, and other ailments. The leaves were sometimes used to treat skin conditions. The flowers were also used to make dyes.

DESCRIPTION: This plant has an erect, woolly stem that reaches 2–7' (60–210 cm) tall. The stem grows from a thick rosette of velvety, basal leaves. Each leaf is 4–16" (10–40 cm) long and ovate, covered with felt-like, gray hair. The leaves become smaller up the stem. The stem leaves are white and woolly. Yellow flowers are tightly packed in a spike-like cluster, opening around dawn and closing around mid-afternoon. The corolla is about ¾–1" (2–2.5 cm) wide and almost radially symmetrical, with five round lobes spreading out flat and five stamens. The top three stamens are covered with yellow hairs on the filaments.

FLOWERING SEASON: June–September.

HABITAT: Fields, roadsides, and waste places.

RANGE: Introduced throughout North America, except the Yukon and Northwest Territories.

SIMILAR SPECIES: Orange Mullein (*V. phlomoides*)—another weedy introduced plant from Eurasia—is similar, but its leaves do not continue down the stem. Its flowers are also wider.

CONSERVATION:
This widely introduced plant is not ranked in North America.

■ NATIVE
■ RARE OR EXTIRPATED
■ INTRODUCED

ALTERNATE NAMES: Devil's-claw, Unicorn Plant, Proboscis Flower, Louisiana Unicorn-plant

The native range of this annual herb is uncertain, but it is believed to originate in the western or south-central United States or Mexico. It is widely introduced throughout North America and in the Old World, where it can be weedy. The common name refers to the fruit, which splits into two horns once it dries. Native Americans used the fruits to make black dye and pickled them to be eaten like okra. The plant was also used to weave baskets. The flowers have an unpleasant scent.

DESCRIPTION: This low, spreading, bushy plant grows 1–2' (30–60 cm) tall and is covered in hairs with oily droplets, which feel wet to the touch. The oil evaporates in the air and gives the plant an unpleasant smell. The leaves are large, long-stemmed, and palmately lobed. The corolla is lavender or yellowish, sometimes with purple spots and yellow nectar guides. This plant is especially known for its fruit. It can produce up to 80 fruit cap-

sules, each of which are about 10 cm long, with a long, curving beak. Once the fruit dries, it splits open into two long horns. The inside contains black or white seeds.

FLOWERING SEASON: June–August.

HABITAT: Riverbanks, meadows, disturbed places, and waste areas.

RANGE: Found throughout much of North America.

CONSERVATION:
Although probably native in parts of North America, this plant's origin is uncertain and its conservation status is not ranked. It is widespread and opportunistic, however.

■ NATIVE
■ RARE OR EXTIRPATED
■ INTRODUCED

ALTERNATE NAMES: False Mint

This plant is a larval host for several butterflies, including Cuban Crescent (*Anthanassa frisia*), Texan Crescent (*Anthanassa texana*), and Crimson Patch (*Chlosyne janais*). Its name refers to the plant's six-angled stems, while the alternate name False Mint refers to the flower's similarity to those in the mint family (Lamiaceae), especially Tropical Sage (*Salvia coccinea*).

DESCRIPTION: This fast-growing, erect perennial has six-angled and multi-branched stems. The leaves are oppositely arranged and simple, egg-shaped, toothless, and pointed. The flowers are bright red and tubular, curved into two lips. It has two stamens with yellow anthers that extend beyond the upper lip. The

fruits, small, green capsules, turn brown at maturity.

FLOWERING SEASON: Nearly year-round, from February–November.

HABITAT: Coastal areas and thickets.

RANGE: In Florida and in Hidalgo County in Texas.

CONSERVATION: G5

Secure, but critically imperiled in its limited range in Texas.

■ NATIVE
■ RARE OR EXTIRPATED
■ INTRODUCED

This perennial herb is native to the southeastern U.S. in the Atlantic coastal plains. Twin Flower grows especially well after wildfires and can be successfully cultivated as an ornamental ground cover outside its native range, including elsewhere in the Southeast and the West Coast. The common name refers to the flowers, which often grow in pairs at the top of the stems. Common Buckeye (*Junonia coenia*) butterflies often use the plant to lay their eggs and the flowers and leaves are often eaten by caterpillars.

DESCRIPTION: This plant grows up to 12" (30 cm) tall, with somewhat hairy stems and leaves. The leaves are oppositely arranged and oval. Trumpet-shaped lavender flowers grow in pairs in the upper leaf axils. The flowers have five lobes and four stamens. Each flower is about 1½" (4 cm) wide, with dark purple spots on the lowermost lobe.

FLOWERING SEASON: April–October; year-round in South Florida.

HABITAT: Pinelands, sand hills.

RANGE: South Carolina to Florida, west to easternmost Alabama.

CONSERVATION:  G4

Apparently secure, but critically imperiled in Alabama.

■ NATIVE
■ RARE OR EXTIRPATED
■ INTRODUCED

ALTERNATE NAMES: Common Water-willow

American Water-willow is a perennial herb native throughout much of eastern North America. This colony-forming plant has showy flowers and underground stems. It is typically an aquatic plant but may be terrestrial as well. Despite its common name and the similarity of their leaves, it is not closely related to willows (*Salix* spp.).

DESCRIPTION: American Water-willow is primarily an aquatic plant with bi-colored flowers. It grows 1–3' (30–90 cm) above water. The leaves are willow-like, about 3–6" (7.5–15 cm) long each, narrow, and oppositely arranged. The flowers grow in dense, spike-like clusters on long, slender stalks rising from the leaf axils. Each flower is ½" (1.5 cm) long, with a calyx shorter than the corolla. The lower corolla lip is three-lobed, white, spotted with purple. The upper corolla lip is pale violet or white, arching over the lower lip. There are two stamens with purplish-red anthers. The fruit is a brown capsule.

FLOWERING SEASON: June–October.

HABITAT: Wet shorelines and shallow water.

RANGE: Ontario, Quebec, and New York south to Florida, west to Iowa, Kansas, and Texas.

SIMILAR SPECIES: Loose-flowered Water-willow (*J. ovata*), a similar but more southern species, has more loosely flowered spikes.

CONSERVATION: G5
Secure, but critically imperiled in Iowa; imperiled in Ontario, Quebec, and Michigan; vulnerable in Louisiana.

■ NATIVE
■ RARE OR EXTIRPATED
■ INTRODUCED

This plant native to the eastern and central United States has many large flowers. The name *humilis* means "low-growing"—a reference to its bushy appearance.

DESCRIPTION: This perennial has multi-branched stems and grows up to 2' (60 cm) tall. The leaves are broadly lanceolate and covered with white hairs on both the upper and lower sides. The leaves also have hairy petioles and are oppositely arranged. Each leaf is about 2½" (6 cm) long and 1" (3 cm) wide. The flowers are purple in color and grow in clusters in the upper leaf axils. Each flower is tubular or bell shaped, with five rounded lobes, about 3" (7.5 cm) long.

FLOWERING SEASON: April–October in the south; June–August in the north.

HABITAT: Gravelly hillsides; dry prairies; open woods.

RANGE: Pennsylvania to Wisconsin and eastern Kansas, south to western North Carolina, Georgia, Florida, and Texas.

CONSERVATION: **G5**

Globally secure but critically imperiled in Maryland, Michigan, North Carolina, Pennsylvania, South Carolina, and West Virginia; imperiled in Wisconsin; and vulnerable in Georgia, Minnesota, and Virginia.

■ NATIVE
■ RARE OR EXTIRPATED
■ INTRODUCED

This native perennial vine climbs via tendrils and can sometimes be found quite high in a tree. The species name *capreolata* means "having tendrils," referring to this plant's tendrils tipped with adhesive pads that allow it to cling to objects such as stone, bricks, trees, and fences. The name Crossvine comes from the pattern revealed in a cross-section of its stems, which have four wedges of tissue (phloem) that form the shape of a cross.

DESCRIPTION: This climbing, woody vine reaches 50' (16.5 m) long. It has showy, orange-red, trumpet-shaped flowers that hang in clusters of two to five. Each flower is about 2" (5 cm) long. Glossy, semi-evergreen leaves change from dark green in summer to reddish-purple in winter. The leaves are oppositely arranged and grow up to 6" (15 cm) long.

FLOWERING SEASON: April–May.

HABITAT: Low, moist woods and fields; dry, open woods.

RANGE: Florida to eastern Texas, north to Maryland and the Ohio River valley.

SIMILAR SPECIES: Trumpet Vine (*Campsis radicans*) is a similar species, but it doesn't grow as long and has more flowers.

CONSERVATION: G5

Globally secure but critically imperiled in Oklahoma and vulnerable in West Virginia.

■ NATIVE
■ RARE OR EXTIRPATED
■ INTRODUCED

ALTERNATE NAMES: Trumpet Creeper, Devil's Shoestrings, Hellvine

This plant is native to parts of eastern North America and naturalized elsewhere. It can be an aggressive weed, giving rise to its colloquial name Hellvine. Trumpet Vine is a fast-growing climber, with stems extending for many feet along the ground. It's known to trip pedestrians, hence its other common name, Devil's Shoestrings.

DESCRIPTION: This woody vine grows up to 20' (6 m) long with reddish or orange trumpet-shaped flowers. The flowers grow in terminal cymes of 4–12. Each flower is 2½" (6.3 cm) long, with a five-lobed corolla. The leaves are pinnately compound, with 7–11 toothed, ovate, pointed leaflets, each about 2½" (6.3 cm) long. The fruit is a capsule, about 6" (15 cm) long.

FLOWERING SEASON: July–September.

HABITAT: Low woods, fallow fields, fencerows, and thickets.

RANGE: Ontario east to New Hampshire, south to Florida, west to Texas, and north to North Dakota.

CAUTION: The sap of this plant can cause skin irritation if touched.

SIMILAR SPECIES: Crossvine (*Bignonia capreolata*) is a similar species, but has fewer flowers—about two to five per cluster.

CONSERVATION:
Secure, but imperiled in Ontario.

■ NATIVE
■ RARE OR EXTIRPATED
■ INTRODUCED

ALTERNATE NAMES: Dwarf Butterwort

Small Butterwort is a native carnivorous plant of the Deep South that grows from a basal rosette of succulent leaves. All *Pinguicula* species have short glandular hairs on the leaf surface that feel clammy. The leaves trap small insects to be consumed and digested by the plant, helping it thrive where nutrients are scarce. The genus name comes from the Latin word *pinguis*, meaning "fat," and references the greasy feeling of the leaf surface.

DESCRIPTION: This carnivorous plant grows 6–8" (15–20 cm) tall, with 1–10 white, pink, lavender, or yellow solitary flowers. The flowers are funnel-shaped, with five lobes. The bottom of the corolla is spurred and reddish. Each flower is about 1" (2.5 cm) wide. Both the sepals and leaves are covered in tiny, glandular hairs. The leaves are yellowish-green and broad, forming a basal rosette.

FLOWERING SEASON: February–May.

HABITAT: Moist, sandy pinelands and savannas.

RANGE: Coastal plain from North Carolina to Texas.

CONSERVATION: **G4**
Apparently secure, but critically imperiled in Alabama and Mississippi and imperiled in the Carolinas.

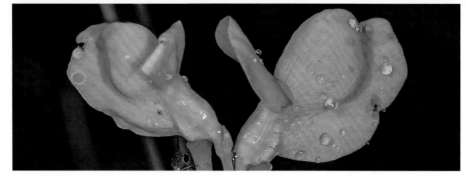

■ NATIVE
■ RARE OR EXTIRPATED
■ INTRODUCED

This is a widespread native carnivorous perennial. Unlike most bladderworts, Horned Bladderwort is primarily a terrestrial rather than aquatic plant, although it may occasionally be submerged in water. It has small bladders on the leaves, which allow it to consume tiny prey, mainly microorganisms. The bladder pores open when a small organism brushes against it, allowing water to rush inside along with the organism.

DESCRIPTION: A terrestrial, carnivorous plant with a few scale-like bracts and one to five yellowish, two-lipped, spurred flowers at the top of a brownish stalk, growing 2–12' (5–30 cm) tall. Each flower is about ¾" (2 cm) long, with a two-lipped corolla. The lower lip is large and helmet-shaped with a backward-projecting spur. It has small, delicate leaves that are mostly underground. The leaves are thread-like, with small bladders.

FLOWERING SEASON: June–September.

HABITAT: Bogs and wet, sandy, muddy, or peaty shorelines.

RANGE: Alberta east to Newfoundland and New England, south to Florida, west to the Great Lakes states and through the Gulf Coast states to Texas.

CAUTION: The seeds and foliage of this plant are poisonous if ingested.

CONSERVATION:  G5
Globally secure but critically imperiled in Alberta and Prince Edward Island in Canada and in Arkansas, Illinois, Indiana, Maryland, North Carolina, Ohio, and Tennessee in the U.S. It is imperiled in Pennsylvania and Rhode Island and vulnerable in Manitoba and Saskatchewan.

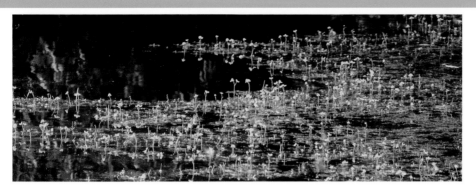

■ NATIVE
■ RARE OR EXTIRPATED
■ INTRODUCED

ALTERNATE NAMES: Small Swollen Bladderwort

This native eastern bladderwort, like most others, doesn't take root and instead floats on the surface of the water with inflated leaves. Several tiny bladders on the leaves are used to trap small organisms, which the plant consumes for nutrients.

DESCRIPTION: This aquatic plant grows 2-8" (5–20 cm) above the water. The lower leaves are alternate and divided many times into hair-like segments, each with many tiny bladders and about 1¼" (3 cm) long. The upper leaves float on the water, just below the flowering stem in whorls of four to seven. The upper leaves are about ⅓–1½" (1–4 cm) long. The flowers are yellow, with two lips, appearing on erect stalks. The lower lip has three lobes. Each flower is about ¾" (2 cm) long.

FLOWERING SEASON: May–October.

HABITAT: Ponds, depression ponds, lakes, and ditches.

RANGE: Primarily in the southeastern and Mid-Atlantic region of the United States. It occurs from Nova Scotia, New Brunswick, and Maine southward to southern Florida, westward to Texas, and disjunct in western Virginia, western Tennessee, and northwestern Indiana.

CAUTION: The seeds and foliage of this plant are poisonous if ingested.

CONSERVATION: G4
Apparently secure. It is critically imperiled in Quebec, Indiana, and Oklahoma; imperiled in New Jersey, New York, Rhode Island, and Vermont; and vulnerable in New Brunswick, North Carolina, and Virginia. It is presumed extirpated in Pennsylvania.

■ NATIVE
■ RARE OR EXTIRPATED
■ INTRODUCED

ALTERNATE NAMES: Greater Bladderwort, *Utricularia macrorhiza*

Common Bladderwort is native to Eurasia; our native North American representative is the subspecies *macrorhiza*, sometimes treated as a separate species (*Utricularia macrorhiza*). Like other *Utricularia* species, Common Bladderwort floats freely in the water thanks to leaves that include bladders, which capture and digest insects and other small organisms. However, communities of microorganisms have been found living inside the bladders, suggesting they may have a mutually beneficial relationship in some cases.

DESCRIPTION: This small, floating plant has 6–20 bilaterally symmetrical, yellow flowers. The flowers are snapdragon-like and grow in an erect raceme at the top of a stout reddish-green stem. The flower stalks rise 2½–8" (6.5–20 cm) above water. The corolla is about ½–¾" (1.5–2 cm) long, with a spur curving forward and downward from near the base. The flowers have a red venation. There are two small, egg-shaped, green sepals behind the flower. The lower sepal has a rounded or notched tip, the upper is slightly larger and more pointed at the tip. The leaves are repeatedly and finely divided into small, hair-like segments, with reddish bladders extending from the leaf filament axils. The bladders are about ⅛" (3 mm) wide. Each leaf is about ½–2" (1.5–5 cm) long.

FLOWERING SEASON: June–August.

HABITAT: Ponds and slow-moving water.

RANGE: Throughout much of North America.

SIMILAR SPECIES: Common Bladderwort has larger flowers than other similar species. Hidden-fruit Bladderwort (*U. geminiscapa*) has similar leaves but favors bogs.

CONSERVATION: G5
Globally secure but critically imperiled on Prince Edward Island in Canada and in Kentucky, North Carolina, South Carolina, and West Virginia in the U.S. It is imperiled in Wyoming and vulnerable in Arizona, Kansas, and Virginia.

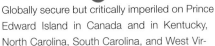

■ NATIVE
■ RARE OR EXTIRPATED
■ INTRODUCED

ALTERNATE NAMES: Sweet William, Rose Vervain, *Verbena canadensis*

This creeping perennial herb is native to the eastern and central U.S. and produces clusters of fragrant flowers at the tips of its branches. The genus name of this showy vervain is from Latin and means "little nuts," referencing the shape of the seedpod.

DESCRIPTION: This is an erect or reclining plant with a hairy stem growing 6–18" (15–45 cm) tall. The leaves are coarsely toothed or lobed, with a partly-winged stalk. Each leaf is 3" (7.5 cm) long, opposite, palmately veined, and ovate to lanceolate. The flowers grow in a dense, terminal, flat-topped cluster. The flowers are tubular and pinkish-lavender or white, with flaring corolla lobes. Each flower is ½–¾" (1.5–2 cm) wide. The corolla has five notched lobes, four stamens, and one pistil with a four-lobed ovary. The fruit, a capsule, separates into four hard, seed-like sections and looks like an acorn.

FLOWERING SEASON: April–October.

HABITAT: Sandy or rocky prairies and roadsides.

RANGE: Minnesota east to Connecticut, south to Florida, west to New Mexico, Colorado, and Nebraska.

SIMILAR SPECIES: Small-flowered Verbena (*G. bipinnatifida*) is a similar species but has stiff hairs on the stem and leaves. The introduced Stiff Vervain (*Verbena rigida*) is also similar, but has a more elongated flower cluster.

CONSERVATION: (G5)

Secure, but critically imperiled in Kentucky and North Carolina.

■ NATIVE
■ RARE OR EXTIRPATED
■ INTRODUCED

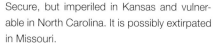

ALTERNATE NAMES: Common Fogfruit, Frogfruit, Matchhead

This native perennial can be found from coast to coast in the southern tier of the United States. This vigorous-spreading plant is often grown as an ornamental or as ground cover and may be a weed in some areas. The flowers resemble a matchstick, giving rise to the colloquial name Matchhead.

DESCRIPTION: This creeping evergreen wildflower grows 1–3" (2.5–7.5 cm) tall, with flower stalks growing up to 6" (15 cm) tall. Opposite, paddle-shaped leaves are toothed above the midpoint. It has many tiny, white or pink flowers that grow around a dark purple center and look like matchsticks.

FLOWERING SEASON: May–October.

HABITAT: Often found in wet or moist habitats, roadsides, lawns, beaches.

RANGE: Across the southern U.S. from southern Virginia west to California.

SIMILAR SPECIES: Lanceleaf Fogfruit *(Phyla lanceolata)* is similar, but Turkey-tangle Fog-fruit has shorter leaves that are often blunt and more rounded.

CONSERVATION: **G5**

Secure, but imperiled in Kansas and vulnerable in North Carolina. It is possibly extirpated in Missouri.

PROSTRATE VERVAIN *Verbena bracteata*

■ NATIVE
■ RARE OR EXTIRPATED
■ INTRODUCED

ALTERNATE NAMES: Large-bract Vervain, Bigbract Verbena, Carpet Vervain

This low-growing, widespread native plant has hairy, squarish stems that grow branched from the plant's base. It readily occupies disturbed sites and spreads rapidly, often becoming weedy. The genus name is Latin for "sacred plant," so named because some European verbenas were thought to be cure-alls among medicinal plants and were used in religious ceremonies.

DESCRIPTION: This sprawling plant has deeply divided opposite leaves with three lobes that are toothed or divided. The stems are somewhat square, with white hairs. The flower stalks are upright, with numerous leaf-like bracts. Dense spikes of pale blue to purple flowers occur at the ends of the branches.

FLOWERING SEASON: May–October.

HABITAT: Weedy, grows well in disturbed sites and cracked pavement.

RANGE: Found throughout most of North America except for the Far North.

CONSERVATION: G5

Globally secure but critically imperiled in North Carolina and vulnerable in Alberta and Manitoba.

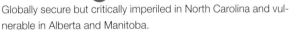

- ■ NATIVE
- ■ RARE OR EXTIRPATED
- ■ INTRODUCED

ALTERNATE NAMES: Simpler's-joy, Blue Verbena

Blue Vervain is a common native plant that is hardy and drought resistant, producing attractive blue-violet flowers in summer. The specific epithet *hastata* means "spear-shaped," a reference to this showy perennial's numerous, pencil-like flower spikes that branch upward.

DESCRIPTION: This clump-forming perennial has a hairy square, grooved stem up to 2–6' (60–180 cm) tall. The leaves are opposite and have a rough texture. Each leaf is about 4–6" (10–15 cm) long, lanceolate, and doubly toothed. The blue-violet flowers are tubular and grow in dense, stiff, pencil-like spikes in candelabra-like clusters. Each flower is about ⅛" (3 mm) wide. Only a few flowers bloom at a time, from the bottom to top of the stem. The corolla has five flaring lobes, four stamens growing in two pairs of different lengths, one pistil and a four-lobed ovary. The fruit separates into four hard, seed-like sections.

FLOWERING SEASON: July–September.

HABITAT: Damp thickets, shorelines, and roadsides.

RANGE: Throughout the United States and Canada, except the Far North. It is most common in the Midwest.

SIMILAR SPECIES: Hoary Vervain (*V. stricta*) and Narrow-leaved Vervain *(V. simplex)* are both similar, but have smaller distribution ranges. Narrow-leaved Vervain's leaves are narrower.

CONSERVATION:
Globally secure but critically imperiled in Prince Edward Island; imperiled in Saskatchewan, North Carolina, and Wyoming; and vulnerable in British Columbia, Nova Scotia, and Montana.

HOARY VERVAIN *Verbena stricta*

■ NATIVE
■ RARE OR EXTIRPATED
■ INTRODUCED

This clump-forming perennial gets its common name from its stiff, white hairs on the leaves and stems. This is an important food plant for many birds, small mammals, and native butterflies, and is often included in butterfly gardens. This plant also has been widely used in traditional medicine. The specific epithet *stricta* means "erect" or "upright."

DESCRIPTION: This plant grows in a narrow clump, about 1–4' (30–120 cm) tall. It has a hairy stem and a terminal cluster of narrow, flowering spikes. The small lavender flowers appear in a ring halfway down the ascending spike. The leaves are covered with fine white hairs, oppositely arranged and about 4" (10 cm) long and 3" (7.5 cm) wide. The leaves are ovate or obovate, and coarsely serrated along the margins.

FLOWERING SEASON: July–September.

HABITAT: Fields, prairies, meadows.

RANGE: Throughout the interior United States and in Ontario and Quebec, and widely naturalized elsewhere. It is most common in the Great Plains and Midwest.

CONSERVATION: (G5)

Secure, but imperiled in Wyoming and Quebec. It is possibly extirpated in West Virginia.

■ NATIVE
■ RARE OR EXTIRPATED
■ INTRODUCED

This attractive native perennial is often grown for its pleasant-smelling flowers. In the wild this plant is an important nectar source for butterflies and of special value to native bees.

DESCRIPTION: This tall, attractive herb grows 2–5' (0.6–1.5 m) tall with a square stem, scattered with short hairs. The leaves are ovate and opposite, with toothed edges and a whitish under-side. Each leaf is about 4" (10 cm) long and has a hairy stem. The flowers are pale purple to purplish-red with a green calyx and grow in a crowded, terminal spike with purple bracts and four long stamens. The lower lip of the flower has three lobes, the middle of which is the biggest. The flower spike is purplish and 1–6" (2.5–15 cm) long.

FLOWERING SEASON: July–September.

HABITAT: Rich woods, woodland edges, thickets, and watery areas.

RANGE: Native from New England to the Midwest south to northernmost Georgia and the Carolinas.

SIMILAR SPECIES: Anise Hyssop (*A. foeniculum*) is similar, but its flowers and leaves are colored differently. The undersides of Anise Hyssop leaves are whitish where Purple Giant Hyssop's are green.

CONSERVATION: G4
Apparently secure but critically imperiled in Ontario, Connecticut, Georgia, Massachusetts, Maryland, Nebraska, Missouri, and Tennessee; imperiled in New Jersey and South Dakota; and vulnerable in North Carolina and West Virginia. It is possibly extirpated in Vermont and Kansas.

■ NATIVE
■ RARE OR EXTIRPATED
■ INTRODUCED

ALTERNATE NAMES: Creeping Bugleweed, Carpenter's Herb, Carpet-bugle

Common Bugle is a spreading plant native to Europe that is often grown in gardens; however it is considered a noxious or invasive weed in some areas, especially the Southeast. It's sometimes called Carpenter's Herb because it was used to stop bleeding after an injury.

DESCRIPTION: This sprawling perennial grows 6" (15 cm) tall with square stems. The basal leaves are whorled and spoon-shaped, while the stem leaves are opposite and green to reddish or purplish in color. It has an elongated spike of blue to purple tubular flowers. Each flower has erect stems and is about ¼" (6 mm) wide, bilater-

ally symmetrical, and tubular. The lobes on the lower lip have dark veins and are longer than the lobes on the upper lip. There are four stamens, two of which are noticeably longer than the corolla and are attached to the tube.

FLOWERING SEASON: Mid- to late spring.

HABITAT: Lawns, cultivated areas, parklands, woods.

RANGE: Introduced throughout North America, especially in the East and Pacific Northwest.

CONSERVATION:

This exotic species is unranked in North America, where it is largely considered a weed.

■ NATIVE
■ RARE OR EXTIRPATED
■ INTRODUCED

ALTERNATE NAMES: Wood Mint, Ohio Horsemint, Pagoda Plant

This is a native, clump-forming woodland perennial herb found in the East. This plant is of special value to bees and supports several beneficial native insects. The Cherokee traditionally used the plant to make a poultice to treat headaches.

DESCRIPTION: This plant has a hairy, stout, four-angled central stem, growing 1–2' (30–60 cm) tall. The leaves have deep venation and are covered in white hairs. The lower leaves have short stout petioles, while the upper leaves are sessile. Each leaf is about 3½" (4.75 cm) long and 1½" (4.25 cm) wide, opposite, and lanceolate. The flowers are white, pink, or lavender, and grow in clusters in the upper half of the central stem. Each cluster is about 2–3" across and round in shape, with the flowers growing in circular rows. The flowers have two lips, with purple spots on the lower one and hairs on the underside.

FLOWERING SEASON: May–August.

HABITAT: Thin woods, granitic or limestone sites, and open areas.

RANGE: New York and Massachusetts west to Wisconsin and Iowa, south to Georgia and East Texas.

SIMILAR SPECIES: Hairy Wood Mint (*B. hirsuta*) is similar but taller, and its stems are covered in longer, spreading hairs.

CONSERVATION: G5

Secure, but critically imperiled in Indiana, Kansas, Massachusetts, New York, and Ontario. It is vulnerable in Maryland, North Carolina, and Wisconsin. It is possibly extirpated in Connecticut, Delaware, New Jersey, Oklahoma, and Vermont.

ALTERNATE NAMES: Oregon-tea, Douglas' Savory, *Satureja douglasii*

This abundant plant is a flat, creeping native perennial of the West Coast that forms mats of bright-green, velvety, rounded leaves. The minty, fragrant leaves have been used to make a mild tea as a beverage and in folk medicine to treat many ailments, including the relief of pain in childbirth.

DESCRIPTION: This creeper has long, slender, trailing stems, about 2' (60 cm) long and 1' (30 cm) tall. The leaves are roundish, about ½–1" (1.5–2.5 cm) long, and opposite. A single white or pale purplish, bilaterally symmetrical flower appears in each upper leaf axil. The corolla is about ¼" (6 mm) long. The upper corolla lip is short, projecting forward, with a shallow notch at the tip. The lower corolla lip is longer, three-lobed and bent downward. There are four stamens.

FLOWERING SEASON: April–October.

HABITAT: Shaded woods.

RANGE: Southern British Columbia, northern Idaho, and south on the western side of the Cascade Range and the Sierra Nevada to Baja California.

SIMILAR SPECIES: The common name Yerba Buena means "good herb" and is given to many species of mint, especially Spearmint (*Mentha spicata*).

CONSERVATION: G5
Secure, but vulnerable in Montana.

■ NATIVE
■ RARE OR EXTIRPATED
■ INTRODUCED

ALTERNATE NAMES: Field Basil, *Satureja vulgaris*

Wild Basil is a native perennial herb that occurs around the globe in suitable habitats, especially in grasslands. The dried leaves can be used as a seasoning and to make herbal tea, although the leaves taste milder than those of commercial basil. The leaves are also used to make brown or yellow dye. The plant was traditionally used as an astringent and antibacterial, among other uses.

DESCRIPTION: This plant has a square, hairy stem and hairy leaves, growing 8–20" (20–50 cm) tall. The leaves are ¾–1½" (2–4 cm) long each and opposite, ovate, mostly untoothed. The flowers are rose-purple and grow in several dense, rounded clusters, with hairy bracts that make the plant look woolly. Each flower is about ½" (1.5 cm) long, with a hairy calyx and four stamens. The corolla has two lips.

FLOWERING SEASON: June–September.

HABITAT: Roadsides, pastures, and thickets.

RANGE: Patchy distribution throughout North America. It was probably introduced from Europe in many parts of its present range; also scattered in the western United States.

CONSERVATION: Ⓖ5

Globally secure but vulnerable in Indiana, New Jersey, Newfoundland, and Quebec.

■ NATIVE
■ RARE OR EXTIRPATED
■ INTRODUCED

ALTERNATE NAMES: Richweed, Stoneroot, Citronella Horse Balm, Canada Horsebalm

Horse Balm is the most widely distributed plant in the *Collinsonia* genus, native to eastern North America. The flowers of this tall wildflower are lemon-scented. Tea can be brewed from the leaves, and the rhizome was formerly used as a diuretic, tonic, and astringent.

DESCRIPTION: This perennial has a stout square stem with loose branching clusters of yellow flowers and grows 2–4' (60–120 cm) tall. The flowers are ⅜–½" (8–13 mm) long each, with a two-lipped corolla. The lower lip is long and fringed. The pistil and two stamens extend beyond the corolla tube. The leaves are opposite, ovate, and sharply toothed, each about 4–8" (10–20 cm) long.

FLOWERING SEASON: July–September.

HABITAT: Rich woods.

RANGE: Ontario and Quebec south to the Gulf Coast

CONSERVATION: G5
Globally secure but imperiled in Vermont, Rhode Island, and Louisiana. It is presumed extirpated in Wisconsin and possibly extirpated in Vermont.

■ NATIVE
■ RARE OR EXTIRPATED
■ INTRODUCED

ALTERNATE NAMES: Creeping Charlie, Gill-over-the-ground

This introduced plant is widely naturalized in North America. In many areas this aggressive plant is considered a lawn weed. It is used as a salad green in many countries, but it can be toxic if consumed in large amounts, especially to livestock.

DESCRIPTION: This plant has weak, creeping stems with four sides that grow 4–16" (10–40 cm) tall. It has opposite leaves that are kidney-shaped and round, about ½–1¼" (1.5–3 cm) long each, with scalloped edges, on long stalks. The flowers are lavender to blue-violet and bilaterally symmetrical, growing in clusters in each leaf axil. Each flower is ½–1" (1.5–2.5 cm) long. The corolla is spotted with purple. The lower corolla lip has three lobes, with a broad central lobe, while the upper corolla lip

arches outward like a hood. It has four stamens, two of which are hiding, and the other two are inside the corolla tube.

FLOWERING SEASON: March–June.

HABITAT: Moist, shaded or sunny areas, roadsides, and lawns.

RANGE: Introduced and naturalized throughout much of North America, except the Far North and the southwestern deserts.

CAUTION: This plant can poison livestock, especially horses, if eaten in large quantities.

CONSERVATION:

This introduced species in North America is unranked.

■ NATIVE
■ RARE OR EXTIRPATED
■ INTRODUCED

ALTERNATE NAMES: Cluster Blushmint

This is a native, shrubby perennial found in the southeastern U.S. The plant has a musky, minty odor, especially when crushed.

DESCRIPTION: This wetland perennial grows 3–4' (1–1.2 m) tall. It has square green or reddish-purple stems. The leaves are opposite, lance-shaped and toothed. The lower leaves are larger than the upper leaves and the upper leaves are stalkless. The flowers are white with pinkish or purple spots and grow in tight, rounded clusters. It has five lobes, the lower of which is yellowish. Only a few flowers appear open at a time. There are four hidden stamens that emerge and deposit pollen when a pollinator lands on the lobe. Fruits are tiny capsules.

FLOWERING SEASON: Year-round.

HABITAT: Wet flatwoods, prairies, and pond margins.

RANGE: Southeastern United States, North Carolina to Texas.

CONSERVATION: (G5)
Secure, but vulnerable in North Carolina.

■ NATIVE
■ RARE OR EXTIRPATED
■ INTRODUCED

ALTERNATE NAMES: Giraffehead, Common Deadnettle

This introduced herb is similar to true nettles (*Urtica*), but it is not related and does not sting. It is widely naturalized in North America. As weedy introduced plants go, this species is less threatening to native ecosystems, and many native birds, pollinators, and browsers readily take to the plant as a food source. The young leaves and shoots of this plant are edible but taste more like kale or celery than mint, and it can be used to make tea.

DESCRIPTION: A small plant with several four-sided stems that grows 3–6" (7.5–15 cm) tall. The leaves are ¾–1¼" (2–3 cm) long, and roundish, scalloped on edges and stalkless. It has a few purplish-pink to reddish-lavender, bilaterally symmetrical flowers in clusters in opposite bracts. The corolla is ½–¾" (1.5–2 cm) long, with a long narrow tube. The lower corolla lip is notched at the tip and strongly constricted at the base. The upper corolla lip is ¼" (6 mm) long, like a hood. It has four hidden stamens. The upper side of the flowers have purple hairs.

FLOWERING SEASON: February–October.

HABITAT: Waste places, fields, and roadsides.

RANGE: Introduced and naturalized throughout North America.

SIMILAR SPECIES: Purple Dead Nettle (*Lamium purpureum*) is similar, but its leaves are stalked.

CONSERVATION:
This introduced species in North America is unranked and largely considered a weed.

PURPLE DEAD NETTLE *Lamium purpureum*

■ NATIVE
■ RARE OR EXTIRPATED
■ INTRODUCED

ALTERNATE NAMES: Red Deadnettle

Purple Dead Nettle is another widespread dead nettle introduced from Eurasia and is often found alongside Henbit (*L. amplexicaule*). Its bright red-purple flowers are frequented by bees. The top leaves of young plants are used in salads and sauces and are high in iron, vitamins, and fiber. It's also used in many folk remedies and as a salve to treat skin conditions.

DESCRIPTION: This mint with four-sided stems grows up to 2' (60 cm) tall. The leaf surface is covered in hairs and tinged with purple. The leaves are oppositely arranged with wavy edges. Small purplish to pink flowers appear amid the leaf whorls toward the top of the stem. The flowers are bilaterally symmetrical. The upper lobe is hoodlike, while the lower lobe is divided in two. The corolla is sometimes hairy at the base of the tube.

FLOWERING SEASON: Year-round.

HABITAT: Turfgrass, landscapes, roadsides, waste places, gardens, and winter grain crops.

RANGE: Introduced throughout much of North America.

SIMILAR SPECIES: Purple Dead Nettle often grows beside Henbit (*L. amplexicaule*). Purple Dead Nettle's leaves have stalks on the stem, while Henbit's leaves are stalkless.

CONSERVATION:
This introduced species is unranked.

■ NATIVE
■ RARE OR EXTIRPATED
■ INTRODUCED

This perennial introduced from Eurasia is largely considered a weed in North America. It has been cultivated medicinally since ancient times, used by herbalists as a stimulant and for menstrual problems, hence the common name Motherwort. The species name *cardiaca* means "for the heart," based on the folklore surrounding this species as a cure for melancholy.

DESCRIPTION: This plant has square stems and grows 2–4' (60–120 cm). The leaves are horizontally held, opposite and lobed. The lower leaves are 2–4" (5–10 cm) long, and palmately cut into three lobes, while the upper leaves are smaller, less deeply cut. Several clusters of pale lavender flowers grow together forming a long terminal spike. Each flower is ½" (1.5 cm) long, ⅜" (8 mm) wide. The calyx has five united sepals, five veins and sharp spines at the tip. The corolla has two lobes. The upper lip is bearded and arching. The lower lip has three lobes. There are four stamens and leafy bracts beneath the cluster.

FLOWERING SEASON: June–August.

HABITAT: Waste places, roadsides, and disturbed areas.

RANGE: Introduced throughout much of North America, except Florida, California, and the Far North.

CAUTION: The foliage can cause skin irritation in some individuals.

SIMILAR SPECIES: Siberian Motherwort (*L. sibiricus*) is similar, but has a 10-veined calyx.

CONSERVATION:
This introduced plant is unranked and largely considered a weed in North America.

■ NATIVE
■ RARE OR EXTIRPATED
■ INTRODUCED

ALTERNATE NAMES: Cutleaf Water Horehound, American Bugleweed

There are 10 species of *Lycopus* in North America, sometimes called bugleweeds because each tiny flower resembles the shape of a bugle. Most of these species are very similar, making identification difficult. All of them grow in moist areas and are non-fragrant mints. Water Horehound is distinguished from other *Lycopus* species by having less coarsely toothed leaves. The genus name is from the Greek *lycos* ("a wolf") and *pous* ("foot"), referring to the leaf's resemblance to a wolf's footprint.

DESCRIPTION: This plant has square stems and grows 6–24" (15–60 cm) tall. The leaves are 1–3" (2.5–7.5 cm) long, lanceolate, coarsely toothed or lobed. The lower leaves are deeply toothed. Tiny white, tubular flowers clustered in dense groups in axils of opposite leaves. Each flower is ¹⁄₁₆" (2 mm) long. The calyx has 5 sharply bristle-tipped lobes. The corollas are 4-lobed and there are 2 stamens.

FLOWERING SEASON: June–September.

HABITAT: Moist sites, shorelines, and wetlands.

RANGE: Throughout North America, except the Far North and Nevada.

SIMILAR SPECIES: Other *Lycopus* species have more coarsely toothed leaves.

CONSERVATION: G5
Globally secure but critically imperiled in Alaska and Georgia; imperiled in North Carolina; and vulnerable in Alberta, Newfoundland, and Wyoming.

■ NATIVE
■ RARE OR EXTIRPATED
■ INTRODUCED

ALTERNATE NAMES: Canadian Mint

This widespread native mint is an aromatic perennial herb. All parts of the plant have a mint scent produced by glands that contain essential oils. Small purple or pink flowers on the stem are almost entirely covered by large, hairy leaves. The leaves are used as flavoring in sauces, jellies, and beverages.

DESCRIPTION: This plant grows upright, about 4–18" (10–46 cm) tall. The flowers are blue, pink, or purple, sometimes with a violet tint, appearing in bunches in the upper leaf axils. The leaves are large, hairy, and oppositely arranged. The fruit is a dry, two-chambered carpel that splits open to reveal two seeds when ripe.

FLOWERING SEASON: July–August.

HABITAT: Wet areas, lake and river edges.

RANGE: Across the northern two-thirds of North America and throughout the West.

CAUTION: The fruit of this plant is toxic and may be fatal (especially to children) if ingested in large quantities.

CONSERVATION: G5
Secure.

■ NATIVE
■ RARE OR EXTIRPATED
■ INTRODUCED

ALTERNATE NAMES: Horsemint, Lemon Mint

The purple flowers of this widespread native annual are highly attractive to hummingbirds, butterflies, and bees. In ideal conditions it can grow aggressively and form large colonies. The name *citriodora* comes from Latin and means "having a citrus scent," referring to the leaves that smell like lemon when rubbed or crushed. The leaves were once used as a balm for bee stings, hence the common name.

DESCRIPTION: This herb has stiff, square stems and grows 12–30" (30–76 cm) tall. The flowers are tubular and lavender to pink, growing in whorls. Two to six flower clusters appear on the top of each stem subtended by white or pink lavender bracts. Narrow, lanceo-late, opposite leaves grow about 1–3" (2.5–7.5 cm) long and ¼–¾" (6–19 mm) wide, with smooth margins.

FLOWERING SEASON: April–July.

HABITAT: Pastures, rocky or sandy prairies, and roadsides.

RANGE: Kentucky, Missouri, and Kansas south to Arkansas, Texas, and New Mexico south into Mexico; it is also introduced elsewhere in North America.

CONSERVATION: **G5**
Secure.

■ NATIVE
■ RARE OR EXTIRPATED
■ INTRODUCED

ALTERNATE NAMES: Oswego Tea, Horsemint, Beebalm

This widespread, showy native perennial is frequently grown in gardens and has fragrant leaves often used in mint tea. Oil from the leaves was also used to treat respiratory issues. The specific epithet *fistulosa* means "hollow," as in a pipe, and refers to the square, hollow stems.

DESCRIPTION: This clump-forming plant forms square stems and grows to a height of 2–4' (60–120 cm). The flowers are pink or purple and tubular, growing in a spherical cluster. Each flower is 1" (2.5 cm) long, with a two-lipped corolla. The upper corolla lip is two-lobed and hairy. The lower lip is broad, with three lobes. There are two stamens that project, with bracts beneath. Leaves are gray-green, opposite, lanceolate, coarsely toothed, about 2½" (6.5 cm) long.

FLOWERING SEASON: June–September.

HABITAT: Dry fields, thickets, and borders; usually common in limestone regions.

RANGE: Throughout the East, except the Maritime Provinces, Newfoundland, and Florida; also in much of the West.

CONSERVATION: G5

Secure, but critically imperiled in Delaware and Utah.

■ NATIVE
■ RARE OR EXTIRPATED
■ INTRODUCED

ALTERNATE NAMES: Horsemint

Punctata is the widespread subspecies of this common native plant; several other subspecies or varieties are found throughout the East. Spotted Beebalm smells like thyme or oregano. The plant contains thymol and has historically been used to treat upset stomachs, colds, kidney disease, and other ailments.

DESCRIPTION: This erect, aromatic perennial grows 6"–3' (15–90 cm) tall. It forms rosettes of tubular, yellowish, purple-spotted flowers that occur in whorls, forming a dense, elongated spike at the end of the stem or from the leaf axils. The whorls occur above whitish, or purple-tinged, leaf-like bracts.

FLOWERING SEASON: June–August.

HABITAT: Open, sandy areas.

RANGE: Vermont west to Minnesota, south to Texas and New Mexico and east to the Atlantic Coast. Isolated populations occur in California.

CONSERVATION: G5

Globally secure, but critically imperiled in Ontario, Ohio, and Vermont. It is possibly extirpated in Pennsylvania.

■ NATIVE
■ RARE OR EXTIRPATED
■ INTRODUCED

ALTERNATE NAMES: Catmint

This plant native to Europe and Asia is introduced in North America, often grown as an ornamental. It attracts butterflies. Cats are also drawn to the plant and often rub against, lick, or chew its foliage. The plant contains the chemical nepetalactone, which has sedative and relaxant properties believed to boost mood and reduce anxiety, and has long been used in traditional medicine, although there isn't enough conclusive medical research to determine the compound's effectiveness in humans. Nepetalactone has been shown to repel pest insects such as mosquitoes. Catnip is also used in herbal teas.

DESCRIPTION: This hairy perennial typically grows in a clump, reaching about 1–3' (30–90 cm) tall, with erect, multi-branched grayish stems. The stems and leaves are downy, giving the plant a gray-green appearance. The leaves are 2½" (6.5 cm) long, opposite, triangular, and coarsely toothed. Clusters of pale white or whitish-lavender flowers with purplish spots grow at the ends of the main stem and branches. Each flower is about ½" (1.5 cm) long. The calyx is hairy and the corolla is tubular, with two lips. There are four stamens.

FLOWERING SEASON: June–September.

HABITAT: Roadsides, waste places, pastures, and barnyards.

RANGE: Introduced throughout much of North America, except the Far North.

CONSERVATION: This exotic species in North America is unranked.

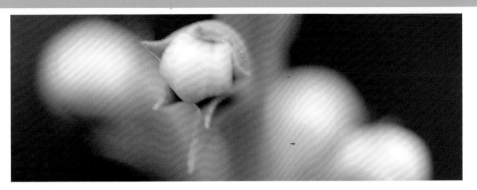

■ NATIVE
■ RARE OR EXTIRPATED
■ INTRODUCED

ALTERNATE NAMES: False Dragonhead

This native perennial herb has attractive flowers that look like those of dragonheads (*Dracocephalum* spp.). It is called Obedient Plant because its flowers bend if pushed and stay in that position temporarily.

DESCRIPTION: This plant has squarish stems and grows 1–4' (30–120 cm) tall. The leaves are narrow, lanceolate, and pointed. Each leaf is about 4" (10 cm) long and opposite, with sharp, incurved teeth, appearing smaller toward the top of the plant. It has opposite pinkish flowers that grow in a spike-like cluster at the top of the stem. Each flower is ¾–1" (2–2.5 cm) long, while the cluster is 4–8" (10–20 cm) long. The calyx has five pointed teeth. The corolla is tubular, with two lips and spotted with purple, enlarging outward. There are four stamens.

FLOWERING SEASON: June–September.

HABITAT: Damp thickets, swamps, and prairies

RANGE: Manitoba, east to New Brunswick, south to Florida, west to Texas, and north to North Dakota.

CONSERVATION: G5
Secure, but imperiled in Vermont.

■ NATIVE
■ RARE OR EXTIRPATED
■ INTRODUCED

ALTERNATE NAMES: Self-heal, Heal-all

This common plant is a perennial herb that is native throughout much of North America and in temperate zones around the world, as well as introduced widely outside its native range. The entire plant is edible and used in salads, soups, and stews. The common names come from this plant's use in herbal medicine to treat throat problems and other ailments.

DESCRIPTION: This plant has square stems, reaching 6–12" (15–30 cm) tall, which sometimes creep and self-root. The leaves are variable and are usually lanceolate to ovate and smooth or obscurely toothed. Each leaf is about 1–3" (2.5–7.5 cm) long and opposite. The flowers are purple or blue and grow in dense, cylindrical, terminal spikes that elongate after flowering. Each flower is about ½" (1.5 cm) long. The calyx has two lips. The upper lip is broader, with three teeth, the lower lip has two deep lobes. The corolla has two lips, the upper of which is arching and the lower lip is fringed and drooping. There are four stamens with greenish, hairy bracts beneath, overlapping the flower spikes.

FLOWERING SEASON: May–September.

HABITAT: Gardens, fields, and roadsides.

RANGE: Throughout much of North America, except the Far North.

CONSERVATION: **G5**
Secure, but imperiled in Saskatchewan and vulnerable in Alberta and Wyoming.

NARROWLEAF MOUNTAIN MINT *Pycnanthemum tenuifolium*

- ■ NATIVE
- ■ RARE OR EXTIRPATED
- ■ INTRODUCED

ALTERNATE NAMES: Slender Mountainmint

This is an herb native to most of eastern North America. As the common name suggests, Narrowleaf Mountain Mint has slender leaves. All parts of the plant have a faint, pleasant, mint-like fragrance when crushed. The genus name *Pycnanthemum* comes from Greek words meaning "dense" and "flower," referring to the densely clustered flowers. This plant, as well as other *Pycnanthemum* species, has been used in teas.

DESCRIPTION: This erect, multi-branched perennial grows 2–3' (60–90 cm) tall, with narrow, simple, needle-like leaves. Dense clusters of small, white flowers grow at the ends of the stems.

FLOWERING SEASON: July–September.

HABITAT: Dry, open rocky woods, dry prairies and fields, roadsides, streams open wet thickets.

RANGE: Eastern North America.

CONSERVATION: G5

Secure, but critically imperiled in Nebraska and vulnerable in Ontario. it is possibly extirpated in Quebec.

■ NATIVE
■ RARE OR EXTIRPATED
■ INTRODUCED

This eastern native perennial herb produces clusters of tiny, fragrant, white flowers that attract a host of native insects ranging from bees to beetles to butterflies. The plant has a strong mint-like fragrance when disturbed or crushed and has been used in teas.

DESCRIPTION: This perennial is multi-branched and grows 24–36" (60–90 cm) tall. The stems are erect, with four sides. The small, white flowers grow in dense clusters at the top of the stem, with only a few blooming at a time. The flowers often have purple spots and grow above white, hairy bracts. The flowers also grow in a round head, about 2" (5 cm) in diameter, with two lips. The leaves are narrow and taper off.

FLOWERING SEASON: July–September.

HABITAT: Moist to wet soils in woodland clearings, prairies, meadows, fields, streamsides, and bluffs.

RANGE: Maine to North Dakota, south to far northern Georgia and Oklahoma.

CONSERVATION: **G5**

Globally secure but critically imperiled in Alabama, Arkansas, Kansas, New Hampshire, North Carolina, and New Brunswick; imperiled in Georgia and Maryland; and vulnerable in North Dakota and Quebec.

■ NATIVE
■ RARE OR EXTIRPATED
■ INTRODUCED

ALTERNATE NAMES: Azure-blue Sage

Native to the eastern and central United States, Blue Sage is related to Garden Sage (*S. officinalis*), the seasoning used to flavor food. The name *Salvia* is derived from a Latin word meaning "to save or heal," referencing the medical properties of plants in the genus. The specific epithet *azurea* means "sky blue" and refers to the attractive blue flowers.

DESCRIPTION: This tall, delicate clump-forming perennial grows 2–5' (60–150 cm) tall with square stems. The stems are hairy or smooth and may flop over from the weight of the flowers. The leaves are opposite and linear to lanceolate on the stem, about 4" (10 cm) long. The flowers are blue to violet and appear in whorls, forming a terminal spike-like cluster. Each flower is ½–1" (1.5–2.5 cm) long. The corolla has two lips and is glandular and hairy outside. The lower lip is much larger than the upper lip. There are two stamens.

FLOWERING SEASON: July–October.

HABITAT: Open or shaded dry prairies and pastures and open pinelands.

RANGE: Wisconsin east to New York, south to Florida, west to New Mexico and Utah.

CONSERVATION: G4
Apparently secure.

- ■ NATIVE
- ■ RARE OR EXTIRPATED
- ■ INTRODUCED

ALTERNATE NAMES: California Sage

A native of the Southwest, Chia was an important food for Native Americans. The seeds were ground into meal and steeped in water to make a thick beverage or mush. The plant was also used as a fiber for building materials. The conspicuous flowers have a distinctive scent and attract birds and butterflies.

DESCRIPTION: This annual has four-sided, hairy stems and grows 4–20" (10–50 cm) tall. The leaves appear dense at the base, with a few stem leaves. Each leaf is about 4" (10 cm) long, oblong, and irregularly divided. It has small, very deep blue to purple, bilaterally symmetrical flowers that grow in round clusters in intervals. There are usually one or two clusters of flowers. The corolla is about ½" (1.5 cm) long, with prominent upper and lower lips. It has two stamens and reddish-purple bracts beneath the flower cluster.

FLOWERING SEASON: March–June.

HABITAT: Open places, deserts.

RANGE: Arizona, California, New Mexico, Nevada, and Utah.

CONSERVATION: 🅖5
Secure, but vulnerable in Utah.

■ NATIVE
■ RARE OR EXTIRPATED
■ INTRODUCED

ALTERNATE NAMES: Cancerweed

Bees favor this eastern native herbaceous perennial. It is sometimes grown as an ornamental for its attractive foliage and flowers, but it can easily escape and become a lawn weed. Native Americans used the root as a salve for sores, and used the whole plant as a tea to relieve colds and coughs. The leaves were once thought to be a cure for cancer, hence the alternate name Cancerweed.

DESCRIPTION: This plant has square stems, reaching 1–2' (30–60 cm) tall. The basal leaves are deeply lobed into rounded segments, about 8" (20 cm) long each, while the stem leaves are much smaller and short stalked or unstalked. The flowers are lavender or blue and grow in whorls of 3–10 in a spike-like cluster. Each flower is about 1" (2.5 cm) long. The corolla has two lips, the lower of which is longer than the upper lip. There are two stamens.

FLOWERING SEASON: April–June.

HABITAT: Sandy, open woods, thickets, and weedy sites.

RANGE: Connecticut and Pennsylvania south to Florida, west to Texas and Kansas.

CONSERVATION: **G5**

Secure, but critically imperiled in Kansas. It is possibly extirpated in New York.

■ NATIVE
■ RARE OR EXTIRPATED
■ INTRODUCED

ALTERNATE NAMES: Hooded Skullcap, Common Skullcap

The genus name *Scutellaria* is from a Latin word meaning "saucer," a reference to the round calyx. The common name references the cap-like shape of the flowers, which are said to look like military helmets worn in the Middle Ages. This plant traditionally has been used in medicinal teas and tablets to ease anxiety.

DESCRIPTION: This plant grows in patches, reaching about 4–31" (10–80 cm) tall. The leaves are oppositely arranged and lanceolate. Each leaf is about ¾–2" (2–5 cm) long with somewhat scalloped edges. One blue to purple, bilaterally symmetrical flower grows in each upper leaf axil. The corolla is ½–1" (1.5–2 cm) long. The upper lip is helmet-shaped, while the lower lip is bent downward.

FLOWERING SEASON: June–September.

HABITAT: Wet meadows, swamps, and along streams at moderate elevations.

RANGE: Throughout North America and much of the Northern Hemisphere.

CONSERVATION: 🅖5
Globally secure but critically imperiled in Delaware, Kansas, Missouri, Virginia, and West Virginia; imperiled in California, Maryland, and Labrador; and vulnerable in Wyoming and Yukon.

MAD-DOG SKULLCAP *Scutellaria lateriflora*

■ NATIVE
■ RARE OR EXTIRPATED
■ INTRODUCED

ALTERNATE NAMES: Blue Skullcap, Side-flowering Skullcap

This wetland perennial herb is native throughout North America, especially in the East. The name *lateriflora* means side-flowering, referring to the snapdragon-like flowers that grow on one side of the stem. The leaves have been used to treat insomnia, epilepsy, anxiety, and other neurological disorders. However, it should not be consumed in high doses because it may cause dizziness or extreme exhaustion. It has also been linked to liver damage.

DESCRIPTION: This slender, heavily branched perennial has square stems that are erect to reclining, reaching 24–32" (60–80 cm) tall. The leaves are dark green on the surface and light green beneath, opposite and heart-shaped at the base. Each leaf is ½–2½" (1.25–6.25 cm) long, with scalloped or toothed edges. The flowers are blue or lavender and grow on just one side of the stem from the upper leaf axils. Each flower is about ⅓" (8.5 mm) long.

FLOWERING SEASON: July–October.

HABITAT: Moist or wet places.

RANGE: Throughout North America, except Montana, Nevada, Utah, Wyoming, Alberta, and the Far North.

CAUTION: This plant may cause liver damage, vertigo, and exhaustion if consumed in high quantities.

CONSERVATION: G5

Globally secure but critically imperiled in Connecticut, imperiled in California, and vulnerable in Newfoundland and Saskatchewan.

- ■ NATIVE
- ■ RARE OR EXTIRPATED
- ■ INTRODUCED

ALTERNATE NAMES: Great Hedge-nettle

This is a perennial herb native to the West Coast from Alaska to California, with hairy and leafy stems often found in swamps. This plant is distinguished from other hedge-nettles by its large flowers and its stalkless leaves.

DESCRIPTION: This plant has four-sided, leafy stems and grows in patches, reaching 2–5' (60–150 cm) tall. All the leaves have stalks and are broadly lanceolate, with blunt teeth, about 2½–6" (6.3–15 cm) long, opposite and hairy on both sides. Red-dish-lavender, bilaterally symmetrical flowers grow in whorls at intervals in a spike at the top. The corolla is ⅝–1" (1.5–2.5 cm) long. The upper lip projects, resembling a hood. The lower lip has three lobes, which are much longer than the upper lips, and bent downward.

FLOWERING SEASON: June–August.

HABITAT: Swamps and moist low ground from sea level to moderate elevations, also common in disturbed sites.

RANGE: West Coast from Alaska to California.

CONSERVATION: **G5**
This plant is secure and common in suitable habitat.

NATIVE
RARE OR EXTIRPATED
INTRODUCED

Native throughout most of North America, Hairy Hedge-nettle is often grown as an ornamental. As its name suggests, the flowers are covered in hairs. Hummingbirds and butterflies visit the flowers for nectar and a variety of birds arrive later to enjoy the juicy (but bitter) berries.

DESCRIPTION: This sprawling or climbing vine can reach as long as 20' (6 m). The upper leaves grow in pairs, fused around the stem. Each leaf is about 12" (30 cm) long, oblong to ovate, and downy. It has pink, tubular flowers that grow in terminal spikes, ½" (12 mm) long, and are covered in glandular hairs, with five protruding stamens and two lips. The upper lip has four lobes. Berries grow in dense, shiny, red clusters, about ⅓" (8 mm) long.

FLOWERING SEASON: May–July.

HABITAT: Canyons, streamsides, woods.

RANGE: Throughout North America except in the Gulf Coast states and the Far North.

CONSERVATION: 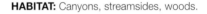 G5

Secure, but critically imperiled in Virginia; imperiled in Vermont; and vulnerable in California, Wyoming, and New Brunswick. It is possibly extirpated in Massachusetts.

■ NATIVE
■ RARE OR EXTIRPATED
■ INTRODUCED

ALTERNATE NAMES: American Germander

This common plant is native to most of North America. The flower's most distinctive feature is an enlarged lower lip that acts as a platform for pollinators to land on. The word "germander" comes from the Greek word *chamaidrys*, or "ground oak." Canada Germander was used by Native Americans to make teas to soothe sores and skin conditions.

DESCRIPTION: A rhizomatous perennial with a downy, square stem that grows 1–3' (30–90 cm) tall. The leaves are densely hairy beneath, about 2–4" (5–10 cm) long, opposite, lanceolate, and toothed. The flowers are pink or lavender and grow in terminal, spike-like clusters. Each flower is ¾" (2 cm) long, with the corolla seemingly one-lipped and with five lobes. The lower lobe is longer, flattened, and lateral. The upper lobes are shorter. There are four projecting stamens.

FLOWERING SEASON: June–September.

HABITAT: Thickets, woods, and shorelines.

RANGE: Throughout the Lower 48 states and southern Canada.

SIMILAR SPECIES: Introduced relatives Wood Sage (*T. scorodonia*), with yellow flowers, and Cut-leaved Germander (*T. botrys*)—an annual with purplish flowers—are both similar species, but are smaller and bushier.

CONSERVATION:  G5

Globally secure but critically imperiled in Wyoming; imperiled in Idaho and Prince Edward Island; and vulnerable in Georgia, British Columbia, Manitoba, New Brunswick, Nova Scotia, and Saskatchewan.

- ■ NATIVE
- ■ RARE OR EXTIRPATED
- ■ INTRODUCED

ALTERNATE NAMES: Bastard Pennyroyal

This annual herb is native to the eastern and midwestern United States and rare in southeastern Canada. Forked Bluecurls have showy blue flowers characterized by their long, curved stamens. They are visited by various native bees and wasps.

DESCRIPTION: This is a small, sticky plant with blue, two-lipped flowers at the tips of short branches rising from leaf axils to a height of 6–30" (15–75 cm). The leaves are ¾–2½" (2–6.5 cm) long, opposite, narrow, and oblong to lanceolate. Each flower is ½–¾" (1.5–2 cm) long and forms a panicle of cymes, which get smaller up the stem. There are also flowers at the tips. The corolla has two lips and five lobes. There are four upper lobes and one lower lobe, curving backward. Four long stamens protrude and curve. The fruit takes the form of two nutlets within the calyx.

FLOWERING SEASON: August–October.

HABITAT: Dry, open, sandy, or sterile sites.

RANGE: Ontario, Quebec, and Maine south to Florida, west to Texas, and north to Iowa, Illinois, and Michigan.

SIMILAR SPECIES: Perennial Bluecurls (*T. suffrutescens*) has tiny, oblong leaves and may not be a distinct species. False Pennyroyal (*T. brachiatum*) has short stamens that don't protrude. Narrowleaf Bluecurls (*T. setaceum*) has narrower leaves.

CONSERVATION: G5

Globally secure but critically imperiled in Ontario, Quebec, and Nova Scotia; imperiled in Michigan; and vulnerable in Illinois. It is possibly extirpated in Iowa.

- NATIVE
- RARE OR EXTIRPATED
- INTRODUCED

This native western plant was formerly called *Mimulus nanus*. It can widely vary in height depending on the amount of rainfall. In poor rainfall, it grows about ¼" (6 mm) tall and it flamboyantly produces one large flower. In years with more moisture, the plants will reach up to 4" (10 cm) tall, with many flowers.

DESCRIPTION: This is a diminutive plant covered with glandular hairs, bearing stems with one or many branches growing up to 4" (10 cm). The leaves are about 1½" (4 cm) long, lanceolate and opposite. Reddish-lavender, bilaterally symmetrical tubular flowers appear on short stalks in the upper leaf axils. Each flower is ½–1" (1.5–2.5 cm) long, with yellow and red lines in the mouth. There are five lobes, three of which are bent down-ward and two are bent upward and marked on the inside with yellow and deep red lines.

FLOWERING SEASON: May–August.

HABITAT: Dry open areas on sagebrush plains and in open pine forests.

RANGE: Central Washington south to northern California and east to northeastern Nevada, northwestern Wyoming, and southwestern Montana.

CONSERVATION: G5

Secure, but critically imperiled in Wyoming and imperiled in Montana.

COMMON MONKEYFLOWER *Erythranthe guttata*

■ NATIVE
■ RARE OR EXTIRPATED
■ INTRODUCED

ALTERNATE NAMES: Seep Monkeyflower, Golden Monkeyflower

This mainly western native species was formerly called *Mimulus guttatus* and still appears under this name in many sources. It can take a wide variety of forms and has been studied extensively as a model organism to explain evolution and ecology. Both annual and perennial forms occur. It is pollinated by bees.

DESCRIPTION: This is a highly variable, leafy plant, which may be spindly and tiny or large and bushy, 2–5' (60–150 cm) tall. The leaves have scalloped edges, and are fleshy, opposite, round to egg-shaped, and petiolate, sessile, or joined at the base at the stem tips. The flowers are yellow and bilaterally symmetrical, appearing on slender stalks in the upper leaf axils. The calyx is swollen and angular with five teeth at the opening, with the upper tooth the longest. The corolla is ½–1½" (1.5–4 cm) long, often with reddish spots near the opening. The upper corolla lip is two-lobed and bent upward. The lower corolla lip is three-lobed and bent downward.

FLOWERING SEASON: March–September. It blooms earlier at lower elevations and later at higher elevations.

HABITAT: Wet meadows and streambanks.

RANGE: Throughout the West from Alaska to the Northwest Territories south to Mexico, from sea level to around 12,000' (3650 m).

SIMILAR SPECIES: Common Monkeyflower is distinguished from the many look-alike monkeyflowers by its longer upper tooth on the angular calyx.

CONSERVATION: G5

Secure, but critically imperiled in Michigan, North Dakota, and Nebraska; imperiled in Saskatchewan; vulnerable in Alberta, Yukon, and New Mexico.

■ NATIVE
■ RARE OR EXTIRPATED
■ INTRODUCED

ALTERNATE NAMES: Musk Monkeyflower

This plant was formerly called *Mimulus moschatus*. As the common name suggests, this plant sometimes has a musky smell. It was grown commercially in the Victorian era for its fragrance, but cultivated specimens seemed to lose their scent in the 1910s. It was speculated throughout the 1920 and 1930s that wild populations of this plant had also mysteriously lost their scent. Botanists at the time speculated about whether the scent had been a recessive trait or cultivated from an abnormal parent plant, or even that humans could no longer detect the smell. Others reported evidence of fragrant wild-growing populations and suggested the loss of scent was merely a myth. It was later determined that some individual muskflowers have a stronger scent than others, and strongly scented plants are actually fairly rare.

DESCRIPTION: This is a sticky, hairy, sometimes slimy, musk-scented plant 8–16" (20–40 cm) tall. The leaves are short-stalked, about 1–2½" (2.5–6.5 cm) long each, opposite, and ovate to lanceolate. Sometimes the leaves are toothed. The flowers are yellow and tubular, appearing in the leaf axils on a weak, ascending stem. Each flower is ¾" (2 cm) long and bilaterally symmetrical, with a flattish face. There are five sepals. The corolla has five lobes, with two lips and an open throat. Four stamens are attached to the petals. There is one pistil. The fruit is a capsule.

FLOWERING SEASON: June–September.

HABITAT: Streambanks and pondsides.

RANGE: In the West, from British Columbia and Alberta south to California, Utah and Colorado. In the East, it occurs (possibly introduced) from Michigan and Ontario east to Newfoundland, south to North Carolina.

SIMILAR SPECIES: There are two similar species, both of which use the common name Yellow Monkeyflower. *E. guttata* is larger and its flowers have closed throats. *E. glabratus* is taller, up to 2' (60 cm), and its flowers are ¼–1" (6–25 mm) long, with wide open throats.

CONSERVATION: G5

This plant is classified as vulnerable in the United States. It is critically imperiled in North Carolina and Massachusetts and vulnerable in Montana.

ALTERNATE NAMES: Yellow Creeping Monkeyflower

This short, mat-forming native perennial is found in the western United States in a variety of mountainous and plateau environments. This plant was formerly called *Mimulus primuloides*. As the common names suggest, the flowers are yellow and primrose-like.

DESCRIPTION: A low, mat-forming perennial, growing 1" (2.5 cm) tall with small, leafless flower stalks. The leaves are typically light green or purplish and grow in dense rosettes. Each leaf is about 2" (5 cm) long. The leaf edges are often lined with white hairs. The flowers are bright yellow and primrose-shaped, with a tubular base, up to 2 cm long, with red-brown spots.

FLOWERING SEASON: June–August.

HABITAT: Moist meadows and wet, grassy banks, mainly between 4000–8000' (1200–2400 m).

RANGE: The U.S. West Coast to southwestern Montana and to Arizona and New Mexico.

CONSERVATION:

Apparently secure, but critically imperiled in New Mexico and Utah; and vulnerable in Montana.

■ NATIVE
■ RARE OR EXTIRPATED
■ INTRODUCED

ALTERNATE NAMES: Allegheny Monkeyflower

This widespread perennial herb is native to much of North America, especially in eastern and central parts of the continent. Some populations, especially in the West, may be introduced. The genus name is from the Latin word *mimus*, meaning "buffoon." Both the common name and scientific name reference the flowers, which look like the face of a monkey with a wide-open mouth.

DESCRIPTION: This plant has a square stem, growing 1–3' (30–90 cm) tall. The leaves are oppositely arranged, clasping the stem. Each leaf is about 2–4" (5–10 cm) long, unstalked, and oblong to lanceolate in shape. Blue flowers rise from the leaf axils, with usually only a few blooming at a time. Each flower is about 1" (2.5 cm) long and the stalks are ⅜–2½" (1–6 cm) high. There are five lobes—the upper corolla lip is erect and has two lobes, while the lower lip has three lobes and a pair of yellow spots inside. The corolla throat is white and nearly closed. There are four stamens. The fruit is a capsule.

FLOWERING SEASON: June–September.

HABITAT: Wet meadows and streambanks.

RANGE: Alberta to Nova Scotia; south to Georgia and northeastern Texas.

SIMILAR SPECIES: Sharp-winged Monkeyflower (*M. alatus*) has shorter flower stalks, about ½" (1.5 cm) long each. The leaves are also stalked and the stem is winged. It is more common farther south and west.

CONSERVATION: G5

Globally secure but critically imperiled in Alberta, Arkansas, Idaho, and Mississippi; and imperiled in Prince Edward Island, Saskatchewan, Louisiana, and Montana. It is possibly extirpated in Colorado.

■ NATIVE
■ RARE OR EXTIRPATED
■ INTRODUCED

ALTERNATE NAMES: American Lopseed

This woodland plant native to eastern and central North America produces downward-hanging fruit, from which it gets its common name. This species is placed in its own family, Phrymaceae. It is similar to members of the mint family (Lamiaceae), but each of its flowers produces only one seed, while those in the mint family produce two to four seeds.

DESCRIPTION: This plant has diverging branches, growing to a height of 1–3' (30–90 cm). The leaves are coarsely toothed. The upper leaves have shorter stalks than the lower leaves. Each leaf is 2–6" (5–15 cm) long and opposite. It has pairs of small, white or pinkish-lavender flowers growing in spike-like clusters. Each flower is ¼" (6 mm) long and the cluster is 6" (15 cm) long. The corolla has two lips, with the lower lip much longer than the upper lip. The upper lip also has a rounded edge and is indented at the center. The lower lip is divided into three lobes. There are four stamens. The fruit is dry and seed-like, enclosed in a calyx and hanging down against the stem.

FLOWERING SEASON: July–September.

HABITAT: Moist woods and thickets.

RANGE: Manitoba south Texas, and east to the Atlantic.

CONSERVATION: G5
Globally secure but critically imperiled in Wyoming; imperiled in New Brunswick; and vulnerable in Manitoba. It is possibly extirpated in Maine.

- ■ NATIVE
- ■ RARE OR EXTIRPATED
- ■ INTRODUCED

ALTERNATE NAMES: Purple False Foxglove

The flowers of this eastern native annual are showy and grow in abundance. It was formerly classified in the genus *Gerardia*, but this species is now in the genus *Agalinis*, while similar species, including Yellow-flowered False Foxgloves (*Aureolaria flava*), were reclassified to the genus *Aureolaria*. However, species in both genera have similar blooms and are difficult to distinguish.

DESCRIPTION: This plant has branching stems and grows 1–4' (30–120 cm) tall. The leaves are opposite and linear, about 1–1½" (2.5–4 cm) long and ⅛" (4 mm) wide. The flowers are pink to rose-colored and bell-shaped on short stalks and arise from the leaf axils at the top of the plant. Each flower is about 1" (2.5 cm) long. The calyx has five fused sepals that are finely hairy. The corolla has five unequal, spreading lobes. The upper two lobes are slightly smaller. The inside of the tube is white with darker reddish-purple spots and usually a pair of pale yellow stripes. There are four hairy stamens, white at the tip, with yellow anthers, attached to the petals, with one pistil. The fruit is a round capsule.

FLOWERING SEASON: July–September.

HABITAT: Moist soil in sandy fields, rocky shores, and serpentine barrens.

RANGE: Ontario and New Hampshire south to the Gulf of Mexico.

SIMILAR SPECIES: Seaside Gerardia (*A. maritima*), found in coastal saltmarshes from New Brunswick and Nova Scotia south to Florida and Texas, is a smaller plant with fleshy, linear leaves and smaller flowers. Small Flower False Foxglove (*Agalinis paupercula*) is also similar and easily confused; it is generally a smaller plant with flowers about half the size and typically lighter pink, with calyx lobes that are shorter or about as long as the calyx tube. Slender-leaved False Foxglove (*Agalinis tenuifolia*) has a longer flower stalk.

CONSERVATION: G5

Globally secure but critically imperiled in Maine; imperiled in Missouri and Nebraska; and vulnerable in Iowa, New York, and West Virginia.

■ NATIVE
■ RARE OR EXTIRPATED
■ INTRODUCED

ALTERNATE NAMES: Naked Broomrape, *Aphyllon uniflorum*

One-flowered Cancer Root is a parasite, obtaining nutrients and water from the roots of nearby plants rather than through photosynthesis.

DESCRIPTION: This widespread native plant uses several kinds of host plants, especially stonecrops (genus *Sedum*). The main stem of this plant is mostly underground. The slender, yellowish-brown flower stalks rise about 1¼–4" (3–10 cm) tall. The leaves are reduced to scales. One purple, lavender, yellowish, or whitish, bilaterally symmetrical flower appears at the tip of each slender stalk. The flowers are about 1" (2.5 cm) long each. The corolla is slightly bent downward in the middle of the tube, with two yellow stripes on the lower side. The lobes are fringed with fine hairs.

FLOWERING SEASON: April–August.

HABITAT: Damp woods and thickets; open places from lowlands to moderate elevations in mountains.

RANGE: Most of North America. Southwestern Canada south throughout West; Ontario east to Newfoundland, south to Florida, west to Texas, and north to South Dakota.

SIMILAR SPECIES: Clustered Broomrape (*O. fasciculata*) is a similar species, but has stouter stems with more flowering pedicels (3–12, usually 5–10) and the stems and pedicels are about equal in length. (The one to three pedicels of *O. uniflora* are longer than the stems.)

CONSERVATION: G5

Globally secure but critically imperiled in Prince Edward Island and Saskatchewan in Canada and Alaska, Arizona, Louisiana, Mississippi, Nebraska, and Oklahoma in the U.S. It is imperiled in New Brunswick, Alabama, Kansas, Minnesota, South Carolina, and South Dakota. It is vulnerable in Alberta, Newfoundland, Delaware, Indiana, New Jersey, North Carolina, Wisconsin, and Wyoming. It is possibly extirpated in North Dakota.

■ NATIVE
■ RARE OR EXTIRPATED
■ INTRODUCED

ALTERNATE NAMES: Littleleaf Mimosa, Catclaw Sensitive-brier

This eastern native perennial is hemiparasitic, attaching to the roots of pines and various hardwoods, most commonly species of oak. It obtains some nutrients from its host in this way but also photosynthesizes. Bumblebees are often attracted to this plant and its yellow flowers.

DESCRIPTION: This plant has weak, arching purplish stems, trailing on the ground and growing to variable lengths. The leaves are opposite and pinnately compound, with many leaflets. The upper leaf surface is deep green with stiff hairs. The undersurface is lighter green and sparsely roughened to glabrous, easily broken or damaged. The flowers are yellow, bell-shaped and short-stalked. The corollas are 1⅓–2⅓" (35–60 mm) long, with

five rounded lobes. There are four stamens, fused at the base of the corolla tube.

FLOWERING SEASON: June–September (year-round in Florida).

HABITAT: Pinelands, grasslands, roadsides, sinkhole ponds.

RANGE: The eastern U.S. west to Texas, Missouri, and Illinois, with limited distribution in Ontario.

CONSERVATION: **G5**

Globally secure, but classified as imperiled in Canada where its distribution is limited. It is critically imperiled in Delaware; imperiled in Ontario, Vermont, and Pennsylvania; and vulnerable in Maryland and North Carolina.

NATIVE
RARE OR EXTIRPATED
INTRODUCED

ALTERNATE NAMES: Applegate's Paintbrush, Wavyleaf Indian-paintbrush

Wavyleaf Paintbrush is a hemiparasitic plant, meaning it obtains nutrients from photosynthesis and by attaching to the roots of host plants, including sagebrushes (*Artemisia* spp.). The modified leaves are vivid red and look like flowers, but they are bracts. The actual flowers are tiny, greenish tubes. There are five subspecies that are sometimes treated as separate species.

DESCRIPTION: This perennial has leafy stems, rising 6–16" (15–40 cm) from a woody crown-root. The leaves are gray-green and usually lance-shaped. The upper leaves are divided into three to five segments, while the lower leaves are undivided. Tubular yellow-green flowers with vibrant pink to scarlet red bracts and sepals appear at the top of erect stems. The bracts and sepals are usually hairy and cup-shaped.

FLOWERING SEASON: April–August.

HABITAT: Dry sagebrush slopes; chaparral; pine forests.

RANGE: Eastern Oregon to Montana, south to southern California and northwestern New Mexico.

SIMILAR SPECIES: Wyoming Indian-paintbrush (*C. linariifolia*) is similar but grows taller, up to 3' (1 m).

CONSERVATION: G5
Globally secure but imperiled in Wyoming.

■ NATIVE
■ RARE OR EXTIRPATED
■ INTRODUCED

ALTERNATE NAMES: Scarlet Indian-paintbrush

This plant has attractive red blooms that look like they've been dipped in paint. It is native across eastern North America but tends to occur in small and scattered populations. Ruby-throated Hummingbirds (*Archilochus colubris*) are a primary pollinator of this annual.

DESCRIPTION: This plant grows 1–2' (30–60 cm) tall. The leaves form a rosette. Those at the base are 1–3" (2.5–7.5 cm) long and elliptical in shape. Those leaves on the stem are stalkless, divided into three to five narrow segments. The flowers are greenish-yellow in color and tubular, appearing hidden in reddish-orange fan-shaped bracts in a dense spike. Each flower is about 1" (2.5 cm) long. The corolla has a long, two-lobed upper lip arching over a shorter, three-lobed lower lip. The styles protrude beyond bracts. The fruit is a capsule.

FLOWERING SEASON: May–July.

HABITAT: Meadows, prairies, and fields.

RANGE: Saskatchewan east to Ontario, New York, and Connecticut, south to Florida, west to Louisiana and Oklahoma, and north to Minnesota.

SIMILAR SPECIES: Purple Painted Cup (*C. purpurea*) is similar, but has purple or violet bracts. Downy Painted Cup (*C. sessiliflora*) has green bracts. Pale Painted Cup (*C. septentrionalis*) has whitish or cream-colored bracts.

CONSERVATION: G5

Globally secure, but its population is scattered, making it rare in some parts of its range. It is critically imperiled in Alabama, Connecticut, Georgia, Kentucky, Maryland, Mississippi, New Jersey, New York, West Virginia, and Saskatchewan; imperiled in Pennsylvania and South Carolina; and vulnerable in Indiana, Iowa, Kansas, North Carolina, and Virginia. It is presumed extirpated in Delaware, Maine, and New Hampshire and possibly extirpated in Louisiana, Massachusetts, and Rhode Island.

NATIVE
RARE OR EXTIRPATED
INTRODUCED

ALTERNATE NAMES: Great Red Indian Paintbrush, Meadow Paintbrush, Greater Red Indian-paintbrush

This native western perennial, like most in its genera, is a partial parasite. Its roots attach to roots of nearby plants for nutrients. The flower clusters look like a ragged red paintbrush.

DESCRIPTION: This leafy plant grows to a height of 1–3' (30–90 cm). The leaves on the stem are lanceolate. Each leaf is about 4" (10 cm) long with thin hairs. Some of the upper leaves and the colorful bracts have three pointed lobes. The flower cluster resembles a red or scarlet paintbrush. The calyx and bracts beneath each flower are brightly colored while the flowers are greenish. This calyx has four pointed lobes. The corolla is bilaterally symmetrical and relatively inconspicuous, about ¾–1½" (2–3.8 cm) long.

FLOWERING SEASON: May–September.

HABITAT: Mountain meadows, thickets, and forest openings.

RANGE: Alaska south and east to Ontario, and south to northern California, Nevada, northern Arizona, northern New Mexico, and North Dakota.

CONSERVATION: G5
Secure, but critically imperiled in Yukon and vulnerable in Alaska and Arizona.

■ NATIVE
■ RARE OR EXTIRPATED
■ INTRODUCED

ALTERNATE NAMES: Great Plains Indian Paintbrush, Downy Painted Cup, Downy Indian-paintbrush

This is a hemiparasitic plant, obtaining water and nutrients through the roots of nearby plants. Host plants include prairie grasses, such as Hairy Grama (*Bouteloua hirsuta*) and June grasses (*Koeleria* spp.), and other wildflowers. This plant can have a different appearance depending on moisture availability and soil quality. In dry areas it is reduced to diminutive tufts, but the plants are taller and more open in better soil.

DESCRIPTION: A hairy plant with clustered stems 4–12" (10–30 cm) tall. The leaves are 1–2" (2.5–5 cm) long. The lower leaves are narrow, while the upper leaves appear in a pair of narrow lobes. It has yellow to pinkish bracts and the corollas protrude like long, pale, curved beaks. The calyx is the same color as the bracts, with four long narrow lobes. Each corolla is 1¼–2½" (3–6.5 cm) long, bilaterally symmetrical, and pale

yellow to pale pink. The upper corolla lip is beak-like, about ½" (1.5 cm) long. The lower corolla lip has two flared lobes.

FLOWERING SEASON: March–September.

HABITAT: Dry, open, rocky or sandy knolls or slopes on plains and among pinyon and juniper.

RANGE: Throughout the Great Plains to western Texas and southeastern Arizona; also in northern Mexico.

CONSERVATION: Ⓖ5

Globally secure, but classified as vulnerable in Canada. It is critically imperiled in Alberta; imperiled in Illinois and Missouri; and vulnerable in Manitoba, Saskatchewan, Iowa, and Wyoming.

■ NATIVE
■ RARE OR EXTIRPATED
■ INTRODUCED

American Cancer-root is a parasitic plant that does not contain chlorophyll and gets its nourishment from the roots of nearby oak trees. The genus name comes from the Greek words *conos* meaning "cone," and *pholos* meaning "scale" and refers to the pinecone-like appearance of this and related plants.

DESCRIPTION: A parasitic plant growing 3–10" (7.5–25 cm) tall. The leaves are reduced to yellow-tan scales that are lanceolate or ovate in shape. Yellowish to cream-colored flowers emerge on the upper part of a fleshy stalk. Each flower is ½" (1.5 cm) long. The upper corolla lip forms a narrow hood over a three-lobed, spreading lower lip. There are four stamens. The fruit is a capsule.

FLOWERING SEASON: May–June.

HABITAT: Woods, under oaks.

RANGE: Throughout much of the East.

SIMILAR SPECIES: Alpine Cancer-root (*C. alpina*) is closely related, but is found in the southwestern United States.

CONSERVATION: G5
Globally secure but critically imperiled in Iowa, Mississippi, Manitoba, and Nova Scotia; imperiled in Delaware and Rhode Island; vulnerable in New Hampshire, Vermont, and Quebec.

- ■ NATIVE
- ■ RARE OR EXTIRPATED
- ■ INTRODUCED

This eastern native annual produces many brown stems with buff-brown or dull magenta flowers. It is a parasitic plant that does not contain chlorophyll and is often found under beech trees, where it receives nourishment from the roots. Dry stalks often grow all winter under the trees.

DESCRIPTION: This plant has unbranched to many-branched, brownish-tan stems growing 6–18" (15–45 cm) tall. The leaves are reduced to scales. The flowers are buff-brown or dull magenta, growing in axils of scattered dry scales. The upper flowers are ½" (1.5 cm) long and tubular, while the lower flowers are ¼" (5 mm) long and bud-like. The flowers never open and are self-pollinated. Small, brown, many-seeded capsules are ¼" (6 mm) long, opening like a clamshell at the top. The capsules open by raindrops.

FLOWERING SEASON: August–October.

HABITAT: Woods, under beech trees.

RANGE: Ontario east to Nova Scotia, south to Florida, west to Texas, and north to Missouri, Illinois, and Wisconsin.

CONSERVATION: (G5)
Secure, but imperiled in Missouri and vulnerable in Quebec and Prince Edward Island.

■ NATIVE
■ RARE OR EXTIRPATED
■ INTRODUCED

ALTERNATE NAMES: American Cow Wheat, Narrowleaf Cow Wheat

This native woodland annual is found across southern Canada and the northern United States, extending down the Appalachian Mountains. Cow Wheat's flowers resemble the shape of a snake's head. This species is partially parasitic, using both photosynthesis and attaching to roots of nearby plants for energy. The genus name is from the Greek words for "black" and "wheat" and refers to the black seeds found in some species.

DESCRIPTION: This is a low, clump-forming plant with upward branches, reaching 6–18" (15–45 cm) tall. There are often two to four bristly teeth at the base. The leaves are ¾–2½" (2–6.5 cm) long, opposite, and linear to lanceolate-ovate in shape. The flowers are small, tubular, and creamy white, with two-lipped flowers on short stalks in the upper leaf axils. Each flower is about ½" (1.5 cm) long. The upper corolla lip is white and arching, with two lobes. The lower lip is yellow and three-lobed. There are four stamens under the upper corolla lip, two of which are shorter. The fruit is a capsule.

FLOWERING SEASON: June–August.

HABITAT: Dry to moist woods, bogs, and rocky barrens.

RANGE: Mostly in southern Canada and the northern United States and south along the Appalachian Mountains to Georgia.

CONSERVATION: **G5**
Secure, but critically imperiled in Indiana and vulnerable in Ohio and Newfoundland. It is presumed extirpated in Illinois.

■ NATIVE
■ RARE OR EXTIRPATED
■ INTRODUCED

ALTERNATE NAMES: Golden-tongue Owl's-clover

This northern and western native plant has eye-like spots on the petals, resembling those of owls. The large corollas also resemble the shape of an owl looking down between the leaves of a tree.

DESCRIPTION: This erect annual grows 4–16" (10–40 cm) with a slender, hairy yellowish-green or purple stem. The leaves are ½–1½" (1.3–3.8 cm) long each and very narrow. The upper leaves are sometimes divided into three narrow lobes. It has many golden-yellow flowers, angled upward and protruding from glandular-hairy bracts near the top. The corolla is bilaterally symmetrical and about ½" (1.3 cm) long, with a pouch on the lower side near the tip. There are three tiny teeth at the end of the pouch and above the pouch. The upper lip forms a short, curving beak. There are four stamens inside.

FLOWERING SEASON: July–September.

HABITAT: Plains and open woods.

RANGE: British Columbia to southeastern California, mostly east of the Cascade Mountains and the Sierra Nevada; east to New Mexico, Nebraska, Minnesota, and Manitoba.

CONSERVATION: G5

Secure, but critically imperiled in the Yukon.

■ NATIVE
■ RARE OR EXTIRPATED
■ INTRODUCED

ALTERNATE NAMES: Great Basin Lousewort

This western native is among the shortest of the louseworts, which are well represented in North America. The flowers of this plant often sit almost upon the ground.

DESCRIPTION: This small perennial grows 4" (10 cm). The leaves are fern-like, about 6" (15 cm) long, pinnately divided into many leaflets, with jagged lobes and teeth. The flowers are pale yellow and bilaterally symmetrical on short racemes. The flowers are club-shaped. The corolla is about ¾" (2 cm) long. The upper corolla lip resembles an overturned canoe. The lower corolla lip is three-lobed, about as long as the upper lip.

FLOWERING SEASON: May–July.

HABITAT: Dry coniferous woods in mountains.

RANGE: Southern Oregon south to southern California and western Nevada.

CONSERVATION: G4

Apparently secure, but imperiled in California.

■ NATIVE
■ RARE OR EXTIRPATED
■ INTRODUCED

ALTERNATE NAMES: Bull Elephant's Head

The pink flowers of this northern and western native plant take the shape of an elephant head—ears, trunk, and all. It is a parasitic plant, obtaining nutrients from the roots of nearby plants.

DESCRIPTION: This plant grows to 28" (70 cm) tall. The leaves are about 2–10" (5–25 cm) long, very narrow, and pinnately divided into sharp-toothed lobes, like ferns. The flowers grow on leafy stems in dense pink racemes that look like elephant heads. Each flower is about ½" (1.3 cm) long and bilaterally symmetrical. The upper lip, or the "trunk," curves forward, well beyond the lower lip. Three lobes form the "ears" and the lower part of the "elephant's head."

FLOWERING SEASON: June–August.

HABITAT: Wet meadows and small cold streams, often at high elevations in the West.

RANGE: Throughout northern North America and in the West, south in the mountains to California, Arizona, and New Mexico.

SIMILAR SPECIES: Little Elephant Heads (*P. attollens*) is similar but shorter, only growing 16" (40 cm) tall. The corolla is white and rose-colored and looks less like an elephant head.

CONSERVATION: G5
Secure, but critically imperiled in Newfoundland; imperiled in Alaska, Saskatchewan, and Yukon; and vulnerable in Manitoba and Labrador.

SICKLETOP LOUSEWORT *Pedicularis racemosa*

■ NATIVE
■ RARE OR EXTIRPATED
■ INTRODUCED

ALTERNATE NAMES: Parrot's-beak, Leafy Lousewort

This western native perennial has a distinctive flower with a beak-like hook, leading some observers to call the plant Parrot's-beak. The undivided leaves separate this plant from other western louseworts. It is often found at the base of spruce trees (*Picea* spp.).

DESCRIPTION: A bushy, leafy plant with several twisted stems growing 6–20" (15–50 cm) tall. The leaves are 2–4" (5–10 cm) long and lanceolate, with many tiny blunt teeth on the edges. The flowers are white or pink and grow on short branches in the upper leaf axils. The corolla is about ½" (1.5 cm) long. The upper corolla lip is narrow and arched. A beak-like hook twists to one side,

touching the lower lip. The lower corolla lip is broad and three-lobed. The central lobe is smaller than the side lobes.

FLOWERING SEASON: June–September.

HABITAT: Coniferous woods and dry meadows in mountains, often associated with spruce trees.

RANGE: British Columbia and Alberta south to the southern Sierra Nevada, Idaho, Utah, and eastern New Mexico, and in the Rocky Mountains from Montana to New Mexico.

CONSERVATION: G5
Secure, but critically imperiled in Alberta; vulnerable in Wyoming.

■ NATIVE
■ RARE OR EXTIRPATED
■ INTRODUCED

ALTERNATE NAMES: Little Yellow Rattle

This annual is mainly a northern species found across the Northern Hemisphere; both native and non-native subspecies may be found in North America. As the common name suggests, the seeds of this plant rattle in the capsule at maturity. Yellow Rattle is partially parasitic, obtaining some of its nutrients from the roots of other plants and some of its nutrients from photosynthesis.

DESCRIPTION: This annual has an erect, sometimes branched stem, growing 4–31" (10–80 cm) tall. The leaves are oppositely arranged and stalkless. Each leaf is about ¾–2½" (2–6.5 cm) long and triangular-lanceolate to oblong in shape, and toothed. Yellow, stalkless flowers appear in a leafy, one-sided spike. The corolla is about ½" (1.5 cm) long. The upper corolla lip is hood-like, with two small round teeth. The lower corolla lip is smaller, with three round teeth projecting forward. The calyx is oval, becoming larger and inflated as the fruit develops, with four teeth at the tip. The fruit is a flattened, circular capsule containing loose, rattling seeds.

FLOWERING SEASON: May–September.

HABITAT: Meadows and fields, thickets, and moist slopes.

RANGE: Throughout the Northern Hemisphere; in the West, south to northwestern Oregon, Arizona, and New Mexico; in the East, south to New York and Connecticut.

CONSERVATION: G5

Globally secure but critically imperiled in Arizona and vulnerable in Newfoundland, Saskatchewan, and Idaho.

■ NATIVE
■ RARE OR EXTIRPATED
■ INTRODUCED

ALTERNATE NAMES: Garden Bellflower, Creeping Bellflower

Common Bellflower is a Eurasian species that has escaped cultivation and become a weedy invasive in North America. It spreads by rhizomes and each plant can produce as many as 15,000 seeds, allowing it to take root rapidly. The genus name comes from the Latin *campana* and means "little bell," a reference to the bell-like flowers.

DESCRIPTION: This plant grows up to 3' (90 cm) tall with smooth to hairy stems containing milky sap. The leaves have hairs on the underside and are alternately arranged. The lower leaves are lance- to heart-shaped, while the upper leaves are lanceolate. Each leaf is about 5" (12 cm) long. The flowers are blue or purplish and bell-like, appearing in a long raceme. Each flower is about 1" (2 cm) long, with five lobes and grows on one side of the thin stem.

FLOWERING SEASON: June–August.

HABITAT: Fields, roadsides, woodlands, disturbed areas.

RANGE: Widespread throughout North America.

SIMILAR SPECIES: Harebell (*Campanula rotundifolia*) has similar flowers and leaves, but is less widespread.

CONSERVATION:

This exotic species is not ranked in North America and is largely considered a weed. It can be highly invasive and is restricted in several states and provinces.

■ NATIVE
■ RARE OR EXTIRPATED
■ INTRODUCED

ALTERNATE NAMES: Bluebells-of-Scotland, Witches' Thimble

This widespread plant is native across the temperate parts of the Northern Hemisphere. It produces bell-shaped, blue-violet flowers singly or in clusters along the top parts of nodding, thread-like stems that grow in small patches. This plant has been mentioned by poets like Emily Dickinson and William Shakespeare.

DESCRIPTION: This delicate perennial grows to a height of 4–40" (10–100 cm). The leaves at the base are round and usually wither before the plant flowers. The leaves along the stem are ½–3" (1.5–7.5 cm) long and very narrow. Blue-violet, bell-shaped flowers appear at the top of long, mostly unbranched stems. The corolla is about ½–1" (1.5-2.5 cm) long, with five pointed lobes somewhat curved back. The style and corolla are about the same length.

FLOWERING SEASON: June–September.

HABITAT: Rocky banks and slopes, meadows, and shorelines.

RANGE: Throughout much of North America, except the southeastern United States.

CONSERVATION:
Globally secure but critically imperiled in Missouri, North Carolina, and Virginia; imperiled in Maryland, Ohio, and Yukon; and vulnerable in Iowa, New Jersey, and West Virginia.

■ NATIVE
■ RARE OR EXTIRPATED
■ INTRODUCED

ALTERNATE NAMES: American Bellflower, *Campanula americana*

This annual is native to eastern and central North America. Despite its name, the flowers are usually flat rather than bell-shaped, unlike other bellflowers. Tall Bellflower was formerly included in the genus *Campanula*; some sources consider *Campanula americana* to be the species' correct name.

DESCRIPTION: This plant has an erect, hairy stem that grows 2–6' (60–180 cm) tall. The leaves are somewhat rough, about 3–6" (7.5–15 cm) long, thin, ovate to lanceolate, and toothed. The flowers are light blue to violet, radially symmetrical, and form an elongated, spike-like cluster. The corolla is ¾–1" (2–2.5 cm) wide and flat, with five deeply cleft lobes. It has a long, curving and recurving protruding style. Bracts beneath the lower flowers are leaf-like. The upper lobes are awl-shaped. Each cluster is 1–2' (30–60 cm) long.

FLOWERING SEASON: June–August.

HABITAT: Rich moist thickets and woods.

RANGE: Ontario and New York south to Florida, west to Louisiana and Oklahoma, and north to South Dakota and Minnesota.

CONSERVATION: **G5**
Secure, but critically imperiled in Louisiana, New York, and South Carolina and vulnerable in Mississippi.

■ NATIVE
■ RARE OR EXTIRPATED
■ INTRODUCED

ALTERNATE NAMES: Bacigalupi's Downinga

This attractive annual is native to the West. It's named after Rimo Bacigalupi (1917–1983), a botanist and former curator of the Jepson Herbarium at the University of California, Berkeley, who was known by the nickname Bach.

DESCRIPTION: This plant has erect, sometimes branching stems, growing 2–12" (5–30 cm) tall. The leaves are small and diamond-shaped or lanceolate, often gone before the flowers appear. One or more tiny bluish flowers appear at the top of each stem. Each flower has two long upper lobes, usually with blue veins, that extend upward like ears. The three lower lobes are spotted with yellow and fused together, creating a cup shape. The center of the flower is usually white.

FLOWERING SEASON: April–August.

HABITAT: Drying mud of vernal pools, muddy margins of lakes, wet meadows, roadsides, irrigation ditches and stream-banks, 3000–6200' (900–1900 m) elevation.

RANGE: California, Idaho, Nevada, and Oregon.

CONSERVATION: G4

Apparently secure, but imperiled in Idaho.

CARDINAL FLOWER *Lobelia cardinalis*

■ NATIVE
■ RARE OR EXTIRPATED
■ INTRODUCED

This widespread native plant with showy red flowers is an important nectar source for butterflies and hummingbirds. It was named after the red robes worn by Roman Catholic cardinals.

DESCRIPTION: This clump-forming perennial has leafy stems and grows 1–3' (30–90 cm) tall. The leaves are dark green, about 2–5" (5–12.5 cm) long each, narrowly lanceolate, and alternate, with fine teeth on the edges. The flowers are tubular and red, white, or pink, growing in clusters. The racemes resemble flaming red spires. The corolla is 1–1½" (2.5–4 cm) long, bilaterally symmetrical and two-lipped. The lower lip is more prominent with three large lobes, while the upper lip has two small lobes.

FLOWERING SEASON: July–September.

HABITAT: Damp sites, especially along streams.

RANGE: Throughout the East from Ontario to New Brunswick, south to Florida and Texas, north to Minnesota; also in the southwestern United States and south into Mexico.

CAUTION: The leaves, seeds, and roots of some plants of the *Lobelia* genus contain poisonous substances and have caused fatalities in humans and animals when ingested. All plants in the genus may contain toxins and should not be ingested.

SIMILAR SPECIES: Sierra Madre Lobelia (*L. laxiflora*) is similar, but the corolla is either red with yellow lobes or entirely yellow.

CONSERVATION: G5
Globally secure but critically imperiled in Nebraska and Nevada, imperiled in Colorado, and vulnerable in New Brunswick.

■ NATIVE
■ RARE OR EXTIRPATED
■ INTRODUCED

ALTERNATE NAMES: Puke Weed

This eastern native contains several toxic alkaloid compounds. Native Americans have used this plant as a purgative to induce vomiting and ceremonially. However, this plant is poisonous and can be fatal if consumed in large quantities. The foliage can be burned to repel insects.

DESCRIPTION: This plant has slightly hairy stems and grows 1–3' (30–90 cm) tall. The leaves are alternate and wavy toothed. Each leaf is about 1–2½" (2.5–6.3 cm) long, thin, and light green. The flowers are lavender or blue-violet, appearing in terminal, leafy, elongated clusters. Each flower is ¼" (6 mm) long, with two lips. The lower lip is bearded. After flowering, the calyx surrounding the fruit becomes inflated and balloon-like, up to ⅓" (8 mm) across. The fruit, a ribbed capsule, is enclosed by the swollen calyx.

FLOWERING SEASON: June–October.

HABITAT: Fields, open woods, and roadsides.

RANGE: Across southern Canada; south to Georgia; west to Arkansas and eastern Kansas.

CAUTION: The leaves, seeds, and roots of some plants of the *Lobelia* genus contain poisonous substances and have caused fatalities in humans and animals when ingested. It can also cause sweating, vomiting, rapid heartbeat, mental confusion, convulsions, hypothermia, and coma, among other adverse effects.

CONSERVATION: G5
Globally secure but critically imperiled in Nebraska, imperiled in Kansas, vulnerable in Louisiana.

■ NATIVE
■ RARE OR EXTIRPATED
■ INTRODUCED

ALTERNATE NAMES: Great Blue Lobelia

This perennial is native to eastern and central North America. Great Lobelia has large flower spikes that are often visited by pollinators, especially bumblebees (*Bombus* spp.). In many respects this plant resembles Cardinal Flower (*L. cardinalis*), except its flowers are blue.

DESCRIPTION: This plant grows 1–4' (30–120 cm) tall with a leafy stem. The leaves are oval to lanceolate in shape and either untoothed or irregularly toothed. Each leaf is about 2–6" (5–15 cm) long. Showy, bright blue flowers grow in the axils of leafy bracts, forming an elongated cluster. Each flower is about 1" (2.5 cm) long, with two lips, and a hairy calyx. The lower lip has a whitish strip. The calyx has five pointed lobes. Five stamens form a united tube around the style.

FLOWERING SEASON: August–September.

HABITAT: Rich lowland woods and meadows, swamps.

RANGE: Western New England south to eastern Virginia and the uplands of North Carolina and Alabama, west to eastern Kansas and Minnesota.

CAUTION: No parts of this plant should be ingested. The root of this species contains alkaloid compounds that cause vomiting if ingested. The leaves and seeds of some *Lobelia* species also contain poisonous substances.

CONSERVATION: G5

Secure, but critically imperiled in Massachusetts, Vermont, and Wyoming; and vulnerable in Georgia and Louisiana.

■ NATIVE
■ RARE OR EXTIRPATED
■ INTRODUCED

ALTERNATE NAMES: Pale-spike Lobelia

This widespread but short-lived native perennial is a highly variable species, with several varieties found across North America. The flowers may be lavender, purplish-blue, or bluish-white and attract a variety of insect pollinators.

DESCRIPTION: This plant grows to a height of 1–4' (30–120 cm). The basal leaves are light green and lanceolate or elliptical in shape. Each leaf is 1–3½" (2.5–9 cm) long, becoming smaller, more widely spaced, and stalkless up the stem. The leaves are so small near the flowers that they're reduced to bracts. The flowers are blue or white and tubular. The corolla is about ⅜–½" (8–13 mm) long, with two lips. The upper corolla lip is two-

lobed, while the lower corolla lip has three lobes. The capsule contains many seeds.

FLOWERING SEASON: June–August.

HABITAT: Fields, woodlands, and rich meadows.

RANGE: Alberta east to Nova Scotia, south to Georgia, west to Texas, and north to North Dakota.

CAUTION: All plants in the genus may contain toxins and should not be ingested.

CONSERVATION: G5

Secure, but critically imperiled in Alberta, Nova Scotia, and Quebec; and imperiled in Delaware and Montana.

■ NATIVE
■ RARE OR EXTIRPATED
■ INTRODUCED

ALTERNATE NAMES: Clasping-leaf Venus' Looking-glass, *Specularia perfoliata*

This native annual produces attractive blue-violet flowers, but some of the flowers are small, inconspicuous, and do not open. Such flowers generally have only three or four sepals and self-pollinate. Native Americans have used this plant for a variety of ceremonial and medicinal purposes, including as treatment for indigestion and to induce vomiting.

DESCRIPTION: This plant has bristly-hairy, leafy stems and grows 6–24" (15–60 cm). The leaves usually have scalloped edges and are ½–1¼" (1.5–3 cm) long each, attached to the stem in a notched base. It has one or a few nearly flat, blue, violet, or white flowers in the axils of upper leaves. The flowers are ½–¾" (1.5–2 cm) wide each, with five narrow sepals, a sharp corolla, usually with five bluntly pointed lobes.

FLOWERING SEASON: April–July.

HABITAT: Dry woods and fields; often in poor soils.

RANGE: Ontario east to New Brunswick, south to Florida, west to Texas, and north to North Dakota; also in much of the West.

CONSERVATION: G5
Secure but imperiled in Colorado and Vermont and vulnerable in Montana and Wyoming.

■ NATIVE
■ RARE OR EXTIRPATED
■ INTRODUCED

ALTERNATE NAMES: Bogbean, Bog Buckbean

This native of North American bogs can also be found in Eurasia. In some locations, it may produce two different types of flowers—some with short styles and long stamens, and others with long styles and short stamens. The common name Bogbean alludes to the plant's habitat and to its small, bean-like seeds, which have been found deeply buried in bogs. The leaves have a strong, bitter taste and were sometimes used in Europe as a substitute for hops in brewing beer and other alcoholic beverages.

DESCRIPTION: This aquatic perennial has leaves and flower stalks reaching 4–12" (10–30 cm) above water. The leaves are 4–12" (10–30 cm) long, with three leaflets, each about 1½–5" (4–12.5 cm) long and broadly lanceolate. The flowers grow in racemes or narrow clusters of white or purple tinged, star-like flowers atop stout stalks, about as high as the leaves. The corolla is about ½" (1.5 cm) wide, forming a tube ¼–¾" (6–9 mm) long (about twice as long as the calyx), with five or six

pointed lobes covered with short hairs. The fruit, a capsule, is covered with many small hard, smooth, shiny seeds.

FLOWERING SEASON: May–August.

HABITAT: Bogs and shallow lakes.

RANGE: Throughout Canada and the northern tier of the United States, especially around the Great Lakes and New England, and widely distributed in the West.

CONSERVATION:

Globally secure but critically imperiled in Arizona, Maryland, Missouri, North Carolina, Ohio, South Dakota, Virginia, and West Virginia. Imperiled in Iowa, Nebraska, New Jersey, North Dakota, Rhode Island, and Utah. Vulnerable in Indiana and Wyoming.

■ NATIVE
■ RARE OR EXTIRPATED
■ INTRODUCED

ALTERNATE NAMES: Big Floatingheart

The common name of this aquatic native plant refers to the floating, heart-shaped leaves. The genus name refers to the plant's similarity to water lilies (family Nymphaea). However, its flowers only have five petals, unlike the many-petaled water lilies.

DESCRIPTION: This aquatic plant rises just above the water. It has floating, heart-shaped, long-stalked leaves. The leaves are 2–8" (5–20 cm) long each and very thick, green above, and veiny beneath. The small white flowers grow in a flat-topped cluster rising just above the leaves. Each flower is ½–¾" (1.5–2 cm) wide, with five petals, nearly separate. The fruit, a capsule, matures underwater, with many rough seeds.

FLOWERING SEASON: July–September.

HABITAT: Ponds and slow streams.

RANGE: Along the coastal plain from Maryland and Delaware south to Florida and west to Texas.

SIMILAR SPECIES: *N. cordata*, also commonly called Floating Heart, is similar but has smaller green leaves mottled with purple above and smooth beneath, not veiny. Yellow Floating Heart (*N. peltata*) has yellow flowers, with fringed corolla lobes.

CONSERVATION: Ⓖ5

Secure but critically imperiled in Maryland and Virginia; imperiled in Mississippi; and vulnerable in North Carolina.

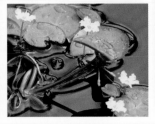

■ NATIVE
■ RARE OR EXTIRPATED
■ INTRODUCED

ALTERNATE NAMES: Common Yarrow, *Achillea lanulosa*

This flowering plant is found throughout the Northern Hemisphere. It is known by many common and colloquial names, including *plumajillo*, or "little feather," in reference to the shape and texture of the leaves. This species has a strong scent that some liken to cabbage.

DESCRIPTION: This perennial plant grows 1–3' (30–90 cm) in height. The flattish cluster of small white flower heads includes 10–30 tiny disk flowers surrounded by three to five white to pinkish ray flowers. The feathery, fern-like leaves grow up to 1½" (4 cm) wide.

FLOWERING SEASON: June–September.

HABITAT: Grows in meadows, fields, and along roadsides.

RANGE: Occurs throughout North America.

SIMILAR SPECIES: Siberian Yarrow (*A. alpina*).

CONSERVATION: G5
This species is globally secure.

■ NATIVE
■ RARE OR EXTIRPATED
■ INTRODUCED

ALTERNATE NAMES: American Trailplant, Trailplant

This woodland plant is native to southern Canada and the northern and western United States. If a hiker walks through a patch of these plants, the whitish undersides of the disturbed leaves become exposed and create a visible path, hence the common name. It is the only species in the genus *Adenocaulon* native to North America.

DESCRIPTION: This flowering plant grows 1–3' (30–90 cm) in height. The white flowers are arranged in tiny clusters at the ends of stem branches, and the large basal leaves are long-stalked, triangular in shape, and white-hairy below.

FLOWERING SEASON: June–October.

RANGE: Occurs from British Columbia and Alberta south to central California, and across the northern United States to Minnesota and Michigan.

SIMILAR SPECIES: As the only member of its genus native to North America, this plant is not likely to be confused with other species.

CONSERVATION: G5

Although globally secure, this species is critically imperiled in Ontario and imperiled in Alberta and Wyoming.

- ■ NATIVE
- ■ RARE OR EXTIRPATED
- ■ INTRODUCED

ALTERNATE NAMES: Richweed, *Eupatorium rugosum*

This poisonous plant is native to eastern North America, where it is most often found in wooded habitats. The genus name *Ageratina* is derived from Greek and means "unaging," referring to the flowers of these plants keeping their color for a long time.

DESCRIPTION: This perennial herb grows 1–3' (30–90 cm) in height. The flower heads are very small and whitish, and the leaves are opposite and sharply toothed, growing 2½–7" (6.5–18 cm) long.

FLOWERING SEASON: July–October.

HABITAT: Grows in woods and thickets.

RANGE: Occurs from Saskatchewan east to Nova Scotia, south to Florida, west to Texas, and north to North Dakota.

CAUTION: This plant is toxic to cattle. Humans who consume the milk or meat of animals that have ingested this plant can become ill with what is known as milk sickness.

SIMILAR SPECIES: Fragrant Snakeroot (*A. herbacea*).

CONSERVATION:  G5
This species is globally secure but critically imperiled in Nova Scotia.

■ NATIVE
■ RARE OR EXTIRPATED
■ INTRODUCED

ALTERNATE NAMES: Pale Goat-chicory, Pale Agoseris

This flowering plant is native to western North America, where it may be found growing along slopes, in mountain meadows, and in open woods. It is quite similar to several other plants, including the introduced Common Dandelion (*Taraxacum officinale*); the thick, beaked fruits help to distinguish Pale Mountain Dandelion from these other species.

DESCRIPTION: This perennial herb grows 4–28" (10–70 cm) in height and features yellow flower heads on leafless stalks. The basal leaves are lance-shaped and range from 2–14" (5–35 cm) long. The fruiting heads are white, round, and fluffy.

FLOWERING SEASON: May–September.

HABITAT: Grows in meadows and along slopes.

RANGE: Occurs from western Canada south through the California mountains and east to New Mexico, South Dakota, and Minnesota.

SIMILAR SPECIES: Prairie False Dandelion (*Nothocalais cuspidata*) and Common Dandelion (*Taraxacum officinale*).

CONSERVATION: G5

Although globally secure, this species is vulnerable in the Northwest Territories.

■ NATIVE
■ RARE OR EXTIRPATED
■ INTRODUCED

ALTERNATE NAMES: Annual Ragweed

Native to North America, this plant occurs in many parts of the world as an invasive weed. The drab flower heads do not attract insects or other pollinators; instead, pollination is accomplished by wind. The wind-dispersed pollen is a common cause of hay fever. The fruits of this plant persist into winter and are readily eaten by a variety of birds.

DESCRIPTION: This hairy-stemmed annual ranges from 1–5' (30–150 cm) in height and features inconspicuous greenish disk flowers in elongated clusters. The light green leaves are highly dissected and grow up to 4" (10 cm) long.

FLOWERING SEASON: July–October.

HABITAT: Grows in fields and alongside roads.

RANGE: Occurs throughout much of North America except in the Far North.

CAUTION: This plant causes hay fever allergies and can irritate the skin if touched.

SIMILAR SPECIES: Great Ragweed (*A. trifida*) and other *Ambrosia* species.

CONSERVATION: G5

Although globally secure, this species is vulnerable in Alberta and Wyoming.

NATIVE
RARE OR EXTIRPATED
INTRODUCED

ALTERNATE NAMES: Giant Ragweed

This giant ragweed is native to North America, where it commonly grows in fields and along roadsides. It occurs in many places outside this continent as an invasive weed. The pollen, dispersed by wind rather than by insects or other pollinators, is a major cause of hay fever.

DESCRIPTION: This hairy annual ranges from 2–15' (60–450 cm) in height. The greenish flowers are arranged in elongated, terminal clusters; some plants can feature hundreds of flower heads. The opposite leaves grow up to 8" (20 cm) long and can be either unlobed or with three to five lobes.

FLOWERING SEASON: June–October.

HABITAT: Grows in fields and alongside roads.

RANGE: Occurs throughout much of North America except in the Far North.

CAUTION: This plant causes hay fever allergies and can irritate the skin if touched. It can be toxic to livestock if ingested.

SIMILAR SPECIES: Common Ragweed (*A. artemisiifolia*) and other *Ambrosia* species.

CONSERVATION: G5
Globally secure but vulnerable in Wyoming.

■ NATIVE
■ RARE OR EXTIRPATED
■ INTRODUCED

This widespread wildflower is native to both North America and Asia. It prefers pastures, roadsides, and other dry, open habitats. The white flowers of this showy species are often used in dried flower arrangements.

DESCRIPTION: This perennial grows 1–3' (30–90 cm) in height and features a white woolly stem and a flat cluster of white ray-less flower heads. The leaves are narrow and usually greenish above and white below, ranging 3–5" (7.5–12.5 cm) in length.

FLOWERING SEASON: June–September.

HABITAT: Grows in open woods, fields, and along roadsides.

RANGE: Occurs across much of Canada and the western and northeastern United States.

SIMILAR SPECIES: May be confused with *Antennaria* species, which are similar but have mostly basal leaves.

CONSERVATION: G5

Globally secure but critically imperiled in Iowa, Nebraska, and Virginia; imperiled in Yukon; and vulnerable in Manitoba, Northwest Territories, Saskatchewan, and Maryland.

■ NATIVE
■ RARE OR EXTIRPATED
■ INTRODUCED

ALTERNATE NAMES: Prairie Everlasting

This small, mat-forming plant is native to Canada and the northern United States, where it grows in dry, open areas. It features fuzzy white flower heads and woolly leaves. This species is a host plant for the American Lady (*Vanessa virginiensis*) butterfly.

DESCRIPTION: This perennial grows 6–10" (15–25 cm) in height and forms dense colonies. The clustered flower heads are fuzzy and white. The basal leaves are arranged in rosettes and feature a single prominent vein, while the smaller stem leaves are linear and widely spaced.

FLOWERING SEASON: April–June.

HABITAT: Grows in dry fields, grasslands, and open woodlands.

RANGE: Occurs throughout much of Canada south to Colorado, Oklahoma, Arkansas, Tennessee, and Virginia.

SIMILAR SPECIES: Pearly Everlasting (*Anaphalis margaritacea*) and other *Antennaria* species.

CONSERVATION: G5

Although globally secure, this species is critically imperiled in Arkansas and Wyoming, and vulnerable in Quebec.

■ NATIVE
■ RARE OR EXTIRPATED
■ INTRODUCED

ALTERNATE NAMES: Ladies' Tobacco

This flowering plant is native to eastern and central North America, where it is widespread and grows in dry and sunny places such as prairies, open woods, and along roadsides. It was named for American botanist John Crawford Parlin, who first characterized this plant as a distinct species.

DESCRIPTION: This native perennial grows 6–17" (15–43 cm) in height and features whitish flower heads arranged in clusters of 4–12. The leaves are basal and alternate, growing 1–3¾" (2.5–9.5 cm) long.

FLOWERING SEASON: April–June.

HABITAT: Grows in dry pastures, prairies, open woods, and along roadsides.

RANGE: Occurs from Manitoba east to Nova Scotia, south to Georgia, and west to Texas.

SIMILAR SPECIES: Pearly Everlasting (*Anaphalis margaritacea*) and other *Antennaria* species.

CONSERVATION: 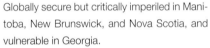 G5

Globally secure but critically imperiled in Manitoba, New Brunswick, and Nova Scotia, and vulnerable in Georgia.

■ NATIVE
■ RARE OR EXTIRPATED
■ INTRODUCED

ALTERNATE NAMES: Mountain Everlasting

This wildflower is a mostly western species that favors meadows and other open areas. A relative of sunflowers and daisies, it is a morphologically diverse species. The flower heads can appear pink, brown, white, or yellow.

DESCRIPTION: This perennial grows 2–12" (5–30 cm) in height. The flower heads, arranged in rounded clusters, are often pinkish but can be brown, white, or yellow. The whitish woolly leaves are mostly basal.

FLOWERING SEASON: May–August.

HABITAT: Grows in meadows and open woods and forests.

RANGE: Occurs in western North America south to California, Arizona, and New Mexico. Rare in eastern Canada.

SIMILAR SPECIES: Pearly Everlasting (*Anaphalis margaritacea*) and Blunt-leaf Rabbit-tobacco (*Pseudognaphalium obtusifolium*).

CONSERVATION: ⒼⒼ5

Although globally secure, this species is critically imperiled in Nova Scotia, Ontario, and Nebraska, and vulnerable in Manitoba and Newfoundland. It is presumed extirpated in Michigan.

■ NATIVE
■ RARE OR EXTIRPATED
■ INTRODUCED

ALTERNATE NAMES: Stinking Chamomile

Native to Eurasia, this bushy plant is widespread across North America as an introduced species. Caution should be exercised around this plant because the foliage may cause skin irritation if handled. This wildflower emits a strong, unpleasant odor, hence the nickname Stinking Chamomile.

DESCRIPTION: This daisy-like plant grows 1–2' (30–60 cm) and features flower heads with white rays surrounding a dome-shaped, yellow central disk. The fern-like leaves are finely dissected and generally grow 1–2½" (2.5–6.5 cm) long.

FLOWERING SEASON: June–October.

HABITAT: Grows along roadsides and other disturbed places.

RANGE: Occurs throughout North America, except the Arctic.

CAUTION: The foliage of this plant may irritate the skin when handled.

SIMILAR SPECIES: Corn Chamomile (*A. arvensis*) is similar but scentless.

CONSERVATION:
The conservation status of this introduced species is not ranked in North America.

■ NATIVE
■ RARE OR EXTIRPATED
■ INTRODUCED

ALTERNATE NAMES: Lesser Burdock

This Eurasian native now grows in old fields and various disturbed areas across much of North America. The prickly flower heads easily catch on fur and clothing, dispersing the seeds far and wide. The similar Greater Burdock (*A. lappa*), also introduced in North America, is a larger plant with bigger flower heads and longer stalks.

DESCRIPTION: This bushy plant grows 1–5' (30–150 cm) in height and bears round, prickly, pink to lavender rayless flower heads. The ovate to heart-shaped leaves are dark green above, woolly below, and grow up to 18" (45 cm) long.

FLOWERING SEASON: July–October.

HABITAT: Grows in old fields and disturbed places.

RANGE: Introduced and occurs throughout much of North America, except the Far North.

SIMILAR SPECIES: Greater Burdock (*A. lappa*).

CONSERVATION:
The conservation status of this introduced species is unranked.

■ NATIVE
■ RARE OR EXTIRPATED
■ INTRODUCED

ALTERNATE NAMES: Heartleaf Leopardbane

This native wildflower may be found in a variety of habitats from lightly shaded mountain woods to moist meadows throughout western North America. It features yellow flowers and heart-shaped leaves, an identifying characteristic from which it gets its common name.

DESCRIPTION: This perennial grows 4–24" (10–60 cm) in height and bears broad yellow flower heads, each with 10–15 rays. The heart-shaped leaves are arranged in pairs and range from 1½–5" (4–12.5 cm) long.

FLOWERING SEASON: April–June.

HABITAT: Grows along roadsides and in mountain woods and meadows.

RANGE: Occurs from Alaska south to New Mexico and westward to the Pacific Coast. Rare in eastern Canada and Michigan.

SIMILAR SPECIES: Spear-leaf Arnica (*A. longifolia*) and Broad-leaf Arnica (*A. latifolia*) have differently shaped leaves.

CONSERVATION: G5

Globally secure but critically imperiled in Manitoba, Ontario, Michigan, and Nebraska, and vulnerable in Saskatchewan and North Dakota.

GROOVE-STEM INDIAN PLANTAIN *Arnoglossum plantagineum*

NATIVE
RARE OR EXTIRPATED
INTRODUCED

ALTERNATE NAMES: Tuberous Indian Plantain, Prairie Indian Plantain

This wildflower is native to the central United States, where it prefers open areas with moist to wet soils. The leaves of this plant resemble those of a plantain. Much of its native prairie habitat has been converted to farmland, leaving its remaining populations mainly in small, isolated colonies.

DESCRIPTION: This native perennial grows 3–4' (90–120 cm) in height. The flowers are white or greenish and arranged in flat-topped clusters. The thick and rubbery leaves are generally 8" (20 cm) long and 4" (10 cm) across and alternately arranged.

FLOWERING SEASON: June–August.

HABITAT: Grows in wet, open areas.

RANGE: Occurs from South Dakota south through Texas, east to Alabama, and north to Michigan and Ontario.

SIMILAR SPECIES: Egg-leaf Indian Plantain (*A. ovatum*).

CONSERVATION: G4

Globally, this species is apparently secure, but is critically imperiled in Alabama; imperiled in Ontario, Minnesota, Ohio, and Tennessee; and vulnerable in Indiana, Michigan, and Wisconsin.

■ NATIVE
■ RARE OR EXTIRPATED
■ INTRODUCED

ALTERNATE NAMES: Riverside Wormwood, Wild Wormwood, Sailor's Tobacco, Old Man

This species, naturalized in North America from Europe, is known by numerous common names. It is highly invasive and difficult to eradicate, commonly occurring along roadsides and in fields and waste places. In some regions of the world this plant is used for medicinal and culinary purposes.

DESCRIPTION: This perennial plant grows 2–4' (60–120 cm) in height and features an extensive rhizome system. The disk flowers, small and dull red, are crowded in elongated panicles. The pinnate leaves are green above with white hairs beneath.

FLOWERING SEASON: July–October.

HABITAT: Grows in fields, along roadsides, and in other weedy places.

RANGE: Introduced and often invasive in North America. Occurs from Alberta east to Newfoundland, south to Florida, west to Louisiana, and north to Kansas, Iowa, and Minnesota. It is also found in much of the northwestern United States and in southern Alaska.

SIMILAR SPECIES: See other *Artemisia* species.

CONSERVATION:
The conservation status of this introduced species is unranked.

■ NATIVE
■ RARE OR EXTIRPATED
■ INTRODUCED

ALTERNATE NAMES: Lawn Daisy

Introduced from Europe, this flowering plant commonly grows in lawns, fields, and other open places in many parts of the United States and Canada. The flowers of this species close at night and open up in the morning, inspiring the Old English name *daeges ege*, meaning "day's eye."

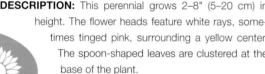

DESCRIPTION: This perennial grows 2–8" (5–20 cm) in height. The flower heads feature white rays, sometimes tinged pink, surrounding a yellow center. The spoon-shaped leaves are clustered at the base of the plant.

FLOWERING SEASON: May–July.

HABITAT: Grows in lawns, fields, and along roadsides.

RANGE: Occurs throughout much of the United States and Canada, especially near the coasts.

SIMILAR SPECIES: Oxeye Daisy (*Leucanthemum vulgare*).

CONSERVATION:

The conservation status of this introduced species is unranked.

■ NATIVE
■ RARE OR EXTIRPATED
■ INTRODUCED

ALTERNATE NAMES: Lyre-leaf Green-eyes, Chocolate Flower

This leafy plant of open fields and roadsides has a limited range in the southwestern United States and northern Mexico. Commonly called Chocolate Flower, it really does smell like chocolate, especially when the rays are plucked from the flower head. This species is cultivated as an ornamental plant.

DESCRIPTION: This perennial grows 1–4' (30–120 cm) in height and features flower heads with yellow rays surrounding

a maroon center. The leaves grow 2–6" (5–15 cm) long and are pinnately lobed or scalloped.

FLOWERING SEASON: April–October.

HABITAT: Grows in fields and along roadsides.

RANGE: Occurs from southeastern Colorado and southwestern Kansas through western Texas, New Mexico, and southeastern Arizona to Mexico.

SIMILAR SPECIES: Florida Greeneyes (*B. subacaulis*) and Texas Greeneyes (*B. texana*).

CONSERVATION:  G5
Although globally secure, this species is critically imperiled in Kansas.

TICKSEED SUNFLOWER   *Bidens aristosa*

- ■ NATIVE
- ■ RARE OR EXTIRPATED
- ■ INTRODUCED

ALTERNATE NAMES: Bearded Beggarticks, Midwestern Tickseed Sunflower

This slender, leafy plant grows in moist or wet places throughout much of the eastern United States. The prickly fruits, commonly called "stickers," often get caught in fur or clothing. It produces numerous yellow flower heads with both ray and disk-shaped florets, which attract a variety of native bees.

DESCRIPTION: This annual grows 1–5' (30–150 cm) in height and features several yellow, daisy-like flower heads. The leaves are pinnately divided into coarsely toothed segments. The fruits are dry achenes, usually with two barbed spines.

FLOWERING SEASON: August–October.

HABITAT: Grows in wet meadows, roadside ditches, and other low, moist places.

RANGE: Occurs from Ontario and the New England states south to Georgia, west to Texas, and north to Nebraska, Iowa, and Minnesota.

SIMILAR SPECIES: Ozark Tickseed Sunflower (*B. polylepis*) is a very similar species that ranges farther west.

CONSERVATION: G5

Although globally secure, this species is imperiled in Iowa.

■ NATIVE
■ RARE OR EXTIRPATED
■ INTRODUCED

ALTERNATE NAMES: Nodding Beggartick

This widespread annual herb prefers swamps and other wet places. The species name *cernua* means "nodding," a reference to the fact that the flower heads are often angled downward. The seeds of this plant are sometimes eaten by ducks, and the flowers attract many species of native butterflies.

DESCRIPTION: This wildflower grows 1–3' (30–90 cm) and features numerous yellow flower heads, each with six to eight petal-like rays and a darker yellow center. The undivided leaves are arranged in opposite pairs and are narrowly lanceolate to elliptical in shape. The seed-like fruits sport two to four barbed spines.

FLOWERING SEASON: August–October.

HABITAT: Grows in swamps, along streambanks, and other wet places.

RANGE: Found throughout much of North America, except southern California, Texas, Mississippi, southern Alabama and Georgia, South Carolina, and Florida.

SIMILAR SPECIES: Smooth Bur Marigold (*B. laevis*).

CONSERVATION: G5

Although globally secure, this species is critically imperiled in Alabama and vulnerable in Arizona, Georgia, North Carolina, Wyoming, and Yukon.

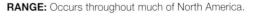

■ NATIVE
■ RARE OR EXTIRPATED
■ INTRODUCED

ALTERNATE NAMES: Pitchfork Weed, Devil's Bootjack

This plant, native to North America, is found in many parts of the world as an introduced species, sometimes becoming an invasive pest. It prefers moist areas with ample sunlight, such as wet fields.

DESCRIPTION: This annual herb grows 8–24" (20–60 cm) and features one or more flower heads with orange disc florets, usually lacking rays. The leaves are divided into lance-shaped leaflets.

FLOWERING SEASON: July–October.

HABITAT: Grows in fields and pastures, along roadsides, and in various open moist places.

RANGE: Occurs throughout much of North America.

SIMILAR SPECIES: Nodding Bur Marigold (*B. cernua*).

CONSERVATION: G5

Globally secure but imperiled in Wyoming and vulnerable in Alberta, Saskatchewan, and Iowa.

■ NATIVE
■ RARE OR EXTIRPATED
■ INTRODUCED

ALTERNATE NAMES: White Doll's-daisy, False Aster

This native plant is found primarily in the central United States and Canada, but also grows in parts of the eastern United States. It favors wet areas such as marshes and ditches. It often escapes from cultivation and establishes itself locally.

DESCRIPTION: This perennial grows 1–5' (30–150 cm) in height, featuring flower heads with white to light purple or pink rays surrounding a yellow, dome-shaped center. The smooth, narrow leaves are usually untoothed. The seed-like fruits usually have two to four short spines and several shorter bristles.

FLOWERING SEASON: July–October.

HABITAT: Grows mostly in moist or wet places.

RANGE: Found from Saskatchewan and Manitoba south to Louisiana, Mississippi, and Alabama, as well as along much of the Atlantic Coast from Maine south to Florida.

SIMILAR SPECIES: Doll's Daisy (*B. diffusa*).

CONSERVATION: **G5**

This species is globally secure but critically imperiled in Pennsylvania; imperiled in Delaware, Michigan, and Manitoba; and vulnerable in Georgia, Ohio, and Virginia.

■ NATIVE
■ RARE OR EXTIRPATED
■ INTRODUCED

ALTERNATE NAMES: Bushy Seaside-tansy, Oxeye Daisy, Sea Oxeye

Native to the United States and Mexico, this salt-tolerant plant grows in a variety of coastal habitats. It often hybridizes with the similar Tree Seaside (*B. arborescens*) where the two species occur together, such as in the Florida Keys.

DESCRIPTION: This bushy, shrub-like perennial grows up to 3' (1 m) tall with ascending branches and attractive foliage. The bright yellow flowers feature petal-like rays surrounding a disk, and the fleshy leaves are arranged opposite on the stem and oval-shaped.

FLOWERING SEASON: July–August.

HABITAT: Grows in saltmarshes, brackish waters, and other coastal areas.

RANGE: Found along the coastal United States from Maryland south to Florida and west to Texas, and in Mexico.

SIMILAR SPECIES: Tree Seaside Oxeye (*B. arborescens*).

CONSERVATION: G5

This species is globally secure, but possibly extirpated in Maryland.

NATIVE
RARE OR EXTIRPATED
INTRODUCED

ALTERNATE NAMES: Bristle Thistle, Nodding Thistle, Nodding Plumeless-thistle

Native to Europe and Asia, this increasingly common invasive plant is now widespread across much of the North American continent. It grows in a variety of disturbed sites. The large flower heads are often nodding, hence the alternate common name Nodding Thistle.

DESCRIPTION: This biennial plant grows 1–9' (30–270 cm) in height and features large pink to purplish-red flower heads. The leaves are dark green and deeply lobed.

FLOWERING SEASON: June–October.

HABITAT: Grows in fields and disturbed places.

RANGE: Introduced, occurring throughout much of North America except the Far North.

SIMILAR SPECIES: Scotch-thistle (*Onopordum acanthium*) and Blessed Milk-thistle (*Silybum marianum*).

CONSERVATION:
The conservation status of this introduced species is unranked.

■ NATIVE
■ RARE OR EXTIRPATED
■ INTRODUCED

ALTERNATE NAMES: Yellow Cockspur, Barnaby Thistle

This thorny plant is native to the Mediterranean region, occurring in North America and many other parts of the world as an invasive species. The bright yellow flowers are frequented by many butterfly species and other pollinators.

DESCRIPTION: This annual plant covered in graying hairs grows 4–39" (10–100 cm) and sports bright yellow flower heads with sharp golden spines. Leaves at the base are 2–3" (5–8 cm) long and deeply lobed, while those on the stem are smaller and not lobed.

FLOWERING SEASON: May–October.

HABITAT: Grows in pastures, dry grasslands and hillsides, and various disturbed areas.

RANGE: Found throughout the western United States and spreading into the East.

CAUTION: This plant is poisonous to horses if continually eaten.

SIMILAR SPECIES: Bachelor's Button (*C. cyaneus*) and Tocolote (*C. melitensis*).

CONSERVATION:
The conservation status of this introduced species is unranked.

■ NATIVE
■ RARE OR EXTIRPATED
■ INTRODUCED

Native to eastern Europe, this slender plant is found in many places throughout the North American continent as an invasive species. It spreads quickly once established in an area, replacing native vegetation and altering the landscape. It is particularly common in the western United States.

DESCRIPTION: This invasive flowering plant grows 12–39" (30–100 cm) and sports many pink to purplish rayless flowers. The leaves are alternate and pinnately divided into narrow lobes.

FLOWERING SEASON: June–September.

HABITAT: Grows in fields, waste places, and along roadsides.

RANGE: Introduced and widespread across much of North America.

CAUTION: This plant may cause skin irritation; it is recommended to wear gloves and long sleeves when handling.

SIMILAR SPECIES: Diffuse Knapweed (*C. diffusa*).

CONSERVATION:
The conservation status of this introduced species is unranked.

■ NATIVE
■ RARE OR EXTIRPATED
■ INTRODUCED

ALTERNATE NAMES: False Yarrow, Pincushion, Hoary Pincushion

Native to western North America, this flowering plant grows in a variety of habitats, often in rocky or disturbed places. Woolly hairs cover its stems and leaves, giving this plant an interesting appearance. Dusty Maiden has been used to treat burns and other ailments.

DESCRIPTION: This flowering herb grows 4–20" (10–50 cm) in height. The flower heads are small, dense, and whitish, arranged in flat clusters. Its woolly leaves are pinnately lobed and fern-like.

FLOWERING SEASON: July–October.

HABITAT: Grows in sandy, open areas.

RANGE: Occurs from British Columbia east to Saskatchewan, south to New Mexico, and west to central California.

SIMILAR SPECIES: Esteve's-pincushion (*C. stevioides*).

CONSERVATION: G5

Although globally secure, this species is imperiled in North Dakota.

- ■ NATIVE
- ■ RARE OR EXTIRPATED
- ■ INTRODUCED

ALTERNATE NAMES: Heath-leaved Chaetopappa, Baby White-aster

This small wildflower is native to the southwestern United States and northern Mexico. The species name *ericoides* means "heath-like," a reference to this plant's leaves.

DESCRIPTION: This perennial herb grows 2–6" (5–15 cm) in height. The yellow disk flowers sport 12–24 white rays, aging pinkish. The leaves are very narrow and untoothed.

FLOWERING SEASON: March–August.

HABITAT: Grows in open places.

RANGE: Occurs from Kansas west to eastern California and south to northern Mexico.

SIMILAR SPECIES: Common Least Daisy (*C. asteroides*).

CONSERVATION: **G5**
Although globally secure, this species is imperiled in Nebraska and Wyoming.

■ NATIVE
■ RARE OR EXTIRPATED
■ INTRODUCED

This wildflower is native to the southeastern United States, where it prefers fields and other open habitats. Broader leaves and larger flowers help to set it apart from other asters. The genus name *Chrysopsis* is derived from Greek words meaning "gold" and "resembling in appearance," referring to the gold-colored flowers of these plants.

DESCRIPTION: This perennial plant grows 6–36" (15–91 cm) in height and features clusters of yellow flowers with petal-like rays. The upper leaves are lance-shaped and the lower leaves, which are larger, are spoon-shaped. The foliage is woolly when young and becomes smoother with age.

FLOWERING SEASON: August–October.

HABITAT: Grows in open woods, fields, and sandy areas.

RANGE: Occurs from New York west to southern Ohio and Kentucky and south to Texas and Florida.

SIMILAR SPECIES: Grass-leaf Golden-aster (*Pityopsis graminifolia*) and other *Pityopsis* species.

CONSERVATION: G5

Globally secure but critically imperiled in Pennsylvania and Rhode Island.

■ NATIVE
■ RARE OR EXTIRPATED
■ INTRODUCED

ALTERNATE NAMES: Blue Sailors

This European introduction is now common and widespread in North America, growing along roadsides and in fields and pastures. The roots of this plant are sometimes roasted and ground as a coffee substitute. It is closely related to Endive (*C. endivia*), a species commonly cultivated for culinary purposes and sometimes called "chicory" in the United States.

DESCRIPTION: This perennial plant grows 1–6" (30–180 cm) in height. It features wiry, branched stems with pale blue flower heads and milky sap. The leaves are lanceolate and pinnately toothed or lobed, edged with small sharp teeth.

FLOWERING SEASON: March–October.

HABITAT: Grows in fields, along roadsides, and in disturbed places.

RANGE: Introduced, occurring throughout much of North America except the Far North.

SIMILAR SPECIES: Endive (*C. endivia*) and *Lygodesmia* species.

CONSERVATION:
The conservation status of this introduced species is unranked.

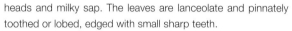

NATIVE
RARE OR EXTIRPATED
INTRODUCED

ALTERNATE NAMES: Canada Thistle

This introduced species is native to Europe, western Asia, and northern Africa; its alternate name refers to its country of origin within North America. A non-spiny stem and smaller flowering heads help to separate this species from the similar Bull Thistle (*C. vulgare*).

DESCRIPTION: This perennial plant grows 1–5' (30–150 cm) and features clusters of fragrant, rayless flower heads that may be pale magenta, lavender, or white in color. The gray-green leaves are spiny, lanceolate, and mostly stalkless.

FLOWERING SEASON: June–October.

HABITAT: Grows in pastures, along roadsides, and in waste places.

RANGE: Found throughout North America, except much of the Far North and southeastern United States.

SIMILAR SPECIES: Bull Thistle (*C. vulgare*).

CONSERVATION: G5

This species is globally secure. It is considered a noxious weed in Canada and the United States.

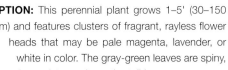

■ NATIVE
■ RARE OR EXTIRPATED
■ INTRODUCED

ALTERNATE NAMES: Yellow Thistle, Horrid Thistle, Bristly Thistle

The common name of this plant reflects its spiny nature. It is often found along the edges of saltmarshes, but also in fields and pastures, especially in the southern part of its range. The similar Prairie Thistle (*C. canescens*) also sports yellow flowers and is common in the western United States.

DESCRIPTION: This biennial herb grows 1–5' (30–150 cm) with a tall branching stem and large, rayless flower heads that are usually yellow but can also be pink, red, purple, or white. The leaves are spiny and pinnately lobed.

FLOWERING SEASON: May–August.

HABITAT: Grows in marshes, fields, and other open places.

RANGE: Occurs in the southeastern United States from southern Maine south to Florida and west to Texas.

SIMILAR SPECIES: Prairie Thistle (*C. canescens*).

CONSERVATION: G5

Globally secure but critically imperiled in Connecticut and Pennsylvania and vulnerable in Maryland and Tennessee. It is possibly extirpated in Maine.

NATIVE
RARE OR EXTIRPATED
INTRODUCED

ALTERNATE NAMES: Nodding Thistle, Gray Thistle

This plant is native to central and western North America but can now be found beyond its natural range as an introduced species. It grows in a wide variety of habitats from prairies to forests. The wavy-edged leaves for which this species is named are a helpful identification feature.

DESCRIPTION: This prickly perennial plant grows 10–40" (25–100 cm) with one or more flower heads that may be whitish or pinkish. The leaves are pinnately lobed and tipped with spines.

FLOWERING SEASON: This prickly perennial plant grows 10–40" (25–100 cm) with one or more flower heads that may be whitish or pinkish. The leaves are pinnately lobed and tipped with spines.

HABITAT: Grows in prairies and grasslands, forests and woodlands, and disturbed areas.

RANGE: Occurs from Michigan, Indiana, and Missouri west to the Pacific Coast and south to Texas.

SIMILAR SPECIES: See other *Cirsium* species.

CONSERVATION: G5

Although globally secure, this species is critically imperiled in Iowa, Missouri, and Manitoba, and vulnerable in Alberta.

■ NATIVE
■ RARE OR EXTIRPATED
■ INTRODUCED

ALTERNATE NAMES: Spear Thistle

This prickly plant, native to Eurasia, is common and widespread throughout North America. It is highly invasive and disruptive, growing in a wide variety of habitats and replacing native vegetation that is more useful to wildlife and livestock.

DESCRIPTION: This biennial plant grows 2–6' (60–180 cm) and has a spiny winged stem and large purplish flower heads surrounded by spiny yellow-tipped bracts. The leaves are coarsely pinnately lobed and spiny.

FLOWERING SEASON: June–September.

HABITAT: Grows along roadsides and in pastures and waste places.

RANGE: Occurs throughout North America, except parts of the Far North.

SIMILAR SPECIES: Creeping Thistle (C. *arvense*).

CONSERVATION:
The conservation status of this introduced species is unranked.

■ NATIVE
■ RARE OR EXTIRPATED
■ INTRODUCED

ALTERNATE NAMES: Blue Boneset, *Eupatorium coelestinum*

This native wildflower grows in low woodlands and along streams in the eastern United States. The bluish flowers, sometimes appearing whitish or pinkish, attract many bees, butterflies, and other pollinators. This plant can sometimes become a pest as it spreads quickly.

DESCRIPTION: This herbaceous perennial grows 1–2' (30–60 cm) tall and spreads by creeping rhizomes. It features dense, bluish flower clusters and covers the ground with a mat of leaves; the leaves are coarsely toothed and triangular.

FLOWERING SEASON: August–October.

HABITAT: Grows in low woodlands, wet meadows, ditches, and along streams.

RANGE: Occurs from Connecticut south to Florida, west to Texas, and north to Illinois, Indiana, and Michigan.

SIMILAR SPECIES: Palm-leaf Mistflower (*C. dissectum*).

CONSERVATION: G5
Globally secure but imperiled in Kansas and vulnerable in New Jersey.

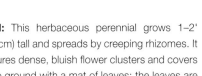

■ NATIVE
■ RARE OR EXTIRPATED
■ INTRODUCED

ALTERNATE NAMES: Large-flower Coreopsis

This vigorous, clump-forming plant is native to the eastern United States, preferring fields and other open areas. The genus name *Coreopsis* is derived from Greek and means "bug-like," referring to the small dry, flat fruits that resemble insects.

DESCRIPTION: This perennial grows 1–3' (30–90 cm) in height and produces yellow to gold daisy-like flowers. The leaves are lanceolate and scattered along the wiry stem.

FLOWERING SEASON: May–June.

HABITAT: Grows in open woods, fields, and along roadsides.

RANGE: Occurs in scattered areas throughout the eastern United States.

SIMILAR SPECIES: Lanceleaf Coreopsis (*C. lanceolata*).

CONSERVATION: G5
This species is globally secure.

- ■ NATIVE
- ■ RARE OR EXTIRPATED
- ■ INTRODUCED

ALTERNATE NAMES: Lanceleaf Tickseed, Sand Coreopsis

This species is native to the eastern and central United States but naturalized in the western United States, Canada, and many other regions of the world. It is commonly cultivated as an ornamental plant and is a useful food source for a variety of native wildlife species.

DESCRIPTION: This perennial plant grows 1–2' (30–60 cm) in height and features yellow daisy-like flower heads on long stalks. The basal leaves are elliptical to linear in shape and occasionally lobed; the upper leaves are linear to oblong. The leaves are generally narrow and lance-shaped.

FLOWERING SEASON: May–July.

HABITAT: Grows in open woodlands, prairies, meadows, and disturbed areas.

RANGE: Native to the eastern United States but also found in much of the western United States and Canada.

SIMILAR SPECIES: Large-flower Tickseed (*C. grandiflora*) and Greater Tickseed (*C. major*).

CONSERVATION: G5
Globally secure but critically imperiled in Wyoming and imperiled in Kansas and Wisconsin.

■ NATIVE
■ RARE OR EXTIRPATED
■ INTRODUCED

ALTERNATE NAMES: Golden Tickseed, Goldenwave, Plains Coreopsis, Calliopsis

This showy wildflower is commonly found throughout much of North America, growing along roadsides, in ditches, and other disturbed areas. It is cultivated extensively as an ornamental plant, hence the common name.

DESCRIPTION: This annual grows 2–4' (60–120 cm) in height and produces daisy-like flower heads with yellow rays surrounding a reddish-purple central disk. The leaves are opposite and dissected into linear segments.

FLOWERING SEASON: June–September.

HABITAT: Grows in open areas, along roadsides, and at disturbed sites.

RANGE: Occurs throughout North America, except parts of the Far North.

SIMILAR SPECIES: Goldenmane Tickseed (*C. basalis*).

CONSERVATION: G5

Globally secure but critically imperiled in Manitoba, imperiled in Wyoming, and vulnerable in Alberta.

TALL TICKSEED *Coreopsis tripteris*

■ NATIVE
■ RARE OR EXTIRPATED
■ INTRODUCED

ALTERNATE NAMES: Tall Coreopsis, Atlantic Coreopsis

Aptly named, this tall plant is native to eastern and central North America, where it thrives in moist habitats. It is sometimes cultivated as an ornamental.

DESCRIPTION: This perennial plant grows 3–9' (1–3 m) in height. The flower heads are large and feature a reddish-brown center surrounded by yellow rays, rounded at the tip. The stalked leaves are often divided into three leaflets.

FLOWERING SEASON: July–September.

HABITAT: Grows in open woods and wet meadows, along streambanks, and in other moist places.

RANGE: Occurs from Ontario east to Quebec, south to Georgia, and west to Texas.

SIMILAR SPECIES: Greater Coreopsis (*C. major*).

CONSERVATION: G5

Although globally secure, this species is critically imperiled in Ontario, Kansas, Maryland, and Texas. It is vulnerable in North Carolina and West Virginia.

- ◼ NATIVE
- ◼ RARE OR EXTIRPATED
- ◼ INTRODUCED

ALTERNATE NAMES: Cosmos, Mexican Aster

This Mexican native, widely cultivated, has escaped from gardens and become common in parts of southern Canada and the northern United States, usually growing in disturbed areas. Its colorful flowers attract birds and butterflies.

DESCRIPTION: This introduced plant may produce maroon, pink, lavender, or white flowers on stalks up to 4' (1.2 m) tall. The large, cup-shaped flowers can make the plants top-heavy. When growing in groups, the plants' bipinnate leaves interlock to support the colony.

FLOWERING SEASON: June–September.

HABITAT: Grows along roadsides and in waste places.

RANGE: Occurs from Ontario east to Quebec and Maine, south to Florida, west to Kansas, and north to Minnesota. It is also found in the Southwest.

SIMILAR SPECIES: Sulphur Cosmos (*C. sulphureus*) is another commonly cultivated, introduced cosmos that has escaped from cultivation in parts of the United States, where it is considered invasive.

CONSERVATION:
The conservation status of this introduced species is unranked.

NATIVE

RARE OR EXTIRPATED

INTRODUCED

ALTERNATE NAMES: Taper-tip Hawksbeard

Native to the western United States, this flowering plant grows in many types of open habitats, such as slopes and meadows. The genus name *Crepis* is derived from the Greek word for "sandal." The reason behind this is unknown.

DESCRIPTION: This perennial plant grows 8–28" (20–70 cm) and features a cluster of numerous narrow flower heads with yellow rays. The leaves are downy and pinnately lobed, edged with teeth.

FLOWERING SEASON: May–August.

HABITAT: Grows in open woods and meadows.

RANGE: Occurs from eastern Washington south to eastern California and east to northern New Mexico, Colorado, and central Montana.

SIMILAR SPECIES: Western Hawksbeard (*C. occidentalis*) and Meadow Hawksbeard (*C. runcinata*).

CONSERVATION: G5

Globally secure but critically imperiled in British Columbia.

■ NATIVE
■ RARE OR EXTIRPATED
■ INTRODUCED

ALTERNATE NAMES: Gray Hawksbeard, Largeflower Hawksbeard

This native flowering plant grows in many types of habitat throughout western North America. It is covered with short, gray, felt-like hairs.

DESCRIPTION: This perennial herb grows 6–18" (15–45 cm) and produces clusters of 2–20 yellow flower heads. The leaves are gray-hairy and mostly deeply lobed.

FLOWERING SEASON: May–August.

HABITAT: Grows in open woods and meadows.

RANGE: Occurs throughout much of western North America.

SIMILAR SPECIES: Hawksbeard (C. *acuminata*) and Meadow Hawksbeard (C. *runcinata*).

CONSERVATION: G5
Globally secure but vulnerable in Alberta, Saskatchewan, and Wyoming.

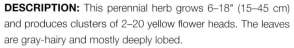

■ NATIVE
■ RARE OR EXTIRPATED
■ INTRODUCED

ALTERNATE NAMES: Parasol Whitetop, *Aster umbellatus*

This conspicuous flowering plant grows in wet areas in Canada and the eastern United States. It is one of the first asters to bloom and has an extremely wide, flat top, for which it is named.

DESCRIPTION: This perennial plant grows 1–7' (30–210 cm) and features a rigid, upright stem bearing a flat-topped cluster of white flower heads. The leaves are stalkless and elliptical, tapering at both ends.

FLOWERING SEASON: August–September.

HABITAT: Grows in woods and thickets, along swamp edges, and in other moist or wet places.

RANGE: Occurs from Alberta east to Newfoundland, south to Georgia and Alabama, and northwest to Missouri, Nebraska, and North Dakota.

SIMILAR SPECIES: Cornel-leaf Whitetop (*D. infirma*).

CONVERSATION: G5

Although globally secure, this species is imperiled in Saskatchewan and vulnerable in Alberta, Delaware, Iowa, and North Carolina.

■ NATIVE
■ RARE OR EXTIRPATED
■ INTRODUCED

ALTERNATE NAMES: Blacksamson Coneflower, *Echinacea pallida* var. *angustifolia*

This species is native to central North America, where it grows in dry habitats. The genus name *Echinacea* is derived from the Greek word meaning "sea urchin" or "hedgehog," referring to the prominent flower heads of these plants.

DESCRIPTION: This perennial herb grows 18–24" (45–60 cm) in height and features large, solitary flower heads with purplish disk flowers. The basal leaves are hairy and elliptical to lance-shaped.

FLOWERING SEASON: June–August.

HABITAT: Grows in dry prairies and plains.

RANGE: Found from Montana, Saskatchewan, and western Minnesota south to Texas.

SIMILAR SPECIES: See other *Echinacea* species.

CONSERVATION: G4

Apparently secure but critically imperiled in Missouri; imperiled in Colorado; and vulnerable in Manitoba, Saskatchewan, Iowa, and Wyoming.

■ NATIVE
■ RARE OR EXTIRPATED
■ INTRODUCED

ALTERNATE NAMES: Eastern Purple Coneflower

This attractive flowering plant is native to eastern North America, favoring dry open woods and prairies. A showy plant that is easy to grow, Purple Coneflower is often cultivated as an ornamental. The flowers of this and other *Echinacea* species are sometimes used to make an herbal tea.

DESCRIPTION: This perennial grows 1–5' (30–150 cm) tall. The flowers feature a spiny, brownish central disk surrounded by purplish, drooping rays. The leaves are rough, long-stalked, and ovately shaped, often with serrated edges.

FLOWERING SEASON: June–October.

HABITAT: Grows in woodlands, thickets, and prairies.

RANGE: Occurs from Quebec and New York south to Florida, west to Texas, and north to Iowa and Wisconsin.

SIMILAR SPECIES: Smooth Purple Coneflower (*E. laevigata*).

CONSERVATION: G4
Apparently secure but critically imperiled in Florida, Kansas, and North Carolina; imperiled in Georgia, Iowa, and Louisiana; and vulnerable in Alabama and Mississippi. It is presumed extirpated in Michigan.

- ■ NATIVE
- ■ RARE OR EXTIRPATED
- ■ INTRODUCED

ALTERNATE NAMES: Goldenhills

This bushy plant is native to the southwestern United States and northwestern Mexico, where it grows on desert slopes and in other dry places. The stems produce a fragrant resin that was burned as incense in early Spanish missions.

DESCRIPTION: This round, leafy bush grows 3–5' (90–150 cm) in height. It produces bright yellow flower heads that bloom in loosely branched clusters. The leaves are silvery gray, hairy, and ovately shaped.

FLOWERING SEASON: March–June.

HABITAT: Grows on dry slopes, flats, and washes.

RANGE: Occurs in southeastern California, southern Nevada, western Arizona, and northwestern Mexico.

SIMILAR SPECIES: Button Brittlebush (*E. frutescens*) and California Encelia (*E. californica*).

CONSERVATION: G5

This species is globally secure but critically imperiled in Utah.

■ NATIVE
■ RARE OR EXTIRPATED
■ INTRODUCED

ALTERNATE NAMES: Eastern Daisy Fleabane, White-top Fleabane, Annual Fleabane

This widespread native annual is a pioneer species that grows in disturbed areas, often able to compete successfully with non-native invasive species that grow in similar habitats. It is named for its daisy-like flowers. The name Fleabane attributed to this and other *Erigeron* species comes from a belief that the dried flower heads could rid a home of fleas.

DESCRIPTION: This plant grows 1–5' (30–150 cm) tall with an erect stem, covered with spreading white hairs. The leaves are lanceolate in shape and hairy. Each leaf at the base is about 5" (12.5 cm) long, 2" (5 cm) wide, and toothed. The leaves on the upper stems are smaller and lack petioles. Small clusters of daisy-like composite flowers occur toward the top of the plant, each about ½–¾" across. The flower buds often have white hairs. The flower heads have 40 or more tightly packed, white to pale pink rays surrounding a yellow central disk. Each flower head is ½" (1.5 cm) wide, with shortish rays. The achenes have tufts of small hairs and are distributed by the wind.

FLOWERING SEASON: June–October.

HABITAT: Fields, roadsides, and waste places.

RANGE: Throughout the East, except the Far North; also in much of the West.

SIMILAR SPECIES: Prairie Fleabane (*E. strigosus*), a very similar plant that's also widespread across North America, was once considered a subspecies of Daisy Fleabane. Prairie Fleabane is more weedy and has narrower leaves, about 1" (2.5 cm) wide. It also has shorter stem hairs.

CONSERVATION: G5
Globally secure but imperiled in North Dakota and Prince Edward Island; vulnerable in Alberta.

■ NATIVE
■ RARE OR EXTIRPATED
■ INTRODUCED

ALTERNATE NAMES: Canadian Horseweed, Canada Horseweed, *Conyza canadensis*

This is an abundant and widespread native annual plant that grows aggressively both in its native range and elsewhere around the world, where it is widely naturalized. Native Americans and early settlers used this annual's leaves in remedies to treat dysentery and sore throat.

DESCRIPTION: A coarse weedy plant with an erect, bristly-haired stem 1–7' (30–210 cm). The leaves are dark green, hairy, linear to narrowly lanceolate in shape. Each leaf is 1–4" (2.5–10 cm) long and toothed. The yellow disk flowers are inconspicuous and numerous. Branching clusters of small, cup-like flower heads rise from the upper leaf axils. This flower head is less than ¼" (6 mm) wide, with white rays. The seed is tiny and one-seeded, with many bristles.

FLOWERING SEASON: July–November.

HABITAT: Fields, roadsides, and waste places.

RANGE: Throughout North America, except parts of the Far North.

CONSERVATION: **G5**
Globally secure. It is widespread in its native North American range and is commonly considered a weed. It has become invasive elsewhere around the world, including temperate Eurasia and Australia.

■ NATIVE
■ RARE OR EXTIRPATED
■ INTRODUCED

ALTERNATE NAMES: Alpine Daisy, Cutleaf Daisy

This dwarf perennial is found in the Arctic areas and at high elevations in the mountains. It's distinguished from other fleabanes by its leaves, which are finely divided.

DESCRIPTION: This cushion-like plant grows 4–10" (10–25.4 cm) tall with compact mounds of woolly, pinnately divided leaves. The stems are red or green and unbranched, usually covered with hairs. Most of the leaves appear at the base. The leaves are spoon-shaped, sometimes divided two or three times into linear lobes, and usually covered with hairs. It usually has one flower head per stem. Each head has 20–60 white, pink, or blue ray florets.

FLOWERING SEASON: May–August.

HABITAT: Gravelly, alpine slopes at high elevations.

RANGE: Alaska and in the Far North, south to Quebec in the East and at high altitudes in the western United States.

CONSERVATION: G5
Globally secure but critically imperiled in Nova Scotia and Newfoundland; vulnerable in Quebec and Saskatchewan.

■ NATIVE
■ RARE OR EXTIRPATED
■ INTRODUCED

ALTERNATE NAMES: Common Fleabane

Philadelphia Fleabane is distinguished from other fleabanes by its clasping leaves and the greater number of ray florets on its flower heads. This is one of the most common fleabanes in North America. Like many of its relatives, this widespread native plant is opportunistic and grows readily in disturbed sites, often leading it to become a problematic weed where it is not native.

DESCRIPTION: This plant grows 6–36" (15–90 cm) tall with a somewhat hairy multiangular stem. The leaves are hairy. Each leaf is 6" (15 cm) long at the base and oblong to narrow in shape. The leaves on the stem are smaller, toothed, and clasping. Small, daisy-like flower heads with 100–300 narrow, white to pink rays surround a large yellow central disk. Each flower head is ½–1" (1.5–2.5 cm) wide. The florets are replaced by achenes with small tufts of white hair, distributed by the wind.

FLOWERING SEASON: April–August.

HABITAT: Rich thickets, fields, and open woods.

RANGE: Throughout the East, except in the Arctic; also in much of the West.

SIMILAR SPECIES: Robin's Plantain (*E. pulchellus*) is similar but slightly smaller and has larger violet flower heads. Robin's Plantain also has fewer stem leaves that do not clasp the stem.

CONSERVATION: G5

Globally secure but critically imperiled in Colorado and Prince Edward Island; imperiled in Nova Scotia, Yukon, and Wyoming; vulnerable in Montana and North Carolina.

■ NATIVE
■ RARE OR EXTIRPATED
■ INTRODUCED

Robin's-plantain has daisy-like flowers that grow in colonies. This widespread eastern native has soft stems that rise from rosettes of paddle-shaped basal leaves appearing along surface runners.

DESCRIPTION: This perennial plant grows 12–18" (30–45 cm) high with a soft, hollow stem covered in hairs. The basal leaves are shaped like a paddle with soft hairs. Lavender-blue to white flower clusters appear at the top of each stem with yellow centers. The flowers somewhat resemble asters but the rays are narrower and numerous. Each cluster has two to six flowers. The flower heads have 50–100 ray florets, densely packed together.

FLOWERING SEASON: April–June.

HABITAT: Rich woods, streambanks, and fields.

RANGE: Maine and Ontario to southeastern Minnesota and Kansas, south to Florida and East Texas.

SIMILAR SPECIES: Philadelphia Fleabane (*E. philadelphicus*) is somewhat similar, but grows taller.

CONSERVATION: **G5**
Globally secure. Braun's Plantain (*E. pulchellus* var. *brauniae*) is an apparently secure, rare variety that occurs from the western panhandle of Maryland—where it is critically imperiled—to eastern Kentucky.

■ NATIVE
■ RARE OR EXTIRPATED
■ INTRODUCED

ALTERNATE NAMES: Prairie Ragwort, Daisy Fleabane

Like most fleabanes, Prairie Fleabane grows in colonies and may be considered weedy. This widespread native annual or biennial herb is also widely naturalized.

DESCRIPTION: This plant grows 1–2' (30–60 cm) tall, with single stems, usually branching above. The stem has white hairs at the base and shorter hairs above. The lower leaves are hairy and oblanceolate in shape, while the upper leaves are elliptical to linear with coarse teeth and mostly hairless. Clusters of yellow, aster-like flowers appear at the top of the stems. Each flower head is about ½" (1.2 cm) wide, with 40–100 white, pink, or purple ray florets surrounding numerous disk florets. The corollas of the disk florets are yellow and tubular in shape, with five lobes, densely packed together.

FLOWERING SEASON: May–June.

HABITAT: Dry prairies and roadsides.

RANGE: Pennsylvania and southern Ontario to Saskatchewan and south-central Montana, south to Louisiana, Texas, and Arizona; widely naturalized elsewhere.

SIMILAR SPECIES: Daisy Fleabane (*E. annuus*) is similar but Prairie Fleabane has fewer and more slender leaves, and the hairs along its middle to upper stems are shorter. Philadelphia Fleabane (*E. philadelphicus*) has larger flower heads with more ray florets (100–300), and wider leaves that clasp the stems. Philadelphia Fleabane's stem hairs are also spreading.

CONSERVATION:

Globally secure but vulnerable in Alberta, Saskatchewan, Texas, and Wyoming.

■ NATIVE
■ RARE OR EXTIRPATED
■ INTRODUCED

ALTERNATE NAMES: Oregon Sunshine, Common Woolly Sunflower

Golden Yarrow's flowers resemble those of sunflowers. This is a common and highly variable species native to western North America. It's covered in white hairs, which reflect heat and reduce air movement on the leaf's surface, allowing it to conserve water and thrive in dry soils.

DESCRIPTION: This grayish, woolly, leafy plant grows 4–24" (10–60 cm) tall. The leaves are 1–3" (2.5–7.5 cm) long and irregularly divided into narrow lobes. Several branched stems terminate in golden-yellow flower heads. Each flower head is 1½–2½" (3.8–6.3 cm) wide, with broadly lanceolate bracts prominently ridged on the back, and 8–12 broad rays, each ½–¾" (1.3–2 cm) long, around the disk. The fruit is narrow, seed-like, and smooth.

FLOWERING SEASON: May–July.

HABITAT: Dry thickets and dry open places.

RANGE: British Columbia to southern California and western Nevada; east to northeastern Oregon and western Montana.

CONSERVATION: G5
Globally secure but critically imperiled in Utah and vulnerable in Wyoming.

■ NATIVE
■ RARE OR EXTIRPATED
■ INTRODUCED

ALTERNATE NAMES: Hyssop-leaf Thoroughwort

This perennial herbaceous plant is native to the eastern United States. It is sometimes planted near crops because it attracts various native birds and beneficial insects. Like other *Eupatorium* species, it has been used externally as medicine to treat rattlesnake and insect bites.

DESCRIPTION: This plant has erect, leafy stems and grows 36" (90 cm) tall. The leaves are very narrow and sometimes linear, and toothed, usually appearing in whorls of three or four (occasionally paired), with groups of smaller leaves in the axils. Flat-topped clusters of tiny, white, rayless flowers appear at the top of the stems. The flowers are fringed and look cloud-like.

FLOWERING SEASON: August–October.

HABITAT: Open sandy areas and roadsides.

RANGE: Coastal plains and piedmont from southern New England and New York to Florida and west to Texas, and inland south and west through Pennsylvania to Ohio, Illinois, and Arkansas.

CONSERVATION: G5
Globally secure but critically imperiled in West Virginia and Ohio; vulnerable in Arkansas.

BONESET *Eupatorium perfoliatum*

■ NATIVE
■ RARE OR EXTIRPATED
■ INTRODUCED

ALTERNATE NAMES: Common Boneset, American Boneset, Feverwort

Boneset is native to eastern and central North America and has a long history of medicinal use. The plant was used to treat pain and fever, including dengue fever in the 18th century. The dried leaves have also been used to make a tonic called boneset tea, thought to be effective in treating colds, coughs, and constipation. It often appears as an ingredient in homeopathic medicine, but the U.S. Food and Drug Administration lists it among its Poisonous Plants Database. The Latin species name refers to the leaves, which unite at the base and circle the stem.

DESCRIPTION: This hairy perennial plant grows 2–4' (60–120 cm). The leaves are oppositely arranged and lanceolate. Each leaf is 4–8" (10–20 cm) long, wrinkled, sessile, and toothed. The leaves are usually united at the base and surround the hairy stem. It has flat-topped clusters of many dull white, fluffy, rayless flower heads. Each flower head is about ¼" (6 mm) long.

FLOWERING SEASON: July–October.

HABITAT: Low woods and wet meadows.

RANGE: Throughout the East, except the Far North.

CAUTION: This plant contains alkaloids that can be potentially toxic in large doses.

SIMILAR SPECIES: Upland Boneset (*E. sessilifolium*) is somewhat similar, but its leaves are not united at the base.

CONSERVATION: **G5**

Globally secure but vulnerable in parts of Canada, including Manitoba, Prince Edward Island, and Quebec.

■ NATIVE
■ RARE OR EXTIRPATED
■ INTRODUCED

ALTERNATE NAMES: Late-flowering Thoroughwort

This perennial is native to the eastern and central United States. Its flowers attract various insects, including butterflies, and birds eat its seeds. Plants in the genus *Eupatorium* have long been used in herbal remedies; Native Americans used this plant to treat typhoid fever.

DESCRIPTION: This plant grows 3–6' (0.9–1.8 m), with stems covered in white pubescent hairs. It appears mostly unbranched below. The leaves are elliptical to ovate and toothed. Most of the leaves are opposite, becoming more alternate at the top of the plant. Each leaf is about 7" (18 cm) long. The top is branched with broad, flat-topped clusters of small white florets with no rays. Each flower head is cup-like and fuzzy, with 12 disk florets, about ¼" (6 mm) long each. It has small achenes with tufts of hair.

FLOWERING SEASON: September–December.

HABITAT: Moist to wet meadows, fields, woodland edges and openings, and bottomlands; near lakes and ponds.

RANGE: Eastern United States from southern New York to Florida, west to Minnesota and Texas.

SIMILAR SPECIES: Late Boneset is distinguished from other bonesets by its leaves. Tall Boneset (*E. altissimum*) has leaves that are more pubescent and narrow. Common Boneset (*E. perfoliatum*) has leaves that wrap around the stem. White Snakeroot (*Ageratina altissima*) is a poisonous plant with similar white flowers but with shorter, broader leaves.

CONSERVATION: G5
Globally secure but vulnerable in Delaware and Iowa.

■ NATIVE
■ RARE OR EXTIRPATED
■ INTRODUCED

ALTERNATE NAMES: Lawn American-aster, *Aster divaricatus*

This perennial native to eastern North America has showy flowers that are attractive to butterflies. The specific epithet *divaricata* means "spreading."

This plant has zigzag stems and grows 12–40" (30–100 cm) tall. The leaves are heart-shaped and coarsely toothed. Each leaf is 2–7" (5–17.5 cm) long and tapering. It has abundant, flat-topped clusters of flower heads with white rays. Each flower head is up to 1" (2.5 cm) wide, 6–10 rays surrounding a yellow or purplish central disk. The bracts are whitish with green tips. The fruit is dry and seed-like, tipped with whitish bristles.

FLOWERING SEASON: July–October.

HABITAT: Dry open woods.

RANGE: Virginia southwest to Tennessee and Alabama, west to Texas, and north to Nebraska.

SIMILAR SPECIES: Largeleaf Aster (*E. macrophylla*) is similar but taller and has lavender rays, not white. Largeleaf Aster's leaves are also coarser and the flower stalks are glandular.

CONSERVATION: G5

Globally secure but classified as a vulnerable species in Canada. It is imperiled in Quebec and vulnerable in Maine and Ontario.

■ NATIVE
■ RARE OR EXTIRPATED
■ INTRODUCED

ALTERNATE NAMES: Flat-top Goldentop, Grass-leaf Goldenrod, *Solidago graminifolia*

This plant is a widespread native perennial mainly found in northern and eastern North America. The plant can appear slender or branching. Some of the common names refer to its narrow, grass-like leaves and rounded, flat-topped flower clusters.

DESCRIPTION: This plant grows to a height of 2–4' (60–120 cm) with a smooth or finely downy stem. The leaves are narrow, elongated, and pointed. Each leaf is 2¾–5" (7–12.5 cm) long, with three to five veins. The stems branch above the midstem. Each branch has one flat-topped cluster of small yellow flower heads. The flower heads are about ¼" (5 mm) long, with 10–20 rays and 8–12 disk flowers.

FLOWERING SEASON: July–October.

HABITAT: Roadsides, fields, and thickets.

RANGE: Alberta east to Newfoundland, south to South Carolina, southwest to Louisiana, and northwest to Oklahoma and North Dakota; also in parts of the West.

SIMILAR SPECIES: Slender Fragrant Goldenrod (*E. tenuifolia*) is similar but smaller and has grass-like leaves with tiny resin dots and one leaf vein. Narrow-leaved Bushy Goldenrod (*E. caroliniana*) is also similar but is smaller and has smaller leaves, about ⅛" (3 mm) wide each.

CONVERSATION: G5

Globally secure but critically imperiled in Idaho, Oklahoma, and Wyoming.

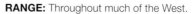
■ NATIVE
■ RARE OR EXTIRPATED
■ INTRODUCED

ALTERNATE NAMES: Western Fragrant Goldenrod, Western Goldentop, Western Goldenrod

This plant is found mostly in western North America. It has long grass-like leaves and showy, bright yellow flowers.

DESCRIPTION: This perennial grows 1½–6' (45–180 cm) tall with many stems covered with waxy, whitish coating. The leaves are long, gray-green, and blade-like with coarse margins. It has numerous bright yellow flower heads in rounded or elongated leafy-bracted clusters.

FLOWERING SEASON: July–October.

HABITAT: Moist or wet places, ditches.

RANGE: Throughout much of the West.

CONSERVATION: G5
Globally secure but imperiled in Wyoming.

- ◼ NATIVE
- ◼ RARE OR EXTIRPATED
- ◼ INTRODUCED

ALTERNATE NAMES: Joe Pye Weed, Spotted Trumpetweed

This native perennial herb thrives in wetlands and other moist habitats. This species is widespread across North America and is the only member of the genus found in the West. Its flowers attract butterflies and bees, and the plant is the larval host of several native butterfly species. Joe-Pye-weeds are named for a Native American healer who, according to folklore, used these plants to treat fevers and typhus. The genus name means "wheel-like" and refers to the whorled leaves, while the specific epithet means "spotted," referencing the purple-spotted stems.

DESCRIPTION: This plant grows 2–6' (60–180 cm) tall with branched, stems spotted with dark purple. Whorls of three to six leaves surround the stem. The leaves are coarsely serrated and lance-shaped. Each leaf is 2½–8" (6.25–20 cm) long. The plant has tiny, light to deep purple flower heads, each ⅜" (8 mm) wide in clusters 4–5½" (10–14 cm) wide.

FLOWERING SEASON: July–September.

HABITAT: Moist, wet places, damp meadows, thickets.

RANGE: Across southern Canada and northern United States; south in mountainous areas to North Carolina, also in western states.

SIMILAR SPECIES: Sweet-scented Joe-Pye-weed (*E. purpureum*) is similar but has a greenish stem and a dome-shaped cluster of dull pink flower heads, which smell like vanilla. Hollow Joe-Pye-weed (*E. fistulosum*) has a hollow stem. Coastal Plain Joe-Pye-weed (*E. dubium*) is a smaller plant with ovate leaves.

CONSERVATION: G5
Globally secure but critically imperiled in Idaho, Montana and West Virginia; imperiled in Arizona, Georgia, Virginia, and Wyoming; and vulnerable in Alberta.

■ NATIVE
■ RARE OR EXTIRPATED
■ INTRODUCED

ALTERNATE NAMES: Sweet Joe-Pye-weed, Gravel Root, Trumpet Weed, Purple Joe-Pye-weed

This eastern native plant is frequently visited by pollinators, such as birds, moths, and bees, which are attracted to its vanilla-scented flowers. Native Americans used this plant to treat fevers.

DESCRIPTION: This plant grows 2–6' (0.6–1.8 m) tall. It has an erect stem with narrow, lance-shaped leaves that grow in a whorl around the stem. Each leaf is about 12" (30 cm) long and slightly wrinkled. Once the flowers bloom, the stem may bend downward from the weight. The flower heads are large and dome-shaped, composed of several branches with small pale pinkish-purple florets.

FLOWERING SEASON: July–September.

HABITAT: Moist prairies, wood edges, and wooded slopes.

RANGE: Ontario and the eastern U.S. west to Nebraska and Oklahoma.

CONSERVATION: G5

Globally secure but critically imperiled in Louisiana and imperiled in Vermont.

- ◼ NATIVE
- ◼ RARE OR EXTIRPATED
- ◼ INTRODUCED

ALTERNATE NAMES: Gaillardia, Brown-eyed Susan

This attractive perennial plant is widely cultivated and may be naturalized rather than native in parts of its range, especially in the East. Some Native Americans used Great Blanketflower to treat wounds and fevers.

DESCRIPTION: This erect, hairy plant grows 2–4' (60–120 cm) tall. It has clasping leaves, which are rough on the surface and lanceolate. Each leaf is about 2–6" (5–15 cm) long and 1" (2.5 cm) wide with edges that may be smooth, toothed or lobed. Daisy-like flower heads appear at the top of the hairy stems with a red center disk, red rays, and yellow flower tips. The seed-like fruit is covered in tufts of hair.

FLOWERING SEASON: July–September.

HABITAT: Plains, prairies, and meadows.

RANGE: Native from British Columbia to Saskatchewan, south to Oregon (mostly east of the Cascades), Utah, Colorado, and Kansas; also in California. It also occurs elsewhere in North America, possibly as an introduced and naturalized species.

CAUTION: The plant's fuzzy hairs may cause skin irritation.

CONSERVATION: **G5**
Globally secure but vulnerable in Minnesota.

FIREWHEEL BLANKETFLOWER *Gaillardia pulchella*

■ NATIVE
■ RARE OR EXTIRPATED
■ INTRODUCED

ALTERNATE NAMES: Firewheel, Indian Blanket, Fire-wheel Blanket-flower, Showy Gaillardia

The showy, vibrantly colored flowers of this species look like pinwheels. There are several varieties often grown in gardens, many of which have entirely yellow flower heads. Arizona Sun Basket Flower *(Gaillardia* x *grandiflora)*, for example, is a popular hybrid with red and yellow florets.

DESCRIPTION: This plant has branched stems and grows 1–2' (30–60 cm) tall, with a leafy base. Each leaf is up to 3" (7.5 cm) long and oblong in shape, with sometimes toothed margins. Showy red flower heads with red rays appear near the leaves at the base and are tipped with yellow—each with 3 teeth. The flower head is

1½–2½" (3.8–6.3 cm) wide with a red, dome-like disk with bristly scales. Each ray is ½–¾" (1.3–2 cm) long. The fruit is seed-like, with whitish scales at the tip.

FLOWERING SEASON: May–July.

HABITAT: Sandy plains and desert, common along roadsides.

RANGE: Arizona to Texas, north to southeastern Colorado and Nebraska, and south into Mexico.

CONSERVATION: G4
Apparently secure, but imperiled in Colorado and Nebraska.

■ NATIVE
■ RARE OR EXTIRPATED
■ INTRODUCED

ALTERNATE NAMES: Broadleaf Gumweed, Compass Plant

This species is toxic but has a long history of medicinal use. Native Americans used the flowers to treat asthma, bronchitis, skin rashes, and other ailments. Some studies have also indicated this plant has a potential as a biofuel crop. The resinous sap that covers the leaves has been used as a substitute for chewing gum. Green and yellow dyes can be made from the yellow flowering heads. This species is sometimes called Compass Plant because its leaves point toward the sun.

DESCRIPTION: This dark green, leafy plant reaches 1–3' (30–90 cm) tall. The leaves are oblong and clasp the stem at the base. Each leaf is 3" (7.5 cm) long, with sharp, pointed teeth at the edges. Many yellow flower heads appear at the top of the plant. The tips at the bracts around each flower head are rolled back.

FLOWERING SEASON: July–September.

HABITAT: Prairies and waste places.

RANGE: Throughout much of North America, except Far North and from North Carolina to Louisiana. It may be naturalized rather than native in parts of its range, especially in the East.

CAUTION: This plant is sometimes used medicinally, but can be mildly toxic to humans and other mammals. The toxicity levels are based on the soil, as the plant absorbs selenium from the ground.

CONSERVATION:
Globally secure. It can invade overgrazed pastures and similar waste places outside its native range.

ALTERNATE NAMES: Fall Sneezeweed, Common Sneeze-wood

Sneezeweed is a widespread native found coast to coast. It flowers in late summer or fall. It gets its common name from the former practice of drying its leaves to make a snuff, which was inhaled to induce sneezing and rid the body of evil spirits and relieve headaches.

DESCRIPTION: This plant grows 2–5' (60–150 cm) tall. The leaves are lanceolate in shape and alternately arranged. Each leaf is about 6" (15 cm) long with toothed margins. The bases form a winged stem with yellow, daisy-like flower heads and fan-shaped, drooping rays. The disk flowers form a greenish-yellow, ball-like structure at the center of the head. Each flower head is about 1–2" (2.5–5 cm) wide. The rays have three lobes.

FLOWERING SEASON: August–November.

HABITAT: Swamps, wet meadows, and roadsides.

RANGE: Throughout North America, except the Maritime Provinces and Far North.

SIMILAR SPECIES: Purple-head Sneezeweed (*H. flexuosum*) is similar but has a purplish-brown ball of disk flowers. Slender-leaved Sneezeweed (*H. amarum*) is also similar but has thread-like leaves on its stems.

CONSERVATION: G5
Globally secure but critically imperiled in Vermont and vulnerable in Alberta, Quebec, Massachusetts, and Wyoming.

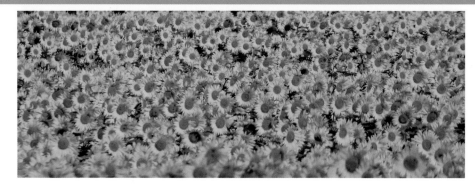

■ NATIVE
■ RARE OR EXTIRPATED
■ INTRODUCED

The seeds and oil of sunflower plants have been cultivated and used in cooking for centuries. Seeds are also cultivated for bird food and livestock. Native Americans produced yellow dye from the flower heads, and a black or dull blue dye from the seeds. Although this plant typically grows up to 13' (4 m) in the wild, the tallest sunflower on record reached 30' (9 m) in Germany in 2014. It was once believed the flowering heads tracked the sun.

DESCRIPTION: This coarse, leafy plant grows 2–13' (0.6–4 m), with a hairy stem. The leaves are very broad and coarsely toothed, appearing either ovate or heart-shaped, becoming smaller and narrower up the stem. The upper half of the plant is usually branched and has several to many yellow flower heads with many bright yellow rays surrounding a dark red central disk. Each flower head is 3–5" (7.5–12.5 cm) wide with disk flowers appearing among stiff scales and ovate bracts that become narrower at the tip. The fruit is seed-like and flattish.

FLOWERING SEASON: June–November.

HABITAT: Prairies, waste places, and roadsides.

RANGE: Throughout much of temperate North America.

CAUTION: The stems and leaves can cause skin irritation in humans and it can be fatal to animals if consumed.

SIMILAR SPECIES: There are many varieties of this plant cultivated in gardens. Prairie Sunflower (*H. petiolaris*) is similar but its scales have white hairs in the central disk.

CONSERVATION: G5
Globally secure but vulnerable in Manitoba.

■ NATIVE
■ RARE OR EXTIRPATED
■ INTRODUCED

ALTERNATE NAMES: Beach Sunflower, Weak Sunflower, East Coast Dune Sunflower

This native perennial spreads low to the ground and is sometimes used as a groundcover in gardens. However, it can grow rapidly and may be a troublesome weed. The flowers attract several native species of beetles and bees.

DESCRIPTION: This plant grows up to 6' (2 m) tall and 1–3' (30–90 cm) long with hairy stems. The leaves are coarse to the touch and heart-shaped, usually arranged in an alternate pattern. Each leaf is about 2–4" (5–10 cm) long and almost twice as wide at its widest and somewhat lobed. The flower stems are brownish and each stem has a single showy yellowish flower head with 20–21 ray florets. The center is surrounded by many red, yellow, or purple disk florets.

FLOWERING SEASON: January–December.

HABITAT: Dunes and disturbed habitats throughout coastal areas and moderately saline environments farther inland.

RANGE: Coastal areas from North Carolina to Florida and Texas; also reported in New Jersey and in New England and Michigan.

CONSERVATION: G5
Globally secure. There are several rare localized and endemic subspecies in Florida, Mississippi, and Texas. It is possibly extirpated in West Virginia.

■ NATIVE
■ RARE OR EXTIRPATED
■ INTRODUCED

ALTERNATE NAMES: Tall Sunflower

Despite this plant's name, the flower heads are relatively small compared to other sunflower species, although its purplish-red stem grows quite tall. It is native to the eastern United States and central and eastern Canada.

DESCRIPTION: This plant has a tall, rough, reddish-purple stem with spreading hairs and grows 3–12' (90–360 cm) tall. The leaves are very coarse, lanceolate in shape, and alternately or oppositely arranged. Each leaf is about 3–7" (7.5–17.5 cm)

long and finely toothed. There are numerous light yellow flower heads. Each flower head is 1½–3" (4–7.5 cm) wide, with 10–20 rays appearing in loose clusters with numerous disk flowers. The bracts are narrow, thin, and green.

FLOWERING SEASON: July–October.

HABITAT: Swamps, wet thickets, and meadows.

RANGE: Alberta east to Nova Scotia, south to Georgia, west to Louisiana, and north to Illinois and Minnesota.

CONSERVATION:  G5
Globally secure but critically imperiled in Illinois, imperiled in Georgia, and vulnerable in Iowa and Manitoba.

■ NATIVE
■ RARE OR EXTIRPATED
■ INTRODUCED

ALTERNATE NAMES: Michaelmas-daisy, Maximillian Sun-flower

This native perennial is frequently visited by pollinators and consumed by livestock and wildlife. Maximilian's Sunflower spreads vigorously and may be weedy in some areas. Its name honors Prince Maximilian of Wied-Neuwied, Germany, who explored parts of western North America in the 1830s.

DESCRIPTION: This plant grows 3–10' (90–300 cm) tall, with rigid, hairy stems. The leaves are rough on both sides and very stiff, often folded lengthwise and curved downward at the tips. Each leaf is 4–6" (10–15 cm) long and alternately arranged. It has yellow flower heads rising from the upper half of the stem. Each flower head is 2–3" (5–7.5 cm) wide, with dark yellow rays at the base.

FLOWERING SEASON: July–October.

HABITAT: Prairies.

RANGE: Alberta east to Quebec, south through Maine to South Carolina and Mississippi, west to Texas, and north to Montana.

SIMILAR SPECIES: Maximilian's Sunflower is generally distinguished from other plants by its folding leaves. Willow-leaved Sunflower (*H. salicifolius*) is similar but has numerous long, drooping leaves covered with soft hairs.

CONSERVATION: G5
Globally secure but imperiled in Wyoming.

■ NATIVE
■ RARE OR EXTIRPATED
■ INTRODUCED

ALTERNATE NAMES: Plains Sunflower

The seeds of this common and widespread native sunflower attract birds. This plant can be weedy and is naturalized in many areas in which it is not native. It probably originated in the dry prairies and plains of central North America and the southwestern U.S. and has expanded its range into the East and the Northwest. The species name is derived from Latin and means "having a petiole."

DESCRIPTION: This showy annual grows 3–5' (1–1.5 m) tall with many-branched, hairy, erect stems. The leaves are dark green, alternatively arranged, and lanceolate in shape. Each leaf is about 2–5" (5–12.5 cm) long, with a rough texture. The flowers are typical of sunflowers, each 1½–3" (4–7.5 cm) wide, with 12–25 yellow rays and a dark brown center disk.

FLOWERING SEASON: June–September.

HABITAT: Open, sandy prairies, plains, and disturbed areas.

RANGE: Great Plains south to Texas, Arizona, and adjacent Mexico; naturalized eastward and occasionally westward.

SIMILAR SPECIES: Prairie Sunflower is distinguished from other sunflowers by its alternate, mostly toothless leaves and flowers that have a brown center disk that's larger than other similar species.

CONSERVATION: G5
Globally secure but vulnerable in Manitoba and Alberta.

■ NATIVE
■ RARE OR EXTIRPATED
■ INTRODUCED

ALTERNATE NAMES: Sunroot, Sunchoke, Wild Sunflower, Earth Apple, *Helianthus tomentosus*

This large, coarse species was cultivated by Native Americans and early settlers. The tubers can be eaten raw, cooked, or pickled, and are highly nutritious. The French explorer Samuel de Champlain consumed the tubers after seeing native people eating them in what is now Massachusetts. In 1805, Lewis and Clark ate the tubers in what is now North Dakota during their famous expedition into the West. Today the tubers are sold in markets and health food stores and can be boiled or roasted like potatoes, and because they contain much less starch than potatoes, they are better tolerated by some people with special dietary needs. The common name comes from the Italian *girasole*, meaning "turning to the sun," because of its similarity to sunflowers.

DESCRIPTION: This stout plant grows to a height of 5–10' (1.5–3 m) with rough, branching stems. The leaves are thick and ovate to lanceolate in shape with rough-toothed margins. Each leaf is about 4–10" (10–25 cm) long, with winged stalks and three veins. The lower leaves are opposite, while the upper leaves are alternate. The stems have large, golden-yellow flower heads. Each flower head is up to 3" (7.5 cm) wide, with 10–20 rays, and narrow, spreading bracts.

FLOWERING SEASON: August–October.

HABITAT: Roadsides, fields, and fencerows.

RANGE: Throughout the East, except in the Far North; also in the northwestern United States.

CONSERVATION: ⓖ5
Globally secure but critically imperiled in Wyoming, imperiled in Saskatchewan, and vulnerable in Manitoba.

◼ NATIVE
◼ RARE OR EXTIRPATED
◼ INTRODUCED

ALTERNATE NAMES: Smooth Oxeye, False Sunflower, Oxeye Sunflower

This common perennial is native to central and eastern North America. Oxeyes, like other plants in the genus *Heliopsis,* resemble and are related to true sunflowers in the genus *Helianthus*—hence the alternate name False Sunflower. Unlike sunflowers, Oxeye's rays persist on the flower heads, while those of sunflowers wither.

DESCRIPTION: This plant grows to a height of 2–5' (60–150 cm) tall with stiff stems. The leaves are opposite and toothed, about 6" (15 cm) long and ovate in shape. The flowers are sunflower-like, with a cone-shaped central disk and persistent rays. Each flower head is 1½–3" (4-8 cm) wide.

FLOWERING SEASON: July–September.

HABITAT: Open woods and thickets.

RANGE: Throughout the East except Nova Scotia and the Far North; west to New Mexico, Colorado, and the Dakotas; also in Washington state.

CONSERVATION: **G5**
Globally secure but critically imperiled in Delaware and vulnerable in Louisiana.

CAMPHORWEED *Heterotheca subaxillaris*

ALTERNATE NAMES: Camphorweed Golden-aster

This widespread native annual or biennial herb is native to the southern two-thirds of the United States but has gradually been advancing farther north.

DESCRIPTION: This plant grows 1–3' (30–90 cm) tall with hairy stems, sometimes appearing lopsided. The leaves are alternately arranged and oblong in shape. The basal leaves are 2–3" (5–7.5 cm) long with stalks, while the upper leaves become smaller and clasp the stem with wavy edges. Both the upper and lower leaf surfaces are green to grayish-green and somewhat hairy. It has yellow, daisy-like flower heads. Each head has 15–35 yellow ray florets surrounding 25–60 disk florets. Each flower head is ½–¾" (1.5–2 cm) wide.

FLOWERING SEASON: July–November.

HABITAT: Prairies, waste places, and roadsides in sandy soil.

RANGE: Iowa east to New York and Connecticut, south to Florida, west to Texas, and north to Nebraska; also in the southwestern United States. It may be introduced in the northernmost parts of its current range.

SIMILAR SPECIES: Prairie Golden-aster (*H. camporum*) and Hairy Golden-aster (*H. villosa*) are both similar, but Camphorweed does not have tufts of hair on its achenes like the other two. Camphorweed also has wider leaves.

CONSERVATION: G5

This species is globally secure. It is possibly extirpated in Utah.

■ NATIVE
■ RARE OR EXTIRPATED
■ INTRODUCED

ALTERNATE NAMES: Hairy False Golden-aster, *Chrysopsis villosa*

This is a variable species found throughout western and central North America, and is particularly abundant in the West. It is typically a round, shrubby plant with showy yellow flower heads in branched clusters. The plant's stem is characteristically covered with rough, grayish hairs.

DESCRIPTION: This plant grows to a height of 8–20" (20–50 cm) with leafy stems, appearing erect or spreading and covered in rough, grayish hairs. The leaves are lanceolate in shape and about ½–1¼" (1.3–3.1 cm) long at the mid-stem. The leaves are alternately arranged and may be very hairy or somewhat hairy. Yellow flower heads appear at the top of branching stems. Each flower head is about 1" (2.5 cm) wide, with 10–35 yellow rays

surrounding a small, yellow disk. The bracts surrounding the base of the flower are variably hairy. The fruit is seed-like, with gray-white bristles.

FLOWERING SEASON: May–October.

HABITAT: Open plains, rocky slopes, cliffs, from low elevations into coniferous forests.

RANGE: From British Columbia to southern California, east to Ontario, Wisconsin, Nebraska, and Texas.

SIMILAR SPECIES: Other *Heterotheca* species; Hairy Golden-aster is distinguished by its rough hairs.

CONSERVATION: G5
Globally secure but critically imperiled in Ontario and Iowa.

NATIVE
RARE OR EXTIRPATED
INTRODUCED

ALTERNATE NAMES: Rattlesnake Hawkweed, Veiny Hawkweed

This native perennial herb is widespread and common in the East, especially in sandy habitats where eastern rattlesnakes live, hence the common name. The plant's leaves have conspicuous red or purple veins.

DESCRIPTION: This plant grows 1–2½' (30–75 cm) tall and exudes milky sap on mostly leafless flower stalks. The basal leaves are elliptical and green, with reddish-purple veins. Each leaf is 1½–6" (4–15 cm) long. It has many yellow, dandelion-like flower heads in open clusters that appear at the top of the stems. Each flower head is ½–¾" (1.5–2 cm) wide. The fruit is seed-like, with yellowish bristles.

FLOWERING SEASON: May–September.

HABITAT: Dry open woods, thickets, and clearings.

RANGE: Ontario and Michigan to New England, south to the Gulf Coast states.

CONSERVATION: G5
This species is globally secure but critically imperiled in Canada. It is critically imperiled in Maine, Mississippi, and Ontario and vulnerable in Indiana, Rhode Island, and Vermont.

■ NATIVE
■ RARE OR EXTIRPATED
■ INTRODUCED

ALTERNATE NAMES: Spotted Cat's-ear

As the common name suggests, the shape of the leaves resemble those of a cat. Both the leaves and flowers are covered in hairs. This ubiquitous perennial introduced from Eurasia is commonly found on North American lawns and largely considered a problem plant.

DESCRIPTION: This plant grows to a height 15" (38 cm). It has wiry, hairy stems covered in hairs which rise from a basal rosette of leaves. The leaves are lanceolate in shape, covered in hairs, and pinnately lobed. The leaves mostly lie flat on the ground. The flowers are yellow and dandelion-like, about 1½" (4 cm) wide each.

FLOWERING SEASON: May–October.

HABITAT: Lawns, fields, and roadsides.

RANGE: Native to Eurasia. It has been introduced and naturalized widely in the East and the West but is largely absent from central North America.

CAUTION: Horses that consume this plant in high amounts may be stricken with stringhalt, a disease causing uncontrolled contractions of the tendons.

CONSERVATION:
The status of this introduced species is unranked. It is widely distributed in North America and considered a noxious weed or a pest plant in several areas.

■ NATIVE
■ RARE OR EXTIRPATED
■ INTRODUCED

ALTERNATE NAMES: Horseheal, Elf Dock

This sunflower-like plant introduced from Eurasia has long been used as a medicinal herb. The rhizomes and roots have been used as a diuretic as well as a treatment for fevers and respiratory ailments. The root is also used to flavor absinthe.

DESCRIPTION: This plant grows to a height of 2–6' (60–180 cm) with a rigid central stem covered in hairs. It has large, rough leaves that are toothed and covered in white woolly hairs beneath. The basal leaves are up to 20" (50 cm) long, with long stalks while the stem leaves are smaller, stalkless, and clasping the stem. Sunflower-like flower heads have long, narrow, straggly rays surrounding a darker central disk. Each flower head is 2–4" (5–10 cm) wide.

FLOWERING SEASON: July–September.

HABITAT: Fields and roadsides.

RANGE: Introduced in the East from the Atlantic west to Manitoba and Missouri; also along the Pacific Coast from British Columbia southward.

CAUTION: This plant has been used medicinally but is considered toxic if consumed in high quantities and may cause drowsiness.

CONSERVATION:

This introduced plant is unranked in North America and is largely considered a weed.

■ NATIVE
■ RARE OR EXTIRPATED
■ INTRODUCED

ALTERNATE NAMES: Flaxleaf Aster, Flaxleaf Whitetop Aster, *Aster linariifolius*

Stiff Aster is a native perennial found throughout the East that is named for its stiff, leafy flower stalks. This attractive plant produces showy blue-purple flowers and grows in a mound.

DESCRIPTION: This plant grows 4–24" (10–60 cm) tall. There are many dark green, linear, needle-like leaves, about 1½" (4 cm) long each. The flower stalk is stiff and has several blue, purple, or pink flower heads with a yellow central disk. Each flower head is about 1" (2.5 cm) wide. The fruit is dry and seed-like, tipped with tawny bristles.

FLOWERING SEASON: August–October.

HABITAT: Dry clearings and rocky banks.

RANGE: Quebec and New Brunswick south to Florida, west to Texas, and north to Iowa.

CONSERVATION:  G5

This species is globally secure but imperiled in Canada. It is critically imperiled in Iowa, Kansas, and Oklahoma and imperiled in Delaware, New Brunswick, and Quebec.

■ NATIVE
■ RARE OR EXTIRPATED
■ INTRODUCED

ALTERNATE NAMES: Two-flower Dwarf-dandelion

This erect native perennial can spread aggressively. The common names refer to the flower stalks, which usually have two flower heads apiece. The genus name honors David Krieg, the German physician who collected this plant in Maryland.

DESCRIPTION: This plant grows to a height of 1–2' (30–60 cm) with a forking stem exuding milky sap. The basal leaves are 2–7" (5–17.5 cm) long each, and elliptical in shape. Most stems have a single leaf (occasionally two) that is small, oval, and clasping the stem. Each plant may have 20 or more yellow-orange, dandelion-like flower heads, often two per stalk, with 25–60 ray flowers.

FLOWERING SEASON: April–September.

HABITAT: Open woods and meadows.

RANGE: Manitoba and Ontario; New York and Massachusetts south to Georgia, west to Oklahoma and Missouri, and north to Minnesota.

SIMILAR SPECIES: Virginia Dwarf-dandelion (*K. virginica*) is smaller and found in sandy soil; it has a single flower head atop each stem and a set of toothed or lobed basal leaves, each less than 1' (30 cm) tall. Potato Dandelion (*K. dandelion*) has a large, solitary flower head on a leafless stem.

CONSERVATION: G5
This species is globally secure but imperiled in Canada. It is critically imperiled in Connecticut and Kansas and imperiled in Colorado, North Carolina, Manitoba, and Ontario. It is possibly extirpated in Delaware.

ALTERNATE NAMES: Dwarf-dandelion

Virginia Dwarf-dandelion is an eastern native annual found mainly in sandy sites. The plant is similar to the familiar Common Dandelion (*Taraxacum officinalis*), but Virginia Dwarf-dandelion has smaller flower heads.

DESCRIPTION: This plant grows up to 12" (30 cm) tall. Both the leaves and flower stalks have a milky sap. The leaves are lanceolate in shape and toothed or lobed, appearing in a rosette at the base of the stem. Each leaf is about 3" (7.5 cm) long. A solitary flower head, about ½" (12 mm) wide, appears at the end of each stalk, composed of numerous bright yellow rays with five teeth at their tips. Each flower head has 9–18 lanceolate floral bracts, about ¼" (6 mm) long. The outer rays have square tips. The fruits are small, bullet-shaped achenes.

FLOWERING SEASON: April–September.

HABITAT: Dry sandy soils in fields, prairies, open woods.

RANGE: Throughout the East, west to Iowa, Kansas, and Texas.

SIMILAR SPECIES: Two-flowered Cynthia *(K. biflora)* is similar, but has two (sometimes up to six) flower heads per stalk and is twice as tall, up to 24" (60 cm).

CONSERVATION: G5

This species is globally secure but critically imperiled in Canada. It is critically imperiled in Maine and Ontario, imperiled in Iowa, and vulnerable in Ohio and Rhode Island.

■ NATIVE
■ RARE OR EXTIRPATED
■ INTRODUCED

ALTERNATE NAMES: Florida Blue Lettuce, Canada Lettuce

This widespread native is a tall plant with small, inconspicuous flowers. Wild Lettuce is related to the commercial staple Garden Lettuce (*L. sativa*) and the flower heads may look similar. Young Wild Lettuce leaves can be used in salads or cooked but taste somewhat bitter. The milky sap produced by the stem, leaves, and roots of this plant has been used externally to treat warts, arthritis, and other ailments.

DESCRIPTION: Wild Lettuce can grow very tall, up to 2–10' (60–300 cm). The leaves exude a milky sap when crushed. Each leaf is about 1' (30 cm) long, nearly toothless, deeply lobed, stalkless, and lanceolate in shape. An elongated cluster of small, pale yellow flower heads

appear at the tips. Each flower head is about ¼" (6 mm) wide with numerous rays. The fruit is flat and one-seeded, with a number of bristles, which aid in seed dispersal.

FLOWERING SEASON: July–September.

HABITAT: Clearings, thickets, and edges of woods.

RANGE: Found throughout much of North America, except the Far North. Its true native range is uncertain, but it probably originated in eastern and central North America and naturalized in the West.

SIMILAR SPECIES: Wild Lettuce is distinguished from other *Lactuca* species by its yellow flowers, its achenes with tufts of hairs, and hairless stems. Prickly Lettuce (*L. serriola*) is similar, but its leaves are spiny. The flower heads of cultivated Garden Lettuce (*L. sativa*) may look similar, but Wild Lettuce tends to grow much taller.

CONSERVATION: G5

Globally secure but critically imperiled in Utah; imperiled in Wyoming; and vulnerable in Manitoba, Prince Edward Island, and Quebec.

■ NATIVE
■ RARE OR EXTIRPATED
■ INTRODUCED

ALTERNATE NAMES: Wild Opium, Milk Thistle

This weedy species is widely introduced across North America. The foliage of Prickly Lettuce has a bitter odor if crushed. A wild-growing relative of cultivated Garden Lettuce (*L. sativa*), its young leaves are edible but taste bitter. It has been used in traditional medicine since ancient times and is sometimes referred to as Wild Opium because the plant's latex contains compounds that are somewhat sedative.

DESCRIPTION: This plant grows to a height of 2–5' (60–150 cm) tall and the foliage exudes milky sap when crushed. The leaves are very prickly, but somewhat soft to the touch, and clasp the stem at the base. Each leaf is 2–12" (5–30 cm) long, oblong in shape, lobed or unlobed, and bristly beneath. It has loose clusters of small, yellow, dandelion-like flower heads on an erect stem. Each flower head is about ¼" (6 mm) wide and composed of 16–24 rays. The fruit is dry, one-seeded and covered in white bristles, which help in seed dispersal.

FLOWERING SEASON: June–October.

HABITAT: Roadsides and waste places.

RANGE: Introduced throughout North America, except the Far North.

CONSERVATION:
The status of this introduced plant is not ranked in North America.

- NATIVE
- RARE OR EXTIRPATED
- INTRODUCED

ALTERNATE NAMES: Glandular Layia, Whitedaisy Tidytips

The seeds of this western native are a favorite among many bird species, and the flowers attract native bees and wasps. As the common name suggests, the showy flowers often are distinctly white.

DESCRIPTION: This small branched annual grows 4–16" (10–40 cm) high and is covered in pubescent hairs. The leaves are thin and oval to linear. The lower leaves are usually lobed, about 4" (10 cm) long. It has showy flowers with several petals. Pure white (sometimes yellow) flower rays surround yellow disks and often fade to rose-purple. The fruit is a hairy achene.

FLOWERING SEASON: March–June.

HABITAT: Dry slopes, mesas.

RANGE: Southern British Columbia to southwestern Idaho, south to New Mexico, Arizona, and California.

CONSERVATION: G4
This species is apparently secure.

■ NATIVE
■ RARE OR EXTIRPATED
■ INTRODUCED

ALTERNATE NAMES: Dog Daisy, Ox-eye Daisy, *Chrysan-themum leucanthemum*

Oxeye Daisy is native to Eurasia and spreads easily by creeping rhizomes and seeds. It is widely introduced and regularly invades lawns and pastures in North America. It was once called *Chrysanthemum leucanthemum* and considered a relative of cultivated chrysanthemums.

DESCRIPTION: This leafy plant grows 8–31" (20–80 cm) tall. All of the leaves have lobed edges. The lower leaves are broadly lanceolate, with long petioles, about 1½–6" (4–15 cm) long each. The upper stem leaves are smaller and lack petioles. A cluster of several stems is topped by mostly leafless branches, each with one flower head and white rays surrounding a yellow central disk. Each flower head is about 3" (7.5 cm) wide, while each ray is about ½–¾" (1.5–2 cm) long. The bracts of the flower heads have a small brown line near the edges. The fruit is seed-like.

FLOWERING SEASON: May–October.

HABITAT: Waste places, meadows, pastures, and roadsides.

RANGE: Introduced throughout North America; less abundant southward.

SIMILAR SPECIES: Another, less widely introduced relative, Max Chrysanthemum (*L. maximum*), is similar but has larger flower rays about ¾–1¼" (2–3 cm) long.

CONSERVATION:

This introduced species is unranked in North America.

■ NATIVE
■ RARE OR EXTIRPATED
■ INTRODUCED

ALTERNATE NAMES: Tall Gayfeather

Rough Blazing Star is native to central and eastern North America. The species is named for its coarse leaves. This plant is taller than other blazing stars, has larger flowers, and prefers drier locations. Native Americans sometimes ate the corms when other food sources were scarce.

DESCRIPTION: This plant grows to a height of 16–48" (40–120 cm) tall with a green or dark red stem covered in short grayish hairs. The leaves are rough and lanceolate to linear in shape, appearing whorled. The lower leaves are 4–12" (10–30 cm) long, becoming smaller up the stem and alternate. Pinkish to lavender (sometimes white) flower heads grow in a spike. Each flower head is about ¾" (2 cm) wide. The bracts are broadly rounded and flaring, with pinkish, translucent edges.

FLOWERING SEASON: August–October.

HABITAT: Open plains and thin woods in sandy soil.

RANGE: Ontario and New York south to Florida, west to Texas, and north to North Dakota.

CONSERVATION: G4

Apparently secure globally but listed as endangered in Canada. It is critically imperiled in North Carolina; imperiled in Ontario; and vulnerable in Georgia, Kentucky, and Virginia.

■ NATIVE
■ RARE OR EXTIRPATED
■ INTRODUCED

ALTERNATE NAMES: Dotted Gayfeather

Dotted Blazing Star is a hardy native perennial that is long-lived and drought-resistant. Deer and elk may browse the foliage and the plants attract native butterflies and birds. Native Americans have used the leaves and roots medicinally to reduce swelling and in teas for stomachaches and bladder and kidney problems.

DESCRIPTION: This plant grows to a height of 6–31" (15–80 cm). The leaves are very narrow and stiff. Each leaf is 3–6" (7.5–15 cm) long and somewhat dotted. Rose-lavender flowers arranged in slender wands appear in crowded heads. Each flower head is about ¾" (2 cm) long, with four to eight flowers. The fruit is seed-like and covered with hairs, with numerous small plumes at the top.

FLOWERING SEASON: August–September.

HABITAT: Dry open places, on plains and among pinyon and juniper, often in sandy soil.

RANGE: Central North America from the eastern base of the Rocky Mountains south to western Texas, east to Manitoba, Michigan, Illinois, and Arkansas. It may be naturalized farther east.

CONSERVATION: G5
This species is globally secure.

SCALY GAYFEATHER *Liatris squarrosa*

NATIVE
RARE OR EXTIRPATED
INTRODUCED

ALTERNATE NAMES: Blazing Star

This native perennial is most common in the southeastern U.S., where it tolerates the rocky or sandy soils. A variety with hairless stems (var. *glabrata*) occurs in the Great Plains and is sometimes treated as a separate species.

DESCRIPTION: This plant grows 10–24" (25–60 cm) with one or several erect and unbranched stems. The lower leaves are 4–6" (10–15 cm) long and about ¼" (6 mm) wide, becoming smaller up the stem and more alternate, linear, and sessile. A few tuft-like, red-violet flower heads appear in a spike along the top of the stem.

FLOWERING SEASON: June–September in the South, June–July in the North.

HABITAT: Dry, sandy prairie; thin woods.

RANGE: Maryland to southern South Dakota and eastern Colorado, south to Florida and eastern Texas.

SIMILAR SPECIES: Other *Liatris* species. This is one of the earliest *Liatris* species to bloom and is distinguished by having fewer leaves.

CONVERSATION: G5

CONSERVATION: G5

Globally secure but critically imperiled in Maryland and vulnerable in Indiana, Iowa, and Ohio. It is presumed extirpated in Michigan and possibly extirpated in West Virginia.

■ NATIVE
■ RARE OR EXTIRPATED
■ INTRODUCED

ALTERNATE NAMES: Skeletonweed

This widespread perennial native to central and western North America is a much-branched plant that is easily overlooked before it blooms. Its soft pink or lavender (rarely white) flowers are visited by native butterflies and bees. It is named for its twiggy, rush-like stems.

DESCRIPTION: This plant grows 6–18" (15–45 cm) with a stiff green heavily branched stem. The lower leaves are stiff and linear, about ½–2" (1.5–5 cm) long, becoming smaller up the stem and reduced to scales on the upper part of the plant. The flowers are pink or lavender, sometimes white, with five petals and small teeth at the tip. Each flower is about ½–¾" (13–19 mm) wide. The bracts are green and have two layers. The outer layer is shorter than the inner layer, forming a column about ¾" (19 mm) long. The fruit is a dry seed with white or brown hairs.

FLOWERING SEASON: June–August.

HABITAT: Prairies, dry fields, and meadows.

RANGE: Wisconsin to Manitoba west to the Pacific Northwest, southward to Missouri and New Mexico.

SIMILAR SPECIES: Annual Skeletonweed (*Shinnersoseris rostrata*) is similar but much less common. Its leaves are larger and its flowers have six or more petals with no teeth at the tips.

CONSERVATION:
Globally secure but critically imperiled in Utah and vulnerable in Iowa, Missouri, and Manitoba.

■ NATIVE
■ RARE OR EXTIRPATED
■ INTRODUCED

This primarily western native plant has a strong, unpleasant tar-like odor, which deters grazing animals. It's covered in bristly hairs and the stem may be branched or unbranched. Native Americans, notably the Cheyenne people, used these plants for food, medicine, and incense.

DESCRIPTION: This erect annual grows from a taproot and reaches up to about 3' (1 m) tall. The leaves grow in an alternate pattern on the stem and are sticky or glandular, covered with rough hairs. Each leaf is about 4" (10 cm) long, narrow, and linear. One to several small clusters of yellow flowers with black tips appear in the upper leaves. Each flower head usually has three to five greenish-yellow ray flowers.

FLOWERING SEASON: May–October.

HABITAT: Disturbed areas, roadsides, open woods.

RANGE: Montane areas of western North America and in the North. It may be introduced in the East.

SIMILAR SPECIES: Little Tarweed (*M. exigua*) is similar but has smaller bracts. Small-head Tarweed (*M. minima*) is also similar but has opposite leaves and smaller bracts.

CONSERVATION: Ⓖ5

Globally secure but vulnerable in Wyoming, North Dakota, and Alberta.

■ NATIVE
■ RARE OR EXTIRPATED
■ INTRODUCED

ALTERNATE NAMES: Smooth Desert-dandelion

In wet years in sandy deserts, this showy wildflower grows in masses with yellow fragrant daisy-like flower heads. The name *glabrata* refers to the mostly hairless leaves. The flower heads are composed of small, strap-shaped flowers called ligules.

DESCRIPTION: This plant grows to a height of 6–14" (15–35 cm) tall. The leaves are 2½–5" (6.3–12.5 cm) long each and divided into a few thread-like lobes. It has a few pinnately divided leaves and bright pale yellow flower heads on branched stems. Pale yellow or sometimes white dandelion-like flower heads appear at the top of the plant. The young flower heads have a reddish color in the middle. The florets are five-lobed and strap-shaped.

FLOWERING SEASON: February–July.

HABITAT: Sandy desert, plains, and washes.

RANGE: Southwestern Idaho and eastern Oregon to southern California, much of Arizona, and southwestern New Mexico.

CONSERVATION: **G5**
Globally secure but vulnerable in Idaho.

■ NATIVE
■ RARE OR EXTIRPATED
■ INTRODUCED

ALTERNATE NAMES: Disc Mayweed, Rayless Chamomile, Pineapple-weed Chamomile, *Matricaria matricarioides, Lepidotheca suaveolens, Chamomilla suaveolens*

This annual, native to northeast Asia, is introduced throughout North America and is considered an invasive species in many areas. Pineapple-weed is closely related to mayweeds and chamomile but its flower heads lack ray florets. The flowers and leaves smell like chamomile or pineapple when crushed, hence the common names. The flowers are also edible and have been used in salads and to make herbal tea. This plant has also been used to treat fevers and wounds.

DESCRIPTION: This much-branched, sprawling plant grows 4–20" (10–50 cm) tall. The leaves have a fruit-like fragrance when crushed and are lacy and bipinnately compound, with tiny, thread-like leaflets. The foliage is deeply divided. Conical flower heads with dense, cup-shaped bracts and tiny, green to yellow, conical, rayless disks appear at the top of the stems.

FLOWERING SEASON: May–September.

HABITAT: River- and streambanks, roadsides, disturbed areas, and fields.

RANGE: Widely naturalized throughout North America.

CONSERVATION: G5
Globally secure. It is considered an invasive plant in many areas.

■ NATIVE
■ RARE OR EXTIRPATED
■ INTRODUCED

ALTERNATE NAMES: Climbing Hempvine

This perennial vine grows to variable heights but often towers over other plants in moist woods and thickets. It is native to the eastern United States and is most prominent in the South and the Atlantic coastal plain.

DESCRIPTION: This twining vine grows to variable heights. The leaves are opposite and ovate to triangular, with wavy-toothed or untoothed margins. Each leaf is 1–3" (2.5–7.5 cm) long. It has flat-topped clusters of white or pinkish flower heads rising from the leaf axils. Each flower head is about ¼" (6 mm) wide, with up to four disk flowers, surrounded by scale-like bracts. The tiny, dried seed-like fruit has a tuft of whitish bristles.

FLOWERING SEASON: July–October; year-round in the South.

HABITAT: Streambanks, moist thickets, swamps, woods.

RANGE: New England south to Florida, west to Texas, and northeast to Illinois and Ohio.

SIMILAR SPECIES: Boneset (*Eupatorium perfoliatum*) has similar flower heads but its leaves grow in pairs and join at the base, surrounding hairy stems. Florida Keys Hempweed (*M. cordifolia*) is also similar, but only occurs from Florida to Texas.

CONSERVATION: 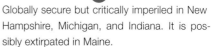 G5
Globally secure but critically imperiled in New Hampshire, Michigan, and Indiana. It is possibly extirpated in Maine.

■ NATIVE
■ RARE OR EXTIRPATED
■ INTRODUCED

ALTERNATE NAMES: Wild Blue Lettuce, Showy Blue Lettuce, Tartarian Lettuce, *Lactuca pulchella*

This delicate native perennial is small and easy to overlook. It was formerly in the *Lactuca* genus and has gone through a number of name changes. It may appear in some sources as *Lactuca pulchella* or as a variety of the Eurasian species *Lactuca tatarica*.

DESCRIPTION: This plant has a deep, creeping root and grows 6"–3' (15–90 cm) tall. The leaves are narrowly lanceolate, about 3–6" (8–15 cm) long and 1½" (4.25 cm) wide, sometimes lobed. The basal leaves are stalked, while the stem leaves are stalkless or clasping and toothless or slightly toothed. The leaves are smooth on the upper surface and sometimes waxy underneath. Its blue flowers look like dandelions. Each flower is whitish at the base, about ½–1" (1.25–2.5 cm) across, with 14–50 petals and five small teeth at the tip. Each ray has a blue style and a blue stamen at the base.

FLOWERING SEASON: June–September.

HABITAT: Meadows, clearings, moist places.

RANGE: Throughout North America except most of New England and most of the southeastern United States.

SIMILAR SPECIES: Chicory (*Cichorium intybus*) has similar but larger flowers and is taller.

CONSERVATION:

Globally secure. It is critically imperiled in Missouri, imperiled in Quebec, and vulnerable in Ontario. It is presumed extirpated in Michigan.

■ NATIVE
■ RARE OR EXTIRPATED
■ INTRODUCED

This perennial herb is native to eastern North America. The Iroquois used roots of this plant to treat rattlesnake bites, hence the common name. It was formerly classified as *Prenanthes altissima*.

DESCRIPTION: This plant grows 1½–7' (46–213 cm) tall, with milk sap in the stems and leaves. The leaves may be heart-shaped and triangular, or lobed or entire. The flowers are greenish-brown or whitish-yellow and pendant-like, hanging in drooping clusters. Each flower head has five or six ray flowers and four to six involucral bracts that are sharply pointed.

FLOWERING SEASON: July–October.

HABITAT: Moist woods.

RANGE: Ontario east to Labrador, south to Georgia, west to Texas, and northeast to Michigan.

SIMILAR SPECIES: Gall-of-the-Earth *(N. trifoliolatus)* is similar, but the flower heads are larger, with 9–20 flowers each. White Lettuce *(N. albus)* is also similar, but has 9–15 flowers.

CONSERVATION: **G5**
Globally secure but critically imperiled in Texas and vulnerable in Louisiana.

WHORLED ASTER *Oclemena acuminata*

■ NATIVE
■ RARE OR EXTIRPATED
■ INTRODUCED

ALTERNATE NAMES: Mountain Aster, Whorled Wood Aster

This perennial herb is native to eastern North America. The common name references the leaves, which grow so close together on the upper part of the stem that they appear whorled.

DESCRIPTION: This plant widely varies in height, growing 6"–3' (15–90 cm) tall. The smaller plants tend to be found in poor soil, typically along the edge of forests or in clearings. It has upright, slightly zigzagged, and somewhat hairy stems. The leaves on the upper part of the stem are larger than those on the lower part and appear whorled. The leaves are thin and toothed, with a midvein. The disk in the center of the flower is yellow or reddish, surrounded by 12–18 white or purplish rays. The ray flowers are bent backward, creating a disheveled-looking appearance.

FLOWERING SEASON: July–October.

HABITAT: Wooded areas.

RANGE: Ontario to Nova Scotia, and New England southwest to northernmost Georgia.

SIMILAR SPECIES: Bog Aster (*Oclemena nemoralis*) is similar but the leaves are entire, not toothed.

CONSERVATION: G5

Globally secure but critically imperiled in Newfoundland and imperiled in Kentucky. It is possibly extirpated in Ohio.

■ NATIVE
■ RARE OR EXTIRPATED
■ INTRODUCED

ALTERNATE NAMES: Hard-leaf Flat-top Goldenrod, Prairie Goldenrod, Stiff Goldenrod

This attractive native perennial is widespread east of the Rocky Mountains. It is a tall, leafy plant with clusters of dark yellow flowers. It is a deep-rooted herbaceous plant that is often found in open, dry areas. Some Native Americans used the plant to treat bee stings and other swellings.

DESCRIPTION: This plant has a tall, coarse, hairy central stem and grows 1–5' (30–150 cm) tall. The basal leaves are semi-evergreen. Each leaf has a rough texture and is elliptical in shape with stalks about 10" (25 cm) long. The upper leaves are alternate and oval, and clasp the stem. The leaves are initially floppy, becoming stiffer and more rigid later in the year. The flowers are dark yellow and bell-shaped, growing in dense, rounded or flat-topped terminal clusters. The flower heads are ⅜" (8 mm) long each, with 7–10 rays and 20–30 disk flowers. The fruit appears as achenes that have small tufts of white or light brown hair. They are distributed by the wind.

FLOWERING SEASON: August–October.

HABITAT: Prairies, thickets, and open woods.

RANGE: Alberta east to Ontario, south through New York to Georgia, west to Texas and the Rocky Mountains.

SIMILAR SPECIES: Other goldenrods. Hard-leaf Goldenrod is distinguished from other goldenrods by the flatter flower heads. It also has a hairy stem, unlike other similar species.

CONSERVATION: **G5**

Globally secure but critically imperiled in Connecticut, Maryland, Pennsylvania, South Carolina, and West Virginia; imperiled in New York and Virginia; and vulnerable in Georgia, Montana, Wyoming, and Ontario. It is possibly extirpated in Massachusetts.

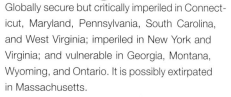

ALTERNATE NAMES: Scotch Cotton-thistle

This tall plant native to Eurasia is widely considered a noxious weed in North America. The seeds can spread easily by attaching to tires, shoes, clothing, and animal fur.

DESCRIPTION: This gray-green plant grows 8' (2.4 m) tall or more and appears velvety white. It has a hairy stem and produces a large rosette of spiny leaves. Each leaf is up to 24" (60 cm) long and lobed. The stem leaves are densely covered in white, woolly hairs. It has large magenta to purple globe-like flower heads that grow upright, about 2" (5 cm) wide each. The bracts are densely spiny.

FLOWERING SEASON: June–September.

HABITAT: Roadsides, waste areas, fencerows, rangelands, pastures, dry areas.

RANGE: Introduced and widely naturalized throughout North America.

SIMILAR SPECIES: Bull Thistle (*Cirsium vulgare*) and Plumeless Thistle (*Carduus acanthoides*) have similar bracts, but the leaves of Scotch Thistle are silvery, making it easy to distinguish.

CONSERVATION:

The status of this introduced species is not ranked in North America. It is largely considered a weed.

■NATIVE
■RARE OR EXTIRPATED
■INTRODUCED

ALTERNATE NAMES: Golden Groundsel, *Senecio aureus*

This eastern native perennial can be somewhat weedy. Its roots colonize an area, creating ground cover over time if left undisturbed. The specific epithet *aurea* means "golden-yellow" and references the flower color.

DESCRIPTION: This smooth plant grows 1–2' (30–60 cm) tall. The basal leaves are heart-shaped, long-stalked, and dark green with a purple tinge on the underside. The basal leaves are ½–6" (1.5–15 cm) long each with rounded teeth at the margins, while the upper leaves are 1–3½" (2.5–9 cm) long and pinnately lobed. The flower heads are yellow and daisy-like, with flat-topped clusters. Each flower head is about ¾" (2 cm) wide with 8–12 rays.

FLOWERING SEASON: April–July.

HABITAT: Wet meadows, swamps, and moist woods.

RANGE: Manitoba east to Newfoundland, south to Florida, west to Texas, and north to Missouri and Minnesota.

CAUTION: Many plants in this genus can cause severe illness and liver damage in humans if consumed.

SIMILAR SPECIES: Roundleaved Ragwort (*P. obovata*) is similar but has spatulate leaves that taper at the base. Prairie Ragwort (*P. plattensis*) grows in prairies and has basal leaves that are woolly on the underside. Woolly Ragwort (*P. tomentosa*) has long, narrow, woolly, basal leaves—especially when young.

CONSERVATION:

Globally secure but critically imperiled on Prince Edward Island; vulnerable in Georgia, Rhode Island, Labrador, and Newfoundland.

■ NATIVE
■ RARE OR EXTIRPATED
■ INTRODUCED

ALTERNATE NAMES: Lobeleaf Groundsel

This native western plant produces a varying number of stems and can range in height depending on elevation. It can have a single stem or a cluster of up to five stems, with single-stemmed plants more likely to occur in drier conditions at lower elevations. It produces showy leaves and an array of flower heads containing many yellow disk florets.

DESCRIPTION: This perennial reaches between 6–24" (15–60 cm) tall. The leaves are mostly basal, up to 5" (12.7 cm) long, 1½" (4.25 cm) wide, and deeply lobed. The basal leaves are spatulate or oblanceolate in shape and widest toward the tip. The leaves are divided into three to six pairs of lobes, which have large, angular teeth along the margins. The upper leaves are stalkless and smaller. The flower heads have 13–21 involucral bracts that are all equal in length, as well as 8–13 yellow ray florets and 40–50 yellow disk florets. The involucral bracts are yellow at the tip and otherwise green.

FLOWERING SEASON: May–July.

HABITAT: Slopes, plains, or open woods.

RANGE: Idaho and Wyoming south through California, Arizona, and New Mexico.

CONSERVATION: **G5**
Globally secure but vulnerable in Wyoming.

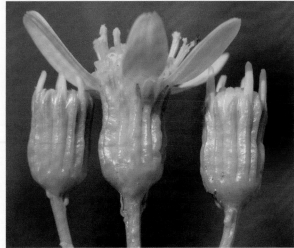

■ NATIVE
■ RARE OR EXTIRPATED
■ INTRODUCED

ALTERNATE NAMES: Northern Ragwort

The leaf shape of Balsam Groundsel can be extremely variable. It produces a cluster of golden-yellow, daisy-like flowers at the apex of the stem.

DESCRIPTION: This perennial grows 6–18" (15–45 cm) tall. The lower leaves are erect and oblong to round or spoon-shaped. Each leaf is 1–2½" (2.5–6 cm) long with sharp teeth along the tip and sides. The stem leaves are widely spaced and become smaller, more lance-shaped, deeply lobed and stalkless, nearly clasping the stem. Balsam Groundsel has an erect, flattish cluster of up to 20 flowers, which are daisy-like with golden centers, with 0, 8, or 13 yellow rays. Each flower is about ¾" (2 cm) wide. Several scale-like narrow bracts are usually widely spaced on the flower stalks and tinged with purple at the tip.

FLOWERING SEASON: May–August.

HABITAT: Ledges or gravel, streambanks.

RANGE: Throughout Canada and the northern tier of the United States, south to Utah and New Mexico in the West and to Florida in the East.

SIMILAR SPECIES: Prairie Ragwort (*P. plattensis*) has similar leaves but is usually covered in dense, woolly hairs and grows in dry areas.

CONSERVATION: G5

Globally secure but critically imperiled in Connecticut, Georgia, and Mississippi; imperiled in New Hampshire, New Jersey, and West Virginia; and vulnerable in Virginia, New York, Ohio, Maryland, Colorado.

ROSY PALAFOX   *Palafoxia rosea*

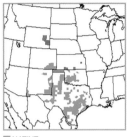

■ NATIVE
■ RARE OR EXTIRPATED
■ INTRODUCED

A number of species of butterflies and moths are attracted to this plant native to the southern Great Plains. The plant typically produces rosy pink flowers, hence its name. Rosy Palafox has been used in folk medicine to treat fever, nausea, and chills.

DESCRIPTION: This native annual has erect, slender stems and grows 4–20" (10–50 cm) tall. The stem is sparsely leaved and branching in the upper part. The lower leaves are opposite, while the upper leaves are alternate, broadly lance-shaped, untoothed, and usually somewhat hairy. Each leaf is 1–2⅓" (3–6 cm) long. A few flower heads appear at the end of the upper branches. The ray florets are reddish to pink and have three narrow lobes.

FLOWERING SEASON: May–October.

HABITAT: Sandy places.

RANGE: Wyoming south to New Mexico and east through Kansas, Oklahoma, and Texas.

CONSERVATION: G5

Globally secure but critically imperiled in Wyoming and imperiled in Colorado.

■ NATIVE
■ RARE OR EXTIRPATED
■ INTRODUCED

ALTERNATE NAMES: American Fever-few

The white flowers of this eastern native perennial attract various bees, wasps, and butterflies. Native Americans used the leaves, which contain tannins, in teas and as a poultice for human and veterinary medical care, including treating burns and dysentery. Today it is often grown as a garden plant and is available in many native plant nurseries.

DESCRIPTION: This clump-forming plant grows 1½–3' (45–90 cm) tall. The basal leaves are coarse, toothed and very large, while the stem leaves are smaller. Clumps of white, button-like ray flowers appear at the top of the plant.

FLOWERING SEASON: June–July.

HABITAT: Mesic to dry prairies and woods.

RANGE: Maryland to southeastern Minnesota, south to Georgia, Alabama, and Arkansas; naturalized to New England, Kansas, and Texas.

CONSERVATION: G5
Globally secure but critically imperiled in Pennsylvania.

NATIVE
RARE OR EXTIRPATED
INTRODUCED

ALTERNATE NAMES: Arctic Sweet Coltsfoot, Arctic Butterbur, *Petasites palmatus*

This cold-hardy plant is native to the northern tier of the U.S. and throughout Canada, including the Arctic. It has attractive, deeply lobed leaves.

DESCRIPTION: This perennial grows 8–24" (20–60 cm) tall. The leaves are palmately lobed at the base of the plant and toothed, appearing below upright flower stalks. The flowers bloom before the leaves appear and grow in round clusters of small heads atop the stems with linear leaf-like bracts. Each flower is ½" (12 mm) wide, with white rays around a white, yellow, or pink disk, sometimes rayless.

FLOWERING SEASON: February–April.

HABITAT: Wet soils in forests.

RANGE: Circumboreal; found throughout Canada and across the U.S. from New England to Washington, and south through Coast Ranges to California.

CONSERVATION: G5

Globally secure but critically imperiled in Idaho and imperiled in North Dakota and Vermont.

■ NATIVE
■ RARE OR EXTIRPATED
■ INTRODUCED

ALTERNATE NAMES: Devil's-paintbrush, Fox-and-cubs

This showy perennial is introduced from central and southern Europe, and today is especially abundant in New England. It's a very colorful plant, but can be a troublesome weed, leading farmers to nickname it Devil's-paintbrush. The genus name is from the Greek word for "hawk" because the Ancient Roman naturalist and author Pliny the Elder believed hawks ate this plant to strengthen their eyesight.

DESCRIPTION: This plant has a slender, usually leafless stalk covered with black hairs growing 1–2' (30–60 cm) tall. All parts of the plant exude milky sap. The leaves grow in a basal rosette and are coarsely hairy. Each leaf is 2–5" (5–12.5 cm) long and elliptical. The flowers have orange or red heads, appearing dandelion-like. Each flower head is about ¾" (2 cm) wide and each ray has five teeth. The bracts are green and covered with black, gland-tipped hairs.

FLOWERING SEASON: June–August.

HABITAT: Fields, clearings, and roadsides.

RANGE: Introduced from Alberta east to Newfoundland, south to Florida, west to Arkansas, and north to South Dakota and Minnesota. It also occurs in scattered naturalized populations farther west.

CONSERVATION:

The status of this introduced plant is not ranked in North America. It is considered an invasive species in many states and provinces, where it is often prohibited and/or targeted by eradication programs.

MOUSE-EAR HAWKWEED *Pilosella officinarum*

■ NATIVE
■ RARE OR EXTIRPATED
■ INTRODUCED

ALTERNATE NAMES: *Hieracium pilosella*

Mouse-ear Hawkweed is native to Eurasia and considered an invasive weed in many parts of North America. It creates dense monocultures, preventing other plants from growing near it by secreting substances that inhibit root growth. It spreads by runners and seeds, often forming dense mats. The leaves are hairy and resemble a mouse's ears, hence the common name.

DESCRIPTION: This small plant grows 3–12' (90–360 cm). The stems and leaves are covered in white or reddish hairs. The entire plant exudes milky sap. The basal leaves are 1–5" (2.5–12.5 cm) long, oblong, and covered with stiff long hairs, appearing white on the underside. It has one yellow or orange dandelion-like flower head on a leafless, hairy, glandular stalk. Each flower head is 1" (2.5 cm) wide. The bracts are covered with black hairs and glands. The small, seed-like fruit has slender bristles.

FLOWERING SEASON: June–September.

HABITAT: Pastures, fields, and lawns.

RANGE: Introduced from Ontario east to Newfoundland and Nova Scotia, south to Georgia, and northwest to Tennessee, Ohio, Michigan, and Minnesota. Also introduced in the Pacific Northwest.

SIMILAR SPECIES: Large Mouse-ear Hawkweed (*H. flagellare*) is similar, but has two to five flower heads and its leaves have green undersides, not white.

CONSERVATION:

The status of this introduced plant is not ranked in North America. This plant is considered a noxious weed in many areas and is often targeted for eradication.

■ NATIVE
■ RARE OR EXTIRPATED
■ INTRODUCED

ALTERNATE NAMES: Silkgrass, Grass-leaf Golden-aster, *Chrysopsis graminifolia*

This perennial is a tough, vigorous evergreen perennial native to the mid-Atlantic and southeastern United States. When not in bloom, it may look like silvery grass. However, it produces daisy-like, bright yellow flowers that attract butterflies in the late summer and fall. The species name is from the Latin words *gramen* and *folius*, meaning "grass" and "leaf."

DESCRIPTION: This plant grows 2–3' tall with mostly basal silvery grass-like leaves covered in fine hairs. The basal leaves grow about 1' (30 cm) long. The stem leaves are few and shorter, appearing alternately arranged. The flowers are bright yellow and daisy-like. Each cluster is about 2" (5 cm) with many orangish-yellow disk florets surrounding ray florets.

FLOWERING SEASON: August–September.

HABITAT: Dry, open sandy sites, scrubby flatwoods.

RANGE: Delaware and Maryland to Ohio, south to Florida and eastern Texas.

CONSERVATION: G5
Globally secure but critically imperiled in Ohio. It is possibly extirpated in Delaware and West Virginia.

■ NATIVE
■ RARE OR EXTIRPATED
■ INTRODUCED

ALTERNATE NAMES: American Star-thistle, *Centaurea americana*

This showy annual is native mainly to the south-central U.S. and is easily cultivated. It looks similar to thistle plants but is not prickly. The common name comes from the stiff, straw-colored bracts beneath the flower head, which have a basketweave pattern.

DESCRIPTION: American Basket-flower grows 1½–5' (45–150 cm) tall. It has a stout, leafy, much-branched stem. The flowers are lavender-pink and filamentous with cream-colored centers. The flower heads are 4–5" (10–12.5 cm) wide each and are subtended by fringed bracts.

FLOWERING SEASON: May–August.

HABITAT: Prairies, pastures, and roadsides.

RANGE: Arizona east to Arkansas and eastern Kansas; widely scattered north to Wisconsin.

CONSERVATION: G5

Globally secure but critically imperiled in Kansas and imperiled in Arizona. It is possibly extirpated in Missouri.

■ NATIVE
■ RARE OR EXTIRPATED
■ INTRODUCED

ALTERNATE NAMES: Sweetscent, Camphorweed, Shrubby Camphorweed, Marsh Fleabane, *Pluchea purpurascens*

Saltmarsh Fleabane is a common plant that adds a showy pink display to marsh areas at the end of the growing season. The plant has a strong, sweet, somewhat camphor-like odor. A medicinal tea made from its leaves is popular in the Caribbean and has been used as a stimulant, diuretic, and antispasmodic.

DESCRIPTION: This erect annual grows 1–5' (30–150 cm) tall. The leaves are short-stalked or stalkless and can appear slightly toothed, scalloped, or smooth. Each leaf is 2–6" (5–15 cm) long and ovate to lanceolate in shape. The flowers appear as flat-topped clusters of pink-lavender, rayless flower heads. Each flower head is about ¼" (5 mm) wide, with pink or purple bracts. The tiny one-seeded dry fruit has a circle of bristles.

FLOWERING SEASON: July–October.

HABITAT: Saline to brackish marshes.

RANGE: Maine south to Florida, largely along the coast, and across the southern tier of the United States.

CONSERVATION:
Globally secure but critically imperiled in Pennsylvania, imperiled in Utah, and vulnerable in North Carolina and Kansas.

■ NATIVE
■ RARE OR EXTIRPATED
■ INTRODUCED

ALTERNATE NAMES: Rabbit-tobacco, Blunt-leaf Rabbit-tobacco, Fragrant Cudweed, *Gnaphalium obtusifolium*

This fragrant annual is native to eastern and central North America and produces sweet-smelling pale flowers. The genus name comes from a Greek word meaning "tuft of wool" and refers to the plant's woolly appearance. The species was formerly classified in the genus *Gnaphalium*.

DESCRIPTION: This white plant has an erect, cottony stem and grows 1–2' (30–60 cm) tall. The leaves are very narrow with whitish woolly hairs beneath. Each leaf is about 1–4" (2.5–10 cm) long, pointed, and stalkless. It has bud-like flowers appearing in branched clusters. Each flower is white or yellowish-white, round, rayless, very fragrant, and about ¼" (6 mm) long. The tiny disk flowers are tubular in shape and covered with bristles. The bracts are white or tinged with yellow and overlapping.

FLOWERING SEASON: August–November.

HABITAT: Dry clearings, fields, and edges of woods.

RANGE: Ontario east to Nova Scotia, south to Florida, west to Texas, and north to Nebraska and Minnesota.

SIMILAR SPECIES: Pussytoes (*Antennaria* ssp.) are similar plants but have mostly basal leaves.

CONSERVATION: G5
Globally secure but critically imperiled in New Brunswick and Prince Edward Island, and vulnerable in Quebec.

■ NATIVE
■ RARE OR EXTIRPATED
■ INTRODUCED

ALTERNATE NAMES: Carolina Desert-chicory

Carolina False Dandelion is native to the eastern and midwestern United States. A common lawn weed of the Southeast, it looks like the familiar Common Dandelion (*Taraxacum officinale*) but is much larger, growing up to knee height. Native Americans ate the taproots.

DESCRIPTION: This erect annual grows to a height of 6–30" (15–76 cm). It has vertically growing, branched flowering stems. The stem leaves are lance shaped, toothed or lobed while the basal leaves are unlobed to deeply lobed, up to 10" (25.4 cm) long. The flowers are yellowish and have five very small teeth, which represent five petals or corolla lobes. Each style is tipped with a Y-shaped stigma covered in fuzz. It has a puffball-like fruiting head that easily disperses via the wind.

FLOWERING SEASON: April–September.

HABITAT: Fields, meadows, roadsides and borders of thin woods, lawns.

RANGE: Maryland south to Florida, west to Nebraska, Oklahoma, and Texas.

CONSERVATION: G5

Globally secure but critically imperiled in Nebraska and New Jersey; vulnerable in West Virginia.

■ NATIVE
■ RARE OR EXTIRPATED
■ INTRODUCED

ALTERNATE NAMES: Upright Prairie Coneflower, Long Headed Coneflower, *Ratibida columnaris*

This colorful perennial is native to open habitats across much of North America. The flower heads are brightly colored and look like a slender sombrero, hence the common name. The flowers often bloom by the thousands. The specific epithet refers to its long, cylindrical central disk on the flowers.

DESCRIPTION: This erect, hairy, clump-forming plant grows 1–4' (30–120 cm) tall. The leaves are pinnately cleft into a few very narrow segments. Each leaf is 1–6" (2.5–15 cm) long. The lower part of the plant has long, leafless stalks bearing flower heads of three to seven yellow or yellow to red-brown drooping rays at the base surrounding a long, red-brown central disk.

FLOWERING SEASON: June–September.

HABITAT: Open areas in limestone soil, roadsides, prairies.

RANGE: Southern Canada south to Arizona, east to Florida, and north to Massachusetts.

SIMILAR SPECIES: Green Prairie Coneflower (*R. tagetes*) has a similar central flower disk, but its leaves grow closer to the flower head.

CONSERVATION: **G5**

Globally secure but vulnerable in Iowa and Mantioba. It is possibly extirpated in West Virginia.

NATIVE

RARE OR EXTIRPATED

INTRODUCED

ALTERNATE NAMES: Gray-head Mexican-hat, Prairie Cone-flower, Yellow Coneflower, Gray Headed Coneflower

This is a rough-looking plant native to the Great Lakes region and the midwestern U.S. It is named for its pinnately divided leaves, and some of its common names reference the dull gray central flower disk. When bruised, the central disk smells like anise. This plant is often grazed by livestock.

DESCRIPTION: This slender, hairy-stemmed perennial grows 1½–5' (45–150 cm) tall. The leaves are about 5" (12.5 cm) long each, compound, and pinnately divided into lanceolate, coarsely toothed segments. The flower heads droop, with yellow rays surrounding a roundish to ellipsoid, dull grayish central disk darkening to brown as rays drop off. The flower rays are 1–2½" (2.5–6.5 cm) long each. The central disk is ⅜–1" (1–2.5 cm) wide.

FLOWERING SEASON: June–September.

HABITAT: Dry woods and prairies.

RANGE: Ontario; Vermont south to Florida, west to Louisiana and Oklahoma, and north to South Dakota and Minnesota.

SIMILAR SPECIES: Mexican Hat (*R. columnifera*) is a similar but shorter plant.

CONSERVATION: G5

Globally secure but vulnerable in Canada. It is critically imperiled in Pennsylvania and South Carolina, imperiled in Florida and Louisiana, and vulnerable in Georgia and Ontario. It is possibly extirpated in West Virginia.

■ NATIVE
■ RARE OR EXTIRPATED
■ INTRODUCED

ALTERNATE NAMES: Hardheads, Mountain Bluet, Turkestan Thistle, *Acroptilon repens*

This plant is native to Eurasia and was introduced to North America in the 19th century. It has an extensive root system, up to 25' (7.6 m) deep, and has become a problem plant in the West. It contains toxins that can suppress the growth of other plant species.

DESCRIPTION: This perennial grows up to 3' (1 m) tall from a basal rosette of leaves. The basal leaves are 2–4" (5–10 cm) long and unlobed or very lobed. The stem leaves are oblong, pinnately lobed to entire and up to 6" (15.2 cm) in length at the base of the stem and become smaller and less lobed toward the top. The

flowers are pink to purple and turn dull yellowish at maturity. The flower heads are approximately ½" wide and urn-shaped.

FLOWERING SEASON: June–September.

HABITAT: Fields, roadsides, disturbed areas.

RANGE: Introduced across much of North America, except in the East.

CAUTION: Russian Knapweed is toxic to horses, with prolonged ingestion known to cause equine nigropallidal encephalomalacia or "chewing disease," a movement disorder similar to Parkinson's disease.

SIMILAR SPECIES: Russian Knapweed resembles Spotted Knapweed (*Centaurea stoebe*) and Diffuse Knapweed (*C. diffusa*), but unlike these, Russian Knapweed has somewhat hairy bracts.

CONSERVATION:
The status of this introduced plant is unranked in North America. It is considered a weed in the West.

■ NATIVE
■ RARE OR EXTIRPATED
■ INTRODUCED

This is a variable species of perennial native to the eastern U.S. This plant tends to grow in clumps. Its flowers are visited by native butterflies and bees, and finches and other birds favor the ripened seeds in winter.

DESCRIPTION: This coneflower grows 1–3' (30–90 cm) tall with scattered stems. The leaves and stems are covered in bristly hairs. The leaves are alternately arranged, dark green, ovate in shape, and sparsely dentate. The upper leaves are sessile on the flowering stems. The yellow-orange flowers have slightly curved petals, each toothed at its apex. The flower heads have black disks and bright yellow-orange ray florets. There is typically one flower head per stem.

FLOWERING SEASON: July–October.

HABITAT: Open woods, meadows, and pastures.

RANGE: Massachusetts west to Wisconsin and Illinois, south to Florida and East Texas.

CONCERVATION:  G5

Globally secure but critically imperiled in Delaware; imperiled in Washington D.C.; and vulnerable in Maryland, Pennsylvania, and West Virginia.

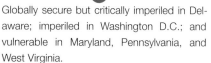

■NATIVE
■RARE OR EXTIRPATED
■INTRODUCED

ALTERNATE NAMES: Yellow Daisy, Brown Betty

This native biennial forms a rosette of leaves the first year, followed by yellowish flowers the second year. Black-eyed Susan is the state flower of Maryland and is widely cultivated. The specific epithet *hirta* is Latin for "hairy" and refers to the stiff hairs on the leaves and stems. The common name refers to the black, round central flower disk.

DESCRIPTION: This upright plant is covered in coarse hairs and grows 1–3' (30–90 cm) tall. Most of the leaves are basal. Each leaf is 2–7" (5–17.5 cm) long, lanceolate to ovate in shape, and roughly hairy. The lower leaves are mostly untoothed, with three prominent veins and winged stalks. It has daisy-like flower heads with showy, golden-yellow rays surrounding a conspicuous brown or black, cone-shaped central disk. The flower head is 2–3" (5–7.5 cm) wide. The fruit is dry and seed-like.

FLOWERING SEASON: June–October.

HABITAT: Fields, prairies, and open woods.

RANGE: Throughout much of the East; sporadic westward.

CAUTION: The stems and leaves can cause skin irritation in humans.

SIMILAR SPECIES: Cutleaf Coneflower (*R. laciniata*) is similar, but the central disk of the flowers is greenish-yellow, not black or brown.

CONSERVATION: G5
Globally secure but vulnerable in Wyoming. It is possibly extirpated in Vermont.

■ NATIVE
■ RARE OR EXTIRPATED
■ INTRODUCED

ALTERNATE NAMES: Green-head Coneflower, Tall Cone-flower, *Rudbeckia ampla*

This widespread native perennial is tall and lanky. It is widely culti-vated in gardens but can spread prolifically and has often escaped outside of its native range. It is somewhat toxic to livestock.

DESCRIPTION: This tall, leafy plant grows 2–7' (60–210 cm) tall with erect branches. The leaves are deeply cut into three, five, or seven variously toothed lobes. The lower leaves have long stalks. Each leaf is 3–8" (7.5–20 cm) long. Large, yellow flower heads with downward-arching rays appear at the ends of the branches. The flower head is 3–6" (7.5–15 cm) wide with a cylindrical or conical brown disk and 6–16 slender rays, about 1–2½" (2.5–6.3 cm) long. The fruit is seed-like and four-sided, with a low crown at the top.

FLOWERING SEASON: July–October.

HABITAT: Moist meadows, slopes, and valleys in the mountains.

RANGE: The Pacific Northwest south through the Rocky Moun-tain region to Arizona and Texas; east to the Atlantic Coast.

CAUTION: This plant is mildly toxic to livestock.

SIMILAR SPECIES: Black-eyed Susan (*R. hirta*) has a similar central flower disk, but its fruit has no crown or ring of scales.

CONSERVATION: G5

Secure but critically imperiled in Nova Scotia and Utah, imperiled in Prince Edward Island, and vulnerable in Manitoba and Louisiana.

■ NATIVE
■ RARE OR EXTIRPATED
■ INTRODUCED

ALTERNATE NAMES: Threadleaf Ragwort

This southwestern native plant has been used in traditional medicine by Native Americans, but it is toxic when consumed in large quantities. This species has several varieties, ranging from hairless annuals to woolly shrubs. It is a fast-growing plant that spreads rapidly over disturbed sites.

DESCRIPTION: This bluish-green, bushy, leafy plant grows to 1–3' (30–90 cm) with wiry stems that arch outward and upward. It's covered with white woolly hairs. The leaves are 1–5" (2.5–12.5 cm) long, divided into a few narrow lobes. The upper leaves are often simple and narrow. Yellow flower heads appear in branched clusters. Each flower head is about 1¼" (3.1 cm) wide, with rays about ½" (1.3 cm) long surrounding a narrow disk. Most bracts are about the same length, lined up side by side and not overlapping. The fruit is seed-like, with a tuft of white hairs at the top.

FLOWERING SEASON: April–September.

HABITAT: Dry rocky plains, deserts, and pinyon-juniper rangeland, open areas.

RANGE: California east to Utah and southern Colorado, and south to western Texas, Arizona, and New Mexico.

CAUTION: Plants of this genus (and herbal remedies derived from them) can cause poisoning and fatal illness in humans. The plant, especially new growth, is highly toxic to livestock.

CONSERVATION: G5
Globally secure.

■ NATIVE
■ RARE OR EXTIRPATED
■ INTRODUCED

ALTERNATE NAMES: Old-man-in-the-spring

Common Groundsel is a widely introduced plant native to Europe. The common name comes from the Anglo-Saxon word *groundeswelge*, meaning "ground swallower," referring to the rapidly spreading, weedy nature of this plant. It has dandelion-like heads and its seeds are spread by the wind. This species has been used in herbal remedies but prenatal or chronic exposure can cause serious illness to humans and other animals.

DESCRIPTION: This weedy plant grows 16–24" (40–60 cm) tall. It has coarse leaves, each about 1–6" (2–15 cm) long and alternately arranged. The lower leaves are stalked while the upper leaves clasp the stem. Green, black-tipped bracts enclose the small cluster of yellow disk flowers. The flowers are rayless, followed by a puff-like head of seeds that resemble a dandelion.

FLOWERING SEASON: Year-round.

HABITAT: Disturbed habitats, such as roadsides, fields, and yards.

RANGE: Widespread, introduced in North America.

CAUTION: All parts of the plant are poisonous to humans and animals if ingested, causing liver disease and sometimes death. The toxins can be passed along in the milk of animals that feed on the plant and in honey made from the flowers. Consumption of hay contaminated with *Senecio vulgaris* plants can poison livestock over time.

SIMILAR SPECIES: Arrowleaf Groundsel (*S. triangularis*) is similar but less common, mostly restricted to temperate regions. Sticky Groundsel (*S. viscosus*) is also similar but has glandular hairs and ray florets.

CONSERVATION:

The status of this introduced plant is not ranked in North America. This plant is largely considered a weed.

■ NATIVE
■ RARE OR EXTIRPATED
■ INTRODUCED

As the common name suggests, the leaves of this plant native to the eastern U.S are conspicuously toothed. A dense cluster of white flowers appears at the top. Unlike other asters, the flowers of this plant have only four to eight rays.

DESCRIPTION: This perennial grows to a height of 6–24" (15–60 cm). It has stiff, erect stems. The basal leaves are spatulate, toothed, and taper to a narrow base without a distinct petiole. Each leaf is about 2–4" (5–10 cm) long and 1–1½" (2.5–4 cm) wide. Dense, flat-topped clusters of white, sometimes pink, rayed florets with a cream-colored center grow at the top of the plant. The fruit is covered with silky hairs.

FLOWERING SEASON: June–October.

HABITAT: Woods, fields, disturbed areas.

RANGE: Michigan east to Maine, south to Florida, and west to Mississippi.

CONSERVATION: G5

Globally secure but critically imperiled in Maine. It is possibly extirpated in Vermont.

■ NATIVE
■ RARE OR EXTIRPATED
■ INTRODUCED

This eastern native perennial is highly variable. The name rosinweed comes from the resinous sap, which Native Americans chewed like gum and used to freshen their breath.

DESCRIPTION: This plant grows up to 5' (1.5 m) tall. The stems are mostly erect and branching near the top. The leaves are usually alternately arranged, but sometimes whorled or opposite, with fine hairs. It has round, yellowish flower heads.

FLOWERING SEASON: July–September.

HABITAT: Prairies, meadows, open woods.

RANGE: Maryland south to Florida, west to Texas, and north to Missouri and Illinois.

CONSERVATION: G5
Globally secure.

- ■ NATIVE
- ■ RARE OR EXTIRPATED
- ■ INTRODUCED

Compass Plant is native to the central and midwestern United States (also native but rare in Ontario). This perennial grows upright and looks like a sunflower.

DESCRIPTION: Its central taproot is woody and large, sometimes extending 15' (4.5 m) underground. These slow-growing plants can live up to 100 years. The common name refers to the plant's deeply incised leaves, which pioneers believed were oriented in a north-south direction. The hardened sap of this plant can be chewed like gum. This plant reaches 3–12' (90–360 cm) tall with a sticky stem that exudes resinous sap. The central stem is thick, light green, and covered in white hairs. The basal leaves also have fine white hairs. Each leaf is 12–24" (30–60 cm) long, broadly lanceolate, and deeply lobed, becoming smaller up the stem. It has yellow flower heads with large, green bracts with hairy edges. Each flower head is about 3" (7.5 cm) wide.

FLOWERING SEASON: July–September.

HABITAT: Prairies.

RANGE: Ontario and New York south to Virginia, Tennessee, and Alabama, west to Texas, and north to North Dakota.

SIMILAR SPECIES: Rosinweed (*S. integrifolium*) is a smaller plant, about 2–5' (60–150 cm) tall. It has opposite, very rough, stalkless, untoothed or slightly toothed leaves. Cup-plant (*S. perfoliatum*) has opposite leaves that enclose its square stem, each leaf forming a "cup" around it. Prairie Dock (*S. terebinthinaceum*) has larger, ovate or heart-shaped basal leaves, about 2' (60 cm) long each.

CONSERVATION: G5
Globally secure but critically imperiled in Canada. It is critically imperiled in Michigan, Ohio, and Ontario; imperiled in Kentucky and Tennessee; and vulnerable in South Dakota. It is possibly extirpated in Colorado.

■ NATIVE
■ RARE OR EXTIRPATED
■ INTRODUCED

This plant native to eastern and central North America is named for the small cup formed by the leaves, which holds water and attracts birds. The cup-like leaves and four-sided stalk distinguish Cup-plant from other *Silphium* species. A blob of resinous sap that exudes from the stalk eventually hardens and can be chewed like gum.

DESCRIPTION: This coarse perennial has a tough, erect, four-sided stem and grows 3–6' (1–2 m) tall. The leaves are coarsely toothed, opposite, and triangular to ovate. Each leaf is about 6–12" (15–30 cm) long and 4–8" (10–20 cm) wide. The upper leaves are stalkless and clasp the stem. Pairs of leaves unite at their bases, and form a cup shape. It has numerous large, yellow daisy-like composite flowers. The flower heads are about 3" (7.5 cm) wide, with 20–40 yellow rays and a dark central flower disk.

FLOWERING SEASON: July–September.

HABITAT: Moist woods, prairies, and low ground.

RANGE: Southern Ontario to North Carolina and Louisiana, west to the eastern Great Plains.

CONSERVATION: G5

Globally secure but imperiled in Canada. It is critically imperiled in Louisiana, imperiled in Michigan and Ontario, and vulnerable in Virginia and West Virginia.

■ NATIVE
■ RARE OR EXTIRPATED
■ INTRODUCED

ALTERNATE NAMES: Late Goldenrod, Canada Goldenrod

This widespread native perennial is among the tallest goldenrods, hence its common name. It can become weedy and has become naturalized in many areas around the world. This plant is part of the *Solidago canadensis* species complex and is sometimes classified as a subspecies of *S. canadensis*.

DESCRIPTION: This plant grows 2–7' (60–210 cm) tall with a grayish, downy stem sometimes covered with light hairs. The upper side of the leaves is rough, while the underside is hairy. All the leaves are lanceolate in shape. The lower leaves are 6" (15 cm) long and become progressively smaller up the stem. Small, yellow flower heads appear on arching branches and form a pyramidal cluster. Each flower head is about ⅛" (3 mm) long.

FLOWERING SEASON: August–November.

HABITAT: Thickets, roadsides, clearings, and disturbed areas.

RANGE: Throughout eastern and central North America, except in the Far North; also in parts of the West.

SIMILAR SPECIES: Late Goldenrod (*S. gigantea*) is similar but the flower heads are larger, about ¼" (6 mm), and white. Canada Goldenrod (*S. canadensis*) has sharply toothed leaves.

CONSERVATION: (G5)

Globally secure but critically imperiled on Prince Edward Island and vulnerable in Maine.

■ NATIVE
■ RARE OR EXTIRPATED
■ INTRODUCED

ALTERNATE NAMES: Broadleaf Goldenrod

This erect woodland perennial native to eastern and central North America is named for its somewhat zigzag stems. The plant's broad, coarsely toothed leaves and axillary racemes of flower heads make it easy to identify. The flowers attract native butterflies and bumblebees (*Bombus* spp.).

DESCRIPTION: This goldenrod is 8–48" (20–120 cm) tall with an unbranched, light green central stem. The stem is usually glabrous to hairy and arranged in a zigzag. The leaves are alternate, oval, and coarsely toothed with hairs on the underside. Each leaf is 2–5" (5–12.5 cm) long, becoming shorter near the flowers. The upper leaves are more lanceolate in shape. The flower heads are small and appear on short axillary stalks from the upper leaves. Each head is about ¼" (6 mm) wide, with three to four yellow rays and four to eight yellow disk florets. The floral bracts are light green.

FLOWERING SEASON: July–October.

HABITAT: Sandy streambanks and rich woods.

RANGE: Quebec to eastern North Dakota, south to Georgia, Louisiana, and northeastern Kansas.

SIMILAR SPECIES: Bluestem Goldenrod (*S. caesia*) and Elm-Leaf Goldenrod (*S. ulmifolia*) are similar, but this species typically occurs in more moist areas than the other goldenrods.

CONVERSATION:

Globally secure but critically imperiled in Kansas, Louisiana, Nebraska, and Rhode Island; imperiled in North Carolina and North Dakota and on Prince Edward Island.

■ NATIVE
■ RARE OR EXTIRPATED
■ INTRODUCED

ALTERNATE NAMES: Salt-marsh Goldenrod, Seaside Goldenrod, *Solidago sempervirens*

Southern Seaside Goldenrod is an eastern perennial primarily native to seashore habitats and introduced farther inland. It is differentiated from other plants in the genus by having toothless, hairless leaves. Many sources treat this species and Northern Seaside Goldenrod (*S. sempervirens*)—which occurs in similar habitat from Virginia northward—as a single species called Seaside Goldenrod (*S. sempervirens*) with northern and southern varieties.

DESCRIPTION: This succulent plant grows 1–8' (30–240 cm) tall. The leaves are fleshy, lanceolate to oblong, and toothless. The upper leaves are 2–8" (5–20 cm) long, and the basal leaves are 1' (30 cm) long. The flower heads are bright yellow and appear in clusters along one side of arching branches. Each head is about ⅜" (8 mm) long, with 7–10 rays.

FLOWERING SEASON: July–November.

HABITAT: Sandy places and edges of saline or brackish marshes.

RANGE: Maryland south to Florida, and west along the Gulf Coast to Texas; mostly introduced inland.

SIMILAR SPECIES: Northern Seaside Goldenrod (*S. sempervirens*) is a closely related plant (often treated as conspecific) that occurs from Virginia northward to Newfoundland and Quebec. Hybrids with Rough-stemmed Goldenrod (*S. rugosa*) are a regular occurrence.

CONSERVATION: G5
Globally secure but critically imperiled in New York.

■ NATIVE
■ RARE OR EXTIRPATED
■ INTRODUCED

ALTERNATE NAMES: Prairie Goldenrod

As the common name suggests, this native perennial is common in Missouri and to the north and west. Missouri Goldenrod is a low-growing plant that thrives in dry soils. It typically blooms earlier than other goldenrods and attracts many native butterflies and bees.

DESCRIPTION: This plant has smooth, reddish stems and grows 1–2' (30–60 cm) tall. The leaves are alternately arranged, up to 5" (12.5 cm) long and ¾" (2 cm) wide, becoming smaller up the stem. Each leaf is narrowly lanceolate with three prominent veins and serrated along the margins with fringed hairs. The flowers are golden-yellow and occur singly or in clusters. Each plant can have up to 210 flower heads. They are arranged along the upper side of the branches, usually forming a nodding plume shape. Each flower is ⅛" (3 mm) wide, with 5–14 ray flowers surrounding 8–20 disk flowers.

FLOWERING SEASON: July–September.

HABITAT: Prairies and other open, dry areas.

RANGE: Southern British Columbia and western Washington to Arizona, east through the Great Plains, to Wisconsin, Indiana, and Arkansas.

SIMILAR SPECIES: Giant Goldenrod (*S. gigantea*) is similar but grows larger — more than 3' (1 m) tall — and lacks basal leaves. It also blooms later in the season. Early Goldenrod (*S. juncea*) is also similar but blooms slightly earlier, forms fewer colonies, and is more common on roadsides rather than in open meadows.

CONSERVATION: G5

Globally secure but imperiled in Ontario. It is presumed extirpated in Michigan.

■ NATIVE
■ RARE OR EXTIRPATED
■ INTRODUCED

ALTERNATE NAMES: Oldfield Goldenrod, Field Goldenrod, Old Field Goldenrod

This is one of the smallest goldenrods. Gray Goldenrod is native across southern Canada and in the U.S. east of the Rocky Mountains. It is a pioneer species that often grows in disturbed areas, such as old agricultural fields, and it can become weedy. It is frequently visited by bees and butterflies, and American Goldfinches (*Spinus tristis*) eat the fruits.

DESCRIPTION: This plant forms clumps of slender, gray-green downy stems and grows up to 2' (60 cm) tall. The basal and lower leaves are oblanceolate in shape and slightly toothed with blunt or rounded edges. The basal leaves taper to long winged petioles while the mid- to upper stem leaves lack teeth and winged petioles and become progressively smaller. Yellow terminal flower plumes grow on one side of the stem in a vase shape. The flower heads are about ¼" (6 mm) wide each.

FLOWERING SEASON: August–October. Individual plants may bloom at different times, extending the flowering season.

HABITAT: Dry, open woods and upland prairies.

RANGE: Nova Scotia to northern Florida, west to British Columbia, Montana, and New Mexico.

CONSERVATION: G5
Globally secure but vulnerable in Wyoming.

■ NATIVE
■ RARE OR EXTIRPATED
■ INTRODUCED

ALTERNATE NAMES: Anise-scented Goldenrod, Blue Mountain Tea

The leaves of this perennial native to the eastern U.S. smell like licorice or anise when crushed, making it easy to identify. A tea can be brewed from its leaves and dried flowers. After the Boston Tea Party and during the American Revolution, American colonists drank tea made from this species, New Jersey Tea (*Ceanothus americanus*), and other wildflowers — the concoction was called Liberty Tea. It has also been used in traditional medicine, notably by the Cherokee.

DESCRIPTION: This slender, upright perennial grows 2–3' (60–90 cm) tall. The stems are downy and green or reddish, and the leaves are lance-shaped and narrow. Each leaf is 1–4" (2.5–10 cm) long, smooth, and stalkless, with small colorless dots (glands). It has crowded, cylindrical clusters of yellow flower heads along one side of slightly arching branches. The flower heads are about ⅛" (4 mm) long each, with three to four ray florets surrounding three or four disk florets. The florets mature into small achenes with tufts of hair.

FLOWERING SEASON: July–September.

HABITAT: Dry fields and open woods.

RANGE: New Hampshire and Vermont south to Florida, west to Texas, and north to Missouri, Kentucky, and Ohio.

CONSERVATION: G5

Globally secure, but imperiled in Ohio. It is possibly extirpated in Vermont.

ROUGH-STEMMED GOLDENROD *Solidago rugosa*

■NATIVE
■RARE OR EXTIRPATED
■INTRODUCED

ALTERNATE NAMES: Wrinkle-leaf Goldenrod, Rough-leaf Goldenrod

This widespread and highly variable goldenrod native to central and eastern North America can form large masses in disturbed sites and old agricultural fields. The Latin name *rugosa* means "wrinkled" and refers to the wrinkled leaves.

DESCRIPTION: This tall plant has a rough, hairy stem and grows 1–6' (30–180 cm) tall. The leaves are very hairy and deeply pinnate, appearing wrinkled. Each leaf is about 1½–5" (4–12.5 cm) long, rough, and sharply toothed. It has arching branches with small, light yellow flower heads mostly on the upper side. Each head is about ⅛" (4 mm) long, with 6–11 rays and four to seven disk flowers.

FLOWERING SEASON: July–October.

HABITAT: Fields, roadsides, and edges of woods.

RANGE: Ontario and the Great Lakes region south to eastern Texas, east to the Atlantic Coast.

SIMILAR SPECIES: Elm-leaf Goldenrod (*S. ulmifolia*) is similar but lacks creeping rhizomes.

CONSERVATION: G5
Globally secure.

- ■ NATIVE
- ■ RARE OR EXTIRPATED
- ■ INTRODUCED

This perennial is one of the showiest goldenrods, producing small, yellow flowers in an erect, pyramidal column. This species is native to most of North America east of the Rockies and is especially common in the East. Unlike other goldenrods, Showy Goldenrod's leaves lack teeth and the flowers appear erect or curved outward.

DESCRIPTION: This plant has a stout stem and grows 2–7' (60–210 cm) tall. The leaves are elliptical and obscurely toothed. The lower and basal leaves are 4–10" (10–25 cm) long and stalked. The upper leaves are much smaller and unstalked.

The leaves are smooth below and rough above. It has a dense, pyramidal or club-shaped, terminal cluster of small yellow flower heads. Each head is about ¼" (6 mm) long.

FLOWERING SEASON: August–October.

HABITAT: Open woods, prairies, and thickets.

RANGE: Ontario and New Hampshire south to Georgia, west to Texas, and north to North Dakota and Manitoba.

CONSERVATION: G5

Globally secure but critically imperiled in Maine; imperiled in Maryland, Ohio, and Wyoming; and vulnerable in Pennsylvania. It is possibly extirpated in Vermont.

■ NATIVE
■ RARE OR EXTIRPATED
■ INTRODUCED

ALTERNATE NAMES: Meadow Goldenrod, California Goldenrod

This western native perennial grows in a variety of habitats. This species now includes several varieties that were long considered separate species, including California Goldenrod (*S. californica*) and Great Basin Goldenrod (*S. sparsiflora*).

DESCRIPTION: This plant grows from creeping rhizomes, up to 5' (1.5 m) tall. Both the leaves and stems are typically covered in stiff, grayish hairs. The leaves are dark green (gray-green at the coast). The lower leaves are oblanceolate in shape, with fine teeth, while the upper leaves become smaller and untoothed. The flowering stalks are showy and wand-like. Spikes of branched clusters containing many tiny yellow flowers appear at the ends of the stalks.

FLOWERING SEASON: July–October.

HABITAT: Moist areas, grasslands, woodland edges, clearings, disturbed areas.

RANGE: Oregon west to South Dakota, south to Southern California and West Texas.

SIMILAR SPECIES: This and other goldenrods are often mistaken for ragweeds (*Ambrosia* spp.), which cause hay fever in people with allergies.

CONSERVATION: G5

Globally secure. It is imperiled in South Dakota and vulnerable in Wyoming.

- ■ NATIVE
- ■ RARE OR EXTIRPATED
- ■ INTRODUCED

ALTERNATE NAMES: Perennial Sowthistle, Field Sowthistle

This widely introduced plant in North America is considered an aggressive weed and can be incredibly difficult to eradicate. Just small pieces of the root system left behind can rapidly colonize disturbed soils. It also spreads by seeds that cling to clothing, hair, and fur and are also carried by wind. The seeds remain viable for many years.

DESCRIPTION: This invasive plant is a tall, patch-forming succulent with milky stems, up to 5' (1.5 m) tall. Most of the leaves grow at the base. The leaves are pinnately lobed with prickly edges, up to 16" (40 cm) long, and clasping the stem. The leaves become rapidly smaller, less lobed, and more widely spaced up the stem. Yellow flowers with many strap-shaped rays, without a central disk, appear at the top. The flower heads are dandelion-like, about 1¼–2" (3–5 cm) wide, and may appear with or without many glandular hairs on the bracts. The single-seed fruit (an achene) is dandelion-like and has long bristles or hairs.

FLOWERING SEASON: July–October.

HABITAT: Ditches, croplands, lake and pond edges, meadows, lawns, mainly in damp soil.

RANGE: Native to Eurasia. Introduced in North America and naturalized widely, except in the Southeast.

SIMILAR SPECIES: Spiny Sowthistle (*S. asper*) is similar but has larger, spiny teeth on the leaves and larger lobes at the leaf base. Common Sowthistle (*S. oleraceus*) is also similar but has more angled lobes on the leaves. Both of these relatives are introduced in North America and have smaller flowers, up to 1" (2.5 cm) wide.

CONSERVATION:

This introduced plant in North America is unranked and is largely considered a weed.

■ NATIVE
■ RARE OR EXTIRPATED
■ INTRODUCED

ALTERNATE NAMES: Stemless Mock Goldenweed

This leafy western native plant grows in dense tufts or mats of dark green, stiff leaves. It produces many nearly leafless stalks, each topped with a single yellow flower head.

DESCRIPTION: Stemless Goldenweed grows ½–6" (1.5–15 cm) long with numerous, nearly leafless stalks. The leaves have three veins and are very stiff and erect, growing ½–2" (1.5–5 cm) long in dense mats or tufts. The stalks are each topped with one yellow flower head, which is about 1½" (4 cm) wide, with 6–15 rays. Each ray is about ½" (1.5 cm) long, surrounding disk flowers. The bracts are pointed and lanceolate in shape. The fruit is seed-like, with many tan bristles at the tip.

FLOWERING SEASON: May–August.

HABITAT: Dry, open places from foothills to mountains, and shrublands.

RANGE: Southeastern Oregon south along eastern slopes of the Sierra Nevada, east to central Idaho and Colorado, and northeast to southwestern Montana.

SIMILAR SPECIES: Thrift Goldenweed (*S. armerioides*), found from Montana south to Arizona and New Mexico and east to Nebraska, is closely similar but has round tips on its bracts, not pointed, and broadly oblanceolate leaves.

CONSERVATION: G5
Globally secure.

■ NATIVE
■ RARE OR EXTIRPATED
■ INTRODUCED

ALTERNATE NAMES: Purple Aster, Pacific Aster, Western Aster, *Aster ascendens*

The Latin name *ascendens* means "rising upwards," referring to the upward-pointing leaves or the erect growth of these plants. Asters have flower heads that resemble those of fleabanes (*Erigeron*). The bracts of aster plants vary in length, however, overlapping like shingles on a roof. This plant, like many asters, has gone through several name changes. Many members of the *Aster* genus, including North America's native asters, were moved to *Symphyotrichum* and a few other genera in the 1990s, but they are still commonly called asters.

DESCRIPTION: This upright perennial grows 8–28" (20–70 cm) tall with a slender stem and ascending branches. The leaves are narrow, about 6" (15 cm) long each. Both the stems and leaves are covered in dense hairs. The flower spikes are composed of showy purple flowers, each about 2" (5 cm) wide surrounding a central golden disk. The dry-seeded fruit is covered with hairs.

FLOWERING SEASON: July–October.

HABITAT: Moist or dry meadows or woodland openings; elevations of 250–7000' (76–2133 m)

RANGE: Saskatchewan to southeastern Washington, south to New Mexico, northern Arizona, and central California.

CONSERVATION: G5
Globally secure but vulnerable in Canada. It is critically imperiled in Nebraska and vulnerable in Alberta and Saskatchewan.

ALTERNATE NAMES: Heartleaf Aster, Common Blue American-aster, Blue Wood-aster, Heart-Leaved Aster, *Aster cordifolius*

This attractive plant native to eastern and central North America produces blue, daisy-like flowers and is often cultivated. Like other asters, its flowers provide nectar for butterflies and other insects.

DESCRIPTION: This upright to arching perennial grows 24–60" (60–150 cm) tall. The leaves are toothed and heart-shaped at the bottom. The lower leaves are up to 5" (12.5 cm) long, becoming smaller and more oval-shaped toward the top of the plant. Each flower is ½–¾" (1.2–2 cm) wide, with blue to purple rays around a reddish to yellow center, appearing in branched clusters atop the stem.

FLOWERING SEASON: August–October.

HABITAT: Woodland edges and clearings.

RANGE: Eastern North America from southeastern Canada south to northern Georgia and Alabama, west to Manitoba, eastern South Dakota, and Kansas.

SIMILAR SPECIES: White Wood Aster (*Eurybia divaricata*) is similar but has white petals in a flat-topped cluster. Bigleaf Aster (*Eurybia macrophylla*) is also similar but has many broadly ovate basal leaves.

CONSERVATION: G5
Globally secure but critically imperiled in Kansas and vulnerable in Delaware and Missouri.

- ■ NATIVE
- ■ RARE OR EXTIRPATED
- ■ INTRODUCED

ALTERNATE NAMES: White Heath American-aster, White Heath Aster, *Aster ericoides*

This widespread native perennial has small flowers that are organized into larger heads, resembling a single, radially symmetrical flower cupped by a ring of green bracts. Its common and Latin names both refer to this plant's resemblance to heaths in the genus *Erica*.

DESCRIPTION: This bushy, grayish plant grows 36" (90 cm) tall. The leaves are small and linear, covered in hairs, and rigid. It has densely clustered white daisy-like flowers. The flowers are about ½" (12 mm) wide, with many white rays surrounding a yellow center, appearing in branched clusters of many dozens.

FLOWERING SEASON: August–November.

HABITAT: Dry fields, prairies, roadsides.

RANGE: Maine to Manitoba and Washington, south to Virginia, Texas, Arizona, and Oregon.

SIMILAR SPECIES: This species is commonly confused with Frost Aster (*S. pilosum*), which occurs in most of the same range, but it has larger flower heads with longer ray petals. Amethyst Aster (*Symphyotrichum* x *amethystinum*) is a naturally occurring hybrid species of Heath Aster and New England Aster (*Symphyotrichum novae-angliae*) that can grow where the two parents are in close proximity.

CONSERVATION: G5

Globally secure but critically imperiled in Georgia and Tennessee; imperiled in Mississippi; and vulnerable in New Jersey, Pennsylvania, and Virginia.

■ NATIVE
■ RARE OR EXTIRPATED
■ INTRODUCED

ALTERNATE NAMES: Alpine Leafy-head American-aster, Alpine Leafy-bract Aster, Leafy-bracted Aster, *Aster foliaceus*

This western native species may vary depending on its habitat and includes four recognized varieties. Bees and butterflies frequently visit the flowers.

DESCRIPTION: This plant grows 8–20" (20–50 cm) tall with leafy stems. The basal leaves are lanceolate in shape, becoming narrower at the base. The leaves at the midstem partially clasp the stem at the bottom. Each leaf is 5–8" (12.5–20 cm) long. Ascending branches are terminated by several flower heads, each with 15–60 narrow, lavender or purple to pink rays. Each flower head is 1–2" (2.5–5 cm) wide, with rays surrounding a yellow central disk. It has overlapping light green or reddish involucre bracts with a whitish base. The fruit is seed-like and smooth or sparsely hairy on the surface.

FLOWERING SEASON: July–September.

HABITAT: Moist areas in the woods, roadsides, and high-elevation meadows.

RANGE: Alaska south to central California, Arizona, and New Mexico.

SIMILAR SPECIES: This species is distinguished from other similar plants by its light green or reddish bracts with a whitish margin.

CONSERVATION: G5
Globally secure but critically imperiled in Yukon.

■ NATIVE
■ RARE OR EXTIRPATED
■ INTRODUCED

ALTERNATE NAMES: Smooth Blue American-aster, Bluebird, *Aster laevis*

As the common name suggests, this widespread native aster has bluish flowers. Its common and Latin names both refer to the foliage that is smooth to the touch. The plants bloom into the fall.

DESCRIPTION: This smooth-leaved perennial grows 2–4' (60–120 cm) tall. The stem produces a light grayish-white bloom. The leaves are thick and slightly toothed, with each leaf about 1–4" (2.5–10 cm) long and elliptical or lanceolate in shape. The lower leaves are stalked and the upper leaves are unstalked, clasping the stem. The flower heads have many lavender-blue rays surrounding a yellow central disk. Each head is about 1"

(2.5 cm) wide and appears in open, panicle-like clusters. The bracts are green at the tip. The fruit is dry and seed-like, often with reddish bristles at the tip.

FLOWERING SEASON: August–October.

HABITAT: Fields, open woods, rocky areas.

RANGE: Throughout much of North America, except the Far North.

CONSERVATION:  G5

Globally secure but critically imperiled in Delaware, North Carolina, Utah and New Brunswick; imperiled in Kentucky, South Carolina, and Vermont; and vulnerable in Georgia, West Virginia, and Wyoming. It is possibly extirpated in Oklahoma.

■ NATIVE
■ RARE OR EXTIRPATED
■ INTRODUCED

ALTERNATE NAMES: White Panicled American-aster, White Panicle Aster, *Aster simplex, Aster lanceolatus*

This widespread native colony-forming perennial is widely cultivated as a garden plant and for cut flowers. Its late-season blooms are important nectar sources for pollinators such as butterflies, moths, and bees. The species can be divided into two subspecies—Swamp Aster (*S. lanceolatum* ssp. *hesperium)* and Lance-leaved Aster (*S. lanceolatum* ssp. *lanceolatum)*—and five varieties. Some of these have only subtle differences among them and overlapping ranges.

DESCRIPTION: This plant grows 2–5' (60–150 cm) tall with fine white hairs on the stem. The leaves are hairless, lanceolate in shape, sharp-pointed and sometimes toothed. The lower leaves, 3–6" (7.5–15 cm) long each with short stalks, become smaller up the stem. A panicle of flower heads with 25–30 white (occasionally violet-tinged) rays appears at the top. Each flower head is ¾–1" (2–2.5 cm) wide, with a yellowish to pinkish central disk. The bracts are narrow and green at the tip. The fruit is dry and seed-like with bristles at the tip.

FLOWERING SEASON: August–October.

HABITAT: Damp areas, meadows, and shorelines.

RANGE: Throughout much of North America, except the Far North.

SIMILAR SPECIES: New York Aster (*S. novi-belgii*) is similar but its ray flowers are blue or violet, not white.

CONSERVATION: G5
Globally secure but imperiled in Newfoundland and North Carolina, and vulnerable in Iowa.

- ■ NATIVE
- ■ RARE OR EXTIRPATED
- ■ INTRODUCED

ALTERNATE NAMES: New England American-aster, *Aster novae-angliae*

New England Aster is a widespread native perennial found across North America but most common in the Northeast and Midwest. This is a highly variable plant, easily distinguished from other asters by its compound flowers, which are larger. It also has more ray flowers than other asters.

DESCRIPTION: This hairy, leafy plant grows 3–7' (90–210 cm) tall. The central stem and side branches are covered with short white hairs, as are the leaves. The leaves are lanceolate in shape and clasp the stem. Each leaf is about 1½–5" (4–12.5 cm) long and toothless, becoming smaller up the stem. Flower heads of bright lavender to purplish-blue rays appear clustered at the ends of the branches. Each head is 1–2" (2.5–5 cm) wide, with 35–45 rays and yellowish disk flowers. The bracts are narrow, hairy, and sticky. The fruit is dry and seed-like, with bristles at the tip.

FLOWERING SEASON: August–October.

HABITAT: Wet thickets, meadows, and swamps.

RANGE: Manitoba east to Nova Scotia, south to Oklahoma and Georgia; also scattered in parts of the western United States.

CONSERVATION: G5

Globally secure but critically imperiled in Saskatchewan, Georgia, South Carolina, and Wyoming; imperiled in Colorado; and vulnerable in North Carolina.

■ NATIVE
■ RARE OR EXTIRPATED
■ INTRODUCED

ALTERNATE NAMES: Purplestem American-aster, Red-stemmed Aster, Swamp Aster, Glossy-leaf Aster, *Aster puniceus*

This aster, which blooms late into autumn, is frequently grown in gardens. It is a variable species with several named varieties. One such variety is sometimes considered a separate species, Shining Aster (*S. firmum*).

DESCRIPTION: This tall, erect plant has a purplish stem and grows to a height of 2–7' (0.6–2.1 m). The leaves are lance-shaped, with toothed margins. Both the stems and leaves are usually rough and bristly. A few flower heads with many violet to bluish rays appear on each branch. The rays surround a yellow disk. Each flower head is about 1½" (4 cm) wide.

FLOWERING SEASON: August–November.

HABITAT: Swamps and wet thickets, frequently with calcareous soil.

RANGE: British Columbia to Newfoundland, south to Georgia, and west to Texas in the southern and North Dakota in the north-ern U.S.

CONVERSATION: **G5**

Globally secure but critically imperiled in Mississippi.

 NATIVE
■ RARE OR EXTIRPATED
■ INTRODUCED

ALTERNATE NAMES: Perennial Saltmarsh American-aster, *Aster tenuifolius*

This native perennial found along the Atlantic and Gulf coasts has few flowers, but is easily noticeable in its habitat because it forms conspicuous masses in brackish tidal marshes.

DESCRIPTION: This straggly plant grows 12–27" (30–70 cm) tall with widely spreading branches. There are few leaves, each about 6" (15 cm) long, fleshy, and narrow, tapering at both ends. A few flower heads with numerous white or pale purple rays appear at the top. Each flower head is ½–1" (1.5–2.5 cm) wide, with a yellowish to pinkish central disk. The fruit is dry and seed-like, with bristles at the tip.

FLOWERING SEASON: August–October.

HABITAT: Salt- or brackish marshes.

RANGE: Maine south along the coast to Florida, and west to Texas.

SIMILAR SPECIES: Annual Saltmarsh Aster (*S. subulatus*) is similar, but has many flower heads and the heads are about half the size of Saltmarsh Aster's, about ¼"–½" (6–13 mm) wide each.

CONVERSATION:

Globally secure but critically imperiled in New Hampshire, imperiled in New York, and vulnerable in North Carolina.

■ NATIVE
■ RARE OR EXTIRPATED
■ INTRODUCED

ALTERNATE NAMES: Golden Button

This European perennial, formerly classified as *Chrysanthemum vulgare*, is introduced in North America. Despite being highly toxic, it was used medicinally for centuries to treat a number of ailments, including plague, colic, digestive issues, fever, and worms. It is also used as an insect repellent. The genus name comes from a Greek word meaning "long-lasting" or "immortal" in reference to the long-lasting flowers, which are often dried and used in winter bouquets.

DESCRIPTION: This erect perennial grows 2–3' (60–90 cm) tall. The leaves have a strong scent and are fern-like, divided into linear, toothed segments. Each leaf is 4–8" (10–20 cm) long. It has flat-topped clusters of bright orange-yellow, button-like flower heads. Each flower head is ½" (1.5 cm) wide, containing disk flowers, with occasional ray-like flowers.

FLOWERING SEASON: July–September.

HABITAT: Roadsides, edges of fields, waste places, and shorelines.

RANGE: Introduced throughout much of North America, except Texas and most of the Deep South, and the Arctic.

CAUTION: The leaves and stem contain thujone, a highly toxic compound that can be fatal to humans and animals if ingested.

SIMILAR SPECIES: Lake Huron Tansy (*T. bipinnatum*) is similar but has long, hairy leaves and larger flower heads. It is native but less widespread in North America, limited to Canada, Alaska, Wisconsin, Michigan, and Maine.

CONSERVATION:
The status of this introduced species is not ranked in North America. This plant has escaped gardens and is largely considered a weed.

■ NATIVE
■ RARE OR EXTIRPATED
■ INTRODUCED

ALTERNATE NAMES: Dandelion

This familiar perennial appears in masses on North American lawns in the early spring and is a common, vigorous weed. At least one subspecies of Common Dandelion is native to North America, but others were introduced from Europe. The word dandelion comes from the French *dent de lion*, for "lion's tooth," a reference to the sharp teeth on the leaves. In some places Common Dandelion is cultivated for food and herbal medicine. The young leaves may be used in salads and soups, and dandelion wine is made from the ray flowers.

DESCRIPTION: This weedy plant grows to a height of 2–18" (5–45 cm) with stems that exude a milky sap when crushed. The basal leaves are 2–16" (5–40 cm) long and deeply toothed or lobed. One flower head appears at the top, composed of numerous yellow rays. Each head is about 1½" (4 cm) wide. The rays have five tiny teeth at the tip and narrow, pointed bracts. The outer bracts bend backward. The fruit is dry and round with a parachute of long, white bristly hairs. The fruit becomes a silky, downy, round head when ripe, and easily disperses via the wind.

FLOWERING SEASON: March–September.

HABITAT: Fields, roadsides, and lawns.

RANGE: Throughout North America, but rarer in the extreme southeastern United States.

CONSERVATION: G5
Globally secure. It is considered a weed in many areas.

■ NATIVE
■ RARE OR EXTIRPATED
■ INTRODUCED

ALTERNATE NAMES: Hopi Tea Greenthread, Navajo Tea

This native perennial is found in the central and southwestern United States. Native American groups — notably the Hopi and Navajo — have used this plant for centuries medicinally, as a source for dye, and for making teas. The attractive flowers are visited by several pollinators, including butterflies, moths, flies, and bees.

DESCRIPTION: This plant grows to a height of 1–3' (30–90 cm). The leaves are green and oppositely arranged. They may be stalkless or have short stalks. Each leaf is 1½–3½" (3.8–9 cm) long, with a few, thread-like lobes. It has slender, erect, smooth, bluish-green stems, which may grow singly or in clusters. Yellow flower heads appear at the tips of branching stems. The bracts surrounding the flower heads are purplish with yellow or white along the margins and fused together.

FLOWERING SEASON: April–October.

HABITAT: Plains, prairies, mesas, and open woods.

RANGE: The western United States from Montana and South Dakota to Arizona and Texas; also in southern California.

CONSERVATION: **G5**
Globally secure but critically imperiled in Utah, and vulnerable in South Dakota and Wyoming.

NATIVE
RARE OR EXTIRPATED
INTRODUCED

ALTERNATE NAMES: *Townsendia florifera*

This is an attractive annual or short-lived perennial that's native to the Mountain West in the United States. It grows from a simple taproot and maintains a low growth form. The foliage is covered in coarse hairs.

DESCRIPTION: This native plant grows 1–6" (2.5–15 cm) tall with one to several erect stems sparsely covered in leaves rising from a simple taproot. The lower leaves are spoon-shaped, about ¾–2¼" (2–6 cm) long each. The upper leaves are smaller and covered in hairs. One to a few flower heads appear at the tips of the stems. Each head has three or four overlapping, green or purple involucral bracts, about 7–10 mm long each, with 13–34 pink or white ray flowers of the same length and yellow disk corollas. The hairy achenes have stiff bristles at the top.

FLOWERING SEASON: May–July.

HABITAT: Gravelly flats, hills, dry, open places.

RANGE: Washington to Montana, south to Nevada and Utah.

CONSERVATION: G5

Globally secure but imperiled in Montana. It is possibly extirpated in Wyoming.

■ NATIVE
■ RARE OR EXTIRPATED
■ INTRODUCED

ALTERNATE NAMES: Yellow Salsify, Meadow Goatsbeard, Western Salsify, Wild Oysterplant

This widely introduced species is native to Eurasia. *Tragopogon* is Greek for "goat's beard" and may refer to the thin, grass-like leaves or the brownish, feathery bristles on the fruit. The basal leaves can be eaten, as can the roots, which are said to taste like oysters. Yellow Goatsbeard hybridizes with a number of other *Tragopogon* species, including North American natives.

DESCRIPTION: This plant has a smooth stem with grass-like leaves and grows 1–3' (30–90 cm) tall. The stems have milky sap and appear swollen just below the flower heads. The leaves are up to 1' (30 cm) long each, and broad at the base where they clasp the stem, becoming narrower at the tip. The plant has one yellow flower head that opens in the morning and closes by noon. Each flower head is 1–2½"

(2.5–6.5 cm) wide and consists of many rays. The bracts are long, pointed, and green. The fruit is seed-like with parachute-like bristles, forming a round, feathery head, about 3" (7 cm) wide.

FLOWERING SEASON: May–August.

HABITAT: Fields and waste places.

RANGE: Introduced throughout much of temperate North America, except the Arctic and parts of the southeastern United States.

SIMILAR SPECIES: Jack-go-to-bed-at-noon (*T. pratensis*) is similar but has yellow flower heads with shorter bracts.

CONSERVATION:

The conservation status of this introduced species is not ranked in North America. This species has a wide global distribution.

■ NATIVE
■ RARE OR EXTIRPATED
■ INTRODUCED

ALTERNATE NAMES: Meadow Salsify, Yellow Goat's Beard, Meadow Goat's-beard

This biennial introduced to North America from Europe grows low with only basal leaves appearing the first year and flowers the second year. The roots can be boiled and eaten like potatoes and the young shoots can be used in salads.

DESCRIPTION: This plant has hollow stems that exude milky sap and grows to a height of 16–32" (40–80 cm). The stems appear branched near the base and have a few long, narrow, tapered leaves. Each leaf is 5–6" (12.5–15 cm) long with parallel veins. The base of the leaves clasp the stem. Both the basal leaves and stem leaves contain milky sap. Bright yellow flower heads appear at the top of the stems. Each head is 1½–2" (3.8–5 cm) wide, while the flowers are about ½" (1.3 cm) long, with 8–13 bracts. The bracts are usually shorter than the flowers. The fruit is brown and seed-like, rough on the surface and tapered at both ends. Pale brown, feathery bristles, nearly 1" (2.5 cm) long, appear at the top of the fruit.

FLOWERING SEASON: May–September.

HABITAT: Roadsides, old lots, and fields.

RANGE: Introduced across much of North America except the Arctic and the southeastern United States.

SIMILAR SPECIES: Oyster Plant (*T. porrifolius*) is similar but has purple flower heads. Yellow Goat's Beard (*T. dubius*) is also similar but has pale yellow flower heads, with about 13 floral bracts that extend beyond the flower head, and green or blue-green foliage.

CONSERVATION:
The conservation status of this introduced species is not ranked in North America.

■ NATIVE
■ RARE OR EXTIRPATED
■ INTRODUCED

This Eurasian perennial is introduced and naturalized in northeastern and north-central North America. The common name refers to the resemblance of the leaves to a horse's hoof. The genus name comes from the Latin word *tussis*, meaning "cough," and references the plant's use as an herbal remedy for coughs. An extract of the leaves can be used to make cough drops or hard candy. The dried leaves can be brewed as tea.

DESCRIPTION: This low-growing, rhizomatous plant reaches a height of 3–18" (8–45 cm), with a scaly stalk. The basal leaves are broad and heart-shaped. Each leaf is 2–7" (5–17.5 cm) long, slightly toothed, and upright, with whitish beneath. It has one yellow flower head at the top of each stem. Each flower head is 1" (2.5 cm) wide, with thin rays surrounding the disk flowers.

FLOWERING SEASON: February–June.

HABITAT: Roadsides and waste places.

RANGE: Introduced in Ontario east to Newfoundland and Nova Scotia, south to North Carolina and Tennessee; also introduced in British Columbia.

SIMILAR SPECIES: Dandelions (*Taraxacum* spp.) have similar flower heads, but Coltsfoot is differentiated by having both ray and disk flowers.

CONSERVATION:
The conservation status of this introduced species is not ranked in North America.

■ NATIVE
■ RARE OR EXTIRPATED
■ INTRODUCED

ALTERNATE NAMES: Cowpen Crownbeard

Golden Crownbeard is an annual herb native to much of the United States and Mexico. This common plant was widely used by Native Americans and early settlers to treat skin ailments. It is often found along roadsides and in other disturbed habitats, where it can sometimes dominate other species.

DESCRIPTION: This heavily branched, grayish-green plant grows 4–60" (10–150 cm) tall. The leaves are almost triangular, oppositely arranged, and coarsely toothed. Each leaf is 1–4" (2.5–10 cm) long. The upper leaves are alternately arranged and the upper surface is covered in hairs or bristles. Yellow or orange flower heads are 1½–2" (3.8-5 cm) wide and ½" (1.3 cm) long and are either three-toothed rays or disk florets. The fruit is

seed-like, and those on the disk flowers have two slender, rigid bristles at the tip.

FLOWERING SEASON: June–September.

HABITAT: Along roads, in pastures and on rangeland, in washes, and at the edges of fields.

RANGE: Montana south through eastern Utah and Colorado to Arizona, New Mexico, Texas, and south into Mexico; west to central California; and east to Kansas and the southeastern United States.

CONSERVATION: **G5**

Globally secure but imperiled in Wyoming and vulnerable in North Dakota.

■ NATIVE
■ RARE OR EXTIRPATED
■ INTRODUCED

ALTERNATE NAMES: Giant Ironweed, *Vernonia altissima*

This upright perennial native to eastern North America was formerly known as *V. altissima*. This is one of the tallest of the ironweeds. The attractive flowers are often visited by butterflies. The species name means "unusually tall" and refers to the tall, tough stem. The flowers and seeds are "iron-like" in their color and durability. The genus name honors English botanist William Vernon, who collected plants throughout North America.

DESCRIPTION: This erect plant grows 3–7' (90–210 cm) tall with stiff, leafy stems that branch at the top. The leaves are thin and pointed, about 6–10" (15–25 cm) long each, lanceolate or ovate in shape and downy beneath. The leaf margins are serrated irregularly. Deep purple-blue composite flower heads appear at the top of the stems in loose clusters. Each flower head is about ¼" (6 mm) wide. There are 13–30 fluffy disk flowers, each with five lobes. The bracts are usually purple and have blunt tips. The fruit is seed-like, with a double set of purplish bristles.

FLOWERING SEASON: August–October.

HABITAT: Meadows, open woods, and pastures.

RANGE: Ontario and New York south to Florida, west to Texas, and north to Iowa.

CONSERVATION: G5
Globally secure but critically imperiled in Canada. It is critically imperiled in New York and Ontario.

■ NATIVE
■ RARE OR EXTIRPATED
■ INTRODUCED

ALTERNATE NAMES: Ironweed

New York Ironweed is a robust, clump-forming perennial native to the eastern United States. This plant has tough stems colored reddish like iron. It tolerates many types of soils and can become an aggressive weed.

DESCRIPTION: This erect plant grows 3–6' (90–180 cm) tall. The stem branches toward the top. Each branch has a cluster of rayless, deep lavender to violet flower heads. Each cluster is 3–4" (7.5–10 cm) wide, while the flower heads are about ⅜" (8 mm) wide. There are 30–50 disk flowers and each flower has five lobes, with erect bracts and long, hair-like tips. The leaves are 4–8" (10–20 cm) long, alternately arranged, finely toothed, lanceolate, and pointed. The fruit is seed-like, with two sets of purplish bristles.

FLOWERING SEASON: August–October.

HABITAT: Moist low ground and streambanks; wet, open, bottomland fields.

RANGE: New Hampshire and New York south to Florida, west to Alabama, Tennessee, and Kentucky.

SIMILAR SPECIES: Tall Ironweed (*V. gigantea*) is similar but usually has fewer flowers per head.

CONSERVATION: G5

Globally secure but imperiled in Rhode Island and vulnerable in Kentucky. It is presumed extirpated in Ohio.

NATIVE
RARE OR EXTIRPATED
INTRODUCED

This western native perennial appears varnished with resin. It produces sunflower-like flower heads, often alongside smaller yellow flowers. The genus is named for Nathaniel Jarvis Wyeth, whose 1834 expedition to explore the West included noted naturalists Thomas Nuttall and John Kirk Townsend and resulted in the collection of more than 100 plant species previously unknown to science.

DESCRIPTION: This clump-forming plant has short, leafy stems and grows to a height of 12–31" (30–80 cm). Its leaves are lanceolate in shape. The basal leaves are 8–24" (20–60 cm) long and short-stalked, while the leaves on the stem are smaller and partly wrapped around the stem. It has large, long-stalked, deep yellow flower heads. The central head is the largest, about 3–5" (7.5–12.5 cm) wide. Each flower head has 13–21 rays. The bases of the disk flowers are surrounded by scales. The bracts are lanceolate in shape, often extending past the top of the central disk. The fruit is very narrow and seed-like, with four sides and a low crown of scales at the tip.

FLOWERING SEASON: May–July.

HABITAT: Open hillsides and meadows, open woods, from foothills to moderate elevations in mountains.

RANGE: Central Washington to western Montana, south to northwestern Colorado, northern Utah, and Nevada.

SIMILAR SPECIES: Arrowleaf plants (*Heterotheca* spp.) are related but have smaller flowers and shorter leaves. All Wyethia species also differ from the similar balsamroots (*Balsamorhiza* spp.) by having leaves on the stem, not just at the base.

CONSERVATION: G4
Apparently secure but vulnerable in Wyoming.

■ NATIVE
■ RARE OR EXTIRPATED
■ INTRODUCED

ALTERNATE NAMES: Lacy Tansy-aster, Cutleaf Ironplant, Spiny Cocklebar, *Machaeranthera pinnatifida*

This western native perennial is highly variable and has been reclassified several times. The genus name *Xanthisma* is from the Greek word *xanth*, meaning "yellow," and refers to the yellow daisy-like flower heads. The species may appear under different Latin names in various sources, having formerly been classified as *Aster pinnatifidus*, *Haplopappus spinulosus*, *Sideranthus spinulosus*, and *Machaeranthera pinnatifida*.

DESCRIPTION: This slender plant has many branched, green stems covered in woolly hairs and grows to a height of 6–14" (15–35 cm). Its small, narrow leaves are about ⅛–¾" (3–20 mm) long each. The lower leaves sometimes have a few lobes, and are angled upward or pressed against the stem. One yellow, daisy-like flower head appears at the tip of each of the upper branches. Each flower head is about 1" (2.5 cm) wide, with 14–60 yellow petals or ray flowers. The rays are each about ⅜" (9 mm) long. The bracts surrounding the base flowers are pointed at the tip with a white bristle and usually covered in woolly hairs, or sometimes hairless or glandular. The fruit is seed-like and densely covered with short hairs, with numerous slender, tan bristles at the tip.

FLOWERING SEASON: August–October.

HABITAT: Open places in arid grasslands and deserts and among pinyon and juniper.

RANGE: Alberta south through the Rocky Mountain region and the Great Plains states to Texas, New Mexico, Arizona, and southern California.

SIMILAR SPECIES: Hairy Goldenaster (*Heterotheca villosa*) has similar yellow flowers, but Yellow Spiny Daisy is distinguished by its divided leaves and bristle-tipped lobes. Slender Goldenweed (*M. gracilis*) is also similar, but is an annual, not a perennial.

CONSERVATION: Ⓖ5
Globally secure but critically imperiled in Utah and vulnerable in Manitoba and Minnesota.

■ NATIVE
■ RARE OR EXTIRPATED
■ INTRODUCED

This is a widespread annual native to North America that has widely naturalized elsewhere around the world. The flowers contain both male and female reproductive organs and it can self-seed. This has led to aggressive growth in some areas. Its burs are covered in short, hooked bristles that attach to clothing and animal fur for seed dispersal. The leaves and seeds are highly toxic to animals, especially when the plant is young.

DESCRIPTION: This plant has a rough stem and grows to a height of 1–6' (30–180 cm) tall. The leaves look like those of maple trees. Each leaf is 2–6" (5–15 cm) long and coarsely toothed, with long stalks. Each raceme produces separate greenish male and female flower heads in the leaf axils. The female flower heads, which produce spiny ovoid burs, appear in the lower part of the racemes. The male flower heads appear on the upper half of the racemes. Each head is ½–1½" (1.3–3.8 cm) long.

FLOWERING SEASON: August–October.

HABITAT: Waste places, roadsides, and low ground.

RANGE: Throughout most of temperate North America.

CAUTION: The seeds and leaves are highly toxic to animals if ingested. Humans should also avoid consuming this plant.

SIMILAR SPECIES: Spiny Cocklebur (*X. spinosum*) is related, but has shiny, veined leaves and distinctive three-branched, orangish spines at the point of each leaf attachment.

CONSERVATION: G5

Globally secure but vulnerable in Alberta and Prince Edward Island.

◼ NATIVE
◼ RARE OR EXTIRPATED
◼ INTRODUCED

ALTERNATE NAMES: American Twinflower

The common name of this widespread native perennial comes from the plant's twinned flowers and leaves. The leaves are paired on opposite sides of the stem, and the fragrant, bell-like flowers also grow in twos. The specific epithet *borealis* means "northern," and refers to this plant's tendency to grow in the north. The genus name honors botanist Carl Linnaeus, who is considered the father of modern taxonomy. This was said to be Linnaeus' favorite flower, and he is often depicted in paintings holding this plant in his hands.

DESCRIPTION: This creeping, somewhat evergreen, matted plant grows less than 4" (10 cm) tall. The leaves are round and leathery, sometimes with shallow teeth on the edges. Each leaf is ¼–1" (6–25 mm) long, and oppositely arranged. Two pink bell-like flowers hang from the tops of erect, short, Y-shaped leafless stalks. Each flower is about ½" (1.5 cm) long. The corolla has five round lobes.

FLOWERING SEASON: June–September.

HABITAT: Cool woods and bogs.

RANGE: Across southern Canada and the northern U.S., and into the southwestern states.

CONSERVATION: G5

Globally secure but critically imperiled in Arizona, Iowa, Pennsylvania, and West Virginia; and vulnerable in Wyoming and Massachusetts. It is presumed extirpated in Indiana, Ohio, and Tennessee.

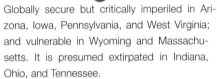

JAPANESE HONEYSUCKLE *Lonicera japonica*

■ NATIVE
■ RARE OR EXTIRPATED
■ INTRODUCED

This woody vine native to eastern Asia is a fast-growing climber, often engulfing native plants and choking trees. It is difficult to eradicate, able to thrive in a variety of habitats, and often damages plant communities by outcompeting native vegetation for light and soil resources. Nectar can be sucked from the base of the corolla, hence the common name.

DESCRIPTION: This climbing or trailing vine grows to 30' (9 m) long with hairy twigs. The leaves are evergreen and covered in hairs. Each leaf is up to 3" (7.5 cm) long and untoothed. Fragrant white and yellow tubular flowers appear in pairs in the leaf axils. Each flower is about 1½" (4 cm) long. The corolla has two lips and five lobes, with stamens extending beyond the inforescence. The fruit is a black berry.

FLOWERING SEASON: April–July, sometimes into the fall.

HABITAT: Thickets, roadsides, woodlands, and open, disturbed areas.

RANGE: Introduced from Ontario east to Maine, south to Florida, west to California, and north to Iowa and Nebraska.

SIMILAR SPECIES: Trumpet Honeysuckle (*L. sempervirens*), native to the southeastern U.S., is similar but the corolla lobes are more spread apart.

CONSERVATION:

The conservation status of this introduced plant is not ranked in North America. This plant is largely considered a weed.

■ NATIVE
■ RARE OR EXTIRPATED
■ INTRODUCED

ALTERNATE NAMES: Coral Honeysuckle, Scarlet Honeysuckle

This southeastern native honeysuckle has attractive trumpet-shaped flowers that are frequently visited by hummingbirds, butterflies, and bees. It is less aggressive than Japanese Honeysuckle (*Lonicera japonica*) but has naturalized north of its native range. The name *sempervirens* is from the Latin words *semper*, meaning "always," and *virens*, meaning "green," and refers to the plant's evergreen habitat, especially in the south.

DESCRIPTION: This perennial vine grows to 16½' (5 m) long. The leaves are deep green or blue and whitish beneath. Each leaf is oblong, oppositely arranged and 1½–3" (4–7.5 cm) long. The uppermost leaf pairs are so united they look like they're attached to the stem. It has large, showy, non-fragrant, trumpet-shaped flowers appearing in several whorled clusters at the tops of the stems. The flowers are red outside and yellowish inside.

Each flower is 1–2"(2.5–5 cm) long. The corolla has five lobes. The fruit is a red berry.

FLOWERING SEASON: April–August.

HABITAT: Woods and thickets.

RANGE: Found from Ontario and Maine south to Florida, west to Texas, and north to Kansas and Iowa; plants in the northern part of the range may have escaped from cultivation.

SIMILAR SPECIES: *L. sempervirens* differs from other similar species by having wide-spreading corolla lobes.

CONSERVATION: G5
Globally secure.

■ NATIVE
■ RARE OR EXTIRPATED
■ INTRODUCED

ALTERNATE NAMES: Longhorn Seablush

White Plectritis is a western native annual herb common in vernally moist habitats. It has pale pink to whitish flowers that occur in dense, head-like clusters.

DESCRIPTION: This erect annual grows 6–24" (15–60 cm) tall. The basal leaves almost surround the smooth stem. Each leaf is ½–2" (1–5 cm) long and up to ⅔" (1.5 cm) wide, paired, and opposite. The leaves appear widely spaced, broadly oval, and stalkless. It has small white or pink spike-like flower clusters at the top, with five lobes.

FLOWERING SEASON: March–June.

HABITAT: Open to partially shaded, seasonally wet areas such as wet meadows.

RANGE: British Columbia south to California, west to Montana and Utah.

CONSERVATION: G5

Globally secure but vulnerable in Montana.

- NATIVE
- RARE OR EXTIRPATED
- INTRODUCED

ALTERNATE NAMES: Feverwort, Tinker's-weed

This eastern native perennial produces a few flowers in small clusters in the axils of the upper leaves. As the common name suggests, the seeds of this plant resemble coffee beans. The fruits can be dried and roasted as a coffee substitute.

DESCRIPTION: This rough plant has hairy, sticky stems and grows to a height of 2–4' (60–120 cm) tall. The leaves are lanceolate to ovate in shape, about 4–10" (10–25 cm) long each and unstalked. The bases of the paired leaves are so united they appear to be fused by the stem. Clusters of two to six small, tubular, red to greenish flowers appear in the upper leaf axils. The flowers are velvety and about ¾" (2 cm) long each. There are five long sepals covered in downy hairs and a five-lobed corolla. The fruit is a hairy, yellow-orange berry with three seeds.

FLOWERING SEASON: May–July.

HABITAT: Open, rocky woods and thickets.

RANGE: Ontario, Quebec, New York, and Massachusetts south to Georgia, west to Oklahoma, and north to Nebraska and Minnesota.

SIMILAR SPECIES: Horse Gentian (*T. aurantiacum*) is closely related but its stem is smooth, not hairy or sticky. Narrow-leaved Horse Gentian (*T. angustifolium*) is also similar but has yellow flowers and leaves about half the size of Wild Coffee's, about 2" (5 cm) long.

CONSERVATION: G5

Globally secure but critically imperiled in Canada. It is critically imperiled in Delaware, Louisiana, Massachusetts, Rhode Island, and Ontario. It is vulnerable in Georgia, Kentucky, New Jersey, New York, North Carolina, and Oklahoma.

EDIBLE VALERIAN *Valeriana edulis*

■ NATIVE
■ RARE OR EXTIRPATED
■ INTRODUCED

ALTERNATE NAMES: Hairy Valerian, Tobacco Root

This perennial is native to western and central North America. As the common name suggests, the large, carrot-like taproot of this plant is edible. This species is polygamo-dioecious, meaning it has bisexual (perfect) flowers with male (staminate) flowers on some plants and bisexual flowers with female (pistallate) flowers on other plants. The flowers provide nectar to a variety of small pollinators like solitary bees, flies, and moths.

DESCRIPTION: This plant grows 1–4" (2.5–10 cm) tall with erect stems. There are multiple stems at the base, covered with a few scattered hairs, often growing in rows. Most of the leaves are basal and spatula-shaped, about 3–12" (7.5–30 cm) long and narrow, with nearly parallel veins. The leaf margins are covered in short, white hairs. There are a few upper leaves that are opposite and smaller than the basal leaves. The upper leaves also have three to nine irregular, pinnate lobes. The flowers are white and grow in vertical panicles with irregular shapes. The flowers have five petals fused together. The lobes are lance- to oval-shaped, at first spreading and then curling back.

FLOWERING SEASON: May–June.

HABITAT: Wet meadows, moist prairies.

RANGE: British Columbia south along the eastern edge of the Cascades and beyond to Arizona and New Mexico, and east through Montana and South Dakota to Iowa and the Great Lakes region.

CONSERVATION: G5

Globally secure but imperiled in Canada. It is critically imperiled in Ontario and imperiled in British Columbia, Iowa, Minnesota.

■ NATIVE
■ RARE OR EXTIRPATED
■ INTRODUCED

ALTERNATE NAMES: Great Angelica, Alexanders

This eastern native perennial was formerly called *Herba angelica*. It is named for its extensive use as a purification herb in many Native American cultures. The plant has been used medicinally to treat various ailments ranging from pain to panic attacks. Many Native American groups also used the plant ceremonially. The young stems and leaf stalks can also be eaten, and parts of the plant can be boiled to make sweets. This plant is a larval host to Short-tailed Swallowtail (*Papilio brevicauda*) butterflies.

DESCRIPTION: This large plant grows 3–10' tall (91–304 cm) with hollow stems that are smooth and purple (or purple spotted), growing from a taproot. The leaves are compound, with toothed leaflets. Each leaf is up to 4" (10 cm) long, on a clasping stalk. The tiny flowers are greenish-white to white, appearing in umbrella-like compound umbels. Each umbel is 4–10" (10–25 cm) wide.

FLOWERING SEASON: June–September.

HABITAT: Wet areas, swamps, and streambanks.

RANGE: Eastern Canada south to Delaware, and west to Iowa; also widely scattered in North Carolina, Kentucky, and Tennessee.

CONSERVATION: G5

Globally secure but critically imperiled in Delaware, Kentucky, Rhode Island, and Tennessee; imperiled in North Carolina and Prince Edward Island; and vulnerable in Iowa and Nova Scotia. It is possibly extirpated in Maryland.

WILD SARSAPARILLA *Aralia nudicaulis*

■ NATIVE
■ RARE OR EXTIRPATED
■ INTRODUCED

This is a widespread native perennial with creeping underground stems. The species name comes from the Latin words *nudus* and *cauli*, meaning "naked" and "stalk," and refers to the leafless flower stalk. At one time the rhizomes of this plant were used as a substitute for sarsaparilla (*Smilax* spp.) to flavor root beer.

DESCRIPTION: This plant grows to a height of 8–20" (20–50 cm) with a leafless stalk. The leaves have long stalks that rise above the flowers. Each leaf is 8–20" (20–50 cm) long and divided into three sections, each with three to five ovate, finely toothed leaflets. A rounded cluster of greenish-white flowers appears at the top beneath one large, umbrella-like leaf. Each flower cluster is 1½–2" (4–5 cm) wide. There are five tiny, flexed petals and five green stamens. Purple-black, clustered berries appear after the flowers.

FLOWERING SEASON: July–August.

HABITAT: Upland woods.

RANGE: Alberta east to Newfoundland, south to Georgia, and northwest to Nebraska and North Dakota; also in the northwestern United States and British Columbia.

SIMILAR SPECIES: American Ginseng (*Panax quinquefolius*) resembles this plant but has palmate leaflets and red berries. Devil's Walkingstick (*A. spinosa*) looks similar but grows as a large shrub or small tree, and its leaves and stems are covered with spines. Bristly Sarsaparilla (*A. hispida*) grows to about half the size of Wild Sarsaparilla and is bristly only at the base. Spikenard (*A. racemosa*) is much larger with more leaflets and it lacks spines.

CONSERVATION: 🅖5

Globally secure but critically imperiled in South Carolina; imperiled in Georgia, Kentucky, Missouri, and Wyoming; vulnerable in Colorado, Nebraska, and Yukon.

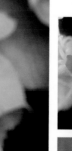

■ NATIVE
■ RARE OR EXTIRPATED
■ INTRODUCED

ALTERNATE NAMES: Wild Chervil

Spreading Chervil is a native of eastern and central North America that produces small white flowers in the spring and early summer. Native Americans used the poisonous roots of this annual to induce vomiting.

DESCRIPTION: This plant grows to a height of 4–16" (10–40 cm) with an erect or drooping main stem that's hairy or shiny. It has fern-like leaves that are alternately arranged and pinnately divided, with oblong leaflet lobes. The flowers are white and very tiny, appearing in compound umbels. Each umbel has about three umbellets, and each umbellet has three to seven flowers. The flowers are about ⅛" (3 mm) wide, with five white petals and five stamens. The fruit is usually glabrous, with ridges that are broad and flat, appearing widest in the middle.

FLOWERING SEASON: May–July.

HABITAT: Open woods, thickets, roadsides, streambanks.

RANGE: Ontario to Florida; New York to Nebraska; North Carolina to Oklahoma, and Washington D.C.

SIMILAR SPECIES: The foliage of Spreading Chervil is similar to that of cultivated Garden Parsley (*Petroselinum crispum*) and Chervil (*Anthriscus cerefolium*), sometimes called French parsley. Cultivated Chervil and the introduced Bur Chervil (*A. sylvestris*) don't have bracts beneath the umbellets. Southern Wild Chervil (*C. tainturieri*) is also similar but the fruit is a different shape — it is widest toward the bottom, whereas Spreading Chervil's fruits are widest in the middle.

CONSERVATION: **G5**

Globally secure but critically imperiled in Canada. It is critically imperiled in New York and Wisconsin, imperiled in Delaware and Georgia, and vulnerable in North Carolina.

WATER HEMLOCK *Cicuta maculata*

■ NATIVE
■ RARE OR EXTIRPATED
■ INTRODUCED

ALTERNATE NAMES: Spotted Water Hemlock, *Cicuta mexicana*

Water Hemlock is usually a large, highly branched plant appearing in wet meadows and swamps, but it can be smaller and found growing among floating vegetation. Water Hemlock is one of the most toxic plants native to North America, and can be fatal to humans and other animals if consumed. Some other hemlocks are said to be edible, but these plants are easily confused, so it is best to avoid all aquatic plants that have white flower clusters.

DESCRIPTION: This smooth, erect, branching plant reaches 3–6½' (90–200 cm) tall with hollow, purple stems. The lower leaves are sharply pointed, about 1' (30 cm) long, and doubly divided. The leaflets are toothed, with veins ending at notches between the teeth. It has dome-shaped, loose compound umbels of small white flowers. Each compound umbel is about 3" (7.5 cm) wide and flattened, without bracts beneath. Each flower is about ⅛" (3 mm) long. The fruit is round and flat, with thick ridges.

FLOWERING SEASON: June–September.

HABITAT: Wet meadows, thickets, and freshwater swamps.

RANGE: Throughout North America.

CAUTION: All parts of this plant are extremely poisonous; consuming only a small quantity can cause death. The roots have been mistaken for parsnips and other common root crops, with fatal results. Cattle, horses, and sheep have also died from grazing on this plant.

CONSERVATION: G5

Globally secure but critically imperiled in Florida; vulnerable in Arizona, Wyoming, and Yukon.

■ NATIVE
■ RARE OR EXTIRPATED
■ INTRODUCED

ALTERNATE NAMES: Deadly Hemlock, Poison Parsley, Spotted Hemlock, European Hemlock, California Fern, Nebraska Fern

This European native is introduced and widespread in North America. This biennial produces fern-like leaves the first year, with flowers appearing in the second year. As the common names suggest, it is highly toxic, killing livestock and other mammals that eat even a small amount of the plant. It can also kill humans or cause serious and permanent illness if eaten. It's believed an extract of this plant was used to execute condemned prisoners in ancient Greece, including famously the philosopher Socrates.

DESCRIPTION: This plant grows 2–10' (60–300 cm) tall with very branched spotted purple stems. The leaves are compound and triangular in shape. The leaves are fern-like and sharply divided with numerous deeply lobed leaflets. Leaf veins extend to the tips of the teeth. The upper leaf stalks are shorter than the lower leaf stalks. Small compound umbels of white flowers appear. Each umbel is about 1½–2" (4–5 cm) wide, with inconspicuous bracts beneath. Each flower is about 1⁄16" (1.5 mm) long, with five petals. The fruit is seed-like, about 1⁄8" (3 mm) long, with a rough surface.

FLOWERING SEASON: June–August.

HABITAT: Waste places, weedy areas, and woodland borders.

RANGE: Introduced throughout most of temperate North America.

CAUTION: All parts of this plant are extremely poisonous and may be fatal if consumed. Handling the plant can also cause skin irritation.

SIMILAR SPECIES: Water hemlocks (*Cicuta* spp.) are similar but their leaves are simply toothed rather than deeply lobed and toothed. The veins on the leaflets of water hemlocks extend only to the notches between teeth rather than running to the tips of the teeth.

CONSERVATION: **G5**

This species is globally secure but it is introduced in North America and considered a noxious weed in many areas.

ALTERNATE NAMES: Wild Carrot

This widely introduced European biennial is related to the cultivated carrot; many botanists believe the cultivated carrot is derived from Queen Anne's Lace. The plant's long taproot can be cooked and eaten in the first year. This plant appears attractive, but is an aggressive weed. When mature, the compound umbel curls inward, resembling a bird's nest. The leaves smell like carrots when crushed.

DESCRIPTION: This plant grows to a height of 1–3½' (30–100 cm). The leaves are deeply lobed and fern-like. Each leaf is 2–8" (5–20 cm) long. Compound umbels of tiny, cream-white flowers appear at the top. Each umbel has a flat top, with 20–90 umbellets and each umbellet has 15–60 flowers with five white petals. There's usually one dark reddish-brown to purplish flower appearing in the middle of each umbel. Each compound umbel has a flat top and is about 3–5" (7.5–12.5 cm) wide, with stiff, leaf-like bracts beneath, divided into three forks. The fruit is bristly and reddish, unlike other members of the carrot family.

FLOWERING SEASON: May–October.

HABITAT: Dry fields and waste places.

RANGE: Introduced throughout most of North America, except the Far North.

CAUTION: Contact with the leaves may cause skin irritation, especially when they are wet.

SIMILAR SPECIES: Small Wild Carrot (*D. pusillus*) is similar but smaller and hairier. The bracts underneath the umbels are also leafier.

CONSERVATION:
The conservation status of this introduced plant is not ranked in North America.

■ NATIVE
■ RARE OR EXTIRPATED
■ INTRODUCED

ALTERNATE NAMES: Button Snakeroot, Button Eryngo

This perennial is native to the central and southeastern U.S. It is distinguished from other members of the carrot family by its basal rosettes of parallel-veined leaves. Its Latin name is a reference to the leaves, which look like those of yuccas (*Yucca* spp.). It was formerly used to treat rattlesnake bites.

DESCRIPTION: This plant has smooth, rigid stems and grows 2–6' (60–180 cm) tall. The leaves are linear, with sharp points, parallel veins, and spiny edges. Each leaf is about 3' (90 cm) long and clasps the stem. The flower heads have many tiny greenish-white flowers appearing in branched clusters. The flowers look bluish at maturity. The flower head is ¾" (2 cm) wide, ovoid, and surrounded by large whitish, pointed bracts.

FLOWERING SEASON: July–August.

HABITAT: Prairies, open woods, and thickets.

RANGE: Minnesota east to Michigan and Virginia, south to Florida, west to Texas, and north to Nebraska; also scattered up the Atlantic Coast to Connecticut.

CONSERVATION:

Globally secure but critically imperiled in Nebraska, imperiled in Michigan and Virginia, and vulnerable in Minnesota and Ohio. It is possibly extirpated in Maryland.

■ NATIVE
■ RARE OR EXTIRPATED
■ INTRODUCED

This Eurasian plant is introduced in the East and on the West Coast, and can be highly weedy and aggressive in some areas. English Ivy can disrupt the natural habitat of other plants. It also carries Bacterial Leaf Scorch (*Xylella fastidiosa*), a plant pathogen that is toxic to elms, oaks, maples, and other native plants. The genus name is from the Latin word for "ivy," while the specific epithet *helix* means "spiral" or "twisted" in Greek, referring to the plant's twisting vine form.

DESCRIPTION: This evergreen vine grows to a height of 65–100' (20–30 m) and may reach 1' (30 cm) wide. It has small rootlike structures, which exude a sticky substance that helps the plant climb surfaces. The leaves are dark green with white veins and alternately arranged.

Each leaf reaches 2–4" (5–10 cm) long and is waxy or leathery, usually with three to five lobes. Unlobed, rounded leaves are often found on mature plants in full sun. The flowers on mature plants are tiny and greenish-white, appearing in small clusters. It has tiny black berries.

FLOWERING SEASON: September–October.

HABITAT: Woodlands, forest edges, fields, hedgerows, coastal areas, saltmarsh edges.

RANGE: Introduced in Ontario and the eastern United States, as well as the western U.S. and British Columbia.

CAUTION: The foliage, sap, and berries are mildly toxic to humans and other mammals if ingested. The sap can cause skin irritation.

CONSERVATION:
The conservation status of this introduced plant is not ranked in North America. This plant is often an aggressive weed and is considered invasive in some areas.

■ NATIVE
■ RARE OR EXTIRPATED
■ INTRODUCED

ALTERNATE NAMES: American Cow Parsnip, *Heracleum lanatum*

Cow Parsnip is a widespread native perennial that is weedy in some areas. Its large flowers provide a good source of nectar. Native Americans ate the young stems, leaf stalks, and roots; however, chemicals in this plant may cause skin irritation. Today, this plant is too easily confused with native and introduced hemlocks, which are fatally poisonous, to consider eating plants collected from the wild.

DESCRIPTION: This tall, leafy plant grows to 10' (3 m) in height. The leaves are round, about 6–16" (15–40 cm) long, and divided into three lobes, with coarsely toothed edges. Large umbels of tiny, white flowers appear at the top of the plant. Each umbel is about 1' (30 cm) wide, and they often grow in groups. There are five petals. The petals at the edge of the umbel are larger, about ¼" (6 mm) long and cleft in the middle. The fruit is oval and has four dark lines extending halfway down, alternating with three ribs.

FLOWERING SEASON: February–September.

HABITAT: Moist areas.

RANGE: Throughout North America except in the extreme southeast and south-central United States.

CAUTION: This plant can cause skin irritation if touched. The flowers look like those of Water Hemlock (*Cicuta maculata*), which is extremely poisonous.

SIMILAR SPECIES: The flowers of this plant look like those of the extremely poisonous Water Hemlock (*C. maculata*).

CONSERVATION: G5

Globally secure but critically imperiled in Georgia and Kansas; imperiled in Tennessee; and vulnerable in Delaware, Maryland, North Carolina, and Labrador. It is possibly extirpated in Kentucky.

MANYFLOWER MARSH-PENNYWORT *Hydrocotyle umbellata*

ALTERNATE NAMES: Dollarweed, Pennywort, Marsh Penny-wort

Manyflower Marsh-pennywort is a vigorous North American native plant that looks like a lily pad and forms dense mats in shallow water, mud, or marshes. It is principally found along the Gulf and Atlantic coasts but can occur inland in suitable habitats. It can become problematic in some aquatic habitats outside its native range, with its floating leaves blocking sunlight and cooling water temperatures.

DESCRIPTION: This aquatic perennial reaches 3–8" (8–20 cm) tall. The leaves are shiny, bright green, circular, and simple, up to 2" (5 cm) wide on petioles up to 6" (15 cm) long. The leaf edges are scalloped. Clusters of 5–10 white, star-shaped flowers appear at the ends of leafless stalks.

FLOWERING SEASON: April–October.

HABITAT: Shallow water, wet ground, often in sandy soil, mud, or clay.

RANGE: Nova Scotia south to Florida, west to Texas, and north to Minnesota. Also in California.

CAUTION: Although purported to be edible in some sources, ingesting the leaves may cause nausea, especially in children. These plants can also absorb toxic substances, such as pesticides and pollutants, from their surroundings.

SIMILAR SPECIES: Water Pennywort (*Hydrocotyle ranunculoides*) is similar but has notched leaves with a distinctive red spot in the center.

CONSERVATION: Ⓖ5
Globally secure but critically imperiled in Canada. It is critically imperiled in Connecticut, Ohio, and Nova Scotia. It is imperiled in New York. It is possibly extirpated in Pennsylvania.

■ NATIVE
■ RARE OR EXTIRPATED
■ INTRODUCED

ALTERNATE NAMES: Chocolate-tips, Fernleaf Biscuitroot

As the common name suggests, this western native perennial has lacy, fern-like foliage. It provides food and nectar for caterpillars and butterflies.

DESCRIPTION: This bushy plant grows to a height of 1–5' (30–150 cm) with purple stems. The leaves are large and fern-like. It has a long, thick taproot. Clusters of purple or yellow flower heads branch from the tops of the stems. The flowers are very tiny and arranged in unevenly branched umbels. Each umbel is about 4" (10 cm) wide. The fruits resemble pumpkin seeds.

FLOWERING SEASON: April–June.

HABITAT: Roadsides, dry slopes.

RANGE: British Columbia to Saskatchewan south to California, Arizona, and Colorado.

CONSERVATION:  G4

Apparently secure but critically imperiled in Saskatchewan and vulnerable in Montana and Wyoming.

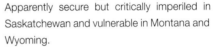

DWARF GINSENG *Nanopanax trifolius*

- ■ NATIVE
- ■ RARE OR EXTIRPATED
- ■ INTRODUCED

ALTERNATE NAMES: *Panax trifolius*

Dwarf Ginseng is a native perennial found in eastern North America. Native Americans brewed this whole plant as a tea to treat indigestion, hives, and other ailments. The root was also chewed for headaches and fainting. Its tubers are edible and can be eaten raw or boiled. The specific epithet *trifolius* means "three leaves" and refers to the compound leaves. The genus *Nanopanax* has been proposed to differentiate this plant from typical ginsengs in genus *Panax*; many sources list this species as *Panax trifolius*.

DESCRIPTION: This plant grows to a height of 4–8" (10–20 cm). A whorl of three compound leaves appears at the bottom of the plant. There are three to five toothed leaflets that are stalkless and ovate in shape. Each leaflet is 1–1½" (2.5–4 cm) long. An umbel of tiny, dull whitish flowers appears at the top. Each flower is about ¹⁄₁₆" (1.5 mm) wide, with five petals. The fruit is yellowish berries clustered together.

FLOWERING SEASON: April–June.

HABITAT: Moist woods and damp clearings.

RANGE: Ontario east to Nova Scotia, south to Georgia, and northwest to Indiana and Minnesota.

SIMILAR SPECIES: American Ginseng (*Panax quinquefolius*) is a larger plant and has five-stalked leaflets.

CONSERVATION: G5

Globally secure but critically imperiled in Georgia; imperiled on Prince Edward Island; and vulnerable in Indiana, North Carolina, Virginia, New Brunswick, and Quebec.

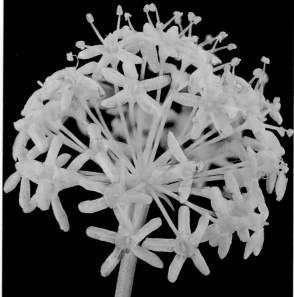

■ NATIVE
■ RARE OR EXTIRPATED
■ INTRODUCED

ALTERNATE NAMES: Smooth Sweet-cicely, Aniseroot

The flowers, leaves, and roots of this widespread native perennial have a strong anise scent, which distinguishes it from other members of the carrot family. Its tuberous root has been used as a substitute for black licorice.

DESCRIPTION: This plant has light green to reddish stems covered in hairs or glabrous, about 1–2½' (30–75 cm) tall. The leaves are hairy, alternately arranged, and compound. Each compound leaf appears divided into three leaflets, which are 1–4" (2.5–10 cm) long, and lanceolate to oval in shape. The margins are coarsely toothed. The lower leaves are 9" (22.5 cm) long, becoming smaller up the stem. White flowers appear in compound umbels on the upper stems. Each umbel has three to six umbellets, about 2" (5 cm) long. The umbellets contain 7–16 flowers clustered together. Each flower has five white petals and five white stamens.

FLOWERING SEASON: April–July.

HABITAT: Moist areas; rich, alluvial woods and wooded slopes.

RANGE: Alberta south to New Mexico and east to the Atlantic.

SIMILAR SPECIES: Sweet Cicely (*O. claytonii*) is similar but has only four to seven flowers per umbellet instead of 7–16; also has shorter fruit.

CONSERVATION:

Globally secure but critically imperiled on Prince Edward Island and in Rhode Island; imperiled in Nova Scotia and New Brunswick; and vulnerable in Alberta, Mississippi, and Wyoming.

NATIVE
RARE OR EXTIRPATED
INTRODUCED

Wild Parsnip is a Eurasian native biennial from which cultivated parsnip—grown as a root vegetable since ancient times—is derived. Wild Parsnip looks and smells like cultivated parsnip. In North America, this highly invasive plant is a problem in fields, roadsides, and disturbed areas where it's invaded native habitats and caused population decline among native plants. The thick taproot can be eaten, but the leaves, stems, and flowers can cause skin irritation, especially when the affected area is exposed to sunlight.

DESCRIPTION: This weedy plant grows to a height of 5' (1.5 m) with ridged stems. The leaves are pinnately compound, with 5–15 ovate, toothed, or lobed leaflets. It produces numerous umbels of tiny yellow flowers that appear in flat, broad clusters with five petals. The fruits are dry, smooth, and somewhat winged.

FLOWERING SEASON: May–June.

HABITAT: Open disturbed habitats, fields, edges of prairies, roadsides, waste places.

RANGE: Eurasia native; introduced in North America and naturalized widely; absent from Georgia to Mississippi and Florida.

CAUTION: This species can be easily confused with the extremely poisonous Poison Hemlock (*Conium maculatum*) and water hemlocks (*Cicuta* spp.); plants collected in the wild should not be eaten. Handling this plant can cause skin irritation. Exposure to sunlight after contact can cause blistering and discoloration of the skin.

SIMILAR SPECIES: Yellow-Pimpernel (*Taenidia integerrima*) is similar, but is less invasive and tends to grow shorter. This species can be confused with the extremely poisonous Poison Hemlock (*Conium maculatum*) and Western Water Hemlock (*Cicuta douglasii*) but Wild Parsnip has a parsnip-like smell and prefers drier soils.

CONSERVATION:
The conservation status of this introduced plant is not ranked in North America. This introduced species is widespread and considered a weed and/or invasive in most places.

■ NATIVE
■ RARE OR EXTIRPATED
■ INTRODUCED

This widespread native perennial has pom-pom-like, greenish-white flower heads that attract various small insect pollinators. Its perfect flowers are nearly stalkless and have a small ovary covered in hooked bristles, with two long styles protruding beyond the bristles.

DESCRIPTION: This perennial grows to a height of 1–4' (30–120 cm). The leaves are alternately arranged on the stem. Both the basal leaves and stem leaves are palmately compound, with five or seven leaflets. Each leaflet is about 6" (15 cm) long and 2" (5 cm) wide, hairless, and deeply toothed or double toothed. The leaves are usually widest in the middle of the stem. The basal and lower stem leaves have long stalks. Pom-pom-like umbels appear at the ends of branching stems. Each umbel has two to four umbellets, about ½" (13 mm) wide, and is made up of 20–60 greenish-white flowers. The flowers have five petals

and protruding stamens with greenish-white tips, turning brown at maturity. The fruit is dry, oval, and covered in bristles.

FLOWERING SEASON: June–July.

HABITAT: Woods, forests, shady areas.

RANGE: Most of North America, except the Far North, the southwestern states, and the south-central U.S. It is rarer in the South and West.

CONSERVATION: G5
Globally secure but critically imperiled in Delaware; imperiled in Kentucky; and vulnerable in Newfoundland, Prince Edward Island, Maryland, Idaho, Washington, and Wyoming. It is possibly extirpated in Louisiana.

■ NATIVE
■ RARE OR EXTIRPATED
■ INTRODUCED

ALTERNATE NAMES: Hemlock Water Parsnip

This is a widespread native perennial with leaves that vary based on its environment. When growing in shallow water, the aquatic leaves grow in clusters; on moist ground the plant grows a basal rosette. The roots of this plant have been boiled and eaten as a cooked vegetable, but because the plant resembles the deadly Water Hemlock (*Cicuta maculata*), it should not be consumed.

DESCRIPTION: This aquatic or wetland plant grows to a height of 2–6' (60–180 cm) with heavily ridged stems. The leaves are pinnately compound and 4–10" (10–25 cm) long each. The basal leaves are often submerged and finely divided into 5–17 lanceolate, toothed leaflets, each about 2½–5½" (6.5–14 cm) long.

The tiny flowers are dull white and grow in compound umbels. Each umbel is about 2–3" (5–7.5 cm) wide, with narrow, leaf-like bracts beneath. The fruit is tiny and ovate with prominent ribs.

FLOWERING SEASON: July–September.

HABITAT: Wet meadows and thickets and muddy shorelines.

RANGE: Throughout North America, except in the Arctic. It is rare in the Southeast.

CAUTION: This species can be easily confused with the extremely poisonous Poison Hemlock (*Conium maculatum*) and water hemlocks (*Cicuta* spp.); plants collected in the wild should not be eaten.

CONSERVATION: G5

Globally secure but critically imperiled in Arizona, Louisiana, Kansas, Texas, and Labrador; imperiled in Arkansas and Wyoming; and vulnerable in Georgia, Kentucky, North Carolina, West Virginia, Newfoundland, and Yukon.

■ NATIVE
■ RARE OR EXTIRPATED
■ INTRODUCED

Yellow-Pimpernel is a widespread native of eastern North America. It has smooth leaves that help distinguish it from other similar plants; unlike other members of the carrot family, this plant's leaflets don't have lobes or teeth. The compound umbels are unusually open and airy in appearance.

DESCRIPTION: This perennial grows 1–3' (30–90 cm) tall with green to brownish stems, sometimes branching, but usually round, hairless, and somewhat glaucous. The leaves are alternately arranged and compound. The base leaves are largest, becoming progressively smaller up the plant. The leaves are often divided two or three times into egg-shaped or elliptical non-toothed leaflets about 1" (2.5 cm) long and ½" (12.5 mm) wide. Some of the upper stems terminate in tiny, yellow flowers in compound umbels. Each umbel is about 4–7" (10–17.5 cm) long and widely spaced. The umbel consists of 12–15 umbellets. Each umbellet has about 12 flowers, which are about ⅛" (3 mm) wide.

FLOWERING SEASON: May–July.

HABITAT: Woods, rocky hillsides, edges of prairies.

RANGE: Eastern North America from Ontario and Quebec south to Georgia, west to easternmost Kansas and Texas.

SIMILAR SPECIES: Wild Parsnip (*Pastinaca sativa*)—an introduced and often invasive plant found on roadsides, fields, and disturbed areas—looks similar but is taller.

CONSERVATION: G5

Secure but critically imperiled in Connecticut, Delaware, Mississippi, Texas, and Quebec; imperiled in Louisiana, Kansas, and Vermont; and vulnerable in Iowa, Minnesota, New Jersey, and North Carolina. It is possibly extirpated in South Dakota.

■ NATIVE
■ RARE OR EXTIRPATED
■ INTRODUCED

ALTERNATE NAMES: Golden Zizia

This plant native to eastern and central North America is distinguished from other members of the carrot family by its lack of a flower stalk on the central flower of each umbel. The species is named for the gold-colored flowers. The genus is named for the 19th-century German botanist Johann Baptist Ziz.

DESCRIPTION: This plant grows 8–24" (20–60 cm) tall. The basal leaves are 1–4" (2.5–10 cm) long and long-stalked. The leaves are ovate in shape and indented at the base with toothed margins. The upper leaves are pinnately divided into three segments with several narrow lobes and toothed edges. Bright yellow flowers in compound umbels appear among leafy stems. The umbel is up to 2½" (6.5 cm) wide. There are five sepals and five petals. The fruit is elliptical and flattish, with low ribs.

FLOWERING SEASON: April–June.

HABITAT: Meadows, shorelines, moist woods, and thickets.

RANGE: Manitoba east to Nova Scotia and south to the Gulf of Mexico.

SIMILAR SPECIES: Prairie Golden Alexanders (*Zizia aptera*) is similar but has simple, heart-shaped basal leaves.

CONSERVATION: G5

Globally secure but critically imperiled in Nova Scotia and Prince Edward Island; imperiled in Delaware, Rhode Island, and Washington D.C.; and vulnerable in Maryland.

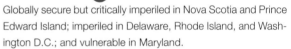

WILDFLOWER FAMILIES

GLOSSARY

INDEX

PHOTOGRAPHY CREDITS

ACKNOWLEDGMENTS

COPYRIGHT

WILDFLOWER FAMILIES

Cabombaceae — Water-shield

This small family includes two genera and six species of aquatic plants formerly in the family Nymphaeaceae.

Nymphaeaceae — Water Lily

This is a family of about 70 aquatic perennials in temperate and tropical climates. Most of the plants float on the surface of water, although some have leaves that grow underwater.

ROCKY MOUNTAIN POND-LILY
(*NUPHAR POLYSEPALA*)

Saururaceae — Lizard's-tail

Saururaceae is a small family of six species native to North America and Southeast Asia. Most of the plants are herbs with creeping rhizomes, generally found in wetlands.

Aristolochiaceae — Birthwort

This family includes more than 500 species, most of which are woody vines that grow in tropical or temperate regions. Some of these plants are used for medicinal purposes. Former families Hydnoraceae and Lactoridaceae are now considered part of Aristolochiaceae.

Acoraceae — Sweetflag

This family contains two species in a single genus, *Acorus*, which was once placed in the Araceae family but was later given its own family. Sweetflags are perennial wetland plants with small flowers. The plants are stalkless with long, narrow leaves that grow from rhizomes.

WATER ARUM (*CALLA PALUSTRIS*)

Araceae — Arum

This family includes about 1,800 species. Most of the plants are small perennial herbs that grow in the water or wetlands, but a few grow on land in tropical areas. The tiny flowers grow in a cluster on a spike, or spadix, sometimes embedded in a leaf-like structure called a spathe.

Alismataceae — Water Plantain

This family includes more than 100 species of flowering plants that are mostly found in aquatic or wetland areas in temperate regions. The leaves may grow out of the water on long stalks or float on the surface. The flowers appear in whorls around the stem, which is usually erect.

WATER-PLANTAIN (*ALISMA TRIVALE*)

CALIFORNIA CORN LILY (*VERATRUM CALIFORNICUM*)

FAIRY SLIPPER (*CALYPSO BULBOSA*)

Melanthiaceae — Bunchflower

The Melanthiaceae family includes about 170 species of large and colorful perennial herbs that grow from bulbs or rhizomes. The leaves are usually clustered at the base of the plant. Species in this family were formerly considered to be part of the family Liliaceae.

Colchicaceae — Flame Lily

The Flame Lily family includes about 285 species of perennial herbs that grow from a corm or a rhizome. These plants grow in temperate and tropical climates and are known to contain alkaloids. Species in this family were formerly considered to be part of Liliaceae.

Smilacaceae — Catbrier

This family includes about 255 species of herbaceous or woody plants that grow in subtropical and tropical areas. This family is similar to the Liliaceae family, but differs in its leaf characteristics. The veins of the leaves are net-shaped, rather than parallel.

Liliaceae — Lily

This family includes just over 600 species of perennials that grow from rhizomes or bulbs. Liliaceae is a smaller family than it used to be as many species formerly considered to be part of the Liliaceae are now placed in other families. The family includes the lilies and tulips that are popular cultivated plants.

Orchidaceae — Orchid

This is one of the largest and most diverse families of flowering plants, with about 26,000 species. Orchids are perennials with upright stems and colorful flowers. They are often grown as ornamentals.

Hypoxidaceae — Star-Grass

There are about 160 species in this family, most of which are perennial herbs with grass-like leaves that grow from an underground stem. This family was once part of Liliaceae.

Iridaceae — Iris

Irises are usually perennial, growing from bulbs, corms, or rhizomes. There are nearly 2,250 known species. Most of these plants have parallel veins on long, straight leaves. The flowers are very showy and colorful. These plants can be found in nearly every type of habitat.

SOUTHERN BLUEFLAG (*LIMNIRIS VIRGINICA*)

WILDFLOWER FAMILIES

Asphodelaceae — Asphodel

This family includes about 900 diverse species found in the tropics and subtropics. Commercially important species in the family include aloe plants, which are used in medicine and beauty products.

Amaryllidaceae — Amaryllis

There are about 1,600 species in this family. Some of these plants are grown as ornamentals. They are typically found in the tropics or subtropics and most plants grow from bulbs.

Alliaceae — Onion

Species in the Onion family are perennial herbs with leaves that grow from a bulb at the base of the plant. The leaves have a characteristic onion-like or garlic-like smell when crushed. The family was formerly part of the Liliaceae family; some sources treat this group as a subfamily within the family Amaryllidaceae.

NODDING ONION (*ALLIUM CERNUUM*)

Asparagaceae — Asparagus

There are about 2,900 known species of Asparagaceae. Most of them are perennial herbs and have leaves with parallel veins. Species in this family were formerly included in the Liliaceae family.

Commelinaceae — Spiderwort

This is a diverse family of about 730 species of annual or perennial herbs, including dayflowers (genus *Commelina*) and spiderworts (genus *Tradescantia*). These species are usually found in the tropics. Many of these plants have soft stems.

OHIO SPIDERWORT (*TRADESCANTIA OHIENSIS*)

Pontederiaceae — Water Hyacinth

This family includes about 34 species, most of which are aquatic herbs that float on water or grow in mud in tropical and subtropical areas.

Typhaceae — Cattail

This family includes about 50 species, some of which were formerly in the Sparganiaceae family. These plants grow in wetland areas and marshes.

Bromeliaceae — Bromeliad

This is a family of about 3,500 diverse species, many of which are found in tropical areas. The family includes both epiphytes like Spanish Moss (*Tillandsia usneoides*) and terrestrial plants, such as Pineapple (*Ananas comosus*).

Xyridaceae — Yellow-eyed Grass

This family of about 400 species grows in wetlands. These plants are distinguished by their long, slender leaves and flowers that appear in a spike at the top of the stem, resembling a cone.

Eriocaulaceae — Pipewort

This is a family of about 1,200 perennial wetland herbs that are widely distributed, particularly in tropical regions. Many of them have grass-like leaves with parallel veins and very small, clustered flowers.

Papaveraceae — Poppy

This family includes 775 species, including several of economic importance. Species in this family grow in temperate and subtropical climates, especially in the Northern Hemisphere. They include herbaceous plants, shrubs, and small trees.

CALIFORNIA POPPY (*ESCHSCHOLZIA CALIFORNICA*)

Menispermaceae — Moonseed

This is a family of about 440 species, most of which are woody climbing plants or vines in tropical areas. Some plants in this family contain alkaloids and are important medicinal plants, used as antispasmodic and anti-inflammatory remedies.

Berberidaceae — Barberry

This is a family of 700 species found throughout subtropical and temperate Africa, Eurasia, and the Americas. These plants are perennial herbs or woody shrubs with alternate leaves. Many barberry species produce small, extremely sour berries that many cultures use for food or medicinally.

Ranunculaceae — Buttercup

This family of about 2,400 species contains mostly herbaceous annuals or perennials, but some are woody climbers and shrubs. Buttercups can be found around the world.

MARSH MARIGOLD (*CALTHA PALUSTRIS*)

Nelumbonaceae — Lotus

This family was once included in the water lily family Nymphaeaceae, but genetic analysis showed these plants did not share closely related ancestors, despite their similarities. Now elevated to its own family, Nelumbonaceae is a small group of aquatic plants, with just three species.

Paeoniaceae — Peony

This family of about 33 species has just one genus — *Paeonia*. Peonies are native to Asia, Europe, and western North America. Most are herbaceous perennials, but some are woody shrubs.

WESTERN PEONY (*PAEONIA BROWNII*)

Saxifragaceae — Saxifrage

Several species formerly thought to be in this family have now been placed in the Hydrangeaceae, Parnassiaceae, or Penthoraceae families. The current Saxifragaceae family includes about 640 known species of herbaceous perennials, mostly in the Northern Hemisphere.

EASTERN FOAMFLOWER (*TIARELLA CORDIFOLIA*)

Penthoraceae — Ditch-Stonecrop

This family includes just two species, both of which are erect herbaceous perennials growing up to 2 feet tall. Species placed in this family were formerly considered to be part of the Saxifragaceae or the Crassulaceae families.

Crassulaceae — Stonecrop

This is a family of about 1,400 diverse species. Most of the plants are herbaceous, but there are some subshrubs and tree-like or aquatic plants.

Zygophyllaceae — Caltrop

This is a family of about 285 species, many of which are trees, shrubs, or herbs. They are often found in dry habitats.

SILVERY LUPINE (*LUPINUS ARGENTEUS*)

Fabaceae — Pea

This is a large family with nearly 20,000 species of vines, herbs, shrubs, trees, and lianas. These plants are widely distributed and easily recognized by their fruit, which is usually a legume.

Polygalaceae — Milkwort

This is a family of about 900 species found all over the world. These plants include herbs, shrubs, and trees.

Rosaceae — Rose

This family includes nearly 3,000 species around the world. Plants in the family include herbs, vines, shrubs, and trees, usually with leaves arranged alternately along the stem. Some species have thorns or prickles. This family includes many edible fruits, such as apples.

PRAIRIE SMOKE (*GEUM TRIFLORUM*)

BOG HEMP (*BOEHMERIA CYLINDRICA*)

Urticaceae — Nettle

This family of about 2,625 species includes mostly lianas and shrubs, and some trees. Most of these plants are found in tropical areas.

Cucurbitaceae — Cucumber

This family includes 965 species, including several used in agriculture. Many are annual vines, but some are woody lianas, thorny shrubs, or trees, found mostly in tropical areas.

Celastraceae — Staff Vine

This is a family of about 1,350 species of herbs, vines, shrubs and small trees. Most of these plants are tropical.

Oxalidaceae — Wood Sorrel

The Oxalidaceae include about 570 species of herbaceous plants, shrubs, and small trees. The leaves of these plants typically open in daylight and close at night.

Hypericaceae — St.-John's-wort

This is a family of about 700 species. These herbs and shrubs are found mostly in tropical areas. Common St.-John's-wort (*Hypericum perforatum*) is a familiar species in this family that is commercially cultivated and has been used as a folk remedy for centuries.

Euphorbiaceae — Spurge

This huge family includes around 7,500 species of mostly shrubs and some trees. Many are found in tropical areas. The family includes several species of economic importance, such as castor beans and rubber trees, as well as several popular ornamental plants.

Violaceae — Violet

This family includes about 980 species. Most of the species are shrubs or small trees found in tropical areas. Violets can be found widely around the world, but some species and genera are specialized to particular habitats and have highly restricted native distribution.

WESTERN BLUE VIOLET (*VIOLA ADUNCA*)

Passifloraceae — Passionflower

This is a family of 980 species, including lianas, trees, and shrubs, found in warm areas. Commercially important passionflowers include Passion Fruit (*Passiflora edulis*) and several popular species of garden plants.

Linaceae — Flax

This family includes 250 species of herbaceous annuals and perennials, as well as woody subshrubs, shrubs, and small trees. Most of these plants are found in temperate and tropical areas.

Geraniaceae — Storksbill

This family includes 830 herbs or subshrubs, some of which are succulent. Most of these plants grow in temperate areas. Geraniums, which are popular garden plants, are included here.

Lythraceae — Loosestrife

This family includes 620 species of mostly herbs, with some trees and shrubs. The trees often have flaky bark. The flower petals appear crumpled. These plants are widely distributed, and many are found in tropical and aquatic habitats.

Onagraceae — Evening Primrose

This family includes 656 species of herbs, shrubs, and trees. Most of these plants are found in tropical areas.

COMMON EVENING PRIMROSE
(*OENOTHERA BIENNIS*)

Melastomataceae — Melastomes

This family includes about 5,115 species of herbs, shrubs, and small trees. These plants are mostly found in tropical areas. A number of species can become invasive when they naturalize outside of their native ranges.

Sapindaceae — Soapberry

This family includes about 1,860 species, some of which were formerly placed in the Aceraceae and Hippocastanaceae families. Most are shrubs, trees, and vines. Many are found in laurel forest habitats.

Malvaceae — Mallow

This family includes species formerly placed in the Tiliaceae family. This is a family of about 4,250 species, most of which are herbaceous perennials, while some are woody.

CHEESES (*MALVA NEGLECTA*)

Cistaceae — Rock-rose

This is a family of 170 species. Many of these plants are herbs or shrubs, noted for having numerous flowers. They tend to be located in temperate areas.

Cleomaceae — Caper

This is a family of about 350 species, some of which were previously included in the family Capparaceae. Most of these are annual, although some are perennial herbaceous plants. This family includes a small number of shrubs or small trees, and some lianas. Most of these plants have glands and contain resins.

GARDEN YELLOWROCKET (*BARBAREA VULGARIS*)

Brassicaceae — Mustard

This family of 3,628 species contains popular cruciferous vegetables, such as broccoli, cabbage, cauliflower, and kale. Most are herbaceous plants, with some shrubs. They are globally distributed, and many commercially important species are widely cultivated.

Santalaceae — Sandalwood

This family of about 1,000 species now includes members of a few former families that were recently combined. These are widely distributed flowering plants, including small trees, shrubs, perennial herbs, and epiphytic climbers. Many members are partially parasitic.

Plumbaginaceae — Leadwort

This is a family of about 725 species that grow in most any environment, from arctic to tropical areas. The plants have perfect flowers and are pollinated by insects.

Polygonaceae — Buckwheat

This is a family of about 1,200 species, most of which are herbs with swollen nodes. There are also some trees, shrubs and vines included here. The family is distributed worldwide.

CUSHION BUCKWHEAT (*ERIOGONUM OVALIFOLIUM*)

Droseraceae — Sundew

This is a family of about 180 species of carnivorous plants. They are mostly perennial herbs with leaves at the base of the plant, with at least one leaf surface containing hairs and mucilage-producing glands at the tip to capture insects or other small prey.

Caryophyllaceae — Carnation

This is a family of 2,625 species. Most of them are annual, biennial or perennial herbs, including popular carnations. Most of these are found in temperate climates, with a few species growing on tropical mountains.

RAGGED ROBIN (*LYCHNIS FLOS-CUCULI*)

Amaranthaceae — Amaranth

This is a family of more than 2,000 species, including the former goosefoot family, Chenopodiaceae. They are mainly annual or perennial herbaceous or woody plants found around the world from tropical to temperate climates. Economically important amaranths include vegetables such as spinach and beets.

Aizoaceae — Ice Plant

This is a family of about 1,900 species, mostly native to southern Africa. They tend to be herbaceous or woody succulents.

Phytolaccaceae — Pokeweed

This family of 33 species includes herbs, shrubs, and trees, mostly native to tropical and subtropical North America and Africa.

Nyctaginaceae — Four-o'clock

This is a family of about 290 species of flowering plants widely distributed in tropical and subtropical regions, with a few plants in temperate regions. This family is distinguished by its fruit, called an "anthocarp," and many genera have extremely large pollen grains.

Montiaceae — Miner's Lettuce

This family of 230 species includes succulent herbaceous plants or shrubs. Representative species can be found around the world. This family was adopted by taxonomists only recently and absorbed several plants that were formerly listed in the purslane family, Portulacaceae.

Portulacaceae — Purslane

This is a family of 115 species with just one genus, *Portulaca*, with a cosmopolitan distribution. This family was formerly much larger, lumping together more than 500 species in about 20 genera, but recent genetic research has split most of these into distinct, more closely related families. These plants are very similar to the Caryophyllaceae family, but the calyx is distinguished by having just two sepals.

Cactaceae — Cactus

About 1,750 species comprise this family. Most cacti live in desert habitats and tolerate extreme drought. Almost all of them are succulents and have spikes in place of leaves.

DESERT FOUR-O'CLOCK (*MIRABILIS MULTIFLORA*)

TREE CHOLLA (*CYLINDROPUNTIA IMBRICATA*)

WILD HYDRANGEA (*HYDRANGEA ARBORESCENS*)

Hydrangeaceae — Hydrangea

Most members of this family of about 223 species are shrubs. Hydrangeas have a wide distribution, particularly in North America and Asia. Species in this family were formerly included in the Saxifragaceae family.

Loasaceae — Stickleaf

There are about 308 known species in this family. Members of the family include annual, biennial and perennial herbaceous plants, as well as a few shrubs and small trees.

Cornaceae — Dogwood

This family includes about 85 species of trees, shrubs, and sometimes herbs found all over the world. Dogwoods often have distinctive and attractive flowers, berries, and bark that make them popular in horticulture. Some of its members were formerly in the Nyssaceae family.

Balsaminaceae — Balsam

This is a family of about 1,000 species of herbaceous plants with showy flowers. It has just two genera — *Impatiens* and *Hydrocera* — with *Hydrocera* containing only one species, found in Southeast Asia. *Impatiens*, however, are found throughout the tropics and the Northern Hemisphere.

Polemoniaceae — Phlox

This is a family of about 400 species including herbs, shrubs, small trees, and vines. Some members of this family are common in deserts and woodlands. They are noted for their attractive flowers that appear for brief periods.

FALSE BABYSTARS (*LEPTOSIPHON ANDROSACEUS*)

Theophrastaceae — Theophrasta

This family contains about seven genera and around 100 species of herbs, shrubs, and trees that are often evergreen. They are found primarily in the American Neotropics. The species now considered to belong to the Theophrastaceae family were formerly treated as part of the Primulaceae. Taxonomists have gone back and forth between lumping these families together and separating them.

Myrsinaceae — Myrsine

This family includes about 50 genera and 1,000 species of herbs, subshrubs, shrubs, and trees that are mainly deciduous. They are distributed worldwide, with herbaceous species found primarily in temperate climates and woody species more prominent in the tropics. As with the related Theophrastaceae family, some sources classify the myrsines as part of the Primulaceae family and others assign them to their own family.

WILDFLOWER FAMILIES

TALL MOUNTAIN SHOOTINGSTAR (*PRIMULA JEFFREYI*)

Primulaceae — Primrose

This family of about 2,800 species consists of herbaceous and woody flowering plants, including some garden plants. Some of the species formerly grouped here are now in the Myrsinaceae and Theophrastaceae families.

Diapensiaceae — Diapensia

This is a family of about 15 species, including evergreen herbs or small shrubs with leaves that may be opposite, alternate, or may grow as a basal rosette.

Sarraceniaceae — Pitcher-Plant

This is a family of 34 species and just three genera, including *Sarracenia* (North American pitcher plants), *Darlingtonia* (the California pitcher plant), and *Heliamphora* (sun pitchers). *Sarracenia* and *Darlingtonia* are native to North America, while *Heliamphora* is native to South America.

Ericaceae — Heath

This is a family of about 4,250 species, including cranberries and blueberries. Species formerly treated in the Empetraceae, Monotropaceae, and Pyrolaceae families are now included here. Most of these plants are herbs or shrubs, often with evergreen leaves.

Rubiaceae — Madder

This is a family of more than 13,500 species that are trees, shrubs, or herbs. Most of the species are found in the tropics or subtropics, including coffee (*Coffea* spp.).

Gentianaceae — Gentian

This is a family of about 1,700 species that consists of trees, shrubs, and herbs, mostly occurring in temperate zones. Some species formerly placed in Gentianaceae are now part of the Menyanthaceae family.

MONUMENT PLANT (*FRASERA SPECIOSA*)

Apocynaceae — Dogbane

This family includes species that were formerly placed in the Asclepiadaceae family. About 5,100 species are included here, most of which are found in tropical areas. These plants include tall trees and perennial herbs or vines.

NARROWLEAF MILKWEED (*ASCLEPIAS FASCICULARIS*)

VIRGINIA BLUEBELLS (*MERTENSIA VIRGINICA*)

Boraginaceae — Borage

There are about 2,000 species in the Borage family found around the world. Most of these plants are annual or perennial herbs with stiff hairs. Species formerly placed in the Hydrophyllaceae family are now included here; however, the placement and circumscription of Boraginaceae is still uncertain.

Hydrophyllaceae — Waterleaf

This group contains about 300 species, including many that are native to western North America. Hydrophyllaceae may be treated as a subfamily of Boraginaceae, but the taxonomy of this group remains uncertain and may be further revised.

THREADLEAF PHACELIA (*PHACELIA LINEARIS*)

Heliotropaceae — Heliotrope

When treated as a distinct family, Heliotropaceae includes roughly 450 species. This group may be treated as a subfamily of Boraginaceae, but the taxonomy of these closely related plants remains uncertain and may be further revised.

SALT HELIOTROPE (*HELIOTROPIUM CURASSAVICUM*)

Convolvulaceae — Morning Glory

These plants were formerly part of the Cuscutaceae family. There are about 1,660 species in this family. Most of these plants are perennial vines, herbs, trees, and shrubs. Some species are parasitic, with leaves that look like scales.

Solanaceae — Nightshade

This family of about 2,600 species includes herbs, vines, lianas, shrubs, and trees. Many members of the family contain alkaloids and may be toxic, but other familiar nightshades—such as tomatoes, eggplants, and potatoes—are grown for food or spices and used in medicine.

Hydroleaceae — False Fiddleleaf

This is a small family of about 12 species. Hydrolea is the only genus of the family. These plants are usually perennial (rarely annual), often semi-aquatic herbs or small shrubs.

Oleaceae — Olive

This family includes close to 800 species, including the widely cultivated Olive (*Olea europaea*) native to the Mediterranean region and southwest Asia. These plants' flowers are often very fragrant. These species are found in most types of environments.

Tetrachondraceae — Juniper-Leaf

This is a small family with three species in two genera — *Polypremum* and *Tetrachondra*. These are perennial (sometimes annual) herbs native to the Americas and New Zealand.

Plantaginaceae — Plantain

This family includes about 1,900 species, some of which were formerly placed in other families, including the Scrophulariaceae. These plants include both terrestrial genera and rooted aquatics. The aquatics usually have whorled or opposite leaves and bisexual flowers.

MOTH MULLEIN (*VERBASCUM BLATTARIA*)

Scrophulariaceae — Figwort

This family includes about 1,700 species. Most of the plants included here are annual and perennial herbs, with some shrubs. They grow around the world, mostly in temperate climates. Scrophulariaceae was once a much larger family of around 5,000 species, but more recent molecular studies have shown that most of the species commonly included in earlier treatments of the family are better placed elsewhere.

Martyniaceae — Unicorn Plant

This is a small family of about 16 species, formerly included in the Pedaliaceae family. These plants have mucilaginous hairs on the stems and leaves, making them feel slimy. The fruits have hooks or horns.

RYDBERG'S PENSTEMON (*PENSTEMON RYDBERGII*)

RAM'S-HORN (*PROBOSCIDEA LOUISIANICA*)

FRINGE-LEAF WILD PETUNIA (*RUELLIA HUMILIS*)

Acanthaceae — Acanthus

This family contains about 2,500 species. Most are tropical herbs, shrubs, or twining vines; some are epiphytes. These plants are found in nearly every habitat, including forests, sea-coasts, and swamps.

Bignoniaceae — Begonias

This family includes about 800 species of mostly tropical trees, lianas, or shrubs with large, showy flowers. Most of the plants are found in subtropical areas. Some are used as ornamentals. The leaves are often opposite or whorled.

TRUMPET VINE (*CAMPSIS RADICANS*)

Lentibulariaceae — Bladderwort

This is a family of carnivorous plants, with about 315 species of aquatic or bog herbs. The genera are diverse, but one similarity is that many plants create mucilage on their leaf surfaces. The mucilage traps insects and absorbs nutrients from the plant's prey.

COMMON BLADDERWORT (*UTRICULARIA VULGARIS*)

Verbenaceae — Verbena

The Verbenaceae family contains about 1,000 tropical flowering plants, including trees, shrubs, and herbs. Most of the plants have square stems with scented flowers, usually arranged in spikes. Some species formerly considered to be part of this family have been moved to the Phrymaceae family.

BLUE VERVAIN (*VERBENA HASTATA*)

COMMON BUGLE (*AJUGA REPTANS*)

Lamiaceae — Mint

There are about 7,500 species in this family. The plants are herbs or shrubs found around the world. Most of them have four-sided stems. Many familiar plants in this group, such as mints (*Mentha* spp.) and Rosemary (*Salvia rosmarinus*), give off a strong scent when the leaves or stems are crushed.

Phrymaceae — Lopseed

This family includes about 200 aquatic or terrestrial plants found all over the world, with large populations in western North America and Australia. Members of this family are found in deserts, riverbanks, and mountains. Some species in this family were formerly considered to be part of the Scrophulariaceae or Verbenaceae.

DWARF PURPLE MONKEYFLOWER (*DIPLACUS NANUS*)

Orobanchaceae — Broomrape

This family includes about 2,000 species of mostly parasitic plants or partially parasitic plants that grow around the world, mostly in temperate regions. Parasitic plants included here have yellow and brown stems with small leaves, while partial parasitic plants have green and leafy stems. Some species were formerly included in the Scrophulariaceae family.

PURPLE GERARDIA (*AGALINIS PURPUREA*)

Campanulaceae — Bellflower

This family includes about 2,300 species of herbaceous plants and shrubs. These plants are found on every continent except Antarctica, with many species native to Hawaii. Campanulaceae grow in diverse habitats, from deserts to rainforests, to sea cliffs, the tropics, and high altitudes.

HAREBELL (*CAMPANULA ROTUNDIFOLIA*)

BUCKBEAN (*MENYANTHES TRIFOLIATA*)

TRUMPET HONEYSUCKLE (*LONICERA SEMPERVIRENS*)

Menyanthaceae — Buckbean

This is a family of about 60 species, including aquatic and wetland plants, that are found worldwide. These plants grow from a rhizome. Some buckbean species have been grown as ornamental plants but can escape and become invasive.

Caprifoliaceae — Honeysuckle

There are an estimated 825 species of Caprifoliaceae worldwide, most of which are woody shrubs or vines. There is a large population in eastern North America. The order Dipsacales to which this family belongs has gone through significant taxonomic reshuffling in recent years. Some former members of Caprifoliaceae are now placed in the family Adoxaceae, and two former families—Dipsacaceae and Valerianaceae—are now included in Caprifoliaceae.

Asteraceae — Aster

Asteraceae is a large family of annuals and perennials, with an estimated 25,000 species. These plants are known to have many tiny flowers that grow into a flower head, resembling a single flower. It's commonly referred to as the aster, daisy, composite, or sunflower family. This family was originally called Compositae, a name given in 1740 and still used synonymously in some sources.

Apiaceae — Carrot

The carrot or parsley family, also called Umbelliferae, includes 3,700 species, most of which are shrubs or small trees. In many species, the leaves give off a strong odor when crushed. Many species in Apiaceae also produce toxic substances called furanocoumarins, which may cause skin irritation. Apiaceae now includes species that were formerly placed in the family Araliaceae.

BLUE MISTFLOWER (*CONOCLINIUM COELESTINUM*)

POISON HEMLOCK (*CONIUM MACULATUM*)

GLOSSARY

A

Achene — A small, dry, hard fruit that does not open and contains one seed.

Adventive — A plant that has been introduced from one region into another but has not yet become fully naturalized. See also *Introduced*.

Air plant — See *Epiphyte*.

Alternate leaves — Arising singly along the stem, not in pairs or whorls.

Annual — Having a life cycle completed in one year or season.

Anther — The sac-like part of a stamen, containing pollen.

Aquatic — A plant growing in water.

Axil — The angle formed by the upper side of a leaf and the stem from which it grows.

B

Banner — The broad, uppermost petal in a pea flower; also called the standard.

Basal leaves — Leaves at the base of the stem.

Bearded — Bearing long or stiff hairs.

Bell-shaped — Flowers with a long tube and a small opening at one end.

Berry — A fleshy fruit with one to many seeds, developed from a single ovary.

Biennial — Growing vegetatively during the first year and flowering, fruiting, and dying during the second.

Bilateral symmetry — In flowers, one that can be divided into two equal halves by only one line through the middle; often called irregular or bilateral.

Bisexual — A flower having both female (pistil) and male (stamen) parts.

Blade — The flat portion of a leaf, petal, or sepal.

Bloom — A whitish, powdery, or waxy covering.

Bowl-shaped — Flowers with a wide opening and petals that flare up, creating a cup or bowl.

Bract — Modified leaves, usually smaller than the foliage leaves, often situated at the base of a flower or inflorescence.

Bud — An undeveloped leaf, stem, or flower, often enclosed in scales; an incompletely opened flower.

Bulb — A short underground stem, the swollen portion consisting mostly of fleshy, food-storing scale leaves.

C

Calyx — Collective term for the sepals of a flower, usually green.

Capsule — A dry fruit with one or more compartments, usually having thin walls that split open along several lines.

Carnivorous — Subsisting on nutrients obtained from the breakdown of animal tissue.

Catkin — A scaly-bracted spike or spike-like inflorescence bearing unisexual flowers without petals, as in willows.

Clasping leaf — A leaf whose base wholly or partly surrounds the stem.

Cluster — A group of several flowers growing together.

Compound leaf — A leaf divided into smaller leaflets.

Cordate — Shaped like a heart, as in some leaves.

Corolla — Collective term for the petals of a flower.

Corona — A crown-like structure on some corollas, as in milkweed flowers.

Creeper — Technically, a trailing shoot that takes root at the nodes; used here to denote any trailing, prostrate plant.

Critically imperiled — A species at very high risk of extinction due to a highly restricted range, very few populations or occurrences, catastrophic declines/population collapse, or other factors.

Cross-pollination — The transfer of pollen from one plant to another.

Cup-shaped — See *Bowl-shaped*.

D

Deciduous — Shedding leaves seasonally, as with many trees, or the shedding of certain parts after a period of growth.

Dehiscent fruits — Fruits that break open at maturity to disperse the seeds.

Dioecious — See *Imperfect flowers*.

Disk flower — The small, tubular flowers in the central part of a floral head, as in most sunflowers.

Dissected leaf — A deeply cut leaf, the cleft not reaching to the midrib; also called a divided leaf.

Drupe — A fleshy fruit with a single seed enveloped by a hard covering (stone); also called a stone fruit.

E

Elliptical — Ellipse-shaped, as in some leaves.

Embryo — The small plant formed after fertilization, contained within the seed and ready to grow with the proper environmental stimulation.

Emergent — An aquatic plant with its lower part submerged and its upper part extending above water.

Endangered — A species in danger of extinction throughout all or a significant portion of its range.

Epiphyte — A plant growing on another plant but deriving no nutrition from it; an air plant.

Escaped — A plant that has spread beyond the confines of a deliberate planting, as from a garden.

Extirpated — A species that no longer exists in the wild in a particular area but still exists elsewhere; a local extinction.

Extinction — The dying out or extermination of a species.

F

Family — A group of closely related genera.

Fertilization — In flowers, occurs when pollen reaches the pistil and goes through a slender tube until it reaches the ovary.

Filament — The slender stalk of a stamen; a thread.

Follicle — A dry fruit developed from a single ovary, usually opening along one line.

Fruit — The ripened ovary or pistil, often with attached parts.

Funnel-shaped — Flowers that are narrow at the base and flare open.

Genus (plural *genera*) — A group of closely related species.

Gland — A small structure usually secreting oil or nectar.

Glandular — Bearing glands.

H

Head — A crowded cluster of flowers on very short stalks or without stalks, as in sunflowers.

Herb — Usually a soft and succulent plant; not woody.

Humus — A brown or black, complex, variable material resulting from partial decomposition of plant or animal matter and forming the organic portion of soil.

I

Imperfect flowers — Flowers that have either a staminate (male) and pistillate (female) reproductive organ, but not both. Also called unisexual or dioecious.

Imperiled — At high risk of extinction due to a restricted range and/or a steep population decline or severe threat.

Indehiscent fruits — Fruits that don't break open and rely on decomposition or some other outside force to release their seeds.

Inflorescence — A flower cluster on a plant or, especially, the arrangement of flowers on a plant.

Involucre — A whorl or circle of bracts beneath a flower or flower cluster.

Irregular flower — A flower with petals that are not uniform in shape but are usually grouped to form upper and lower "lips"; generally bilaterally symmetrical.

L

Lanceolate leaf — Lance-shaped, much longer than wide and pointed at the end.

Leaflet — One of the leaf-like parts of a compound leaf.

Legume — A dry fruit developed from a single ovary, usually opening along two lines, as in the pea family (Fabaceae).

Linear — Long and narrow with parallel sides, as in the leaf blades of grasses.

Lip petal — The lower petal of some irregular flowers, often elaborately showy, as in orchids.

Lobed — Indented on the margins, with the indentations not reaching to the center or base.

Local — A plant occurring sporadically but sometimes common where found.

M

Monoecious — See *Perfect flowers*.

N

Naturalized — A plant that has been introduced from one region into another, where it has become established in the wild and reproduces as though native.

Node — The place on the stem where leaves or branches are attached.

Nuts — Fruits with thick, durable outer walls, such as acorns.

O

Oblong — Long and oval, as in some leaves.

Opposite leaves — Occurring in pairs at a node, with one leaf on either side of the stem.

Orbicular — Shaped like a circle, as in some leaves.

Ovary — The swollen base of a pistil, within which seeds develop.

Ovate leaf — Roughly egg-shaped leaves, pointed at the top and broader near the base or sometimes broadest at the middle.

Ovule — The immature seed in the ovary that contains the egg.

P

Palmate leaf — Having three or more divisions or lobes, looking like the outspread fingers of a hand.

Panicle — A branched, open inflorescence in which the main branches are again branched.

Parasite — A plant deriving its nutrition from another organism.

Pedicel — The stalk of an individual flower.

Peduncle — The main flower stalk or stem holding an inflorescence.

Perennial — A plant that lives for more than two years, usually producing flowers, fruits, and seeds annually.

Perianth — The calyx and corolla or, in flowers without distinct sepals and petals, simply the outer whorl, as in the buckwheat family (Polygonaceae) and others.

Perfect flowers — Flowers that contain both staminate (male) and pistillate (female) reproductive parts. Also called bisexual or monoecious.

Petal — The basic unit of the corolla; modified leaves that are flat, usually broad, and brightly colored.

Petaloid — Petal-like, usually describing a colored sepal.

Petiole — The stalk-like part of a leaf, attaching it to the stem.

Photosynthesis — The process by which a plant converts sunlight, water, and carbon dioxide into nutrients like sugars and starches.

Pinnate leaf — A compound leaf with leaflets along the sides of a common central stalk, much like a feather.

Pistil — The female organ of a flower, consisting of an ovary, style, and stigma.

Pistillate flower — A female flower, having one or more pistils but no functional stamens.

Pith — A spongy material present in the center of stems of certain plants.

Pod — A dry fruit that opens at maturity.

Pollen — Spores formed in the anthers that produce the male cells.

Pollen sac — The upper portion of the stamen, containing pollen grains; the anther.

Pollination — The transfer of pollen from an anther to a stigma.

Pubescent — Covered with hairs.

R

Raceme — A long flower cluster on which individual flowers each bloom on small stalks along a common, larger, central stalk.

Radial symmetry — In flowers, one with the symmetry of a wheel; often called regular.

Ray flower — The bilaterally symmetrical flowers around the edge of the head in many sunflowers; each ray flower resembles a single petal.

Receptacle — The base of the flower where all flower parts are attached.

Regular flower — With petals and/or sepals arranged around the center, like the spokes of a wheel; always radially symmetrical.

Reinform — Kidney-shaped, as in some leaves.

Rhizome — A horizontal underground stem, distinguished from roots by the presence of nodes, often enlarged by food storage.

Rhomboid — Star- or diamond-shaped, as in some leaves.

Root — A specialized structure that absorbs water and nutrients from the soil and transports them to the plant's stem.

Rose hip — A smooth, rounded, fruit-like structure consisting of the cup-like calyx enclosing seed-like fruits.

Rosette — A crowded cluster of leaves; usually basal, circular, and appearing to grow directly out of the ground.

Runner — A stem that grows on the surface of the soil, often developing leaves, roots, and new plants at the nodes or tip.

S

Samaras — Fruits with wing-like structures that catch the wind, causing the seed to spin and float aloft.

Sap — A general term for the liquid contained within the parts of a plant.

Saprophyte — A plant lacking chlorophyll and living on dead organic matter.

Saucer-shaped — Flowers that are flatter than bowl-shaped flowers and have a wide, round opening.

Scale — A small, flattened, thin, usually green structure; the scales of a grass spikelet, among which the flowers develop, or the scales (much reduced leaves) in a flower cluster.

Sepal — A basic unit of the calyx, usually green, but sometimes colored and resembling a petal.

Sessile — Without a stalk.

Sessile leaf — A leaf that lacks a petiole, the blade being attached directly to the stem.

Sheath — A more or less tubular structure surrounding a part, as the lower portion of a leaf surrounding the stem.

Shoot — A young stem or branch with its leaves and flowers not yet mature.

Shrub — A woody, relatively low plant with several branches from the base.

Simple leaf — A leaf with an undivided blade.

Solitary flower — A single flower growing from a single stem.

Spadix — A dense spike of tiny flowers, usually enclosed in a spathe, as in members of the arum family (Araceae).

Spathe — A bract or pair of bracts, often large, enclosing the flowers.

Spatulate — Spatula or spoon-shaped, with a rounded tip and tapering to the base.

Species — A fundamental category of taxonomic classification, ranking below a genus; individuals within a species usually reproduce among themselves and are more closely related than they are to individuals of other species.

Spherical — Flowers that are round, shaped like a ball.

Spike — An elongated flower cluster, each flower of which is without a stalk.

Spur — A slender, usually hollow projection from a part of a flower.

Stalk — A general and less precise term for the stem, used to describe a supporting structure, such as a leaf stalk or flower stalk.

Stamen — The male organ of a flower, composed of a filament topped by an anther; usually several occur in each flower.

Staminate flower — A male flower, having anthers and no pistils.

Stem — The main axis of a plant and its branches, responsible for supporting the leaves and flowers.

GLOSSARY

Stigma — The tip of the pistil where the pollen lands.

Stipules — Small appendages, often leaf-like, on either side of some petioles at the base.

Stolon — A stem growing along or under the ground; a runner.

Style — The narrow part of the pistil, connecting the ovary and stigma.

Succulent — Fleshy and thick, storing water; a plant with fleshy, water-storing stems or leaves.

T

Tendril — A slender, coiling structure that helps support climbing plants.

Threatened — A species that is likely to become endangered in the foreseeable future in all or a significant portion of its range.

Toothed — Having a sawtooth edge.

Trumpet-shaped — Flowers that are narrow at the base and flare open, but the petals curl back.

Tuber — A fleshy, enlarged part of an underground stem, serving as a storage organ, as in a potato.

Tubular — Flowers with united petals that form a long tube.

U

Umbel — A flower cluster in which the individual flower stalks grow from the same point, like the ribs of an umbrella.

Undulate — Having a wavy edge.

Unisexual — A flower of one sex only, either pistillate (female) or staminate (male).

V

Vulnerable — A species at moderate risk of extinction due to a limited range or a widespread decline or threat.

W

Whorled — A circle of three or more leaves, branches, or pedicels at a node.

Wing — In plants, a thin, flat extension found at the margins of a seed or leaf stalk or along the stem; the lateral petal of a pea flower.

A

Abronia fragrans	425
Abutilon	
parvulum	337
theophrasti	338
Achillea millefolium	697
Acmispon americanus	206
Aconitum columbianum	151
Acorus calamus	29
Actaea rubra	152
Adam's Needle	124
Adenocaulon bicolor	698
Agalinis purpurea	671
Agastache	
scrophulariifolia	635
Agave	
deserti	111
parryi	112
Agave, Parry's	112
Ageratina altissima	699
Agoseris glauca	700
Agrimonia gryposepala	263
Agrimony	263
Agrostemma githago	402
Ajuga reptans	636
Alexanders, Golden	874
Alfalfa	240
Alisma triviale	35
Alliaria petiolata	354
Allionia incarnata	426
Allium	
acuminatum	106
cernuum	107
schoenoprasum	108
tricoccum	109
vineale	110
Allotropa virgata	495
Alplily	59
Alumroot	188
Pink	189
Amaranthus retroflexus	417
Ambrosia	
artemisiifolia	701
trifida	702
American-aster, Western	829
Amorpha fruticosa	207
Amphicarpaea bracteata	208
Amsinckia menziesii	540
Amsonia	
tabernaemontana	526
Anagallis arvensis	479
Anaphalis margaritacea	703
Androsace occidentalis	486
Anemonastrum	
canadense	153
Anemone	
cylindrica	154
quinquefolia	155
Anemone	
Rue	183
Wood	155
Angelica atropurpurea	857
Angelica, Purple-stem	857

Anglepod, Common	537
Antennaria	
neglecta	704
parlinii	705
rosea	706
Anthemis cotula	707
Anticlea elegans	38
Apios americana	209
Apocynum cannabinum	527
Aquilegia	
caerulea	156
canadensis	157
formosa	158
Arabis pycnocarpa	355
Aralia nudicaulis	858
Arbutus, Trailing	498
Arceuthobium douglasii	379
Arctium minus	708
Arethusa bulbosa	65
Argemone polyanthemos	137
Argentina anserina	264
Arisaema triphyllum	30
Armeria maritima	381
Arnica cordifolia	709
Arnica, Heartleaf	709
Arnoglossum	
plantagineum	710
Arrowhead	
Broadleaf	37
Narrowleaf	36
Artemisia vulgaris	711
Artichoke, Jerusalem	766
Arum	
Arrow	33
Water	31
Asarum	
canadense	27
caudatum	28
Asclepias	
amplexicaulis	528
fascicularis	529
incarnata	530
latifolia	531
speciosa	532
subverticillata	533
syriaca	534
tuberosa	535
Aster	
Common Blue Wood	830
Flat-topped White	738
Heath	831
Leafy	832
New England	835
Panicled	834
Purple-stemmed	836
Saltmarsh	837
Smooth Blue	833
Stiff	773
Toothed White-top	814
White Wood	752
Whorled	790
Astragalus	
agrestis	210
canadensis	211

Aureolaria flava	673
Avens, White	269

B

Bacopa monnieri	586
Balm, Horse	640
Baneberry, Red	152
Baptisia	
alba	212
bracteata	213
Barbarea vulgaris	356
Basil, Wild	639
Basket-flower, American	802
Bassia scoparia	418
Beardtongue	
Beaked	606
Broad-beard	597
Cobaea	598
Foxglove	599
Hairy	602
Large-flower	601
Lilac	600
Pale	604
Rocky Mountain	609
Royal	608
Bedstraw	
Fragrant	509
Northern	508
Bee Plant, Yellow	352
Beebalm	
Lemon	648
Spotted	650
Beechdrops	679
Beetleweed	492
Beggartick, Devil's	716
Bellflower	
Common	686
Tall	688
Bellis perennis	712
Bellwort	
Large-flower	47
Sessile	48
Bergamot, Wild	649
Berlandiera lyrata	713
Bidens	
aristosa	714
cernua	715
frondosa	716
Bignonia capreolata	624
Bindweed	566
Hedge	564
Bistort, Western	383
Bistorta bistortoides	383
Bittercress, Heartleaf	362
Bitterroot, Alpine	435
Bittersweet, American	286
Bladderpod, Fendler's	374
Bladderwort	
Common	629
Horned	627
Little Floating	628
Blanketflower	
Firewheel	758

Great	757
Blazing Star	
Dotted	781
Rough	780
Scaly	782
Blazingstar	
Giant	453
Ten-petal	452
Bleeding Heart, Western	142
Blephilia ciliata	637
Bloomeria crocea	113
Blue-eyed Grass	
Idaho	101
Strict	102
White	99
Maiden	588
Bluebells	
Mountain	550
Virginia	551
Bluecurls, Forked	664
Bluedicks	117
Blueflag, Southern	98
Bluestar, Eastern	526
Bluets, Small	510
Boechera stricta	357
Boehmeria cylindrica	280
Boerhavia coccinea	427
Boltonia	717
Boltonia asteroides	717
Boneset	750
Hyssop-leaf	749
Late	751
Borrichia frutescens	718
Bouncing Bet	409
Boykinia occidentalis	186
Boykinia, Coast	186
Brassica nigra	358
Brittlebush	741
Brooklime, American	612
Brookweed, Seaside	478
Buckbean	695
Buckwheat	
Cushion	385
Sulphur	386
Buffalo Bur	582
Bugle, Common	636
Bunchberry	454
Bur-reed, Giant	132
Burdock, Common	708
Burning Bush	418
Bush-clover	
Creeping	231
Round-head	230
Butter-and-eggs	593
Buttercup	
Common	174
Creeping	178
Early	177
Kidneyleaf	173
Swamp	179
Water	176
Water-plantain	175
Butterfly-pea, Climbing	214
Butterfly Weed	535

Butterweed, Basin 794
Butterwort, Small 626
Buttonweed 506

C

Cabbage, Skunk 34
Cabomba caroliniana 22
Cactus, Desert Christmas 441
Cakile edentula 359
Calandrinia menziesii 430
Calicoflower, Bach's 689
Calla palustris 31
Callirhoe involucrata 339
Calochortus
    elegans 50
    macrocarpus 51
Calopogon tuberosus 66
Caltha palustris 159
Caltrop, Warty 204
Calypso bulbosa 67
Calyptridium umbellata 431
Calystegia
    sepium 564
    soldanella 565
Camas, Common 114
Camassia quamash 114
Campanula
    rapunculoides 686
    rotundifolia 687
Campanulastrum
    americanum 688
Camphorweed 769
Campion
    Bladder 414
    Starry 413
    White 412
Campsis radicans 625
Cancer-root
    American 678
    One-flowered 672
Candystick 495
Candytuft, Wild 373
Capnoides sempervirens 138
Capsella bursa-pastoris 360
Cardamine
    bulbosa 361
    cordifolia 362
    diphylla 363
Cardinal Flower 690
Cardiospermum
    halicacabum 336
Carduus nutans 719
Carnegiea gigantea 438
Carpobrotus chilensis 422
Carrionflower 49
Castilleja
    applegatei 674
    coccinea 675
    miniata 676
    sessiliflora 677
Cat's Ears 50
Cat's-ear, Hairy 771

Catchfly
    Cardinal 411
    Sticky 410
Catnip 651
Cattail, Broadleaf 133
Caulophyllum
    thalictroides 148
Celandine 139
    Lesser 169
Celastrus scandens 286
Centaurea
    solstitialis 720
    stoebe 721
Centaurium pulchellum 513
Centaury, Branched 513
Centrosema virginianum 214
Century Plant 111
Cerastium
    fontanum 403
    strictum 404
Cevallia sinuata 451
Cevallia, Stinging 451
Chaenactis douglasii 722
Chaerophyllum
    procumbens 859
Chaetopappa ericoides 723
Chamaecrista fasciculata 215
Chamaenerion
    angustifolium 318
Chamaepericlymenum
    canadense 454
Chamaesaracha sordida 575
Checker-mallow, New
    Mexico 347
Cheeses 346
Chelidonium majus 139
Chelone glabra 587
Chenopodium album 419
Chervil, Spreading 859
Chestnut, Water 317
Chia 657
Chickweed
    Common 416
    Field 404
    Mouse-ear 403
Chicory 725
Chimaphila
    maculata 496
    umbellata 497
Chives, Siberian 108
Chlorogalum
    pomeridianum 115
Cholla
    Buckhorn 439
    Tree 440
Chrysopsis mariana 724
Chrysosplenium
    americanum 187
Cichorium intybus 725
Cicuta maculata 860
Cinquefoil
    Common 278
    Dwarf 274
    Marsh 265

Rough-fruit 277
Slender 275
Tall 266
Circaea alpina 319
Cirsium
    arvense 726
    horridulum 727
    undulatum 728
    vulgare 729
Clammyweed,
    Red-whisker 353
Clarkia amoena 320
Claytonia
    lanceolata 432
    perfoliata 433
    virginica 434
Cleavers 507
Clematis
    crispa 160
    hirsutissima 161
    occidentalis 162
    virginiana 163
Clinopodium
    douglasii 638
    vulgare 639
Clintonia
    borealis 52
    uniflora 53
Clitoria mariana 216
Clover
    Cow 254
    Rabbit-foot 250
    Red 252
    Spanish 206
    White 253
    White Sweet 241
Cluster-vine, Hairy 573
Cocklebur, Rough 850
Coeloglossum viride 68
Coffee, Wild 855
Cohosh, Blue 148
Collinsia parviflora 588
Collinsonia canadensis 640
Collomia linearis 456
Coltsfoot 844
    Palmate 798
Columbine
    Colorado Blue 156
    Red 158
    Wild 157
Comandra umbellata 380
Comarum palustre 265
Commelina
    communis 127
    erecta 128
Compass Plant 816
Coneflower
    Cutleaf 811
    Narrowleaf Purple 739
    Orange 809
    Pinnate Prairie 807
    Purple 740
Conium maculatum 861
Conoclinium coelestinum 730

Conopholis americana 678
Convolvulus arvensis 566
Coptis trifolia 164
Corallorhiza
    maculata 69
    wisteriana 70
Coralroot
    Spotted 69
    Spring 70
Coreopsis
    grandiflora 731
    lanceolata 732
    tinctoria 733
    tripteris 734
Coreopsis
    Garden 733
    Lanceleaf 732
Corn Lily
    California 45
    Green 46
Corncockle 402
Corydalis aurea 140
Cosmos bipinnatus 735
Cosmos, Garden 735
Cow Wheat 680
Cranberry, Small 505
Cranesbill 313
Creamcups 145
Crepis
    acuminata 736
    occidentalis 737
Cress, Spring 361
Crinklemat, Woody 556
Crinkleroot 363
Crocanthemum
    canadense 350
Cross, Widow's 202
Crossvine 624
Crotalaria sagittalis 217
Crown-vetch 245
Crownbeard, Golden 845
Cryptantha ambigua 541
Cryptantha, Basin 541
Cucumber
    Creeping 284
    Indian 63
    One-seeded Bur 285
    Wild 283
Culver's Root 615
Cup-plant 817
Cuscuta gronovii 567
Cylindropuntia
    acanthocarpa 439
    imbricata 440
    leptocaulis 441
Cynanchum laeve 536
Cynoglossum officinale 542
Cynthia, Two-flowered 774
Cypripedium
    acaule 71
    parviflorum 72

D

Daisy
English 712
Oxeye 779
Sea 718
Showy Townsend 841
Yellow Spiny 849
Dalea
candida 218
purpurea 219
Dame's Rocket 368
Dandelion
Common 839
Desert 785
Dasylirion wheeleri 116
Datura stramonium 576
Daucus carota 862
Dawnflower, Southern 574
Dayflower
Asiatic 127
Erect 128
Daylily, Orange 103
Dead Nettle, Purple 644
Deathcamas, Mountain 38
Decodon verticillatus 315
Decumaria barbara 449
Delphinium
carolinianum 165
scaposum 166
tricorne 167
Descurainia sophia 364
Desert-parsley, Fernleaf 867
Desmanthus illinoensis 220
Desmodium
canadense 221
glutinosum 222
paniculatum 223
Diamond-flowers 512
Dianthus armeria 405
Dicentra
cucullaria 141
formosa 142
Dicliptera sexangularis 620
Dielsiris missouriensis 96
Digitalis purpurea 589
Diodia virginiana 506
Diplacus nanus 665
Dipterostemon capitatus 117
Dock
Curly 398
Mexican 399
Doellingeria umbellata 738
Downingia bacigalupii 689
Draba
cana 365
verna 366
Draba
Lanceleaf 365
Spring 366
Dragon's Mouth 65
Drosera
capillaris 400
rotundifolia 401

Drymocallis arguta 266
Dudleya cymosa 198
Dudleya, Canyon 198
Dutchman's Breeches 141
Dwarf Mistletoe,
Douglas-fir 379
Dwarf-dandelion, Virginia 775
Dyer's Woad 369
Dyschoriste oblongifolia 621

E

Echinacea
angustifolia 739
purpurea 740
Echinocereus
engelmannii 442
viridiflorus 443
Echinocystis lobata 283
Echium vulgare 543
Elecampane 772
Elephant's Head 683
Encelia farinosa 741
Enemion biternatum 168
Epifagus virginiana 679
Epigaea repens 498
Epilobium
densiflorum 321
leptophyllum 322
Epipactis
gigantea 73
helleborine 74
Erigeron
annuus 742
canadensis 743
compositus 744
philadelphicus 745
pulchellus 746
strigosus 747
Eriocaulon decangulare 136
Eriogonum
annum 384
ovalifolium 385
umbellatum 386
Eriophyllum lanatum 748
Erodium cicutarium 312
Eryngium yuccifolium 863
Erysimum capitatum 367
Erythranthe
guttata 666
moschata 667
primuloides 668
Erythronium
americanum 54
grandiflorum 55
Eschscholzia californica 143
Escobaria vivipara 444
Eucrypta
chrysanthemifolia 544
Eulobus californicus 323
Eupatorium
hyssopifolium 749
perfoliatum 750
serotinum 751

Euphorbia
albomarginata 293
corollata 294
cyathophora 295
cyparissias 296
marginata 297
Eurybia divaricata 752
Eustoma exaltatum 514
Euthamia
graminifolia 753
occidentalis 754
Eutrochium
maculatum 755
purpureum 756
Evening Primrose
Common 325
Cutleaf 329
Hooker's 327
Narrowleaf 328
Pale 330
Showy 332
Tansyleaf 334
Tufted 326
Everlasting
Pearly 703
Sweet 804
Evolvulus nuttallianus 568

F

Fairybells, Wartberry 64
Fairy Slipper 67
Fallopia scandens 387
False Babystars 462
False Buckwheat,
Climbing 387
False Dandelion, Carolina 805
False Fiddleleaf,
One-flower 583
False Foxglove, Smooth
Yellow 673
False Lily-of-the-Valley,
Feathery 119
Fanwort, Carolina 22
Farewell-to-spring 320
Featherbells 39
Ficaria verna 169
Fiddleneck, Menzies' 540
Fiesta Flower, Blue 562
Figwort, Lanceleaf 616
Fire-on-the-mountain 295
Fireweed 318
Five-eyes, Hairy 575
Flag, Yellow 97
Flax
Blue 309
Large-flower Yellow 311
Stiff Yellow 310
Fleabane
Daisy 742
Dwarf Mountain 744
Philadelphia 745
Prairie 747
Saltmarsh 803

Flixweed 364
Floating Hearts 696
Flower-of-an-hour 343
Foamflower, Eastern 196
Fog-fruit, Turkey-tangle 631
Foldwing, Six-angle 620
Forget-me-not
Spring 553
True 552
Four-o'clock
Desert 429
Trailing 426
White 428
Foxglove 589
Fragaria
vesca 267
virginiana 268
Frasera speciosa 515
Fritillaria
affinis 56
atropurpurea 57
pudica 58
Fritillary, Spotted 57
Frostweed 350

G

Gagea serotina 59
Gaillardia
aristata 757
pulchella 758
Galax urceolata 492
Galium
aparine 507
boreale 508
triflorum 509
Gaping Bush 590
Garlic
False 105
Field 110
Gaultheria procumbens 499
Gaura, Scarlet 333
Gentian
Catchfly Prairie 514
Closed Bottle 516
Explorer's 517
Fringed 520
Northern 519
Star 525
Gentiana
andrewsii 516
calycosa 517
saponaria 518
Gentianella amarella 519
Gentianopsis crinita 520
Geranium
carolinianum 313
maculatum 314
Geranium, Wild 314
Gerardia, Purple 671
Germander, Canada 663
Geum
canadense 269
fragarioides 270

triflorum	271
Ghost Pipe	502
Gilia	
capitata	457
tricolor	458
Gilia	
Bird's-eye	458
Globe	457
Ginger, Wild	27
Ginseng, Dwarf	868
Glandularia canadensis	630
Glasswort	
Common	421
Perennial	420
Glechoma hederacea	641
Globe-mallow	
Scarlet	348
Small-leaf	349
Streambank	344
Glycyrrhiza lepidota	224
Goat's-rue	248
Goatsbeard, Yellow	842
Golden Smoke	140
Golden-aster	
Hairy	769
Maryland	724
Goldenrod	
Gray	822
Hard-leaf	791
Lanceleaf	753
Missouri	821
Rough-stemmed	824
Showy	825
Southern Seaside	820
Sweet	823
Tall	818
Three-nerve	826
Western Grass-leaf	754
Zigzag	819
Goldenstars, Common	113
Goldenweed, Stemless	828
Goldthread	164
Gonolobus suberosus	537
Goodyera	
oblongifolia	75
pubescens	76
Greeneyes	713
Greenthread, Slender	840
Grindelia squarrosa	759
Groundcherry, Clammy	578
Groundnut	209
Groundsel	
Balsam	795
Common	813
Threadleaf	812
Gumweed, Curlycup	759

H

Habenaria repens	77
Hackelia	
floribunda	545
virginiana	546
Harebell	687

Harlequin, Rock	138
Harvestbells	518
Hat, Mexican	806
Hawksbeard	736
Western	737
Hawkweed	
Mouse-ear	800
Orange	799
Heart's-delight	425
Heath	
Beach	351
Rose	723
Hedera helix	864
Hedge-nettle	
Coastal	661
Hairy	662
Hedgehog Cactus	
Engelmann's	442
Simpson's	448
Hedysarum boreale	225
Helenium autumnale	760
Helianthus	
annuus	761
debilis	762
giganteus	763
maximiliani	764
petiolaris	765
tuberosus	766
Heliopsis helianthoides	767
Heliotrope, Salt	563
Heliotropium	
curassavicum	563
Helleborine	74
Giant	73
Hemerocallis fulva	103
Hemlock	
Poison	861
Water	860
Hemp, Bog	280
Hempweed, Climbing	787
Henbit	643
Hepatica americana	170
Hepatica, Round-lobed	170
Heracleum maximum	865
Hesperis matronalis	368
Hesperochiron pumilus	547
Hesperochiron, Dwarf	547
Heterotheca	
subaxillaris	768
villosa	769
Heuchera americana	188
Heuchera rubescens	189
Hibiscus	
denudatus	340
laevis	341
moscheutos	342
trionum	343
Hibiscus, Rock	340
Hideseed, Spotted	544
Hieracium venosum	770
Hoarycress	370
Hoffmannseggia glauca	226
Hogpeanut, American	208

Honeysuckle	
Japanese	852
Trumpet	853
Honeyvine	536
Horehound, Water	646
Horsenettle	579
Horseweed	743
Houndstongue	542
Houstonia pusilla	510
Hudsonia tomentosa	351
Hydrangea arborescens	450
Hydrangea, Wild	450
Hydrocotyle umbellata	866
Hydrolea uniflora	583
Hydrophyllum	
occidentale	557
virginianum	558
Hylotelephium telephium	199
Hymenocallis occidentalis	104
Hypericum	
anagalloides	289
gentianoides	290
perforatum	291
virginicum	292
Hypochaeris radicata	771
Hypopitys americana	500
Hypoxis hirsuta	95
Hyptis alata	642
Hyssop	
Purple Giant	635
Water	586

I

Iliamna rivularis	344
Impatiens capensis	455
Indian Plantain,	
Groove-stem	710
Indian-hemp	527
Indigo	
Bastard	207
Cream False	213
White Wild	212
Inula helenium	772
Ionactis linariifolius	773
Ipomoea	
coccinea	569
hederacea	570
lacunosa	571
purpurea	572
Ipomopsis	
aggregata	459
longiflora	460
Iris, Rocky Mountain	96
Ironweed	
New York	847
Tall	846
Isatis tinctoria	369
Isotria verticillata	78
Ivesia gordonii	272
Ivesia, Gordon's	272
Ivy	
English	864
Ground	641

J

Jack-go-to-bed-at-noon	843
Jack-in-the-pulpit	30
Jacquemontia tamnifolia	573
Jeffersonia diphylla	149
Jewelweed, Orange	455
Jimsonweed	576
Joe-Pye-weed	
Spotted	755
Sweet-scented	756
Juniper-leaf	585
Justicia americana	622

K

Kallstoemia parviflora	204
Keckiella breviflora	590
Knapweed	
Russian	808
Spotted	721
Knotweed	
Douglas'	395
Japanese	396
Prostrate	394
Kosteletzkya	
pentacarpos	344
Krigia	
biflora	774
virginica	775
Kudzu	244

L

Lactuca	
canadensis	776
serriola	777
Ladies'-tresses	
Great Plains	93
Hooded	94
Nodding	92
Lady's Slipper	
Large Yellow	72
Pink	71
Lady's-thumb	391
Lamb's-quarters	419
Lamium amplexicaule	643
Lamium purpureum	644
Langloisia setosissima	461
Langloisia, Great Basin	461
Laportea canadensis	281
Larkspur	
Bare-stem	166
Carolina	165
Dwarf	167
Lathyrus	
japonicus	227
latifolius	228
palustris	229
Layia glandulosa	778
Leather Flower, Swamp	160
Leek, Wild	109
Leonurus cardiaca	645

Lepidium	
draba	370
virginicum	371
Leptosiphon	
androsaceus	462
nuttallii	463
Lespedeza	
capitata	230
repens	231
Lettuce	
Blue	788
Miner's	433
Prickly	777
Wild	776
Leucanthemum vulgare	779
Lewisia pygmaea	435
Liatris	
aspera	780
punctata	781
squarrosa	782
Licorice, American	224
Lilium	
canadense	60
columbianum	61
philadelphicum	62
Lily	
Bluebead	52
Canada	60
Checker	56
Columbia	61
Glacier	55
Trout	54
Wood	62
Limniris	
pseudacorus	97
virginica	98
Limonium carolinianum	382
Linanthus pungens	464
Linanthus, Nuttall's	463
Linaria	
dalmatica	592
vulgaris	593
Linnaea borealis	851
Linum	
lewisii	309
medium	310
rigidum	311
Liparis loeselii	79
Lithophragma parviflorum	190
Lithospermum	
canescens	548
occidentale	549
Lizard's-tail	26
Lobelia	
cardinalis	690
inflata	691
siphilitica	692
spicata	693
Lobelia	
Great	692
Spiked	693
Locoweed, Purple	242
Lomatium dissectum	867

Lonicera	
japonica	852
sempervirens	853
Looking-glass, Venus'	694
Loosestrife	
Fringed	481
Purple	316
Swamp	315
Whorled	484
Lopseed	670
Lotus corniculatus	232
Lotus, American	184
Lousewort	
Dwarf	682
Sickletop	684
Love-in-a-puff	336
Ludwigia alternifolia	324
Lupine	
Blue-pod	237
Dwarf	234
Kellogg's Spurred	235
Seacoast	236
Silky	238
Silvery	233
Lupinus	
argenteus	233
caespitosus	234
caudatus	235
littoralis	236
polyphyllus	237
sericeus	238
Lychnis flos-cuculi	406
Lycopus americanus	646
Lygodesmia juncea	783
Lysichiton americanus	32
Lysimachia	
borealis	480
ciliata	481
maritima	482
nummularia	483
quadrifolia	484
terrestris	485
Lythrum salicaria	316

M

Madia glomerata	784
Maianthemum	
canadense	118
racemosum	119
Maiden, Dusty	722
Malacothrix glabrata	785
Malaxis unifolia	80
Mallow	
Dwarf Indian	337
Seashore	345
Malva neglecta	346
Marbleseed, Soft-hair	549
Marigold	
Marsh	159
Nodding Bur	715
Mariposa Lily, Sagebrush	51
Marsh-pea	229

Marsh-pennywort, Manyflower	866
Matricaria discoidea	786
Maurendella antirrhiniflora	594
May-apple	150
Mayflower, Canada	118
Mayweed	707
Meadow Beauty, Virginia	335
Meadow-rue	
Early	181
Purple	180
Western	182
Medeola virginiana	63
Medic, Black	239
Medicago	
lupulina	239
sativa	240
Melampyrum lineare	680
Melilotus alba	241
Melothria pendula	284
Menispermum	
canadense	147
Menodora scabra	584
Menodora, Rough	584
Mentha canadensis	647
Mentzelia	
decapetala	452
laevicaulis	453
Menyanthes trifoliata	695
Mertensia	
ciliata	550
virginica	551
Micranthes	
pensylvanica	191
rhomboidea	192
virginiensis	193
Microsteris gracilis	465
Mikania scandens	787
Milk-vetch	
Canada	211
Field	210
Milkweed	
Broad-leaf	531
Clasping	528
Common	534
Narrowleaf	529
Poison	533
Showy	532
Swamp	530
Milkwort	
Purple	260
White	259
Whorled	261
Mimosa, Prairie	220
Mimulus ringens	669
Mirabilis	
albida	428
multiflora	429
Mistflower, Blue	730
Mitchella repens	511
Mitella diphylla	194
Miterwort	194
Mock Vervain, Rose	630
Moehringia lateriflora	407

Monarda	
citriodora	648
fistulosa	649
punctata	650
Moneses uniflora	501
Moneywort	483
Monkeyflower	
Common	666
Dwarf Purple	665
Primrose	668
Square-stemmed	669
Monkshood, Western	151
Mononeuria groenlandica	408
Monotropa uniflora	502
Monument Plant	515
Moonseed, Common	147
Morning Glory	
Beach	565
Common	572
Ivy-leaf	570
Shaggy Dwarf	568
Small Red	569
Small White	571
Moss, Spanish	134
Motherwort	645
Mountain Dandelion, Pale	700
Mountain Mint	
Narrowleaf	654
Virginia	655
Mountain-trumpet, Narrowleaf	456
Mugwort	711
Mule's-ears, Northern	848
Mulgedium oblongifolium	788
Mullein	
Common	618
Moth	617
Muskflower	667
Musky-mint	642
Mustard	
Black	358
Garlic	354
Hedge	377
Tumbling	376
Myosotis	
scorpioides	552
verna	553

N

Nabalus altissimus	789
Nanopanax trifolius	868
Nasturtium officinale	372
Navarretia	
intertexta	466
squarrosa	467
Navarretia, Needle-leaf	466
Nelumbo lutea	184
Nemophila parviflora	554
Nemophila, Small-flower	554
Neogaerrhinum filipes	595
Neottia convallarioides	81
Nepeta cataria	651
Nettle, California	282

INDEX

Nettle, Wood 281
Nicotiana attenuata 577
Nightshade
    Climbing 580
    Common 581
    Enchanter's 319
Noccaea fendleri 373
Nolina microcarpa 120
Nolina, Beargrass 120
Nothoscordum bivalve 105
Nuphar
    advena 23
    polysepala 24
Nuttallanthus canadensis 591
Nymphaea odorata 25
Nymphoides aquatica 696

O

Obedient Plant 652
Obolaria virginica 521
Oclemena acuminata 790
Oenothera
    biennis 325
    caespitosa 326
    elata 327
    fruticosa 328
    laciniata 329
    pallida 330
    serrulatus 331
    speciosa 332
    suffrutescens 333
Oligoneuron rigidum 791
Onion
    Hooker's 106
    Nodding 107
Onopordum acanthium 792
Opuntia
    cespitosa 445
    fragilis 446
    phaeacantha 447
Orchid
    Alaska Rein 90
    Blunt-leaf 87
    Bracted 68
    Green Adder's-Mouth 80
    Green Bog 82
    Lesser Purple Fringed 89
    Pale Green 85
    Ragged Fringed 86
    Roundleaf Rein 88
    Water-spider 77
    White Rein 86
    Yellow Fringed 83
Ornithogalum
    umbellatum 121
Orobanche uniflora 672
Orthocarpus luteus 681
Osmorhiza longistylis 869
Owl's-clover, Yellow 681
Oxalis stricta 288
Oxeye 767
Oxyria digyna 388
Oxytropis lambertii 242

P

Packera
    aurea 793
    multilobata 794
    paupercula 795
Paeonia brownii 185
Pagoda-plant, Downy 637
Paintbrush
    Giant Red 676
    Great Plains 677
    Wavyleaf 674
Painted Cup, Scarlet 675
Palafox, Rosy 796
Palafoxia rosea 796
Pansy, Field 299
Papaver californicum 144
Parnassia glauca 287
Parnassia, Fringed 287
Parsnip
    Cow 865
    Water 872
    Wild 870
Parthenium integrifolium 797
Partridge-pea 215
Partridgeberry 511
Pasqueflower
    American 171
    Western 172
Passiflora incarnata 308
Passionflower 308
Pastinaca sativa 870
Pathfinder 698
Pea
    Beach 227
    Butterfly 216
    Everlasting 228
    Mountain Golden 249
Pedicularis
    centranthera 682
    groenlandica 683
    racemosa 684
Pediocactus simpsonii 448
Pediomelum argophyllum 243
Peltandra virginica 33
Pencil-flower 247
Pennywort 521
Penstemon
    albidus 596
    angustifolius 597
    cobaea 598
    digitalis 599
    gracilis 600
    grandiflorus 601
    hirsutus 602
    linarioides 603
    pallidus 604
    procerus 605
    rostriflorus 606
    rydbergii 607
    speciosus 608
    strictus 609
Penstemon
    Littleflower 605

Rydberg's 607
    Toadflax 603
    White 598
Penthorum sedoides 197
Peony, Western 185
Pepperweed, Virginia 371
Peritoma lutea 352
Periwinkle 538
Persicaria
    amphibia 389
    hydropiperoides 390
    maculosa 391
    pensylvanica 392
    punctata 393
Petasites frigidus 798
Petrophyton caespitosum 273
Phacelia
    hastata 559
    linearis 560
    sericea 561
Phacelia
    Silky 561
    Threadleaf 560
    Whiteleaf 559
Phemeranthus parviflorus 436
Phlox
    diffusa 468
    drummondii 469
    glaberrima 470
    longifolia 471
    paniculata 472
    pilosa 473
    stansburyi 474
    subulata 475
Phlox
    Annual Garden 469
    Cold-desert 474
    Downy 473
    Garden 472
    Granite Prickly 464
    Longleaf 471
    Slender 465
    Smooth 470
    Spreading 468
Pholistoma auritum 562
Phryma leptostachya 670
Phyla nodiflora 631
Physalis heterophylla 578
Physaria fendleri 374
Physostegia virginiana 652
Phytolacca americana 424
Pickerelweed 131
Pigweed 417
Pilosella
    aurantiaca 799
    officinarum 800
Pimpernel, Scarlet 479
Pineapple-weed 786
Pinesap 500
Pineweed 290
Pinguicula pumila 626
Pink
    Deptford 405
    Grass 66

    Moss 475
Pipewort 136
Pipsissewa 497
Pitaya, Green 443
Pitcher-plant
    Northern 494
    Yellow 493
Pityopsis graminifolia 801
Plagiobothrys hispidulus 555
Plantago
    lanceolata 610
    major 611
Plantain
    Common 611
    English 610
Platanthera
    aquilonis 82
    ciliaris 83
    dilatata 84
    flava 85
    lacera 86
    obtusata 87
    orbiculata 88
    psycodes 89
    unalascensis 90
Platystemon californicus 145
Plectocephalus
    americanus 802
Plectritis macrocera 854
Plectritis, White 854
Pluchea odorata 803
Podophyllum peltatum 150
Pogonia ophioglossoides 91
Pogonia
    Large Whorled 78
    Rose 91
Pokeweed 424
Polanisia dodecandra 353
Polemonium
    occidentale 476
    reptans 477
Polemonium, Western 476
Polygala
    alba 259
    sanguinea 260
    verticillata 261
Polygala, Fringed 262
Polygaloides paucifolia 262
Polygonatum biflorum 122
Polygonum
    aviculare 394
    douglasii 395
Polypremum
    procumbens 585
Pond-lily
    Rocky Mountain 24
    Yellow 23
Pontederia cordata 131
Popcorn-flower, Harsh 555
Poppy-mallow, Purple 339
Poppy
    Annual Prickly 137
    California 143
    Fire 144

Wood	146
Portulaca oleracea	437
Potato, Hog	226
Potentilla	
canadensis	274
gracilis	275
indica	276
recta	277
simplex	278
Prairie Clover	
Purple	219
White	218
Prairie Star	190
Prickly-pear	
Desert	447
Eastern	445
Fragile	446
Primrose, Parry's	489
Primula	
jeffreyi	487
meadia	488
parryi	489
pauciflora	490
tetrandra	491
Prince's-plume, Desert	378
Proboscidea louisianica	619
Prosartes trachycarpa	64
Prunella vulgaris	653
Pseudognaphalium	
obtusifolium	804
Puccoon, Hoary	548
Pueraria montana	244
Pulsatilla	
nuttalliana	171
occidentalis	172
Puncturevine	205
Purslane, Common	437
Pussypaws, Mt. Hood	431
Pussytoes	
Field	704
Parlin's	705
Rosy	706
Pycnanthemum	
tenuifolium	654
virginianum	655
Pyrola	
asarifolia	503
elliptica	504
Pyrrhopappus	
carolinianus	805

Q

Queen Anne's Lace	862
Queen's Cup	53
Quinine, Wild	797

R

Ragweed	
Common	701
Great	702
Ragwort, Golden	793
Ram's-horn	619

Ranunculus	
abortivus	173
acris	174
alismifolius	175
aquatilis	176
fascicularis	177
repens	178
septentrionalis	179
Raspberry, Dwarf Red	279
Ratibida	
columnifera	806
pinnata	807
Rattle, Yellow	685
Rattlebox	217
Rattlesnake Master	863
Rattlesnake	
Plantain	75
Downy	76
Rattlesnake Root, Tall	789
Rattlesnake-weed	293
Rattlesnake Weed	770
Redmaids	430
Reynoutria japonica	396
Rhaponticum repens	808
Rhexia virginica	335
Rhinanthus minor	685
Rhodiola integrifolia	200
Robin, Ragged	406
Robin's-plantain	746
Rockcress	
Drummond's	357
Hairy	355
Rockjasmine, Western	486
Rockmat	273
Rorippa palustris	375
Rose-gentian, Marsh	523
Rose-mallow,	
Halberd-leaf	341
Rose-pink	522
Rosemallow,	
Crimson-eyed	342
Rosinweed, Starry	815
Rubus pubescens	279
Rudbeckia	
fulgida	809
hirta	810
laciniata	811
Rue Anemone, False	168
Ruellia humilis	623
Rumex	
acetosella	397
crispus	398
triangulivalvis	399

S

Sabatia	
angularis	522
dodecandra	523
stellaris	524
Sabatia, Saltmarsh	524
Sage	
Blue	656
Lyre-leaf	658

Sagittaria	
cuneata	36
latifolia	37
Saguaro, Giant	438
Salicornia	
ambigua	420
depressa	421
Salvia	
azurea	656
columbariae	657
lyrata	658
Samolus valerandi	478
Sand-Spurry, Salt	415
Sandwort	
Grove	407
Mountain	408
Sanicula marilandica	871
Saponaria officinalis	409
Sarracenia	
flava	493
purpurea	494
Sarsaparilla, Wild	858
Saururus cernuus	26
Saxifraga caespitosa	195
Saxifrage	
American Golden	187
Diamond-leaf	192
Early	193
Swamp	191
Tufted	195
Scaldweed	567
Scrophularia lanceolata	616
Scurfpea, Silverleaf	243
Scutellaria	
galericulata	659
lateriflora	660
Sea Fig	422
Sea Lavender, Carolina	382
Sea-milkwort	482
Sea-Purslane, Slender	423
Searocket, American	359
Securigera varia	245
Sedum	
lanceolatum	201
pulchellum	202
ternatum	203
Seedbox	324
Selfheal, Common	653
Senecio	
flaccidus	812
vulgaris	813
Senna marilandica	246
Senna, Wild	246
Sericocarpus asteroides	814
Sesuvium maritimum	423
Shepherd's Purse	360
Shinleaf	504
Shootingstar	
Alpine	491
Eastern	488
Few-flower	490
Tall Mountain	487
Sicyos angulatus	285
Sidalcea neomexicana	347

Silene	
caroliniana	410
laciniata	411
latifolia	412
stellata	413
vulgaris	414
Silkgrass, Narrowleaf	801
Silphium	
asteriscus	815
laciniatum	816
perfoliatum	817
Silverweed	264
Sisymbrium	
altissimum	376
officinale	377
Sisyrinchium	
albidum	99
angustifolium	100
idahoense	101
montanum	102
Sium suave	872
Skeleton Plant, Rush	783
Skullcap	
Mad-dog	660
Marsh	659
Skunk Cabbage, Yellow	32
Skunkweed	467
Skyrocket	459
Sleepy Dick	121
Smartweed	
Dotted	393
Pennsylvania	392
Swamp	390
Water	389
Smilax herbacea	49
Smoke, Prairie	271
Snakeroot	
Maryland Black	871
White	699
Sneezeweed	760
Snow-on-the-mountain	297
Soap-plant, Wavyleaf	115
Solanum	
carolinense	579
dulcamara	580
emulans	581
rostratum	582
Solidago	
altissima	818
flexicaulis	819
mexicana	820
missouriensis	821
nemoralis	822
odora	823
rugosa	824
speciosa	825
velutina	826
Solomon's Seal, Smooth	122
Sonchus arvensis	827
Sorrel	
Mountain	388
Sheep	397
Sotol	116
Sow-thistle, Field	827

Sparganium eurycarpum 132
Speedwell
  Corn 613
  Thyme-leaf 614
Spergularia marina 415
Sphaeralcea
  coccinea 348
  parvifolia 349
Spiderlily, Carolina 104
Spiderling, Scarlet 427
Spiderwort
  Ohio 129
  Virginia 130
Spinystar 444
Spiranthes
  cernua 92
  magnicamporum 93
  romanzoffiana 94
Springbeauty
  Narrowleaf 434
  Western 432
Spurge
  Cypress 296
  Flowering 294
St.-John's-wort
  Common 291
  Marsh 292
Stachys
  chamissonis 661
  pilosa 662
Stanleya pinnata 378
Star-grass, Yellow 95
Starflower 480
Stellaria media 416
Stenanthium gramineum 39
Stenaria nigricans 512
Stenotus acaulis 828
Stickseed
  Many-flower 545
  Virginia 546
Stonecrop
  Ditch 197
  Lanceleaf 201
  Ledge 200
  Woods 203
Storksbill 312
Strawberry
  Barren 270
  Indian 276
  Wild 268
  Woodland 267
Stylisma humistrata 574
Stylophorum diphyllum 146
Stylosanthes biflora 247
Sugarbowls 161
Sunbright 436
Suncup, California 323
Sundew
  Pink 400
  Roundleaf 401
Sundrops, Yellow 331
Sunflower
  Common 761
  Cucumber-leaf 762

  Giant 763
  Maximilian's 764
  Prairie 765
  Tickseed 714
Susan, Black-eyed 810
Swallow-wort, Black 539
Swamp Candles 485
Sweet-vetch, Northern 225
Sweetflag 29
Sweetroot, Longstyle 869
Swertia perennis 525
Symphyotrichum
  ascendens 829
  cordifolium 830
  ericoides 831
  foliaceum 832
  laeve 833
  lanceolatum 834
  novae-angliae 835
  puniceum 836
  tenuifolium 837
Symplocarpus foetidus 34

T

Taenidia integerrima 873
Tanacetum vulgare 838
Tansy, Common 838
Taraxacum officinale 839
Taraxia tanacetifolia 334
Tarweed, Mountain 784
Tephrosia virginiana 248
Teucrium canadense 663
Thalictrum
  dasycarpum 180
  dioicum 181
  occidentale 182
  thalictroides 183
Thelesperma
  megapotamicum 840
Thermopsis montana 249
Thimbleweed
  Long-head 154
  Roundleaf 153
Thistle
  Bull 729
  Creeping 726
  Horrible 727
  Musk 719
  Scotch 792
  Wavyleaf 728
  Yellow Star 720
Thrift, California 381
Tiarella cordifolia 196
Tick-Trefoil
  Large 222
  Panicled 223
  Showy 221
Tickseed
  Large-flower 731
  Tall 734
Tidytips, White 778
Tillandsia usneoides 134
Tinker's Penny 289

Tiquilia canescens 556
Toadflax
  Bastard 380
  Blue 591
  Dalmatian 592
Tobacco
  Coyote 577
  Indian 691
Townsendia florifer 841
Tradescantia
  ohiensis 129
  virginiana 130
Tragopogon
  dubius 842
  pratensis 843
Trapa natans 317
Trefoil
  Bird's-foot 232
  Hop 251
Tribulus terrestris 205
Trichostema dichotomum 664
Trifolium
  arvense 250
  campestre 251
  pratense 252
  repens 253
  wormskioldii 254
Trillidium undulatum 40
Trillium
  cernuum 41
  erectum 42
  grandiflorum 43
  ovatum 44
Trillium
  Large-flower 43
  Nodding 41
  Painted 40
  Red 42
  Western 44
Triodanis perfoliata 694
Triosteum perfoliatum 855
Trumpet Vine 625
Trumpets, Pale 460
Turtlehead, White 587
Tussilago farfara 844
Twayblade
  Broadleaf 81
  Loesel's 79
Twin Flower 621
Twinflower 851
Twining Snapdragon 594
  Yellow 595
Twinleaf 149
Typha latifolia 133

U

Urtica gracilis 282
Utricularia
  cornuta 627
  radiata 628
  vulgaris 629

Uvularia
  grandiflora 47
  sessilifolia 48

V

Vaccinium oxycoccos 505
Valerian
  Edible 856
  Greek 477
Valeriana edulis 856
Velvetleaf 338
Veratrum
  californicum 45
  viride 46
Verbascum
  blattaria 617
  thapsus 618
Verbena
  bracteata 632
  hastata 633
  stricta 634
Verbesina encelioides 845
Vernonia
  gigantea 846
  noveboracensis 847
  americana 612
  arvensis 613
  serpyllifolia 614
Veronicastrum virginicum 615
Vervain
  Blue 633
  Hoary 634
  Prostrate 632
Vetch
  American 255
  Cow 256
  Spring 257
Vicia
  americana 255
  cracca 256
  sativa 257
Vinca minor 538
Vincetoxicum nigrum 539
Viola
  adunca 298
  bicolor 299
  canadensis 300
  incognita 301
  missouriensis 302
  pedata 303
  pedatifida 304
  pubescens 305
  purpurea 306
  sororia 307
Violet
  Bird's-foot 303
  Canada 300
  Common Blue 307
  Downy Yellow 305
  Goosefoot 306
  Larkspur 304
  Missouri 302
  Sweet White 301

Western Blue 298
Viper's-bugloss,
    Common 543
Virgin's Bower
    Blue 162
    Eastern 163

W

Wallflower, Contra Costa 367
Water-lily, Fragrant 25
Water-plantain 35
Water-willow, American 622
Watercress, True 372
Waterleaf
    Virginia 558
    Western 557
Wild Ginger, Long-tailed 28
Wild Mint, American 647
Wild Petunia, Fringe-leaf 623
Wild-buckwheat, Annual 384
Willowherb
    Bog 322
    Dense-flower 321
Wintergreen 499
    Bog 503
    Spotted 496
Wisteria frutescens 258
Wisteria, American 258
Witch's Moneybags 199
Wood Nymph 501
Wood Sorrel, Yellow 288

Woodvamp 449
Wyethia amplexicaulis 848

X

Xanthisma spinulosum 849
Xanthium strumarium 850
Xyris difformis 135

Y

Yarrow 697
    Golden 748
    Yellow Bell 58
Yellow-eyed Grass, Bog 135
Yellow-Pimpernel 873
Yellowcress, Common 375
Yellowrocket, Garden 356
Yerba Buena 638
Yucca
    arkansana 123
    flaccida 124
    schidigera 125
    torreyi 126
Yucca
    Arkansas 123
    Mojave 125
    Torrey 126

Z

Zizia aurea 874

Please note that we have provided full names of the photographers where possible; some contributors are known by usernames, which we also wish to credit.

A. Drauglis
A. Palu
Aaron Carlson
Aaron Liston
aarongunnar
abbamouse
Accuruss
Acer Hwang
Adam Humphreys
Adam Peterson
Adam Skowronski
Adelaide Pratt
Adrián Pablo Rodríguez Quiroga
AfroBrazilian
Agnes Monkelbaan
Agnieszka Kwiecień, Nova
Aiko, Thomas & Juliette+Isaac
Aimee Lusty
ajari
AJC1
aka CJ
Akos Kokai
Al
Alan Levine
Alan Prather
Alan Schmierer
Alastair Rae
Albert Bussewitz
Alejandro Bayer Tamayo
Aleksandrs Balodis
Alex Lomas
Alex Popovkin
Alex Zelenko
Alexandr frolov
Alexandre Dulaunoy
Alexis
Algirdas
alh1
Alinja
Alison Day
Alison Northup
Allefant
Allen Gathman
Allison Cox
Alpsdake
Altairisfar
Alun Williams333
Alvesgaspar
Alvin Kho
Amada44
Amanda Fisher
Amazonia Exotics U.K
Ambientalista e fotógrafo amador
Amos Oliver Doyle
Amy Washuta
André-Philippe D. Picard
Andrea Romero
Andrea Westmoreland

Andreas Eichler
Andreas Rockstein
Andrew Butko
Andrew C
Andrew Cannizzaro
Andrew Sebastian
Andrew Weitzel
Andrey Zharkikh
Andy Blackledge
Andy Hawkins
Andy Melton
Andy Reago & Chrissy McClarren
Andy Wraithmell
anechaffin
AnemoneProjectors
AnnaFialkoff
Anne Adrian
Anne Reeves
Anneli Salo
Annika Lindqvist
anoldent
AnRo0002
Antennaria neglecta
Antepenultimate
Anthony Zukoff
Aphidoidea
Arches National Park
Arnstein Rønning
Arria Belli
Atrian
Atsuko-y
Auckland Museum
Audrey
AudreyMuratet
Aung
Austin R. Kelly
Auvo Korpi
Averater
avogel_schweiz
AydarNabiev
Ayla
B A Bowen Photography
B. Domangue
bambe1964
Bambizoe
barbarab
Barnes Dr Thomas G, U.S. Fish and Wildlife Service
Bartosz Cuber
benet2006
Benjamin Burgunder
Benjamin Zwittnig
Bering Land Bridge National Preserve
Bernard Dupont
Bernd Haynold
Bernd Thaller
Bernhard Friess
Bewareofdog

Bff
Bill Bouton
Bill Keim
Biosthmors
Björn S.
Blake Bringhurst
BlindGoofy
BLM Nevada
Blondinrikard Fröberg
bluefootedbooby
Bob
Bob Danley
Bob Greenburg
Bob Peterson
Bob Richmond
bobistraveling
bobkennedy
Bogdan Giuşcă
Böhringer
Böhringer Friedrich
Boiron Belgium
Bonnie James
Bonnie Semmling
Borealis55
born1945
Borrichia frutescens
BotanischerVerein Sachsen-Anhalt
botany08
BotBln
Brett Francis (Oort)
brewbooks
Brian Gratwicke
Brocken Inaglory
brokinhrt2
Bruce Kirchoff
brybrysciguy
Buddha Dog
Buendia22
BuhaM
Bureau of Land Management, Christine Williams, Mackenzie Cowan, Sandra Miles, Sally Villegas, and West Eugene Wetlands staff
Bureau of Land Management Alaska
Bureau of Land Management Oregon and Washington
Burkhard Mücke
bwinesett
C T Johansson
CAJC- in the PNW
Calibas
California Department of Fish and Wildlife
candiru
CanyonlandsNPS
Carl Lewis
carlfbagge

CarlsbadCavernsNPS
carmona rodriguez.cc
CARNIVORASLAND
Carol VanHook
cassi saari
Cassondra Skinner
Cathie Bird
CayteMcDonough
Cbaile19
ceasol
Cédric Buffler
Cephas
Cette photo a été prise par André Alliot
Charles de Mille-Isles
Charles T. Bryson, USDA Agricultural Research Service
Charles T. Bryson, USDA Agricultural Research Service, Bugwood.org
Charlie Hohn
Charlie Jackson
Checkermallow
Chelsea Monks, Black Hills National Forest
Chihiro H
ChildofMidnight
Chinasaur
Chloe & Trevor Van Loon
Chmee2
Choess
Chris Fannin
Chris Hartman
Chris Light
Chris M Morris
Chris Meloche
Chris Parker
Christian Ferrer
Christian Fischer
Ciar
Cichorium intybus
CK Kelly
Claire Houck
Clinton & Charles Robertson
cmadler
Cody Hough
Colin Croft
color line
Congaree National Park
Consultaplantas
CostaPPPR
Cptcv
Craig Martin
cristina.sanvito
cultivar413
Curt Kline
Curtis Clark
Cyndy Sims Parr
D. Gordon E. Robertson

Daderot
Dakota Duff
Dalgial
Damiano Pappadà
Dan Jaffe
Dan Johnson
Dan McKay
Dandy1022
Daniel
Daniel Ballmer
Daniel J. Layton
Daniel Jolivet
Daniel Schwen
Daniel X. O'Neil
Danny S.
Dante Hinson
Dave Bonta
Dave Powell, USDA Forest
  Service
davecz2
Davefoc
David Abercrombie
David Brossard
David Eickhoff
David Evans
David J. Stang
David Prasad
David Rasp
David Reber
David Stang
David Wipf
David~O
DavidFrancis34
Dawn
Dawn Endico
Dax Ledesma
DCHNwam
Dcrjsr
Dean Morley
Dean Wm. Taylor
Deanna Phillips
Deb Nystrom
Debbie Ballentine
Decumaria barbara
Dehaan
delirium florens
Denali National Park and
  Preserve
DenesFeri
denisbin
Dennis Lamczak
Dennis Rex
Derek Ramsey
Derek Winterburn
Derell Licht
desultrix
Dewhurst Donna, U.S. Fish
  and Wildlife Service
Dick Culbert
Diego Delso
Dileep Eduri
Dinesh Valke
Dinkum
DM
Dmitri Popov
Dmitry Makeev

docentjoyce
Dominicus Johannes
  Bergsma
Don Henise
Don Loarie
Donald Hobern
Doppelbrau
Dornenwolf
Dorothy Long
Doug Goldman
Doug McGrady
Doug Murphy
Doug Suitor
Dr. Alexey Yakovlev
Dr. Boli
Dr. Hans-Günter Wagner
Dr. Thomas Barnes, Universtiy
  of Kentucky, U.S. Fish and
  Wildlife Service
Dulup
dw_ross
Earl McGehee
Earthdirt
echoe69
ecov ottos
Ed Ogle
EDevost1
Edwards allanb
EHM02667
Ejohnsonboulder
El Grafo
Elaine with Grey Cats
Elias
Elise Smith, U.S. Fish and
  Wildlife Service
Elizabeth Axley
Emilie Chen
Emma Forsberg
Engler, Mark A. Publisher-
  U.S. Fish and Wildlife
  Service
Enrico Blasutto
er-birds
Eric Hunt
Eric in SF
Erutuon
Esin Üstün
Espirat
Ethan Rose
Etienne Falquet
Ettore Balocchi
Eugene Zelenko
Evan M. Raskin
Evelyn Simak
Everglades NPS
Ezra S F
F Delventhal
F. D. Richards
f99aq8ove
Fabian Horst
Falcoperegrinus (Matthieu
  Gauvain)
Famartin
Felix
Ffaarr
Fice

Florian Grossir
Florida Fish and Wildlife
Flowersinmyyard
Fluff Berger at iNaturalist
Follavoine
Forest and Kim Starr
Forest Service Alaska Region,
  USDA
Forest Service Northern
  Region
Forest Service, Eastern
  Region
Fornax
Fralambert
Francis Bourgouin
Francisco Emilio Roldán
  Velasco
Franco Folini
Frank Kovalchek
Frank Mayfield
Frank Vincentz
FrankBramley
Franz Franzen
Franz Xaver
Fredlyfish4
FredrikLähnn
Fritz Flohr Reynolds
Fritz Geller-Grimm
Fritz Hochstaetter
Fritzflohrreynolds
Fungus Guy
GA-Kayaker
gailhampshire
garmonb0zia
Gary A. Monroe
Gary Chang
Gary Ford
Gary J. Wood
Gauna - USDA FS Modoc NF
Gazamp
Geoff Gallice
Geoffrey.landis
Georg Slickers
George Chernilevsky
George F Mayfield
George Williams
Gerard
Gertjan van Noord
Géry Parent
Gh5046
Giles Watson
Gilles San Martin
GlacierNPS
gmayfield10
Godpasta
Gordon Leppig & Andrea J.
  Pickart
graibeard
Grand Canyon National Park
gravitat-OFF
Grrewa
Guettarda
Gunera
Guy Waterval
Gzen92
H. Zell

Hajotthu
Halpaugh
Hana Oshima
Hanna Zelenko
Hannah E. Miller
Hans Hillewaert
Hans Kylberg
Hans Stieglitz
Haplochromis
Harald Henkel
Harald Loss
Hardyplants
Harlan B. Herbert
harrier
Harry Rose
harum.koh
hedera.baltica
Hedwig Storch
Heiditoronto
Helen Simonsson
Heliopsis helianthoides
Hellerhoff
Hideyuki Kamon
HIM Nguyen
Hiroyuki Takeda
Homer Edward Price
houroumono
Hovenweep National
  Monument
Howcheng
Hugh Knott
Hüseyin Cahid Doğan
I, BS Thurner Hof
I, DL.
I, NobbiP
I, Rolf Engstrand
icosahedron
Idéalités
Illustr
ilya_ktsn
InAweofGod'sCreation
incidencematrix
Intermountain Forest Service,
  USDA Region 4
Ipomopsis longiflora
Irene Grassi
Ivan Radic
IvanTortuga
Ivar Leidus
Izawa Ryu
J Brew
J. Schmidt
J.F. Gaffard Jeffdelonge
J7uy5
Jac. Janssen
jacinta lluch valero
Jacob W. Frank
JacobEnos
Jaknouse
jam343
Jamain
James St. John
James Steakley
Jamesgowld
Jamie Richmond
Jan Helebrant

Jane Nearing
Jane S Richardson
Jane Schlossberg
janet graham
Jason Hollinger
Jason Ksepka
Jay Horn
Jay Sturner
Jayesh Patil
JD
je_wyer
Jean
Jean and Fred
Jean-François Roch
Jean-Jacques Milan
Jean-Marie Van der Maren
JefferyRayCoffman
Jefficus
Jeffrey J. Witcosky
Jena Fuller
Jennifer Anderson
Jerry Friedman
Jerry Yates
Jerzy Opioła
Jerzy Strzelecki
Jesse Rorabaugh
Jessica
Jesús Cabrera
Jill Lee
jilllybean
Jim Duggan
Jim Evans
Jim Morefield
Jim Pisarowicz
Jim Staley
jimduggan24
Jinx McCombs
JKehoe_Photos
JLPC
jmaley
JMK
Joachim Lutz
Joan Simon
Joe Blowe
Joe Decruyenaere
Joe Mabel
Joe Passe
joergmlpts
John B.
John Brew
John D. Byrd, Mississippi
    State University
John Flannery
John Game
John Hayes
John Lodder
John Lynch
John Lynch & Michael
    Piantedosi
John Newton
John Pavelka
John Rusk
John Tann
Johnathan J. Stegeman
    (Midimacman)
johndal

Johnoliverwatts
johnyochum
Jomegat
Jonathan Hover
Jonmallard
Jordan Meeter
Jörg Hempel
Josep Gesti
Joseph Gage
Joseph Hubbard
Josh Graciano
Joshua Mayer
Joshua Tree National Park
Jotterand
Jsegraves99
Juan Carlos Fonseca Mata
Judy Gallagher
Just a Prairie Boy
Justin Meissen
Justina Madrigal
JW Stockert
k yamada
k.draper
Kaden Stebbins
Kaemat
Kagor
Kalathalan
Kaldari
kallerna
Karelj
Karen Roussel
karen_hine
Kate Mostad
Katja Schulz
Katya
Kazuhiro Tsugita
Ken Gibson
Ken-ichi Ueda
KENPEI
Kenraiz
Kevin Faccenda
Kevin Gepford
Kevin Gessner
Kevin Kenny
Kevin Thiele
khteWisconsin
Killarnee
Kimblad
kirybabe
Kleuske
Koerner Tom, U.S. Fish and
    Wildlife Service
Koichi Oda
Kolforn (Kolforn)
Kor!An (Андрей Корзун)
Krazytea
Kristian Peters - Fabelfroh
Kristina Ugrinović
Kristine Paulus
Kruczy89
Krzysztof Golik
Krzysztof Ziarnek, Kenraiz
ksblack99
Kunal Mukherjee
kylerossner
laikolosse

Laineypaige
Lake Mead NRA Public Affairs
Lamiot
Land Between the Lakes KY
laogooli
Larry Lamsa
Larry Smith
Larry Trekell
Laslovarga
Laura Camp
Laura Gaudette
Laurel F
Lauren
laurent houmeau
Lawrence Newcomb
Lazaregagnidze
lazarus
Lazarus000
Lee Bonnifield
Leila Dasher
Len Worthington
Leo-setä
Leonardo DaSilva
Leonhard Lenz
Leonora (Ellie) Enking
Leslie J. Mehrhoff
Leslie Seaton
LesMehrhoff
LiCheng Shih
liesvanrompaey
Light in Colors
Lihtsaltmaria
Lillie
Linaria canadensis
Lisa King
liz west
LizaGreen
lostinfog
Louisiana Sea Grant College
    Program Louisiana State
    University
Luca 4891
Lucash
Luis Fernández García
Lukas Riebling
Łukasz Rawa
Lumaca
Lutz Blohm
Lydia Fravel
M van Ree
M. O'Hearn, J. Carstens, L.
    Pfiffner, USDA, ARS
M. Readey
m.shattock
M's photography
Madeleine Claire
magnolia1000
Magnus Manske
Maja Dumat
Malcolm Manners
Malte
Manuel Freiría
manuel m. v.
maplegirlie
Marcus Winter

Margo Akermark
Marianne Serra
Mariejanelle
Mariia Zykova
Marina Ziabchenkova
Mark
Mark A. Garland. USA, NC,
    Jackson Co., Balsam,
    Daytona Road, cove forest
Mark AC Photos
Mark Freeth
Mark Marathon
Marko Vainu
Markus
Marshal Hedin
Martin Bravenboer
Martin Cooper
Marvin Smith
Mary Krieger
Mason Brock
MathewTownsend
MathieuMD
Matt Berger
Matt Lavin
Matt Tillett
Matthew Dillon
Matthew T Rader
Maurice Flesier
mauro halpern
Max071086
Mbdortmund
Megan Hansen
Melanie Shaw
Melinaguene
Melissa McMasters
Meneerke bloem
mfeaver
Michael Figiel
Michael Goodyear
Michael Gras, M.Ed.
Michael Hodge
Michael Jutzi
Michael Mueller
Michael Piantedosi
Michael Wolf, Penig
MichaelPiantedosi
Michał Strzelecki
Michel Langeveld
Michele Dorsey Walfred
Michele Ursino
Michelle
MichielSt
Mick Talbot
Midnight Runner
Miguel Andrade
Mike Cline
Mike Finn
Mike Michael L. Baird
Mike Ostrowski
Mike Procario
Mike Steinhoff
Millie Basden
Miltos Gikas
Mimi Cummins
Mimulusmoschatus
Mirror

Miwasatoshi
mjpapay
Mokkie
Monarda citriodora
MONGO
MostlyDross
Mount Rainier National Park
Mrs. Gemstone
Mttswa
MurielBendel
Murray Foubister
Mustang Joe
My-Lan Le
Mykola Swarnyk
Nadiatalent
Naoki Takebayashi
Natasha de Vere & Col Ford
National Archives at College Park
National Park Service, Alaska Region
Nature at La Tuilerie Western France
NatureServe
NatureShutterbug
NC Orchid
NC Wetlands
Neelix
neil
Nicholas A. Tonelli
nicholas_gent
Nicholas Tippery
Nicholas_T
Nick Doty
nick fullerton
Nick Varvel
Nicolas Torquet
Nicolas Weghaupt
Niepokój Zbigniew
Nina
NMSU IPM
NoahElhardt
Noj Han
Nolan Exe
Nonenmac
Nordhage
Norio Nomura
Northeast Coastal & Barrier Network
NotAnonymous
Notjake13
NPS
NPS / Jacob W. Frank- Denali National Park and Preserve
NPS Staff
NPS Staff - NPGallery
NPS / Diane Renkin
Nucatum amygdalarum
NY State IPM Program at Cornell University
OakleyOriginals
Obsidian Soul
Old Photo Profile
Oleg Kosterin
Olive Titus
oliveoligarchy

oliver.dodd
Olivier Pichard
Orchi
Orest Ukrainsky
Over The Arroyo Gang
Oxfordian Kissuth
Øyvind Holmstad
Pacific Southwest Forest Service, USDA
Pacific Southwest Region 5
Paisley Scotland
Panegyrics of Granovetter
Park staff - NPGallery
Pat Farris
Patrice78500
Patrick Alexander
Patrick Hacker
Patrick Standish
Paul
Paul Asman and Jill Lenoble
Paul Prior
Paul Smith (Romfordian)
Paul Sullivan
Paul van de Velde
Paul VanDerWerf
peardg
peganum
Peggy A. Lopipero-Langmo
pellaea
Petar Milošević
pete beard
Peter Abrahamsen
Peter Cooper
Peter Corbett
Peter coxhead
Peter D. Tillman
Peter O'Connor aka anemoneprojectors
Peter Stenzel
Peter Stevens
Peter Toporowski
Peterwchen
Pethan
Petrified Forest National Park, Lauren Carter
Petrified ForestNPS
Petritap
petur r
peupleloup
Peyman Zehtab Fard
pfly
Pgrunow
Phil Sellens
Philip Stepnowski
PhilippeGerbet
PhotoDoc
photogramma1
pietila4
Plant Image Library
Plant Right
PlayMistyForMe
Pmau
pointnshoot
Pom'
Potomacpalms
Priit Tammets

prilfish
PROPOLI87
ptrktn
Puddin Tain
PumpkinSky
Quartl
Quinn Dombrowski
Qwert1234
R Stringham
R-E-AL
R. A. Nonenmacher
R6, State & Private Forestry, Forest Health Protection
Radio Tonreg
Radu Chibzii
Raffi Kojian
Ragesoss
Ragnhild&Neil Crawford
Rain0975
Ramón Portellano
Randal
Randi Hausken
Randy A. Nonenmacher
Rasbak
Raulbot
Reaperman
Reggaeman
Reinhold Möller
Renee Grayson
Reuven Martin
rhonddawildlifediary
Richard Bartz
Richard Thomas
Richtid
Rickard Holgersson
Rictor Norton & David Allen
RiverBissonnette
Rob Bertholf
Rob Hille
Rob Hodgkins
Rob Routledge, Sault College
Robb Hannawacker
Robert
Robert Flogaus-Faust
Robert H. Mohlenbrock
Robert Lafond
Robert Nunnally
Robert S Remie
rockerBOO
Rocky Mountain National Park
Rolf Dietrich Brecher
Rolf Engstrand
Ron Conlon
Roozitaa
Rorolinus
Rosa-Maria Rinkl
Roy Cohutta
Rüdiger Stehn
Rudolphous
Ruppia2000
Ruth Hartnup
Ruthven
Ryan Elliott
Ryan Hagerty
Ryan Hodnett
Ryan Kitko

Ryan McMinds
Ryan Somma
Ryan Watson
Rylan Sprague, Black Hills National Forest
Ryusuke Seto
S. Rae
SABENCIA Guillermo César Ruiz
Sabina Bajracharya
Sabrina Setaro
Sage Ross
Saguaro National Park
saiberiac
Saivann
Salicyna
Sam Fraser-Smith
Samantha Forsberg
San Bernardino Nat'l Forest
SanctuaryX
Sandy Wolkenberg
Sanja565658
Sasata
Satoru Kikuchi
SB Johnny
schizoform
Scott
Scott Darbey
Scott Loarie
Scott Wilson
Seiya Ishibashi
Sergey Stefanov
sergio niebla
sfbaywalk
Shadowmeld Photography
sharloch
Shauna Littlewood, Jeremydwade
Shenandoah National Park
Shishirdasika
siamesepuppy
Silk666
Simon
Sixflashphoto
Smith Elise, U.S. Fish and Wildlife Service
Solidago sempervirens
Sönke Haas
sonnia hill
SoulRider.222
SriMesh
Stan Shebs
stanze
Staudengärtnerei Forssman
Stefan.lefnaer
Stemonitis
Sten
Sten Porse
Stephan van Helden
Stephanie Harvey
StephanieFalzone
Stephen Hornbeck
Stephen Horvath
Steve Berardi
Steve Cyr
Steve Fung

Steve Redman (MORA)
Steven Katovich, USDA Forest Service
StevenZiglar
Stickpen
stinger
StingrayPhil
Stolz Gary M, U.S. Fish and Wildlife Service
Storye book
Stosh Morency
Stoutcob
Strobilomyces
Sue Ann (Suna) Kendall
Suhaib Sarabta
sunoochi
Superior National Forest
Susan Blayney
Susan Elliott
Susan Young
Svarði2
Swallowtail Garden Seeds
Syrio
T. Kebert
T.K. Naliaka
Taenidia integerrima
Tanaka Juuyoh
Tarciso Leão
Tarry Edington
Tatters
Ted Bodner, Southern Weed Science Society
Teresa Grau Ros
Tero Karppinen
Tero Laakso
Terry Lucas
TeunSpaans
Texas Sea Grant
Thayne Tuason
The Cartographer
The Cosmonaut
The Forest Vixen's CC Photo Stream
The Marmot
The Plantography Project
The World Through Athene's Eyes
The_Gut
Thegreenj
Theodore Webster, USDA Agricultural Research Service
Thibault Lefort
Thierry Caro
Thierry Ploquin
Thomas Griesohn-Pflieger
Thomas R. Koll
Thomas Shahan
Tim Binns
Tim Green
Tim Ross
Tim Sackton
Ting Chen
titanium22
Tj Campbell
Tom Hilton

Ton Rulkens
Tony Alter
Tony Fischer
Tracie Hall
troy mckaskle
tsiegretlop
Tubifex
Tucker SP
Tulipasylvestris
Tyrrhium
U.S. Department of Agriculture
U.S. Fish & Wildlife Service Southwest Region
U.S. Fish and Wildlife Service Headquarters
U.S. Fish and Wildlife Service Southeast Region
u278
Udo Schmidt
Uleli
UliLorime
UliLorimer
ume-y
Under the same moon...
Uoaei1
Ural-66
USDA NRCS
USDA NRCS Montana
USFWS - Pacific Region
USFWS Midwest Region
USFWS Mountain
USFWS Mountain-Prairie
USFWSmidwest
USGS Bee Inventory and Monitoring Lab
Utricularia cornuta
Uwe W.
V Maslyak
Vahe Martirosyan
VanLap Hoàng
VasenkaPhotography
vastateparksstaff
Vengolis
Veronica arvensis
Vijay Barve
Vinayaraj
Vinilka
Virginia Oakes
Virginia State Parks
Vivek Yadav
VP
W.carter
Wallace Keck
Walter Siegmund
Wanderland
Wasp32
Wasrts
Wayne National Forest
Wendell Smith
Wendy Cutler
Wendy McCrady
weta2000nz
WideClyde
Wilhelm Zimmerling PAR
Will Brown

Will McFarland
Will Pollard
Will Simpson
Willamette Biology
William Farr
William Herron
William L. Farr
WilliamLarkin
Willthomas
Wilson Bilkovich
Wilson44691
Wing-Chi Poon
Wintertanager J.T. Storey
Wlodzimierz
Wouter Hagens
Wsiegmund
xulescu_g
y_egan
yamatsu
Yasuo Kida
Yellowstone National Park
yewchan
yguaba
Yoko Nekonomania
zabdiel
Zen sutherland
Zeynel Cebeci
Zihao Wang
ZooFari
Ανώνυμος Βικιπαιδιστής
Σ64
Аимаина хикари
Александр Б.
Анна Митрошенкова
Ассан Аринец
Качанов Сергій
Марина Емельянова
Національний природний парк "Великий Луг"
Савин Игорь Игоревич
Сарапулов
СССР
Хомелка
Эльвира Клищенко
Юлия
松岡明芳 - 松岡明芳
阿橋 HQ

Any work of this scope takes a dedicated team of talented experts, advisors, and editors to bring it to life. It also takes an inspirational spark, and for that we want to thank the National Audubon Society, which since 1905 has championed the conservation of bird habitats and the natural world throughout North America. **www.audubon.org**

We are particularly grateful to Native Plant Trust for their counsel and direction throughout this project. Their breadth of knowledge and nuanced understanding of the native plant species of North America is unparalleled and served as a vital foundation. We especially thank Senior Research Botanist Arthur Haines for his taxonomic expertise which shaped the species representation that is the backbone of this work.

The images used throughout this book reflect countless hours in the field on the part of scores of photographers and the painstaking curation of those images. We particularly want to thank Liz Green of Native Plant Trust for her dedication to this project, helping us bring these species to the printed page with immediacy and intimacy.

We are grateful to the following collaborators whose expertise informed the taxonomic data, the species representation we have chosen, and articles in the beginning of this book. We appreciate their work to further the cause of botanical research, conservation, and education:

Ted Elliman, Vegetation Ecologist at Native Plant Trust

Jessamine Finch, Ph.D., Research Botanist at Native Plant Trust

Dan Jaffe Wilder, Ecological Horticulturist, Norcross Wildlife Sanctuary

Warm thanks to Jane Cirigliano and her design team for rendering every page of this book with precision and elegance; to Katy Savage, Kyle Carlsen, and Ben Shuman for their contributions to the writing of species profiles; and to Heather Rowland of Add+Water for the book design. Very special thanks to our publishing partners at Alfred A. Knopf for their ongoing encouragement, guidance, and support; most notably Andy Hughes and his editorial staff.

We also thank the Biota of North America Program (BONAP) for supplying the range maps for the book using the most current data from its online database of the native and naturalized species of plants growing in North America.

At the core of this effort is the team at Fieldstone Publishing, whose devotion to this project imbues every page of this book:

Shyla Stewart, President & CEO
Andrew Stewart, Publisher Emeritus
Jim Cirigliano, Editor-In-Chief
Katy Savage, Editor, Production Coordinator
Heather Coon, Finance Director
Lawrence Farr, Digital Development
Shavar Dawkins, Photo Editor / Research
Sophia Foster, Project Management

We thank all of the Audubon experts who contributed their knowledge to ensure this book represents the organization's mission: Audubon's Development Division, including Sean O'Connor, Chief Development Officer; Kevin Duffy, Vice President of Corporate, Public, Foundation, and Legacy Giving; Julisa Colón, Director of Brand Marketing; Holly Fairall, Manager of Brand Marketing; Audubon's Science and Network divisions, including John Rowden, Senior Director of Bird-Friendly Communities; Marlene Pantin, Partnerships Manager of Plants for Birds; Audubon's Content Division, including Jennifer Bogo, Vice President of Content; Kristina Deckert, Art Director; Sabine Meyer, Photography Director; Melanie Ryan, Assistant Art Director; Camilla Cerea, Photo Editor and Photographer; and Alex Tomlinson, Graphic Designer.

THIS IS A BORZOI BOOK PUBLISHED BY ALFRED A. KNOPF

Copyright © 2023 by Fieldstone Publishing, Inc.

All rights reserved. Published in the United States by Alfred A. Knopf, a division of Penguin Random House LLC, New York, and distributed in Canada by Penguin Random House Canada Limited, Toronto.

www.aaknopf.com

Knopf, Borzoi Books, and the colophon are registered trademarks of Penguin Random House LLC.

Audubon™ is a licensed, registered trademark of the National Audubon Society. All rights reserved.

Based on *National Audubon Society Field Guide to North American Wildflowers: Eastern Region* (Alfred A. Knopf 2001) and *National Audubon Society Field Guide to North American Wildflowers: Western Region* (Alfred A. Knopf 2001)

For more information about Audubon, including how to become a member, visit www.audubon.org or call 1-844-428-3826.

Library of Congress Cataloging-in-Publication Data
Names: Cirigliano, Jim, [date] editor.
Title: National Audubon Society wildflowers of North America / edited by Jim Cirigliano.
Description: First edition. | New York : Alfred A. Knopf, 2023. | Series: National Audubon Society field guides | Includes index.
Identifiers: LCCN 2022004081 (print) | LCCN 2022004082 (ebook) | ISBN 9780593319949 (hardcover) | ISBN 9780593319956 (ebook)
Subjects: LCSH: Wild flowers—North America—Identification.
Classification: LCC QK110 .N38 2023 (print) | LCC QK110 (ebook) | DDC 582.13097—dc23/eng/20220421
LC record available at https://lccn.loc.gov/2022004081
LC ebook record available at https://lccn.loc.gov/2022004082

Jacket image: Columbine or Granny's Bonnet (*Aquilegia*) by Jacky Parker Photography/Getty Images
Cover design by Linda Huang
Frontispiece by PaulSat - stock.adobe.com

Manufactured in China
First Edition